Numerical Methods and Computer Simulations in Energy Analysis, 2nd Edition

Numerical Methods and Computer Simulations in Energy Analysis, 2nd Edition

Topic Editors

Marcin Kamiński
Mateus Mendes

MDPI

Basel • Beijing • Wuhan • Barcelona • Belgrade • Novi Sad • Cluj • Manchester

Topic Editors

Marcin Kamiński
Department of Structural Mechanics
Lodz University of Technology
Lodz
Poland

Mateus Mendes
Coimbra Institute of Engineering
Polytechnic University of Coimbra
Coimbra
Portugal

Editorial Office

MDPI AG
Grosspeteranlage 5
4052 Basel, Switzerland

This is a reprint of the Topic, published open access by the journals *Energies* (ISSN 1996-1073), *Applied Sciences* (ISSN 2076-3417), *Mathematics* (ISSN 2227-7390), *Entropy* (ISSN 1099-4300) and *Computers* (ISSN 2073-431X), freely accessible at: https://www.mdpi.com/topics/numerical_methods_ii.

For citation purposes, cite each article independently as indicated on the article page online and as indicated below:

Lastname, A.A.; Lastname, B.B. Article Title. *Journal Name* **Year**, *Volume Number*, Page Range.

ISBN 978-3-7258-4169-1 (Hbk)
ISBN 978-3-7258-4170-7 (PDF)
https://doi.org/10.3390/books978-3-7258-4170-7

Cover image courtesy of Marcin Kamiński

Contents

About the Editors vii

Preface ix

Shuangshuang Fan, Ying Wang, Kun Yao, Jiakui Shi, Jun Han and Jie Wan
Distribution Characteristics of High Wetness Loss Area in the Last Two Stages of Steam Turbine under Varying Conditions
Reprinted from: *Energies* **2022**, *15*, 2527, https://doi.org/10.3390/en15072527 1

David Watling, Patrícia Baptista, Gonçalo Duarte, Jianbing Gao and Haibo Chen
Systematic Method for Developing Reference Driving Cycles Appropriate to Electric L-Category Vehicles
Reprinted from: *Energies* **2022**, *15*, 3466, https://doi.org/10.3390/en15093466 14

Gang Xu, Kegang Li, Mingliang Li, Qingci Qin and Rui Yue
Rockburst Intensity Level Prediction Method Based on FA-SSA-PNN Model
Reprinted from: *Energies* **2022**, *15*, 5016, https://doi.org/10.3390/en15145016 42

Lomena Mulenda Augustin, Sumuna Temo Vertomene, Ndaye Nkanka Bernard, Amsini Sadiki and Mbuyi Katshiatshia Haddy
A New Perspective on Cooking Stove Loss Coefficient Assessment by Means of the Second Law Analysis
Reprinted from: *Entropy* **2022**, *24*, 1019, https://doi.org/10.3390/e24081019 61

Marcin Kamiński and Alberto Corigliano
Shannon Entropy in Stochastic Analysis of Some Mems
Reprinted from: *Energies* **2022**, *15*, 5483, https://doi.org/10.3390/en15155483 88

Ho Jin Park, Mohammad Alosaimi, Seong-Ah Yang, Heejeong Jeong and Sung Hoon Choi
McCARD Criticality Benchmark Analyses with Various Evaluated Nuclear Data Libraries
Reprinted from: *Energies* **2022**, *15*, 6852, https://doi.org/10.3390/en15186852 102

Lei Jiang, Minjun Diao and Chuan'ai Wang
Investigation of a Negative Step Effect on Stilling Basin by Using CFD
Reprinted from: *Entropy* **2022**, *24*, 1523, https://doi.org/10.3390/e24111523 119

Paweł Łojek, Ireneusz Czajka and Andrzej Gołaś
Numerical Study of the Impact of Fluid–Structure Interaction on Flow Noise over a Rectangular Cavity
Reprinted from: *Energies* **2022**, *15*, 8017, https://doi.org/10.3390/en15218017 137

Jarosław Artyszuk and Paweł Zalewski
Power Consumption and Effectiveness of Azimuth Stern-Drive (ASD) Tug While Assisting at Ship's Bow
Reprinted from: *Energies* **2022**, *15*, 9519, https://doi.org/10.3390/en15249519 159

Marcin Kamiński and Michał Strąkowski
An Application of Relative Entropy in Structural Safety Analysis of Elastoplastic Beam under Fire Conditions
Reprinted from: *Energies* **2023**, *16*, 207, https://doi.org/10.3390/en16010207 176

Roberto Franco Cisneros and Freddy Jesús Rojas
Determination of 12 Combustion Products, Flame Temperature and Laminar Burning Velocity of Saudi LPG Using Numerical Methods Coded in a MATLAB Application
Reprinted from: *Energies* **2023**, *16*, 4688, https://doi.org/10.3390/en16124688 . . . . **197**

Sebastian Łapczyński, Michał Szulborski, Łukasz Kolimas, Przemysław Sul, Maciej Owsiński, Przemysław Berowski, et al.
Electrodynamic Forces in Main Three-Phase Busbar System of Low-Voltage Switchgear—FEA Simulation
Reprinted from: *Energies* **2024**, *17*, 1891, https://doi.org/10.3390/en17081891 . . . . **220**

Nasrin Kalamian, Masoud Soltani, Fariba Bouzari Liavoli and Mona Faraji Niri
Observer-Based Suboptimal Controller Design for Permanent Magnet Synchronous Motors: State-Dependent Riccati Equation Controller and Impulsive Observer Approaches
Reprinted from: *Computers* **2024**, *13*, 142, https://doi.org/10.3390/computers13060142 . . . . . . **240**

Kaiyuan Sun, Kun Chen, Haifeng Cen, Fucheng Tan and Xiaohui Ye
Study of a Numerical Integral Interpolation Method for Electromagnetic Transient Simulations
Reprinted from: *Energies* **2024**, *17*, 3837, https://doi.org/10.3390/en17153837 . . . . **257**

Hui Zhang and Fajian Nie
Study of the Impact of Acquisition Parameters on Fault Feature Identification Based on Magnetotelluric Modeling
Reprinted from: *Appl. Sci.* **2024**, *14*, 9720, https://doi.org/10.3390/app14219720 . . . . **274**

Zahid Mehmood, Zhenyu Wang, Xin Zhang and Guiying Shen
Aerodynamic Performance and Numerical Validation Study of a Scaled-Down and Full-Scale Wind Turbine Models
Reprinted from: *Energies* **2024**, *17*, 5449, https://doi.org/10.3390/en17215449 . . . . **286**

Yueyang Li, Enli Zhang, Ping Yue, Han Zhao, Zhiwei Xie and Wei Liu
Simulating Water Invasion Dynamics in Fractured Gas Reservoirs
Reprinted from: *Energies* **2024**, *17*, 6055, https://doi.org/10.3390/en17236055 . . . . **308**

Xingping Liu, Xingxing Huang, Weijiang Chen, Zhengwei Wang
Flow-Induced Strength Analysis of Large Francis Turbine Under Extended Load Range
Reprinted from: *Appl. Sci.* **2025**, *15*, 2422, https://doi.org/10.3390/app15052422 . . . . **318**

About the Editors

Marcin Kamiński

Marcin Kamiński has been a Full Professor in the Department of Structural Mechanics, Łódź University of Technology, since 2015. He was the Head of the Discipline Civil Engineering, Geodesy & Transportation in the Faculty of Civil Engineering, Architecture & Environmental Engineering, in the period of 2019–2024. His main research includes the stochastic finite element method, the numerical modeling of random composites including homogenization theory, reliability assessment and optimization of civil engineering structures, stochastic aging processes, and recently, an application of probabilistic entropy in engineering computations. He spent a postdoctoral study at Rice University in Houston, USA, and was a visiting professor at Leibniz Institute of Polymer Research in Dresden, Germany, Politecnico di Milano, and also University of Naples Federico II in Italy. He has authored more than 200 research papers in various journals including the *International Journal for Numerical Methods in Engineering*, *Composite Structures*, *International Journal of Solids & Structures*, and also *Computers & Structures*. He has published the two following monographs: "Computational Mechanics of Composite Materials", Springer, 2005, and "The Stochastic Perturbation Method for Computational Mechanics", Wiley, 2013. He serves as the Associate Editor of *Mechanics Research Communications* and is a member of the Editorial Board of *Acta Mechanica*, *Computers & Structures*, *Journal of Composites Science*, and *SCI*. He has been the recipient of fellowships from the Foundation for Polish Science, the J. Argyris Award from ECCOMAS and Elsevier, the J.T. Oden Scholarship at The University of Texas at Austin, and also some Polish national or academic awards including the Bronze Cross of Merit (1998), Silver Cross for the Longstanding Service (2017), Medal of Commission of Education (2022), and also The Research Prize of Ministry of Education & Science in 2023. He has recently been recognized (since 2021) as the World's Top 2% Scientists by Stanford University, CA, USA.

Mateus Mendes

Mateus Mendes received his BSc, MSc, and PhD from the University of Coimbra, Portugal. He is currently a Coordinator Professor at the department of Computer Science of the Institute of Engineering of the Polytechnic University of Coimbra. He is also an integrated researcher at the RCM^{2+}—Research Centre for Asset Management and Systems Engineering. He is also a member of the Data Science Portuguese Association and president of the Executive Commission of the PAMDaS international conference on physical asset management and data science. His areas of research include mainly applications of artificial intelligence techniques, with a focus on classical and deep machine learning methods, and data analysis and simulation processes for different industrial problems. He has developed different algorithms for quality control, fault detection, and predictive maintenance in paper, transport, plastic injection, and other industrial settings. He has also been involved in creating intelligent simulators for different forest engineering processes. He has authored more than 100 research publications, including books, journal papers, and conference publications. He has also been the Guest Editor of different Special Issues for the journals *Algorithms* and *Energies*.

Preface

This collection reflects a deep need for numerical simulations in modern society and demonstrates how they are a fundamental part of the scientific research focused on energy analysis. It is widely understood that modern energy resources and energy transformation, harvesting, and transmission in all their aspects still demand more efficient numerical simulations. This need is also driven by expanding computer technologies, software, and numerical methods in the context of forthcoming quantum computing and the exponentially increasing role of artificial intelligence. The contributions collected in this book represent quite different points of view and various aspects of energy analysis, which is natural for a variety of both computed methods, software, and research disciplines. This could be beneficial for scientists, practicing engineers, and students at different levels of academic education to whom the book is addressed. Modern computer simulation includes a variety of methods with different efficiencies and possible numerical errors, models (multiscale, multi-physical, stochastic, etc.), discretizations based on meshes or grids, and implementations (traditional, parallel, or grid computing). The papers collected in this reprint enable us to look to the future of both simulations and energy analysis, where the applicability of artificial intelligence tools and applications must naturally increase. The first author would like to acknowledge the financial support of the National Science Center in Cracow for the Research Grant OPUS no. 2021/41/B/ST8/02432 "Probabilistic entropy in engineering computations".

Marcin Kamiński and Mateus Mendes
Topic Editors

energies

MDPI

Article

Distribution Characteristics of High Wetness Loss Area in the Last Two Stages of Steam Turbine under Varying Conditions

Shuangshuang Fan [1,2], Ying Wang [3], Kun Yao [1,*], Jiakui Shi [4], Jun Han [5] and Jie Wan [1,4,*]

1 Laboratory for Space Environment and Physical Sciences, Harbin Institute of Technology, Harbin 150001, China; 2201800327@neepu.edu.cn
2 College of Energy and Power Engineering, Northeast Electric Power University, Jilin 132000, China
3 College of Power and Energy Engineering, Harbin Engineering University, Harbin 150009, China; gecx@harbin-electric.com
4 School of Electrical Engineering and Automation, Harbin Institute of Technology, Harbin 150001, China; shijiakui@stu.hit.edu.cn
5 School of Energy Science and Engineering, Harbin Institute of Technology, Harbin 150001, China; hanjun@hit.edu.cn
* Correspondence: 19b353009@stu.hit.edu.cn (K.Y.); wanjie@hit.edu.cn (J.W.); Tel.: +86-18846178483 (K.Y.); +86-13936411359 (J.W.)

Abstract: Wetness loss of a steam turbine seriously affects the security of the unit when operating in deep peak regulation. To obtain the distribution characteristics of the high wetness loss area under different working conditions, especially low-load conditions, the last two stages of the low-pressure cylinder (LPC) of a 600 MW steam turbine were simulated using the non-equilibrium condensation model proposed in this study. The nucleation rate distribution, supercooling degree, and steam velocity droplet were analyzed. Consequently, the diameter distribution of coarse water droplets under 100%, 50%, 40%, 30%, and 20% THA conditions and the distribution of the thermodynamic loss and water droplet resistance loss were obtained. Thermodynamic loss mainly occurred at the front end of second-stage stator blades and trailing end of the last-stage stator blades. The water droplet resistance loss mainly occurred at 40% of the blade height and at the tip of the last-stage stator blades. Moreover, with a reduction in the unit load, the thermodynamic loss continued to decrease, but the water droplet resistance loss continued to increase.

Keywords: last two stages; steam turbine; thermodynamic loss; water droplets resistance loss; wetness loss

1. Introduction

The generation of wet steam is inevitable in the process of steam turbine running in a wide range of working conditions [1]. In particular, the last two stages of LPC are in a wet steam environment for a long time. During the flow of wet steam, the generation and development of small droplets leads to the wetness loss, which not only reduces the efficiency of the unit, but even threatens the operation safety [2,3]. The area of the distribution of high wetness loss, is accurately obtained. There is a certain sense of improving the operational safety and economy of the steam turbine.

Wet steam condensation flow is a long-standing problem. An accurate stepwise description of the flow characteristics is carried out, from the initial one-dimensional nozzle structure [4] and two-dimensional cascade structure [5], to the current three-dimensional multi-stage structure [6,7]. Simultaneously, an empirical estimation method for wetness loss was proposed, where a 1% average humidity corresponding to a 1% drop in the turbine efficiency [8] was obtained. However, some researchers have found that this estimation method is not completely accurate. This is because the value of the humidity and deposition distribution of water droplets should be considered in the process of estimating wetness loss [9–11]. Therefore, as indicated by references [12], the surface pressure and humidity distribution in the wet steam stage undergo significant changes. Thus, by comparing the two different flows, the non-design loss and the non-equilibrium loss was roughly evaluated. Simultaneously, some scholars began to study the value of the wetness loss using a semi-analytical method [13] and numerical simulation calculations [14], which laid the foundation for the subsequent three-dimensional simulation calculation of the wetness loss. The influence of the nozzle and cascade on the thermodynamic loss was studied in reference [15]. In [16], condensation loss was analyzed from two aspects: multi-dispersed droplets and cascade trailing edge reduction. To reduce the condensation loss, the idea of turbulent flow disturbance on the suction side to reduce the liquid phase fraction was proposed in reference [17], and reference [7] took the initiative of injecting droplets.

In summary, the current research on wetness loss is limited to the quantitative estimation of empirical algorithms, where the distribution of wetness loss in steam turbines is not given under different operating conditions.

Therefore, the following issues need to be discussed:

- How are the thermodynamic losses distributed in the last two stages of steam turbine LPC under different working conditions?
- How are the water droplet resistance losses distributed in the last two stages of steam turbine LPC under different working conditions?
- What factors are related to the above two losses?

In order to study the high wetness loss area (thermodynamic and water droplet resistance losses) in the last two stages of the LPC under different working conditions of steam turbine, corresponding measures are taken to reduce the wetness loss. Therefore, efficiency and safety are guaranteed. Thus, in this study, the last two stages of a 600 MW steam turbine LPC were considered as the research object. The distribution characteristics of the high wetness loss area under different working conditions, which is of great significance in improving the design level of the steam turbine, were analyzed using simulations.

The paper is organized as follows: Section 2 introduces the physical model, numerical methods, mathematical model verification, and composition of high wetness loss. In Sections 3.1 and 3.2, the thermodynamic loss and the water droplets resistance loss are analyzed, and their distributions in the last two stages of the low-pressure cylinder of the steam turbine are summarized in Section 3.3. In Section 3.4, the methods to control wetness loss is proposed, and Section 4 provides the concluding remarks.

2. Models and Numerical Methods

2.1. Physical Model and Boundary Conditions

A 600 MW steam turbine low-pressure cylinder with the last two stage blades was taken as the research object, where the calculated single-flow channel structure is shown in Figure 1. According to the thermodynamic characteristics specification, the steam turbine speed was 3000 r/min, the inlet temperature of the last second stage was 357.45 K, the inlet humidity was 1.5%, the mass flow rate was 1.305 kg/s, and the outlet pressure of the last stage was 4.9 kPa. The frozen rotor method is used for the interface [18,19], and the convergence residual limit was set to the order of 10^{-5}.

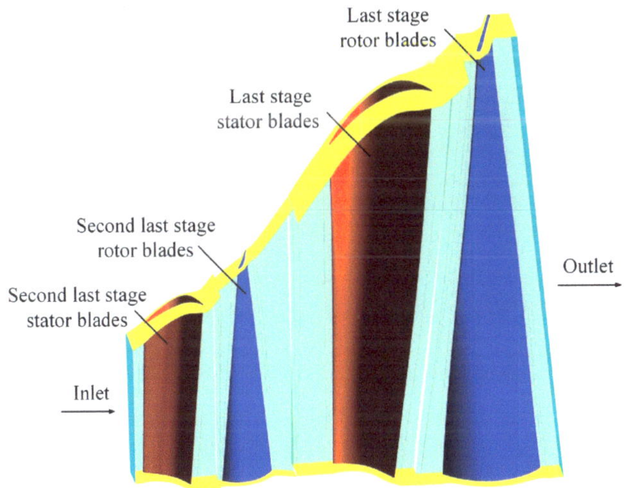

Figure 1. Last two stages of the single-channel structure in the low-pressure cylinder of a steam turbine.

2.2. Grid Independence Verification

The ANSYS Turbogrid in ANSYS is used to mesh the physical model, whose grid around the blade is refined at the same time. The model grid is shown in Figure 2. The grid independence has been verified in reference [20], and the number of grids is chosen to be 6 million.

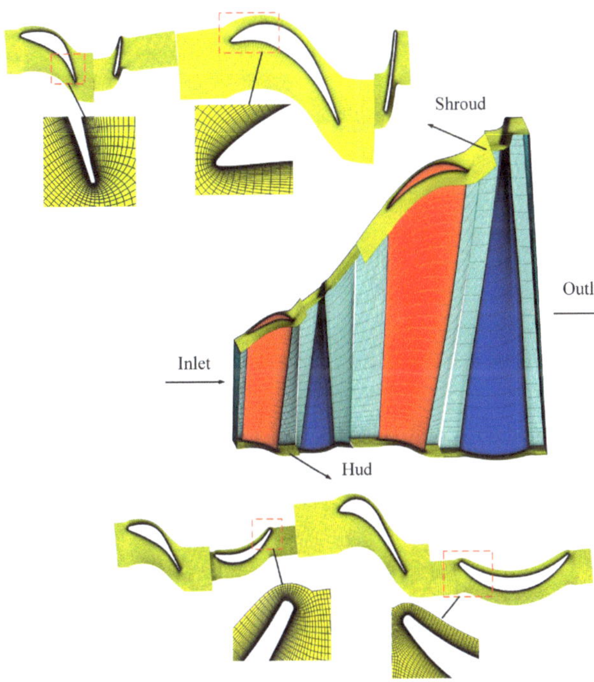

Figure 2. Last two stages of the structure grid and local grid enlargement in the low-pressure cylinder of steam turbine.

2.3. Wet Steam Condensation Flow Model

The wet steam condensation nucleation model adopts the Kantrowitz model modified by the non-isothermal effect [21,22]. The nucleation equation is shown in Equation (1):

$$J_m = \frac{q_c}{1+\theta}\frac{\rho_c^2}{\rho_d}\sqrt{\frac{2\sigma}{\pi m^3}}exp\left(-\frac{4\pi r^{*2}\sigma}{3kT_c}\right) \quad (1)$$

where J_m is the nucleation rate of the droplets, q_c is the condensation coefficient, ρ_c and ρ_d are the densities of the vapor and liquid phases, respectively, σ is the surface tension, m and r^* represent mass and radius of the droplet, respectively, k is the Boltzmann constant, T_c is the gas phase temperature, and θ is the non-isothermal correction coefficient. The subscripts c and d are the vapor phase and the liquid phase, respectively.

2.4. Mathematical Model Verification

In order to ensure the accuracy of results, the mathematical model used in this paper is verified. The simulation results in this paper are compared with the experimental data in reference [23]. First, the same physical model is built as the experimental object. The span, chord length, pitch, axial chord length, and inlet flow angel are 76, 35.76, 18.26, 25.27, and 0, respectively. Then, the same boundary conditions as the experimental conditions were set, and used the mathematical model for calculation in this paper. The inlet pressure and temperature were 172 kPa and 380.7 K, respectively. The outlet and inlet pressure ratio was 0.48.

The static pressure distributions on the suction side and pressure side of the blade obtained by numerical simulation are compared with the experimental results, as shown in Figure 3. It can be seen that the numerical results are basically consistent with the experimental results. Simultaneously, non-equilibrium condensation was used as the numerical calculation method, whereby a large number of studies have verified its accuracy [18,24]. Therefore, the mathematical model used is accurate in this paper, and the simulation results are also credible.

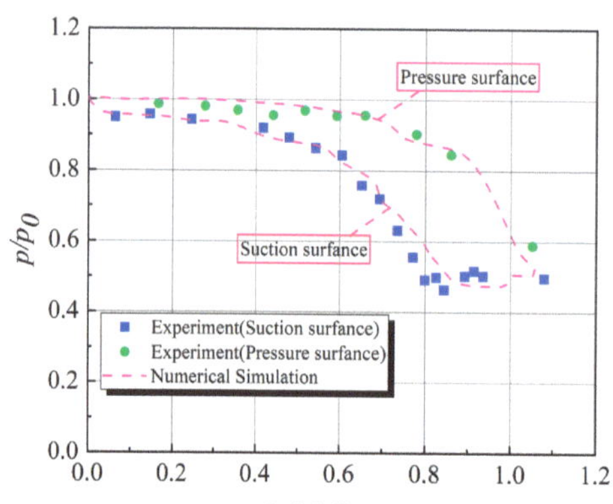

Figure 3. Comparison between numerical simulation results and experimental results.

2.5. Wetness High Loss Composition

According to the different formation mechanisms, wetness loss can be divided into five types: thermodynamic, hydrophobic, water droplet, droplet impact, and centrifugal losses. In the process of steam flow, the proportion of water droplet resistance loss is negligible because the diameter of the primary water droplet is less than 1 μm. The hydrophobic and centrifugal losses can be ignored to account for a small proportion of the wetness loss, which has little impact on the unit [25–27]. Therefore, only two high wetness losses, thermodynamic loss and droplet impingement loss, were analyzed in detail.

2.5.1. Thermodynamic Loss

In the process of the condensation of wet steam in the steam turbine cascade channel, supercooling is caused by steam flow at a high velocity. After non-equilibrium condensation, nucleation leads to thermodynamic losses, which are inseparable from the supercooling value. The formula is shown in Equation (2) [27]:

$$\Delta P_h = \int_{Z_1}^{Z_2} L\frac{dQ_1}{dZ}\frac{\Delta T}{T_s}dZ \tag{2}$$

where Z_1 is the axial starting point coordinate of the turbine stage, Z_2 is the axial endpoint coordinate of the turbine stage, L is the latent heat, Q_1 is the mass flow rate, ΔT is the degree of supercooling, and T_s is the saturation temperature; 1 and 2 indicate start and end, respectively.

2.5.2. Water Droplet Resistance Loss

Under the action of steam flow, the water film deposited on the trailing edge of the stationary blade is torn and converges into larger diameter water droplets called coarse. These cannot efficiently pass through the cascade channel with the steam and impinge on the rotor blades. The rotor blades are inhibited, thereby consuming the mechanical work of the steam turbine and causing losses, known as water droplet impact losses. The formula is shown in Equation (3) [27]:

$$\Delta P_s = M\omega\int_{R_1}^{R_2} P_w l_R R(1-\alpha_2)dR \tag{3}$$

where M and ω represent the number and angular velocity of rotor, respectively, R_1 is the radial coordinates of the blade root, R_2 is the radial coordinate of the blade tip, l_R is the axial direction in the radial direction brake length, and $(1-\alpha_2)$ and P_W represent the volume fraction and hammer pressure of water, respectively.

3. Computational Results and Analysis

3.1. Thermodynamic Loss Distribution Characteristics under Different Steam Turbine Working Conditions

3.1.1. Distribution of the Supercooling Degree under Different Working Conditions

Figure 4 shows the distribution of the degree of supercooling under different working conditions. It can be seen from the figure that irrespective of the blade tip, 50% blade height, and blade root, the extreme value of the supercooling degree, whose value is 20 K, first appears at the trail end of the second-stage stator blades under the same working conditions. Subsequently, the supercooling value began to gradually decrease until it reached 0 K. Before reaching the extreme supercooling degree, the steam expands continuously in the low-pressure cylinder channel and gradually approaches the equilibrium point. The supercooling degree reaches the maximum value because the unbalanced degree of the steam reaches the extreme value. Therefore, this is an important factor affecting the thermodynamic loss, that is, the position where the thermodynamic loss is the largest. Subsequently, the steam expands rapidly at the large local curvature of the trailing edge blade on the coarse last pressure surface. The trailing edge of the pressure surface becomes the peak nucleation area under the action of the violent expansion of the steam flow. A large number of water droplets are generated in the flow channel with steam flow, which weakens the continuous accumulation of the imbalance. At this time, the supercooling value began to decrease until the supercooling value reached 0. With the flow of steam, the supercooling degree reaches a maximum at the trail end of the last-stage stator blades. However, different operating conditions corresponded to different extreme supercooling values. The steam reaches supersonic flow, which accelerates its own expansion rate, resulting in secondary nucleation of the last-stage stator blade, and the supercooling degree increases from zero to the peak again.

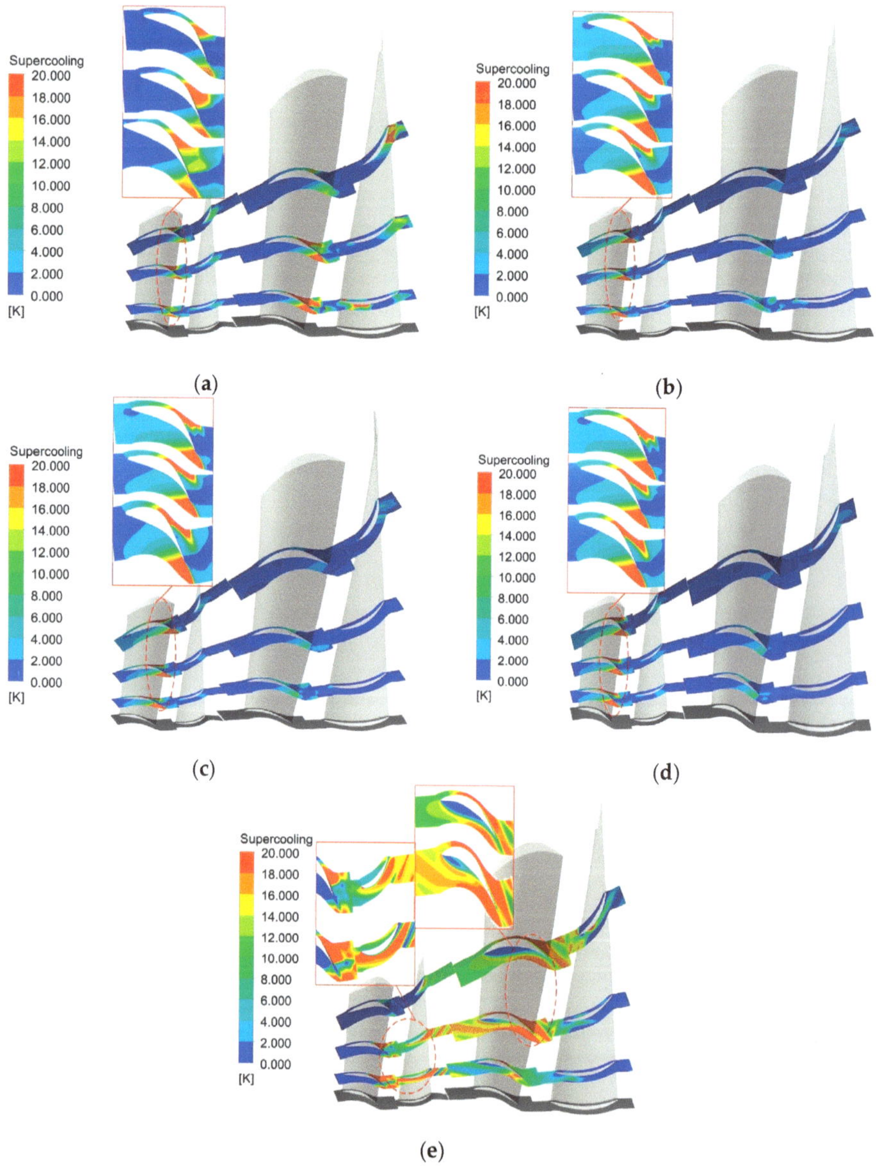

Figure 4. The distribution of supercooling degree along the blade height of steam turbine under different working conditions. (**a**–**e**), respectively, represent 100%, 50%, 40%, 30%, and 20% THA.

3.1.2. Nucleation Rate Distribution for the Last Two Stages of the Steam Turbine under Different Working Conditions

Figure 5 shows the distribution of the nucleation rate along the blade height of the steam turbine under different working conditions. From the above analysis, we know that the supercooling degree gradually increases starting from the second-last-stage stator blades and reaches the maximum value at the trail end of the second-last-stage stator blades. Therefore, a larger supercooling degree leads to the nucleation process of steam from the second-last-stage stator blades, where the peak position of the nucleation rate is also advanced. With an increase in the degree of supercooling, the area of the nucleation rate gradually increases, and the nucleation phenomenon becomes more intense. For five different working conditions, regardless of the blade tip, 50% leaf height, and blade root, the nucleation rate value at the second-last stage of the stator blade is the largest. The corresponding extreme values of the nucleation rate were 30 m^{-3} s^{-1}, 27 m^{-3} s^{-1}, 21 m^{-3} s^{-1}, 9 m^{-3} s^{-1}, and 6 m^{-3} s^{-1} under five working conditions. Subsequently, the supercooling degree begins to gradually decrease to 0 with a decrease in the nucleation rate. Simultaneously, the nucleation rate at the last stage of the stator blade increases again to 30 m^{-3} s^{-1} under the 20% THA condition. This is also because the value of the supercooling degree increases again to a peak value of 20 K under the 20% THA conditions. With a gradual decrease in the unit load, the peak value of the nucleation rate also gradually decreased by approximately five times from 30 m^{-3} s^{-1} in the 100% THA condition to 6 m^{-3} s^{-1} in the 20% THA condition. This is because the steam flow is not sufficiently expanded in the process of spontaneous condensation, which leads to the degree of supercooling not reaching the condensation condition. In addition, the nucleation rate decreased sharply with a decrease in the supercooling degree.

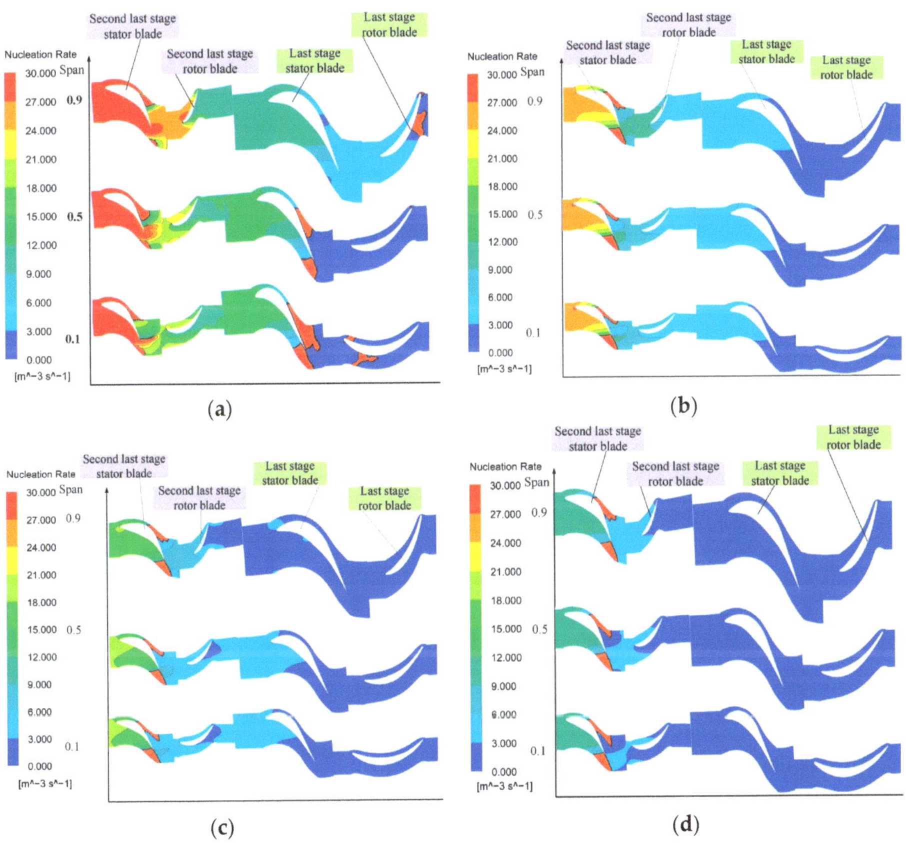

Figure 5. *Cont.*

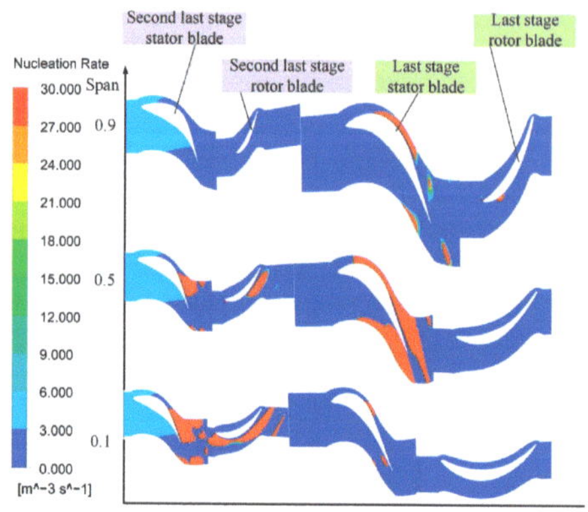

(e)

Figure 5. Distribution of nucleation rate along the blade height of steam turbine under different working conditions. (**a**–**e**), respectively, represent 100%, 50%, 40%, 30%, and 20% THA.

The above analysis of the distribution of the supercooling degree and nucleation rate shows that the maximum of the supercooling degree occurs at the trail end of the second-last-stage stator blades and the last-stage stator blades. This means that the thermodynamic losses in these two places are relatively large, owing to the large influence of the degree of subcooling on the thermodynamic loss. Combined with the analysis of the distribution of the nucleation rate, it was further proved that the thermodynamic losses in these two locations are the largest. Meanwhile, a decrease in the unit load resulted in continuous decrease in the supercooling degree and nucleation rate. Therefore, the thermodynamic losses also decreased as the unit load decreased.

3.2. Water Droplet Resistance Loss under Different Steam Turbine Working Conditions

The diameter distribution of the coarse water droplet has a great influence on the water droplets resistance loss. Therefore, it is necessary to obtain the diameter distribution of the coarse water droplets along the leaf height direction. The windage phenomenon occurs in the steam turbine under the working condition of 20% THA, which causes the temperature to rise in the last two stages. It could not be deposited on the trail end of the blade by a water film.

Hence, only the 100%, 50%, 40%, and 30% THA conditions were analyzed for water droplet resistance loss.

The maximum coarse water droplet diameter was calculated using Equation (4):

$$d_{\max} = \frac{W_e \cdot \sigma}{\rho_g (u - u_p)^2} \tag{4}$$

where W_e is the critical Weber number, ρ_g and u represent the density and velocity of steam, u_p is the movement velocity of the water droplet, and $d_{\max}$ is the maximum coarse water droplet diameter. The subscript g stands for steam.

According to the results of the experimental measurement, the diameter of the coarse water droplets approximately satisfies the normal distribution law; thus, half of the maximum diameter is taken as the average diameter of the coarse water droplets [28].

The calculation formula is shown in Equation (5):

$$\overline{d} = \frac{1}{2} d_{\max} \tag{5}$$

where $\overline{d}$ is the average diameter.

3.2.1. Velocity Distribution of Steam under Different Working Conditions

It can be seen from Equation (4) that the steam velocity needs to be obtained. As shown in Figure 6, it is the distribution of steam velocity in the last two stages of the steam turbine under different working conditions. For four working conditions, the value of steam velocity decreases gradually from the blade root to the blade tip, and the value ranges between 100 m/s and 500 m/s.

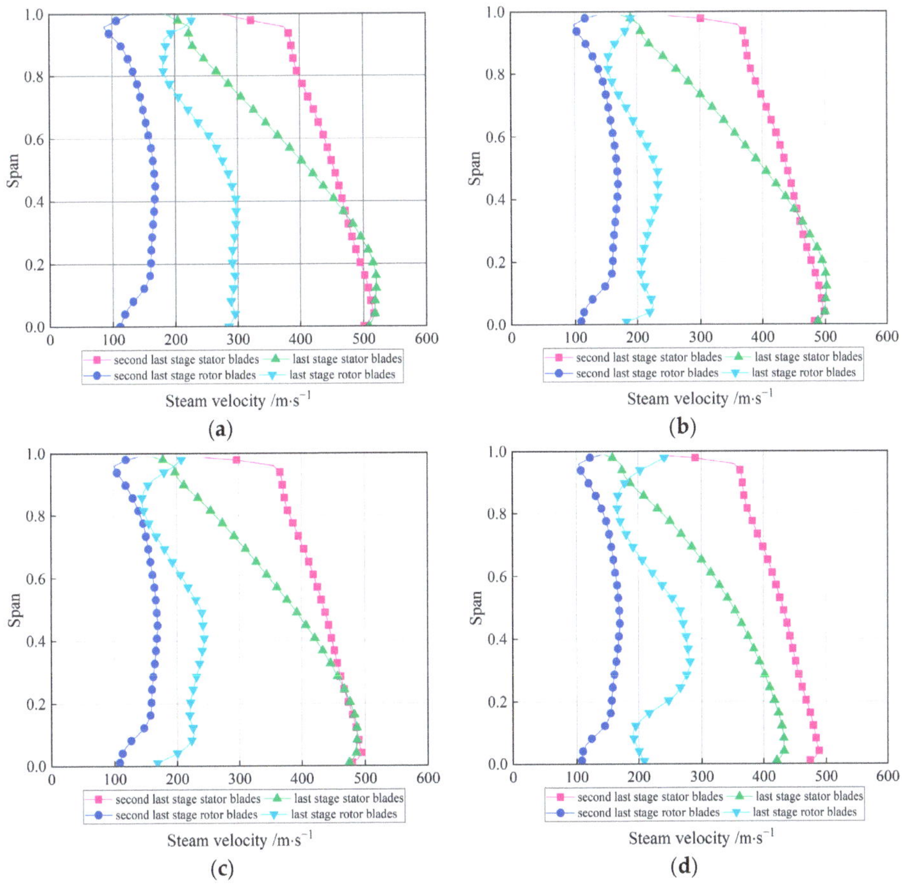

Figure 6. Steam velocity under different working conditions. (**a**–**d**), respectively, represent 100%, 50%, 40%, and 30% THA.

3.2.2. Diameter Distribution of Coarse Water Droplets in the Last Two Stages under Different Steam Turbine Working Conditions

When wet steam flows in the passages of the last two stage blades of the turbine, the velocity of the water droplets on the trailing edges of the last two stage stator blades is 0, and the velocity on the trailing edges of the last two stage rotor blades is equal to the rotation velocity of the blades. The diameter distribution of coarse water droplets can be obtained using Equations (4) and (5), as shown in Figure 7.

The water film deposited on the trail end of the blade begins to tear under the action of steam, and then converges to form coarse water droplets, whose diameter is larger, which has a greater impact on the turbine blades and is the key factor causing the water droplets resistance loss. Figure 6 shows the water droplets diameter distribution along the blade height with different working conditions. For the same working condition, the diameter of the stator blades increases from the root to the tip, whereas the diameter of the rotor blade increases from the root to 40% of the blade height; and from the tip to 40%, the leaf height is reduced.

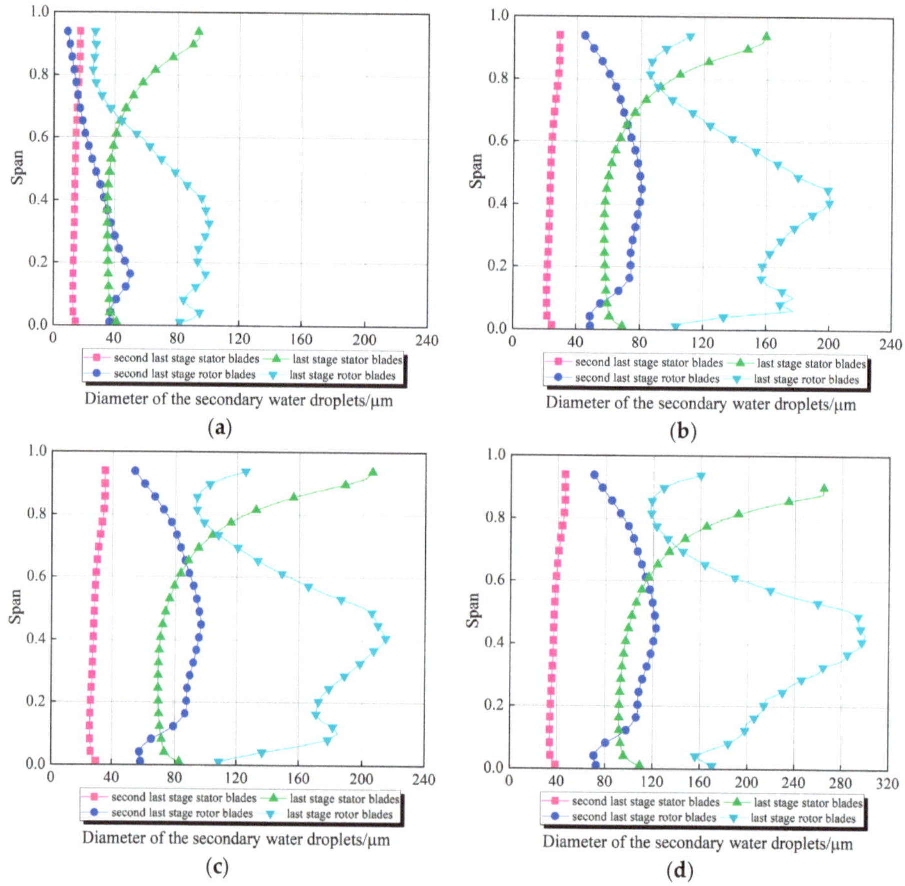

Figure 7. Diameter distribution of coarse water droplets along the blade height under different working conditions. (**a**–**d**), respectively, represent 100%, 50%, 40%, and 30% THA.

Based on the above analysis, the maximum diameter of the coarse water droplets occurs at the tip of the last-stage stator blade and at 40% of the blade height of the rotor blade, which indicates that the water droplet resistance loss is the largest at these two locations. Simultaneously, as the unit loads decreased, the diameter of the coarse water droplets gradually increased. In other words, the corresponding water droplet resistance loss increased.

3.3. Analysis of High Wetness Loss Areas under Different Working Conditions

Figure 8 shows the distribution of the high wetness loss area of the steam turbine under different working conditions. The two major losses that make up the wetness loss are thermodynamic loss and water droplet resistance loss. The figure above shows the areas where these two major losses occur under different working conditions of the steam turbine. It can be seen from the figure that thermodynamic loss mainly occurs at the leading edge of the second-stage stator blades and the trail end of the last-stage stator blades. The water droplet resistance loss mainly occurs at 40% of the blade height and at the tip of the last-stage stator blades. Simultaneously, with a reduction in the unit load, the thermodynamic loss continues to decrease, but the water droplet resistance loss continues to increase.

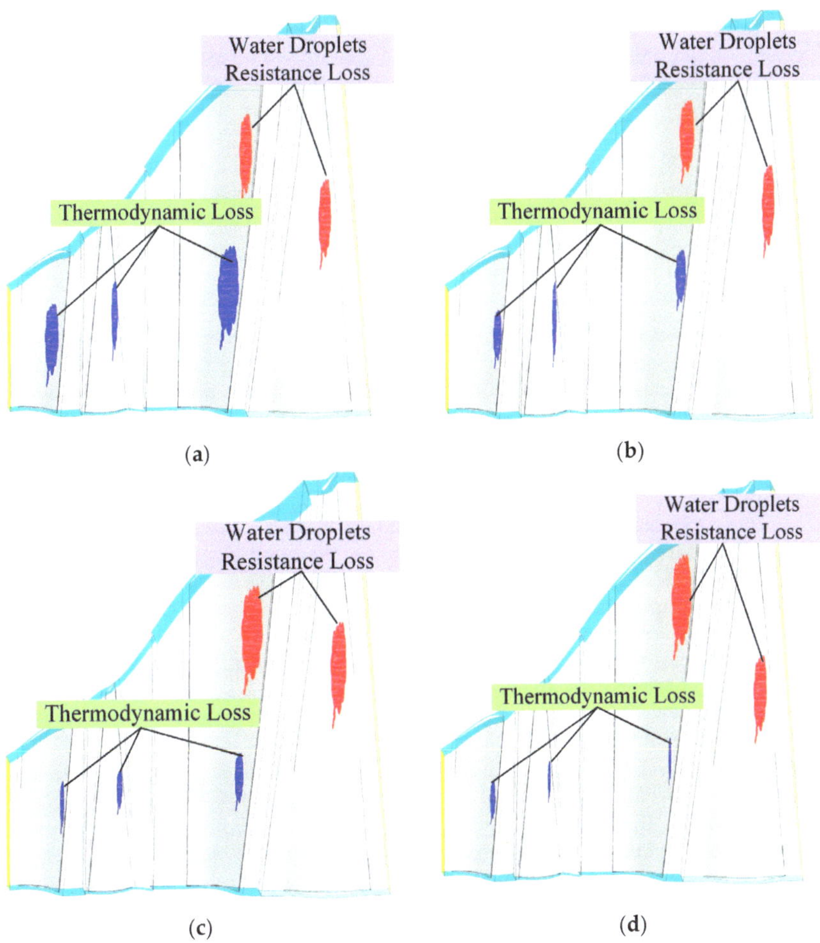

Figure 8. Distribution of high wetness loss area under different working conditions. (**a**–**d**), respectively, represent 100%, 50%, 40%, and 30% THA.

3.4. Method of Controlling High Wetness Loss

After obtaining the distribution characteristics of the thermodynamic and droplet impingement losses, an effective method is proposed to control these two high moisture losses. Surface heating treatment is performed on the dynamic and static blade cascades in two stages at the end of the low-pressure cylinder of the steam turbine [29]. In this way, the growth rate of droplets is gradually reduced, and the condensation nuclei require a longer time to form droplets, thereby reducing the thermodynamic loss. At the same time, the reduction in the growth rate of water droplets leads to a smaller droplet radius, which in turn reduces the diameter of the secondary water droplet, and finally reduces the droplet impact loss.

The degree of supercooling and nucleation rate were analyzed. Consequently, the distribution of thermodynamic losses was obtained in the last two stages of the LPC under different working conditions of the steam turbine; the distribution of water droplet resistance losses was acquired by studying the diameter distribution of the coarse water droplets. The two regions of high wetness losses were analyzed and the following results were obtained: thermodynamic loss mainly occurred at the front end of second-stage stator blades and trailing end of the last-stage stator blades. The water droplet resistance loss mainly occurred at 40% of the blade height and at the tip of the last-stage stator blades. Moreover, with a reduction in the unit load, the thermodynamic loss continued to de-crease, but the water droplet resistance loss continued to increase.

4. Conclusions

Taking a 600 MW steam turbine as the research object, the nucleation phenomenon of wet steam was simulated. The distributions of the degree of supercooling, nucleation rate, and steam velocity were obtained, and the diameter distribution of the coarse water droplets was calculated. Finally, the distributions of the two wetness high-loss regions, namely, the thermodynamic loss and water droplet resistance loss, were obtained. The conclusions are as follows.

(1) Regardless of the blade tip, 50% blade height, and blade root, the extreme value of the supercooling degree with a value of 20 K first appears at the trailing edge of the second-last-stage stator blades under the same working conditions. Subsequently, the supercooling value began to gradually decrease until it reached 0 K. The value of the nucleation rate at the second-last-stage stator blade was the largest. The corresponding extreme values of the nucleation rate were 30 $m^{-3}\ s^{-1}$, 27 $m^{-3}\ s^{-1}$, 21 $m^{-3}\ s^{-1}$, 9 $m^{-3}\ s^{-1}$, and 6 $m^{-3}\ s^{-1}$ under the five working conditions. Subsequently, the value of the nucleation rate begins to decrease. However, the nucleation rate at the last stage of the stator blade increased again to 30 $m^{-3}\ s^{-1}$ under the 20% THA condition.

(2) For four working conditions, the value of steam velocity decreases gradually from the blade root to the blade tip, and the value ranges between 100 m/s and 500 m/s. The maximum diameter of the coarse water droplets occurs at the tip of the last-stage stator blade and 40% of the blade height of the rotor blade. At the same time, as the unit loads decrease, the diameter of the secondary coarse water droplets gradually increases.

(3) Thermodynamic loss mainly occurs at the leading edge of the second-stage stator blades and trail end of the last-stage stator blades. The water droplet resistance loss mainly occurs at 40% of the blade height and at the tip of the last-stage stator blades. Moreover, with a reduction in the unit load, the thermodynamic loss continued to decrease, but the water droplet resistance loss continued to increase.

Author Contributions: All authors contributed equally for the investigation task and presented the article in its final format as full research work outcome. All authors have read and agreed to the published version of the manuscript.

Funding: This work was supported by the National Key R&D Program of China (Grant No. 2017YFB0902100) and the China Postdoctoral Science Foundation (Grant No. 2021T140154).

Informed Consent Statement: Not applicable.

Conflicts of Interest: The authors declare no conflict of interest.

References

1. Wen, C.; Yang, Y.; Ding, H.; Sun, C.; Yan, Y. Wet steam flow and condensation loss in turbine blade cascades. *Appl. Therm. Eng.* **2021**, *189*, 116748. [CrossRef]
2. Gribin, V.G.; Tishchenko, A.A.; Tishchenko, V.A.; Gavrilov, I.Y.; Sorokin, I.Y.; Alexeev, R.A. Experimental Study of the Features of the Motion of Liquid-Phase Particles in the Interblade Channel of the Nozzle Array of a Steam Turbine 1. *Power Technol. Eng.* **2017**, *51*, 82–88. [CrossRef]
3. Tishchenko, V.A.; Alekseev, R.A. Numerically Simulating the Formation and Motion of Water Films and Erosion-Hazardous Droplets in Flow-Through Parts of Steam Turbines. *Therm. Eng.* **2019**, *66*, 830–838. [CrossRef]

4. Li, L.; Feng, Z.; Li, G. Numerical Simulation of Wet Steam Condensing Flow with an Eulerian/Eulerian Model. *Chin. J. Mech. Eng.* **2003**, *16*, 156–159. [CrossRef]
5. Senoo, S.; White, A.J. Numerical simulations of unsteady wet steam flows with non-equilibrium condensation in the nozzle and the steam turbine. In Proceedings of the Fluids Engineering Division Summer Meeting, Miami, FL, USA, 17–20 July 2006; Volume 47500, pp. 757–767.
6. Han, X.; Han, Z.; Zeng, W.; Li, P.; Qian, J. Numerical simulation of wet steam transonic condensation flow in the last stage of a steam turbine. *Int. J. Numer. Methods Heat Fluid Flow* **2018**, *28*, 2378–2403. [CrossRef]
7. Aliabadi, M.A.F.; Zhang, G.; Dykas, S.; Li, H. Control of two-phase heat transfer and condensation loss in turbine blade cascade by injection water droplets. *Appl. Therm. Eng.* **2021**, *186*, 116541. [CrossRef]
8. Baumann, K. Recent Developments in Steam Turbine Practice. *J. Inst. Electr. Eng.* **1912**, *48*, 768–842. [CrossRef]
9. Jaiswal, A.K.; Khandekar, S. Dynamics of a Droplet Impacting a Sessile Droplet on a Superhydrophobic Surface: Role of Boundary Conditions during Droplet Placement. *J. Flow Vis. Image Process.* **2021**, *28*, 69–89. [CrossRef]
10. Zhan, Y.; Kuwata, Y.; Okawa, T.; Aoyagi, M.; Takata, T. Experimental Study on Coarse Droplets Produced During Liquid Jet Impingement onto A Horizon Solid Surface. *Exp. Therm. Fluid Sci.* **2021**, *120*, 110249. [CrossRef]
11. Lopez-Gavilan, P.; Velazquez, A.; García-Magariño, A.; Sor, S. Breakup Criterion for Droplets Exposed to the Unsteady Flow Generated by An Incoming Aerodynamic Surface. *Aerosp. Sci. Technol.* **2020**, *98*, 105687. [CrossRef]
12. Li, L.; Li, Y.; Wu, L.; You, W.; Li, J.; Feng, Z. Numerical Study on Condensing Flow in Low Pressure Cylinder of A 300MW Steam Turbine. In Proceedings of the Turbo Expo: Power for Land, Sea, and Air, Glasgow, UK, 14–18 June 2010; Volume 44021, pp. 2289–2296.
13. Kawagishi, H.; Onoda, A.; Shibukawa, N.; Niizeki, Y. Development of Wetness Loss Models in Steam Turbines. *Heat Transf.—Asian Res.* **2013**, *42*, 651–664. [CrossRef]
14. Starzmann, J.; Casey, M.; Sieverding, F. Non-Equilibrium Condensation Effects on the Flow Field and the Performance of a Low Pressure Steam Turbine. In Proceedings of the Turbo Expo: Power for Land, Sea, and Air, Glasgow, UK, 14–18 June 2010; Volume 44021, pp. 2199–2208.
15. Joseph, J.; Subramanian, S.; Vigney, K.; Prasad, B.V.S.S.S.; Biswas, D. Thermodynamic wetness loss calculation in nozzle and turbine cascade: Nucleating steam flow. *Heat Mass Transf.* **2018**, *54*, 2521–2531. [CrossRef]
16. Aliabadi, M.A.F.; Lakzian, E.; Jahangiri, A.; Khazaei, I. Numerical investigation of effects polydispersed droplets on the erosion rate and condensation loss in the wet steam flow in the turbine blade cascade. *Appl. Therm. Eng.* **2020**, *164*, 114478. [CrossRef]
17. Yang, Y.; Peng, H.; Wen, C. A Novel Dehumidification Strategy to Reduce Liquid Fraction and Condensation Loss in Steam Turbines. *Entropy* **2021**, *23*, 1225. [CrossRef] [PubMed]
18. Cao, L.; Wang, J.; Luo, H.; Si, H.; Yang, R. Distribution of condensation droplets in the last stage of steam turbine under small flow rate condition. *Appl. Therm. Eng.* **2020**, *181*, 116021. [CrossRef]
19. Laksana, A.R.; Kokong, A.M.; Sigit, P.A.; Widjajanto, T.; Setiawan, H.; Djunaedi, I. Flutter Analysis of Last Stage Steam Turbine Power Plant Blade Through Transient Blade Row Simulation. In *IOP Conference Series: Materials Science and Engineering*; IOP Publishing: Atlanta, GA, USA, 2021; Volume 1096, p. 012093.
20. Fan, S.; Wang, Y.; Yao, K.; Fan, Y.; Wan, J.; Gu, W. Research on Water Droplet Movement Characteristics in the Last Two Stages of Low-pressure Cylinder of Steam Turbine under Low Load Conditions. *Front. Energy Res.* **2021**, *9*, 798305. [CrossRef]
21. Kantrowitz, A. Nucleation in Very Rapid Vapor Expansions. *J. Chem. Phys.* **1951**, *19*, 1097–1100. [CrossRef]
22. Gyarmathy, G. *Grundlagen Einer Theorie Der Nassdampfturbine*; Jurisv-Erlag: Zürich, Switzerland, 1962.
23. Ding, H.; Li, Y.; Lakzian, E.; Wen, C.; Wang, C. Entropy generation and exergy destruction in condensing steam flow through turbine blade with surface roughness. *Energy Convers. Manag.* **2019**, *196*, 1089–1104. [CrossRef]
24. Han, Z.; Zeng, W.; Han, X.; Xiang, P. Investigating the dehumidification characteristics of turbine stator cascades with parallel channels. *Energies* **2018**, *11*, 2306. [CrossRef]
25. Li, L.; Li, Y.; Xie, X.; Li, J. Quantitative evaluation of wetness losses in steam turbines based on three-dimensional simulations of non-equilibrium condensing flows. *Proc. Inst. Mech. Eng. Part A J. Power Energy* **2014**, *228*, 708–716. [CrossRef]
26. Petr, V.; Kolovratnik, M. Wet steam energy loss and related Baumann rule in low pressure steam turbines. *Proc. Inst. Mech. Eng. Part A J. Power Energy* **2014**, *228*, 206–215. [CrossRef]
27. Starzmann, J.; Casey, M.M.; Mayer, J.F.; Sieverding, F. Wetness loss prediction for a low pressure steam turbine using computational fluid dynamics. *Proc. Inst. Mech. Eng. Part A J. Power Energy* **2014**, *228*, 216–231. [CrossRef]
28. Yang, B.; Xiang, Y.; Cai, X.; Zhou, W.; Liu, H.; Li, S.; Gao, W. Simultaneous measurements of fine and coarse droplets of wet steam in a 330 MW steam turbine by using imaging method. *Proc. Inst. Mech. Eng. Part A J. Power Energy* **2017**, *231*, 161–172. [CrossRef]
29. Han, X.; Zeng, W.; Han, Z. Investigating the dehumidification characteristics of the low-pressure stage with blade surface heating. *Appl. Therm. Eng.* **2020**, *164*, 114538. [CrossRef]

energies

MDPI

Article

Systematic Method for Developing Reference Driving Cycles Appropriate to Electric L-Category Vehicles

David Watling [1,*], Patrícia Baptista [2], Gonçalo Duarte [2,3], Jianbing Gao [4] and Haibo Chen [1]

1 Institute for Transport Studies, University of Leeds, Leeds LS2 9JT, UK; h.chen@its.leeds.ac.uk
2 Centre for Innovation, Technology and Policy Research (IN+), Associação para o Desenvolvimento do Instituto Superior Técnico, Universidade de Lisboa, Avenida Rovisco Pais, 1049-001 Lisboa, Portugal; patricia.baptista@tecnico.ulisboa.pt (P.B.); goncalo.duarte@isel.pt (G.D.)
3 Mechanical Engineering Department, Instituto Superior de Engenharia de Lisboa (ISEL), Rua Conselheiro Emídio Navarro 1, 1959-007 Lisboa, Portugal
4 School of Mechanical Engineering, Beijing Institute of Technology, Beijing 100081, China; gaojianbing@bit.edu.cn
* Correspondence: d.p.watling@its.leeds.ac.uk

Abstract: Increasingly, demanding environmental standards reflect the need for improved energy efficiency and reduced externalities in the transportation sector. Reference driving cycles provide standard speed profiles against which future developments and innovations may be tested. In the paper, we develop such profiles for a class of electric L-category vehicles, which are anticipated to play an increasing future role in urban areas. While such driving cycles exist for regular L-category vehicles, these may not be suitable in the case of electric vehicles, due to their power output limitations. We present a methodology for deriving these new driving cycles, developed from empirically deduced power relationships, before demonstrating their application under different assumptions on the terrain and vehicle characteristics. The applications demonstrate the feasibility of the method in developing appropriate driving patterns for alternative real-world contexts. On flat terrain, the adjustments made to cope with the power limitations of L-EV do not introduce significant differences in energy consumption, suggesting that the certification does not require extensive modification. However, when considering road slope, differences of up to 5% in energy use and up to 10% in regenerated energy were observed, showing the importance of the developed method for assessing vehicle performance in real-world driving.

Keywords: vehicle specific power; driving cycle; regenerative braking; powered light vehicle; e-bike; micro-mobility

1. Introduction

Due to the severe problems, such as global warming and harm to human health, caused by internal combustion engine-powered vehicles, electric vehicles have been attracting increasing attention [1–3]. They enable higher energy efficiency without exhaust emissions, and the use of electricity incorporating higher percentages of renewable energy resources plays a significant role in enabling improved life-cycle impacts compared to conventional technologies [4,5]. The inventory of electric vehicles continues to increase significantly, with over 7 million vehicles worldwide in 2019 [6]. Furthermore, their purchase cost barriers are expected to be alleviated by the anticipated evolution in battery technology [6,7]. Consequently, electric vehicles are expected to be the main mode of personal passenger transport in the future.

Furthermore, urban mobility brings additional challenges in terms of use of space, parking requirements, and the associated drive cycle, bringing opportunities for small electric vehicles, which have the particular benefits of small size, low cost, and low energy consumption for short-distance travelling [8]. While many different names and classifications exist for such vehicles around the world, we refer to them as *L-category electric vehicles*, following the terminology used in Europe and based on the class of vehicles that is specifically used in our case study [9]. This class of vehicles is framed within the L7 category for quadricycles, whose unladen mass is not more than 400 kg (not including the mass of batteries in the case of electric vehicles) and whose maximum net engine power does not exceed 15 kW [10]. These small vehicles are generally within the M1 category according to the UNECE global technical regulations [11,12]. However, this classification changes over the geographical area. In India, quadricycles must comply with the maximum permissible kerb weight of 450 kg in case of a passenger vehicle, up to 15 kW maximum power and a maximum speed of 70 km/h [13]. In Korea, these vehicles are included within the concept of micro-mobility, defining maximum allowable dimensions (length up to 3.6 m, width up to 1.5 m, height up to 2.0 m), as well as a maximum power of 15 kW and a maximum mass of 600 kg [14]. In the U.S., quadricycles fall within the class of Low-Speed Vehicles, whose speed attainable in 1.6 km (1 mile) is more than 32 km/h and lower than 40 km/h, also limiting Gross Vehicle Weight Rating up to 1134 kg due to safety purposes [15,16]. Elsewhere, other regional rules can apply, however it can be stated that, generally, quadricycles represent a mobility opportunity due to their small size, low power and weight, energy efficiency, and suitability for all types of drivers.

With their more efficient use of space due to their small size, the increased use of L-category electric vehicles has the potential to reshape urban mobility. One of the main applications is as the first-mile or last-mile access mode for a longer journey [17]. Especially under epidemic situations, L-category electric vehicles could be a good option to decrease infection spread by reducing public transport use for those who cannot afford conventional electric vehicles. The relatively low purchase and operating costs of such vehicles also provide them with the potential to better address concerns over social inequalities concerning energy poverty [18].

Thus, although the current market for L-category vehicles is relatively small (Santucci et al., 2016, estimated around 10,000 had been sold within EU countries by that date), their great potential in benefiting our lives suggests they deserve greater attention. Indeed, as the market analysis model forecasts have shown [19], their sales volume is likely to continue to grow for the foreseeable future. One of the first examples of these L-category electric vehicles was the Renault Twizy [20], launched in 2012, with a more recent (2020) example the Citröen Ami [21], suggesting that this type of vehicle is seen as an opportunity by major vehicle manufacturers. According to the plan approved by the European Commission for achieving clean urban transport, the goal is that "by 2050 nearly all cars, vans, buses as well as new heavy-duty vehicles will be zero-emission" [22], while at the same time encouraging the use of smaller, lighter and more specialized road passenger vehicles [8], namely focused on micro-mobility.

The popularization of L-category electric vehicles will most probably be fostered by the increasingly accepted concept of micro-mobility and mobility as a service, but its adoption relies on the existence of a correlated test procedure necessary to ensure their production consistency and to meet the requirements of consumers, namely regarding energy consumption. For M-category (light-duty vehicles, passenger cars and vans, carrying passengers) electric vehicles, certification driving cycles and procedures are well established. However, their representativeness can be discussed, since differences have been found between energy use within certification cycles and real-world conditions [23–26]. Consequently, even for M-category vehicles, the pursuit of representative drive cycles for alternative vehicle technologies based on real-world use has also started to be explored. For example, Zhang et al. developed tailored driving cycles for electric vehicles based on a sample of 40 electric taxis, by developing a Markov Monte Carlo method considering the driving features of different roads [27]. Defining a driving cycle for a test procedure is the main basis for assessing vehicle performance [28,29], particularly for electric vehicles where regeneration plays an important role in the gross energy consumption.

A detailed test procedure is available for a different type of L-category vehicle to that considered in the present paper, namely those powered by an internal combustion engine [30]. However, since L-category vehicles have power and weight limits (without a battery), additional mass associated with energy storage could limit the power-to-weight ratio, possibly leading to differences in their driving capabilities, as has been observed from comparisons of conventional L-category vehicles and electric ones [31,32]. Problems may therefore arise if the driving cycle of a conventional L-category vehicle is applied to the performance assessment of electric modes, such as the electric ones being unable to follow the driving cycle due to their small-rated power-to-weight output.

Due to the short running distance of L-category electric vehicles, they are more suitable for urban utilization. In real-world driving, traffic conditions are much more complex than the ones considered in driving cycles [33]. One study quantified that traffic conditions and driving behaviour may increase energy consumption by up to 40% and 16%, respectively, compared to the worst performance condition [34]. Regarding road grade, ascending roads (with 3% grade) increased energy consumption by 50% while descending roads (−3% grade) decreased energy consumption by 80%, due to the presence of regenerative braking systems. These facts may lead to more difficulty for L-category electric vehicles in real driving, such as on hilly terrain where acceleration is limited due to their power output.

Additionally, in urban driving, vehicles are periodically under stop-and-go situations [35] (e.g., resulting from many traffic lights), which will cause much energy loss if the energy regeneration technology is missing [25]. With regards to the L-category electric vehicles on the market, most of them do not include energy recovery devices leading to lower energy efficiency performance. Energy loss will be significant if the vehicle is travelling downhill where brake actions are necessary to ensure safety and to be within the speed limit. Furthermore, a better understanding of real-world performance in specific routes, acknowledging the influence of road grade and traffic, may also contribute to better route optimization and charging optimization systems [36,37].

Therefore, it has been shown that M-category vehicles have well-established certification procedures but do not incorporate real driving energy consumption in a generalized way. For L-category vehicles, the certification procedures are mostly suited to conventional propulsion technologies and do not consider real driving energy use, which gains importance due to their limited power-to-weight ratio or the possibility of regeneration. Taking into consideration the emergence of L-category electric vehicles in an urban context and their power-to-weight limitations, this paper proposes an innovative approach for evaluating current certification drive cycles available for this vehicle class and compares them with real-world driving cycles. The impacts of regenerative capabilities and road grade on the L-category vehicles' energy performance are also assessed, paving the way for improved future energy characterization of a growing vehicle market.

The structure of the paper is as follows. Firstly, in Section 2 we describe the previously-reported experimental data that we will re-purpose for the study, and begin to explain the specific motivation for our work. In Section 3 we present the details of a generic methodology for modifying a given reference speed profile to apply to a vehicle with generally lower power capabilities. This methodology is applied in Section 4 to a particular case of electric L-category vehicles, where the sensitivity of the method is explored in tests on fictitious gradient profiles. In Section 5, it is applied to a realistic case of road profiles from the city of Lisbon. Finally, conclusions and directions for further research are described in Section 6.

2. Experimental Data Description and Motivation

2.1. Empirical Evidence of L-Category Electric Vehicle Characteristics

The method to be described requires knowledge of the characteristics of the electric vehicle under study, in particular the relationship between velocity, acceleration, and *vehicle specific power* [38]. Here, we draw on the relationships derived in a real-world study of electric quadricycles [23]. The general specifications of the quadricycles used as a basis for this work, are presented in Table 1 and can be found in more detail in reference [23].

Table 1. Specification of test vehicles.

Vehicle	Dimensions (m) Length; Width; Height	Total Weight w/Batteries (kg)	Max. Power (kW)	Motor Technology	Battery Technology
Type 1	1.74; 1.03; 1.57	370	4	DC	Pb-Acid
Type 2	2.40; 1.03; 1.50	400	4	DC	Pb-Acid
Type 3	2.63; 1.32; 1.51	565	8	AC induction	Pb-Acid

Let $f(v, a; \boldsymbol{\theta}(x))$ denote the function whose output is the instantaneous Vehicle Specific Power (VSP) (in W/kg) corresponding to a velocity of v (m/s) and acceleration a (m/s^2), given a parameter vector $\boldsymbol{\theta}(.)$ that defines the relevant properties of the vehicle type under consideration and the local road surface at a given location x. Alves et al. [23] demonstrate how a simplified form of this relationship may be derived for $f(.)$, of the form:

$$f(v, a; \boldsymbol{\theta}(x)) = (1.1a + 9.81\theta_1(x) + \theta_2)v + \theta_3 v^3 \tag{1}$$

where $\theta_1(x)$, θ_2 and θ_3 are respectively the grade (in m/m) at location x, coefficient of rolling resistance (in N/kg) and coefficient of aerodynamic drag (in Ns^2/m^{2}kg).

In addition, Alves et al. posited a relationship for the function $e(p; \boldsymbol{\beta})$, denoting the instantaneous Energy Consumption Rate (ECR, in Wh/s) corresponding to an instantaneous VSP of p (W/kg), given parameter vector $\boldsymbol{\beta}$ (which differs by vehicle type). They supposed $e(.)$ to be a continuous, piecewise-linear function of the form:

$$e(p; \boldsymbol{\beta}) = \begin{cases} \beta_1 p + \beta_2 & p < -3 \\ \beta_3 p + \beta_4 & -3 \leq p < 0 \\ \beta_5 p + \beta_6 & 0 \leq p < 3 \\ \beta_7 p + \beta_8 & p \geq 3 \end{cases}. \tag{2}$$

Note that this energy relationship explicitly models negative instantaneous events arising from negative VSP, arising from the regenerative braking capability of the vehicles considered. While Equation (1) is the main element required to apply our method, Equation (2) will prove useful in our sensitivity analyses, to judge the consequences of different assumptions on overall energy consumption.

Relationship (1) will be used later in the paper as the basis for developing relevant speed profiles, and relationship (2) will be used to assess the energy impacts of alternative speed profiles. In fact our methodology is rather generic, in that if desired alternative power and energy relationships may be implemented using the same techniques. Before continuing we provide some reflection on our choice of (1) and (2) as the basis of our numerical demonstrations, and discuss how alternative approaches may be implemented:

1. In our analysis we suppose a single value for the coefficient of rolling resistance, θ_2 in (1). The methodology used is based on already published work, which follows established methods, namely the Vehicle Specific Power (VSP) methodology [23,38–43]. The simplicity and accuracy of the VSP which is followed here, with coefficients provided from already published work [23] is one of the main strengths of this method. In reality θ_2 may vary over time depending, for example, on whether road surface conditions are wet or dry. The method we subsequently present is sensitive to the value of θ_2, and so if different values of θ_2 were known relevant to the prevailing weather conditions, then different speed profiles may be deduced relevant to the situation under study. In addition our method allows sensitivity testing of the impact of deriving the speed profile with (say) a wet weather θ_2 when the weather is in fact dry; this would follow the same process as we follow in Section 4.2 for testing

a different parameter, when instead we test the effect of using a profile derived for flat terrain on a hilly terrain. A simple extension of this would be to represent the impact of modeller uncertainty by assuming a distribution of true θ_2 values, randomly sampling from this distribution, and then evaluating the performance of a speed profile derived from some given $\theta_2 = \hat{\theta}_2$.

2. Our methodology and derived speed profiles do not depend on (2); however, this equation is used to evaluate the energy impacts of the speed profiles, and so this evaluation is sensitive to the assumed value of the $\boldsymbol{\beta}$ parameter. The detailed process for estimating the values in $\boldsymbol{\beta}$ we shall assume is described in the source reference [23], based on 1 Hz data in order to represent a large range of driving conditions. Again, this is based on the scientifically accepted VSP methodology, which uses representative coefficients obtained from the characteristics of the vehicles studied. For instance, rolling resistance is based on the tyre properties, and the aerodynamic coefficient is based on the aerodynamic resistance coefficient and frontal area. Since a modal analysis is made, the numerical value calculated for VSP based on the coefficients is the same as that collected under real-world conditions. Consequently, a VSP value calculated for a speed cycle (based only on speed, acceleration and topography) corresponds to a certain amount of energy used (monitored, measured) or harvested under real-world conditions with a similar VSP value. These variables have physical meaning and are adapted for bicycles, motorcycles, quadricycles, light-duty and heavy-duty vehicles. However, they do not interfere deeply in the implementation of VSP, since measured energy consumption values are associated to driving conditions, allowing the estimation of energy used on a driving cycle without effectively running that driving cycle with a particular vehicle. If it is desired to test the sensitivity with respect to $\boldsymbol{\beta}$ then we note that this is even more easily achieved than testing the impact of alternative power assumptions, since the derived speed profiles do not depend on the relationship between energy and VSP, only on the relationship between VSP and speed, acceleration and the features of the vehicle and physical environment.
3. The remarks in the two points above are examples of a wider issue, whereby it could be argued that all of the parameters in (1) and (2) have uncertainty associated with them, either because they vary in the real-world (within or between vehicles, or due to environmental conditions), or because of modeller uncertainty. Our methodology is readily applied to study such impacts by the kinds of methods already described in the two points above. In our later analysis (Section 4) we study one such example of real variation by exploring the impact of different types of vehicle, as well as one such example of modeller uncertainty in terms of road topography. Such an analysis could be repeated with a range of randomly sampled values, to represent either true variation or modeller uncertainty, and could be repeated for the whole range of parameters $\boldsymbol{\beta}$, $\theta_1(x)$, θ_2 and θ_3. On the other hand, when considering VSP in particular, the use of fixed values for the parameters of the VSP relationship has been widely studied, with calibration results leading to errors below 5% for electric vehicles [23]. Consequently, our own subsequent analysis based on fixed parameter values is unlikely to lead to significant error.
4. Alternative non-linear forms to the ECR relationship (2) may also be considered; the piecewise-linear form was motivated in the original source paper by considering the 'breakpoints' at which different energy consumption states arise. By combining the above equations for VSP and ECR, the basic assumption is that the instantaneous energy consumption at time t is related to the vehicle velocity and acceleration *at the same time* t. Our ultimate objective will be to adapt a reference speed profile (developed for a different vehicle technology) to conform to the physical constraints of a new technology, as captured in the above equations. Alternative methods exist for more sophisticated analysis of such temporal profiles, such as wavelet transformations, and these would be particularly suited to future research focused on forecasting [44] or feature extraction [45].

In their empirical study [23], Alves et al. estimated the parameters of Equation (1) for the quadricycle class to be:

$$\theta_2 = 0.159 \quad \theta_3 = 0.984 \times 10^{-3} \tag{3}$$

and estimated the parameters of Equation (2) to differ by the specific type of quadricycle monitored, as in Table 2.

Table 2. Estimated coefficients for relationship between ECR and VSP for three types of quadricycle.

	β_1	β_2	β_3	β_4	β_5	β_6	β_7	β_8
Type 1	0.12067	0.25192	0.04765	0.11759	0.02083	−0.02056	0.02311	0.02083
Type 2	0.1101	0.26834	0.04526	0.11621	0.02778	−0.02039	0.02379	0.02778
Type 3	0.19095	0.16002	0.05973	0.10772	0.05556	−0.03427	0.04107	0.05556

A final important estimated element was the maximum VSP. Rather than a physical limit of the battery/vehicle, the estimated value was more of a 'behavioural limit', namely a value below which 99% of the measured VSPs fell in the real-world driving experiments. These 99th percentile VSPs, which will be used as the maximum VSPs in the methodology presented below, were estimated (when rounded to the nearest integer) as 9, 9, and 12 for vehicle types 1, 2, and 3 respectively.

2.2. *Reference Driving Cycles for Conventional L-Category Vehicles*

Independently from the work described in Section 2.1, the World Harmonised Motorcycle Test Cycle (WMTC) is a system of driving cycles, specifically for measuring environmental emissions [46], and which are intended to represent real-world driving conditions. These are used as part of the detailed technical requirements and test procedures for the approval of L-category vehicles in the EU. While not developed with the requirements of *electric* L-category vehicles in mind, they nevertheless represent an important standard, and specifically, the two cycles for urban driving conditions (where electric L-category vehicles will likely operate) represent important reference points for any work designed at developing new driving cycles. These two cycles will be referred to subsequently in this paper as (WMTC) Reference Cycles 1 and 2. The speed and acceleration profiles are displayed later in this report, as comparators as we seek to adjust them.

2.3. *Issues with Adopting Reference Cycles for Electric L-Category Vehicles*

As motivation for our later work to be presented, we illustrate the problem of directly combining the evidence and relationships described in Section 2.1 with the reference cycles in Section 2.2. In Figure 1 WMTC Reference Cycle 2 is illustrated, and in Figure 2, the implied time-profile of vehicle specific power, if an electric quadricycle (with the characteristics defined in Section 2.1) were to follow the speed/acceleration profile in Figure 1 over a flat terrain.

Note that, as intended with the power model developed in [23], both positive and negative VSP values arise, the negative values corresponding to intervals of regenerative braking (corresponding to 15.65% of time in WMTC Reference Cycle 2), where the vehicle's kinetic energy is converted back to electrical energy. On the other hand, peaks of VSP, corresponding to acceleration events in the reference cycle, give rise to VSP values that exceed the maximum of 9 estimated for vehicle type 1 or type 2, with the VSP exceeding the maximum for 5.16% of the time duration of the driving cycle. In this respect, the reference cycles are thus not realistic for the performance or driving characteristics of electric L-category vehicles.

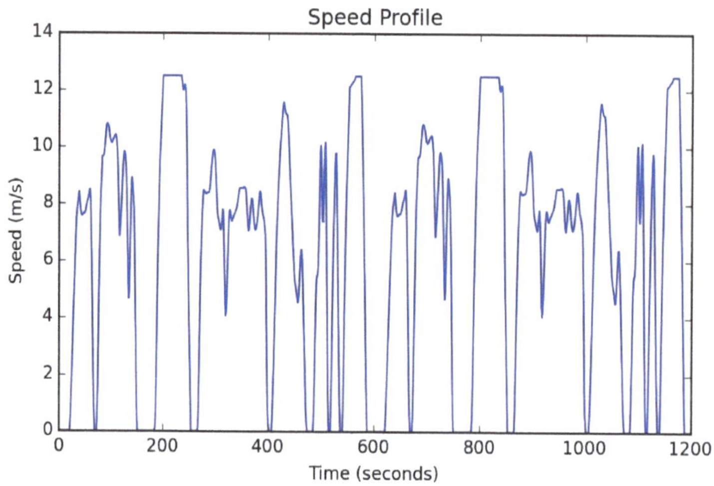

Figure 1. Speed profile from WMTC Reference Cycle 2.

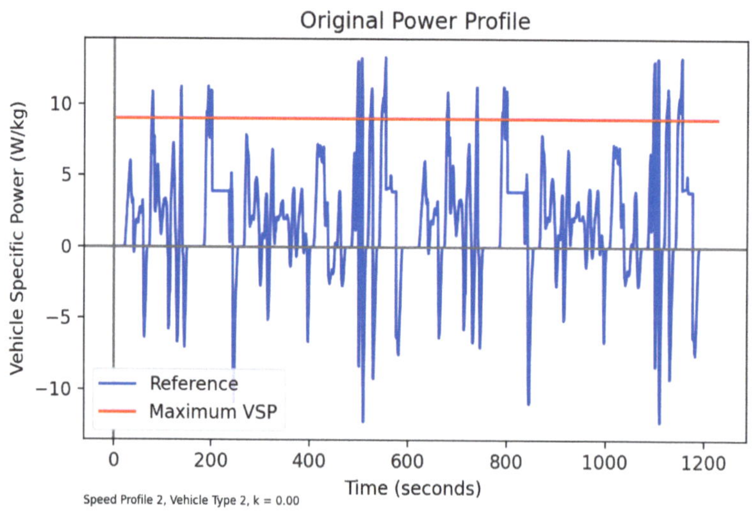

Figure 2. Implied Vehicle Specific Power for an electric quadricycle following speed profile stipulated in WMTC Reference Cycle 2 (Flat terrain).

Typically, reference cycles are developed assuming a flat terrain and we are not aware of corresponding cycles taking gradient into account. As an experiment, though, we repeat the same test as above, but now instead of assuming a hilly terrain, using the gradient function described later (see Section 4) which is uphill for the first half of the road section and downhill for the second, but with the start and endpoints on the same elevation (i.e., zero net gain in elevation over the whole profile).

In Figure 3, we can observe more pronounced peaks of the VSP in the first half of the cycle (on the uphill section of the route), and a lesser violation of the maximum on the second (downhill) half. This balances out to in fact a lesser overall time (4.83%) exceeding the VSP bound than the flat-terrain case, and a greater time (17.65%) in negative power states. More relevant than these summary measures is, however, the fact that when it does exceed the bound it exceeds it by more, making it even more physically infeasible.

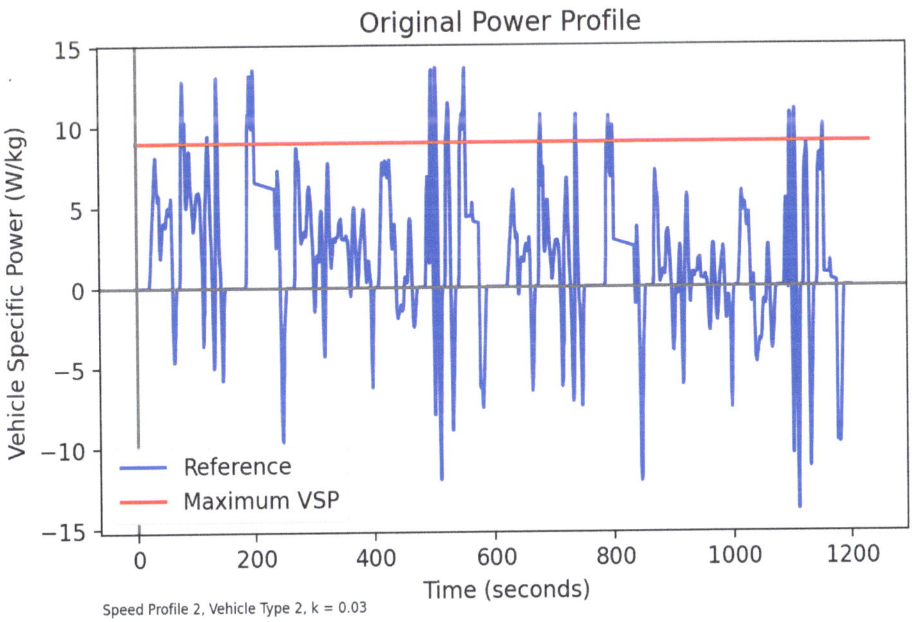

Figure 3. Implied vehicle-specific power for an electric quadricycle following the acceleration profile stipulated in WMTC Reference Cycle 2 (Hilly terrain, $k = 0.03$).

From these initial experiments, we can highlight some key, distinctive features:

- it is important that the positive impact on energy consumption of regenerative braking, as uniquely present in electric vehicles, can be captured in any reference profiles;
- the speed/acceleration profiles developed for regular L-category vehicles may not be attainable by electric L-category vehicles, particular when speed/acceleration is high;
- any adjusted driving cycle for electric L-category vehicles should consider the sensitivity to the gradient of the terrain on which they are driven.

Addressing these features will serve as the main motivation for the present paper.

3. Methodology for Creating Adjusted Speed Profiles

3.1. Goals of Methodology

An overview of the methodology to be applied is presented in Figure 4. It assumes as inputs (a) a given reference speed profile, and (b) a relationship between VSP and (speed, acceleration, gradient) for a given vehicle type, along with a maximum VSP. Equivalently to working with the reference speed profile in (a) we may derive from it a reference acceleration profile and initial speed at the entry to the road stretch; in addition, a reference distance-time profile may be deduced. The goal of the adjustment process is to derive an adjusted acceleration profile that does not violate the given maximum VSP.

At first sight, since high VSP values are typically associated with high acceleration events, this might seem to be a quite trivial task of simply bounding the acceleration profile at some given maximum. However, on closer inspection, there are several complications to this process, and it is these considerations that motivate our method (to be subsequently described):

1. From Equation (1), it can be seen that high VSP values arise from a combination of speed, acceleration, and gradient; it is not simply that there is a maximum acceleration, regardless of speed or gradient.

2. Given speed and gradient, Equation (1) is easily rearranged to make acceleration a function of VSP, and a maximum acceleration is then deduced given the maximum VSP. If done in continuous time, then this would indeed impose the maximum VSP constraint. However, the reference profiles are specified at a given time discretization. If we compute the maximum acceleration, given the speed at the *start* of a time increment, so as to be bounded by the maximum VSP, then by the end of the time increment the vehicle (since accelerating) will be going faster, and so may violate the maximum VSP by the end of the time increment.
3. In general, the adjustment process will result in lower acceleration and lower speeds. As a result, downstream from any adjustment made during an acceleration event, we may have a deceleration event with physically impossible consequences, i.e., a negative speed.
4. As noted, the adjustment process will generally result in lower speeds. The reference speed profiles are intended to represent travel over a road stretch of a given length, but are specified by a speed profile over a given time period. If the adjusted speed profile is applied for the same time period, then it will not represent the same length of road, since by the end of the time period, the vehicle (travelling at lower speeds) will not have reached the end of the road stretch. Thus, in tandem with adjusting speeds, a logical method is needed for extending the time period over which the driving cycle applies, so that it applies to the same road length.

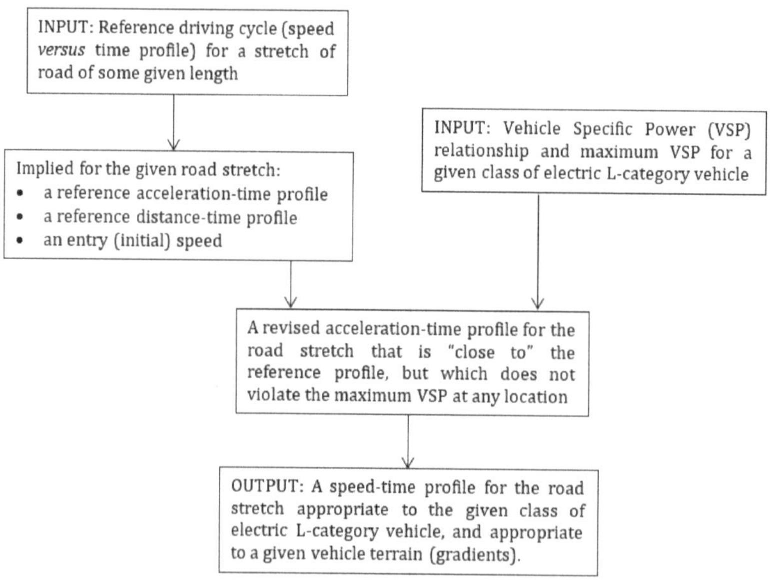

Figure 4. Schematic representation of the applied methodology.

In brief, points 1 and 2 are addressed by a numerical search (Newton) method at each time increment, which deduces the maximum acceleration possible at the current location, given the current gradient, and given the consequential effect of that acceleration on the speed at the end of the time increment. Point 3 is handled by bounding acceleration so that the consequential speed at the end of the increment is at least zero. Point 4 is dealt with essentially by assuming that speed reduction will mean that vehicles will tend to cruise for longer; it is as if there is a vehicle following the unadjusted speed profile, with which our adjusted vehicle tries wherever possible to 'catch up' in terms of the distance traversed. Taken together, these three adjustment processes are thus concerned with acceleration, deceleration, and cruising respectively. The three processes are first described in more detail in Section 3.2, Section 3.3, and Section 3.4 respectively, before presenting the overall algorithm in Section 3.5.

3.2. Acceleration Steps: Bounding of Acceleration

Consider a particular increment of time of length Δ seconds, and suppose that we are given the speed u m/s and location from some origin x metres of a vehicle at the start of the time increment. The aim will be to determine the constant acceleration a m/s^2 during that increment, and thereby the resulting speed v m/s at the end of the time increment. Clearly by standard equations of motion:

$$v = u + a\Delta \tag{4}$$

and the location y of the vehicle at the end of the time increment is given by:

$$y = x + u\Delta + \frac{a\Delta^2}{2}. \tag{5}$$

Then, according to Equation (1), we may associate a VSP with any (location, speed, acceleration) triple (y, v, a). Since we are imagining time increments in the order of seconds, then the road gradient will hardly change between location y and x, so we approximate $\theta_1(y) \simeq \theta_1(x)$ for some simplification. Then the VSP at the end of the time increment $f(v, a; \Theta(x))$ is obtained by combining Equations (1), (4) and (5) to give:

$$f(u + a\Delta,\ a; \Theta(x)) = (1.1a + 9.81\theta_1(x) + \theta_2)(u + a\Delta) + \theta_3(u + a\Delta)^3. \tag{6}$$

Thus, given the (location, speed) (x, u) at the start of the time increment, the VSP at the end of the time increment may be thought of as a function only of the acceleration a, parameterized by (x, u):

$$\phi(a; x, u) = (1.1a + 9.81\theta_1(x) + \theta_2)(u + a\Delta) + \theta_3\ (u + a\Delta)^3. \tag{7}$$

By inspection, ϕ is monotonically increasing in a, and so its inverse function exists. Let $\psi(p; x, u)$ denote this inverse function for a VSP p given (x, u), i.e., such that: $a = \psi(\phi(a; x, u); x, u)\ \forall a \in \mathbb{R}$.

Now let $p_{\max}$ denote the assumed maximum VSP of a vehicle under consideration. We may then, by the reasoning above, calculate a maximum acceleration $a_{\max}$ corresponding to the maximum VSP $p_{\max}$ according to $a_{\max} = \psi(p_{\max}; x, u)$. Note that this maximum acceleration is not constant, depending through ψ on the current speed u and local conditions in terms of gradient through x; this is implicitly handled through the method described (so we could write $a_{\max}(x, u)$ to denote this dependence).

Since the inverse function is not available in analytic form, the procedure is to instead numerically estimate the acceleration a that satisfies $\phi(a; x, u) = p_{\max}$ or equivalently $\phi(a; x, u) - p_{\max} = 0$. This is done via a Newton scheme whereby the $(k + 1)^{\text{th}}$ iteration is given by:

$$a^{(k+1)} = a^{(k)} - \frac{\phi\left(a^{(k)}; x, u\right) - p_{\max}}{\phi'\left(a^{(k)}; x, u\right)} \tag{8}$$

where $\phi(a; x, u)$ is given by (7) and:

$$\phi'(a; x, u) = 1.1(u + a\Delta) + \Delta(1.1a + 9.81\theta_1(x) + \theta_2) + 3\theta_3\Delta(u + a\Delta)^2. \tag{9}$$

3.3. Deceleration Steps: Bounding of Deceleration

The adjustments described in Section 3.2 will in general decrease (or leave unaltered) accelerations and speeds, relative to the reference values. Suppose that we are considering a particular time-step, and that in a *previous* time-step the bounding of acceleration has indeed resulted in a speed reduction relative to the reference speed. Thus, even before we consider what might happen in the current time-step, the vehicle may be travelling at a lower initial speed than the reference speed at the start of the time-step. Now consider the current reference acceleration; if it is negative, then there is no guarantee that such a deceleration will be feasible from the initial speed, as it may result in a negative speed by the end of the time increment. In practice, this is quite a rare occurrence, but it can occur and lead to physically impossible profiles. Therefore, a simple adjustment is made for deceleration events, namely that given any time-period (length Δ) and an initial speed at the start of that time period u, then the acceleration a is bounded below so as the speed at the end of the time-interval v is no less than zero:

$$v = u + a\Delta \geq 0 \tag{10}$$

implying:

$$a \geq -\frac{u}{\Delta} \tag{11}$$

3.4. Cruising Steps: "Catching Up" Distance

The third and final type of step is the cruising one. This arises as we aim to specify a speed profile for a road segment of a given *length*. The main speed adjustment steps (as described in Section 3.2) will mean that in the adjusted profile the speeds will generally be lower than in the reference profile. If we specify the adjusted profile for only the same number of time steps as the reference profile, a vehicle following that trajectory will then clearly only traverse part of the road segment by the end of the time period. For a slower vehicle to traverse the same length as a vehicle following the reference profile, it must travel for longer in time. The question is: what logic might we use to add additional time periods?

The approach is based on the premise that the variations in speed/acceleration in the reference profile are associated with spatial (as opposed to temporal) locations on the route, e.g., intersections, locations where queuing occurs. Therefore, the adjusted speed profile should, where possible, aim to track the spatial variations in the reference speed profile. It can be imagined that a vehicle following the (slower) adjusted profile is trying to 'catch up' the distance travelled by a vehicle following the reference profile, and so each time it falls behind it will cruise at the current speed for several time steps. The effect of this, as seen in Section 4, is that the peaks of the adjusted profile both lag behind those of the reference profile in time, as well as being somewhat extended; this seems logical behaviorally, that someone travelling at a slower speed than another vehicle will just maintain that slower speed for longer to reach the same point.

Suppose then, having applied a step from Section 3.2 or Section 3.3 as appropriate, to a given reference time increment, the total distance travelled on the adjusted profile so far is x metres, and the distance travelled using the reference speeds by the end of this reference time interval is x_{REF}. Then the cumulative lost distance relative to the reference profile is $D = x - x_{\text{REF}}$. Suppose that when applying the methods of Section 3.2 or Section 3.3 an adjusted speed of $\widetilde{v}$ m/s was determined for the end of the last time increment. If $\widetilde{v} = 0$ then skip this step; so let us assume now that $\widetilde{v} > 0$. Firstly, we determine whether the vehicle is in a 'cruising mode', and if it has reached some desired speed. This is indicated either by the reference acceleration for the previous period a_{REF} being equal to zero, or by the fact that the reference acceleration is about to "pass through" zero if it were not for discretization. This latter state is indicated by a positive reference acceleration a_{REF} for the previous period and a negative reference acceleration a^{+}_{REF} for the following period.

Hence overall our test is:

$$\widetilde{v} > 0 \text{ and } \left\{a_{\text{REF}} = 0 \text{ or } \left\{a_{\text{REF}} > 0 \text{ and } a^{+}_{\text{REF}} < 0\right\}\right\} \tag{12}$$

We now check whether additional time increments can be inserted with the vehicle cruising at a speed of $\widetilde{v}$, without the distance for these increments exceeding D. Due to the discretization, we are unlikely to be able to catch up all the distance in D (though that is the goal), so instead, we insert as many time increments as possible in order to travel as close as possible to D metres. Since by cruising at a speed of $\tilde{v}$ m/s we will travel $\Delta\tilde{v}$ metres in one time increment of Δ seconds, the number of time increments we will be able to insert is:

$$\text{number } n \text{ of cruising time intervals} = \left\lfloor \frac{D}{\Delta\tilde{v}} \right\rfloor \tag{13}$$

where for any real number x the notation $\lfloor x \rfloor$ denotes the largest integer $\leq x$. These time increments are then inserted, before moving on to consider the next time increment from the reference speed profile.

Having applied this process to the full time period of the reference speed profile, it is likely that there will be a small amount of cumulative lost distance remaining at the end. This is accounted for by rewinding to the last time at which cruising increments were inserted, and inserting additional increments at that time to account for as much of the remaining cumulative lost distance as is possible within the error of discretization.

3.5. Overall Solution Process

The solution process makes use of three *methods* corresponding to the techniques described in Sections 3.2–3.4. A method in this sense is a computational process with given inputs and outputs as follows:

$$(\tilde{a},\tilde{v}) = \text{BoundAcceleration}\ (x, u, \Delta, p_{\max}, a_{\text{REF}}) \tag{14}$$

$$(\tilde{a},\tilde{v}) = \text{BoundDeceleration}\ (u, \Delta, a_{\text{REF}}) \tag{15}$$

$$n = \text{CalculateCruisingTime}\ (\tilde{v}, D, \Delta). \tag{16}$$

The method BoundAcceleration, applied only when $a_{\text{REF}} > 0$, first estimates an upper bound on acceleration over a time interval of length Δ seconds given an initial speed of u m/s and given a bound on vehicle specific power of $p_{\max}$ W/kg. The acceleration bound will vary with both speed u and location x (the latter dependence due to variations in gradient). The upper bound is estimated by the iterative process (8) based on (7) and (9), with the converged solution denoted $a_{\max}$ (m/s^2). The method then provides as output the bounded acceleration $\tilde{a}$ (in relation to the input reference/default acceleration a_{REF}) and the corresponding speed $\tilde{v}$ according to:

$$\tilde{a} = \min(a_{\text{REF}}, a_{\max}) \qquad \tilde{v} = u + \tilde{a}\Delta\,. \tag{17}$$

The method BoundDeceleration, applied only when $a_{\text{REF}} < 0$, bounds acceleration from below to avoid physically impossible (negative) speed on exit from the time increment. It provides as output the bounded acceleration $\tilde{a}$ (in relation to a_{REF}) and corresponding speed $\tilde{v}$ according to:

$$\tilde{a} = \max\left(a_{\text{REF}}, -\frac{u}{\Delta}\right) \qquad \tilde{v} = u + \tilde{a}\Delta. \tag{18}$$

The method CalculateCruisingTime is applied after either of the two methods above, and only when three conditions are all met: (i) the adjusted speed $\tilde{v} > 0$, (ii) the reference acceleration is indicative of cruising ($a_{\text{REF}} = 0$ or $\{a_{\text{REF}} > 0$ and $a^{+}_{\text{REF}} < 0\}$), and (iii) the cumulative lost distance $D > 0$. It provides as output the maximum number n of time intervals of length Δ seconds at which a vehicle may cruise at $\tilde{v}$ m/s in order to recover the most lost distance, according to (13). An implementation issue to note is that since a_{REF} is stored as a real number, the practical test of "zero acceleration" applied is $|a_{REF}| < \epsilon$ for some small $\epsilon > 0$.

With these three methods as the building blocks, it is then possible to describe the overall adjustment process. Suppose that a vehicle is moving in time and space along a stretch of road of given length L metres, and that time is discretized into increments of length Δ seconds. Let $\{v_0^{\mathrm{REF}}, v_1^{\mathrm{REF}}, \ldots, v_m^{\mathrm{REF}}\}$ denote the speeds over time increments in a given reference driving cycle, with v_i^{REF} m/s the reference speed at the end of the ith time increment (i.e., so at continuous time $i\Delta$). Assuming uniform acceleration within a time increment (between the exit speed of the last time increment and the exit speed of the present one), Equation (4) may be applied to additionally infer a reference acceleration profile $\{a_1^{\mathrm{REF}}, a_2^{\mathrm{REF}}, \ldots, a_m^{\mathrm{REF}}\}$. In addition, Equation (5) may be used to deduce a reference distance-time profile $\{x_1^{\mathrm{REF}}, x_2^{\mathrm{REF}}, \ldots, x_m^{\mathrm{REF}}\}$. The basic logic of the adjustment method is that a vehicle will attempt to follow the acceleration/deceleration patterns in the reference acceleration profile, as far as possible; and that at the same time the distance lag relative to the reference speed profile will be used to indicate when additional cruising time steps should be inserted.

- The solution process to deduce the adjusted speed profile $\mathbf{v}^{\mathrm{ADJ}} = \left\{v_0^{\mathrm{ADJ}}, v_1^{\mathrm{ADJ}}, \ldots\right\}$ and acceleration profile $\mathbf{a}^{\mathrm{ADJ}} = \left\{a_1^{\mathrm{ADJ}}, a_2^{\mathrm{ADJ}}, \ldots\right\}$ is defined in Algorithm 1.

Algorithm 1. Overall solution process.

1. Initialise the vector $\mathbf{v}^{\mathrm{ADJ}}$ with the single element $v_0^{\mathrm{ADJ}} = v_0^{\mathrm{REF}}$, and initialise $\mathbf{a}^{\mathrm{ADJ}}$ as an empty vector. Set $u = v_0^{\mathrm{ADJ}}$. Set $x = 0$.
2. For $j = 1, 2, \ldots, m$ do the following steps in sequence of increasing j:
   a. If $a_j^{\mathrm{REF}} > 0$, apply: $(\widetilde{a}, \widetilde{v}) = \mathrm{BoundAcceleration}\left(x, u, \Delta, p_{\max}, a_j^{\mathrm{REF}}\right)$.
   b. If $a_j^{\mathrm{REF}} < 0$, apply: $(\widetilde{a}, \widetilde{v}) = \mathrm{BoundDeceleration}\left(u, \Delta, a_j^{\mathrm{REF}}\right)$.
   c. If $a_j^{\mathrm{REF}} = 0$, set $\widetilde{a} = 0$ and $\widetilde{v} = u$. Set $d^+ = 0$.
   d. Append $\mathbf{v}^{\mathrm{ADJ}}$ with $\widetilde{v}$ and $\mathbf{a}^{\mathrm{ADJ}}$ with $\widetilde{a}$.
   e. If $\widetilde{v} > 0$ and $(\left|a_j^{\mathrm{REF}}\right| < \epsilon$ or $(a_j^{\mathrm{REF}} > 0$ and $a_{j+1}^{\mathrm{REF}} < 0))$ and $D = x_j^{\mathrm{REF}} - x > 0$ apply: $n = \mathrm{CalculateCruisingTime}(\widetilde{v}, D, \Delta)$. Otherwise set $n = 0$.
   f. Append $\mathbf{v}^{\mathrm{ADJ}}$ with n additional elements of $\widetilde{v}$, and $\mathbf{a}^{\mathrm{ADJ}}$ with n additional elements of 0.
   g. Set $x = x + \frac{1}{2}(u + \widetilde{v})\Delta + n\widetilde{v}\Delta$.
   h. Set $u = \widetilde{v}$

4. Application of Adjustment Process and Discussion

The adjustment process described in Section 3 will now be applied to the WMTC driving cycles for motorcycles described in Section 2.2, adapted for L-category electric quadricycles based on the empirical evidence described in Section 2.1. As noted earlier in Figure 2, following the WMTC driving cycles gives rise to a problem for the electric quadricycles under study, as the speed/acceleration profiles assumed would give rise to violations of the limits on Vehicle Specific Power (VSP). In the first part of the analysis, we study the adjusted profiles assuming that the stretches of road are completely flat; note that the original WMTC profiles do not refer to a gradient. In the second part of the analysis, we perform a sensitivity analysis, exploring the impact of including gradients in the correction process.

When summarising the kinetic properties of the various driving cycles, the following statistics are calculated for the whole profile:

- mean and maximum speed;
- mean acceleration considering only positive acceleration events;
- maximum acceleration, estimated as the mean of the ten highest acceleration events;
- mean deceleration considering only negative acceleration events;
- maximum deceleration, estimated as the mean of the ten highest deceleration events;
- total idling time, estimated as the number of time increments with vehicle speed $< 10^{-9}$ m/s;
- total travel time to traverse the road segment.

4.1. Adjusted Speed Profiles: Flat Road Stretches

The adjustment process was applied to the two WMTC driving cycles and the three vehicle types, with the results for WMTC Cycle 1 given in Table 3 (kinetic summary statistics) and Table 4 (corresponding power/energy statistics), and the results for WMTC Cycle 1 in Tables 5 and 6.

Table 3. Kinetic summary statistics of adjusted drive cycles (WMTC Cycle 1).

Profile	Speed (m/s)		Positive Acceleration (m/s^2)		Negative Acceleration (m/s^2)		Total Idling Time (s)	Total Travel Time (s)
	Mean	Max	Mean	Max	Mean	Max		
Reference	4.89	6.94	0.60	1.64	−0.44	−1.66	185.0	1201.0
Type 1 Adjusted	4.87	6.94	0.60	1.64	−0.44	−1.66	185.0	1206.7
Type 2 Adjusted	4.87	6.94	0.60	1.64	−0.44	−1.66	185.0	1206.7
Type 3 Adjusted	4.87	6.94	0.60	1.64	−0.44	−1.66	185.0	1206.7

Table 4. Energy-related summary statistics of adjusted drive cycles (WMTC Cycle 1).

Vehicle	Profile	Total Energy (Wh)	Regenerated Energy (Wh)	% Time in Negative Energy State	% Time in Positive Energy State
Type 1	Reference	296.19	21.93	8.49	91.42
	Adjusted	294.68	21.52	8.46	91.46
Type 2	Reference	293.95	21.03	7.99	91.92
	Adjusted	292.60	20.63	7.96	91.96
Type 3	Reference	361.29	34.47	7.99	91.92
	Adjusted	361.29	34.47	7.99	91.92

Table 5. Kinetic summary statistics of adjusted drive cycles (WMTC Cycle 2).

Profile	Speed (m/s)		Positive Acceleration (m/s^2)		Negative Acceleration (m/s^2)		Total Idling Time (s)	Total Travel Time (s)
	Mean	Max	Mean	Max	Mean	Max		
Reference	6.33	12.50	0.46	1.64	−0.55	−1.74	227.0	1201.00
Type 1 Adjusted	6.20	11.87	0.44	1.64	−0.54	−1.74	233.0	1226.14
Type 2 Adjusted	6.20	11.87	0.44	1.64	−0.54	−1.74	233.0	1226.14
Type 3 Adjusted	6.32	12.50	0.46	1.64	−0.55	−1.74	227.0	1203.22

Table 6. Energy-related summary statistics of adjusted drive cycles (WMTC Cycle 2).

Vehicle	Profile	Total Energy (Wh)	Regenerated Energy (Wh)	% Time in Negative Energy State	% Time in Positive Energy State
Type 1	Reference	460.65	52.86	16.65	83.26
	Adjusted	456.46	47.57	15.82	84.09
Type 2	Reference	445.23	50.76	15.65	84.26
	Adjusted	442.64	45.65	14.85	85.07
Type 3	Reference	587.83	83.28	15.65	84.26
	Adjusted	586.59	82.38	15.63	84.29

Tables 3 and 4 suggest that, for WMTC cycle 1, the modifications made to the speed profile to make it compatible with low-power electric vehicles do not generate large differences in the drive cycle statistics (only −0.4% change in average speed). This is also reflected in a small decrease in total energy consumption (−0.3%) and regenerated energy

(−1.3%). The same trend was found for WMTC Cycle 2 (Tables 5 and 6), where the maximum differences were found for Vehicle Type 1 and Type 2, with adjustments to average speed of −2%, maximum speed of −5%, average positive acceleration of −4% and average negative acceleration of −2%. These modifications to the driving cycle also have an impact on total energy consumption (−0.6%) and particularly on regenerated energy (−7%).

As anticipated, the least adjustment required is for Vehicle Type 3, with the largest maximum power, with no adjustment required to Cycle 1 and a minimal adjustment to Cycle 2. For the other vehicle types, small but appreciable differences can be observed. A general pattern is that the adjusted profiles are associated with slightly less overall energy consumption (up to 0.9% decrease in energy consumption), but also significantly less regenerated energy (up to 10% decrease). In all cases where adjustments are made, the time to traverse the given section of road is longer than in the reference case (with travel time increasing from 0.5 to 2.1%), reflecting the expected reduction in speed. Consequently, these results indicate that the modifications introduced in the cycles do not translate into large modifications in total energy consumption, which provides a good baseline for comparison, even though the modifications introduced reduce acceleration (and consequently deceleration), penalizing the positive impact of regenerative braking.

While the summary measures provide some indication of the effects, they are better appreciated by exploring plots of individual cases. Unless otherwise stated we will henceforth focus on Vehicle Type 2 as an example (since it is one of the types with a lower maximum VSP of 9, and so requires more significant adjustments). The adjusted speed profiles for the two reference driving cycles are illustrated in Figures 5 and 6. With generally lower speeds the adjustments to Cycle 1 are relatively smaller. The effects of the adjustment—though evident in both profiles—are more easily appreciated with Cycle 2, and so we shall henceforth focus on examining Cycle 2.

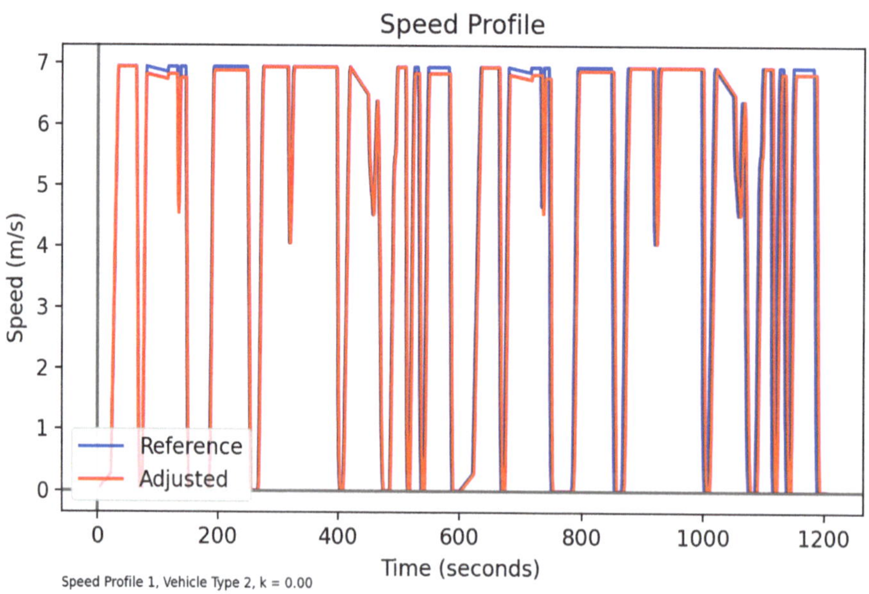

Figure 5. Adjusted speed profile for electric quadricycle in comparison with reference cycle (WMTC Reference Cycle 1, Vehicle Type 2, Flat terrain assumed in adjustment).

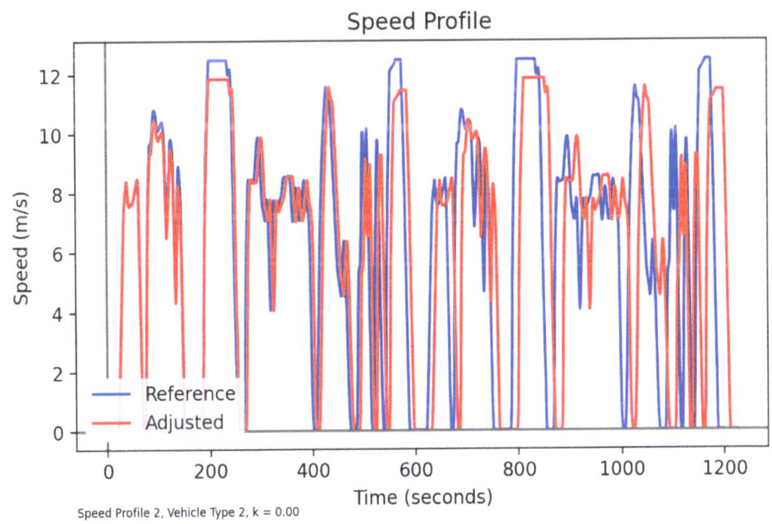

Figure 6. Adjusted speed profile for electric quadricycle in comparison with reference cycle (WMTC Reference Cycle 2, Vehicle Type 2, Flat terrain assumed in adjustment).

In particular, in Figure 6, in the earlier stage of the profile (up to about time 250), a general reduction in peak speed at the same time increment can be observed (due to the acceleration capping process). After that time, it is increasingly evident that the peaks of the adjusted profile are time-lagged behind those of the reference profile (due to the process of "catching up" distance), and so then any comparison can be made allowing for this growing time-lag during the period represented.

For the case of Reference Cycle 2, the effect on the distance-time profile (Figure 7) confirms the increasing time-lag of the adjusted profile behind the reference profile, along the length of the route profile.

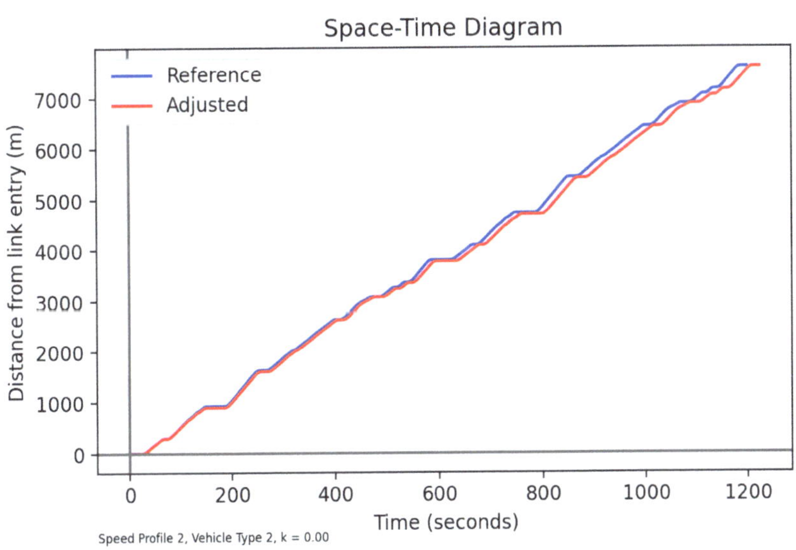

Figure 7. Distance travelled as a function of time for adjusted and reference speed profiles (WMTC Reference Cycle 2, Vehicle Type 2, Flat terrain assumed in adjustment).

In Figure 8, the limiting effect of the adjustment is evident, with the peaks of the reference profile eliminated so that the adjusted profile is within the maximum VSP bound. However, when we correlate these adjustments in time with the acceleration profile (Figure 9) it is evident that these adjustments are not simply about curbing high acceleration events; VSP is a combination of speed and acceleration. In the early part of the profile there is practically no adjustment (so the two acceleration profiles overlap), but then similarly to the adjustments to speed, an increasing time lag of the adjusted profile can be seen in the acceleration peaks.

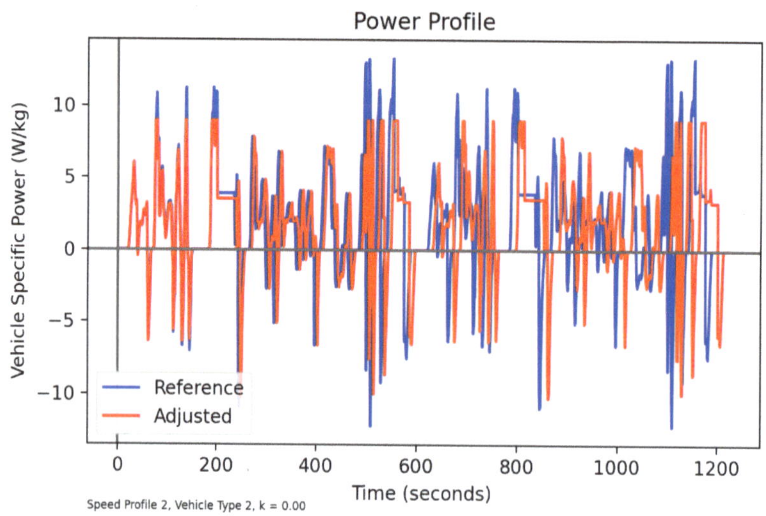

Figure 8. Vehicle Specific Power profile for adjusted and reference speed profiles (WMTC Reference Cycle 2, Vehicle Type 2, Flat terrain assumed in adjustment).

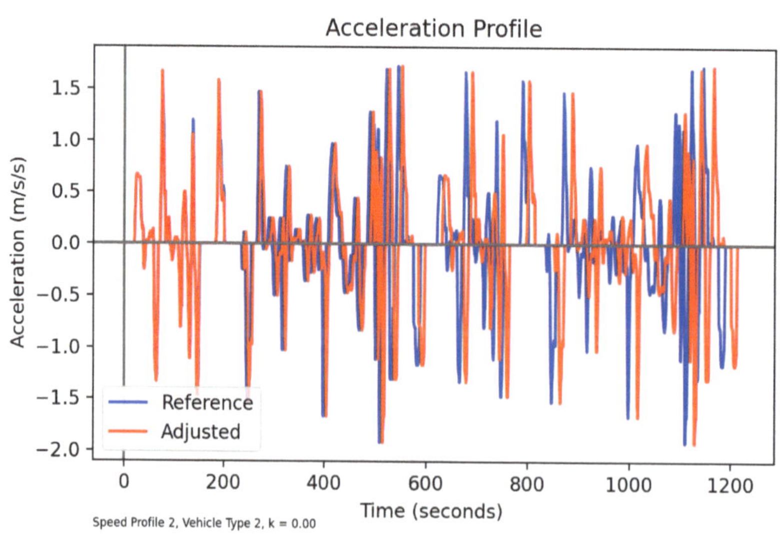

Figure 9. Acceleration profile for adjusted and reference speed profiles (WMTC Reference Cycle 2, Vehicle Type 2, Flat terrain assumed in adjustment).

4.2. Adjusted Speed Profiles: Sensitivity Analysis to Gradient

In the analysis in Section 4.1, a flat terrain was assumed. In practice, there may be hilly terrain that could particularly affect the performance of L-category EVs. This is especially likely to be true when accelerating on uphill terrain, with now a triple dependence to capture between velocity, acceleration, and gradient. In addition, EVs have a particular feature of regenerative braking, which is likely to be especially relevant in downhill sections.

We are not aware of any equivalent reference profiles to the WMTC reference profiles for use in undulating terrain. However, the approach described in Section 3 is able to nevertheless create an adjusted profile for undulating terrain, given a reference profile for flat terrain. This is because VSP depends on all three of (velocity, acceleration, gradient), and so bounding VSP will automatically allow for changes in gradient. As an initial illustration of the performance of the method on such terrain, and to easily perform sensitivity tests, the approach taken is to hypothesize a smooth, symmetric hill/valley, and to vary the gradient profile. The equation of the height of the hill/valley in metres as a function of the distance along the route profile is given by:

$$g(x) = \frac{k}{L}x(L-x) \qquad (0 \le x \le L;\ L > 0; -\infty < k < \infty) \tag{19}$$

where L metres is the length of the route profile, and where $k > 0$ corresponds to a hill and $k < 0$ to a valley. Note that this route profile has the start and endpoints at the same elevation (i.e., zero net gain in elevation over the whole profile). The implied gradient function, as is needed for (1), is given by:

$$\theta_1(x) = g'(x) = \frac{k}{L}(L-2x) \qquad (0 \le x \le L). \tag{20}$$

From (20), it is evident that the parameter k denotes the gradient on entry to the route profile as well as the negative of the gradient on exit from it.

Figure 10 compares the adjusted speed profile obtained for $k = 0.03$ in Equations (19) and (20), with that obtained on a flat terrain ($k = 0$; the 'Adjusted' profile from Figure 6). For $k > 0$, the maximum positive gradient in the profile (19) is at the start of the route, gradually decaying to zero at half-way along the route. Thus, the VSP-based adjustment, in taking into account speed, acceleration, and gradient, makes the maximum impact in the first part of the route, as can be seen from the comparison in Figure 10. This in turn increases the time-lagged effect of subsequent peaks, since the vehicle has travelled more slowly on the first part of the journey. The second part of the route is characterized by an increasingly downhill section, and in that case, less adjustment is needed to speed than on the flat, explaining why some comparable peaks are higher for $k = 0.03$ than for $k = 0$.

We now turn to a different kind of comparison and aim to answer the question of how much gradient matters. In order to do this:

- Firstly, adjusted speed profiles are calculated (i) assuming a flat terrain ($k = 0$), and then (ii) assuming a given value of k ($k = k_{\text{TRUE}}$). We shall refer to (i) as the flat-adjusted profile, and to (ii) as the gradient-adjusted profile.
- The impacts of following each of these profiles on an undulating terrain are then evaluated by running each of the speed profiles to compute the VSPs on a terrain with $k = k_{\text{TRUE}}$.
- Considering the typical, ultimate application of reference speed profiles in making assessments of overall energy/environmental impacts, we then examine how much gradient influences the overall energy consumption along the route profile, by computing the energy expended following the two speed profiles, under a common assumption of $k = k_{\text{TRUE}}$.
- Recalling that from Section 2.1, both the adjustment (through the maximum VSP) and the energy model depend on vehicle type, the comparisons above are repeated for different vehicle types.

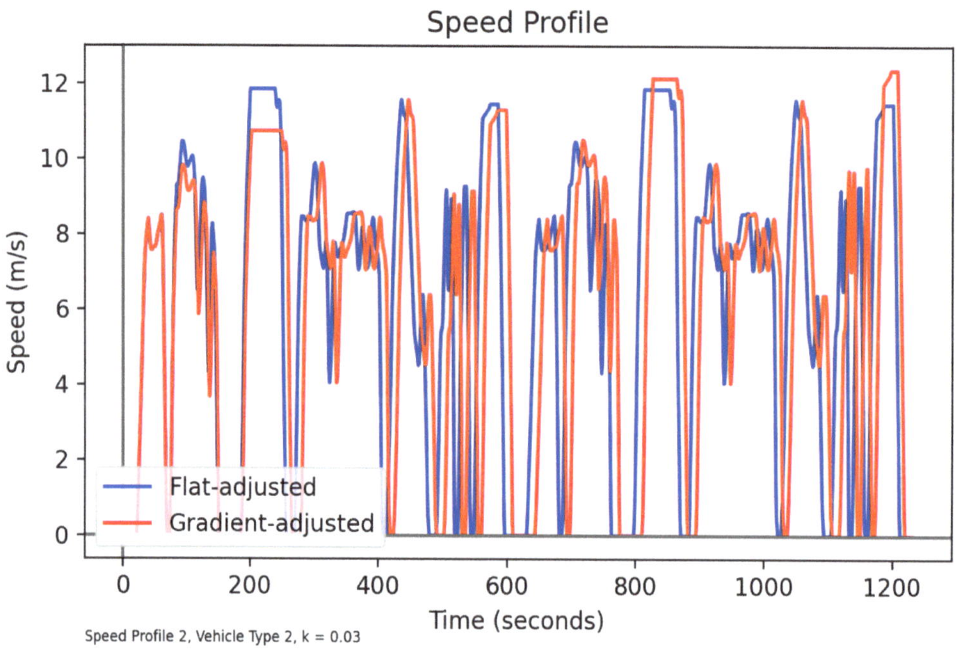

Figure 10. Comparison of adjusted speed profiles for Flat ($k = 0$) and Gradient ($k = 0.03$) cases (Both adjusted from WMTC Reference Cycle 2; Vehicle Type 2).

In Figure 11, we compare the VSP profiles for the undulating terrain under the two assumed speed profiles. Since the 'Flat' adjusted speed profile was computed by neglecting the impact of gradient, it no longer is able to guarantee that the maximum VSP of 9 is not violated; there are instances in the figure illustrated (notably at the beginning of the route, where the uphill gradient is highest) where the 'Flat' VSP does indeed violate this bound.

In Figure 12, the consequential effects on differences in cumulative energy consumed are displayed. The pattern for the first part of the route section (the uphill part) is that following the speed profile computed for the true undulating profile gives rise to slightly less energy consumed than when following a profile computed for a flat profile, but that for the second (downhill) part this is gradually reversed, and in fact at the end of the route, the speed profile for a flat route gives slightly less energy. This might seem counter-intuitive—surely a profile that is calculated knowing the true undulating profile should be better than one that does not? It should be recalled that the adjustments being made are not intended to represent some kind of optimal strategy for minimizing VSP or energy. Instead, we are attempting to follow as closely as possible a reference speed pattern that is intended to be a realistic behavioural representation. The objective of developing the adjusted speed profile might be described as "attempt to follow the reference speed profile, and when this is not possible go as fast as possible to approach the reference profile within the limits of vehicle power". Since there is a complex relationship between speed/acceleration/gradient and energy consumption, there is nothing in the developed methodology that ensures the developed profiles are maximally efficient for energy consumed. In this case, we are seeing the effect of the pattern observed in Figure 10, where the higher speeds on the later downhill parts of the route have resulted in higher energy consumption in the second part of the journey.

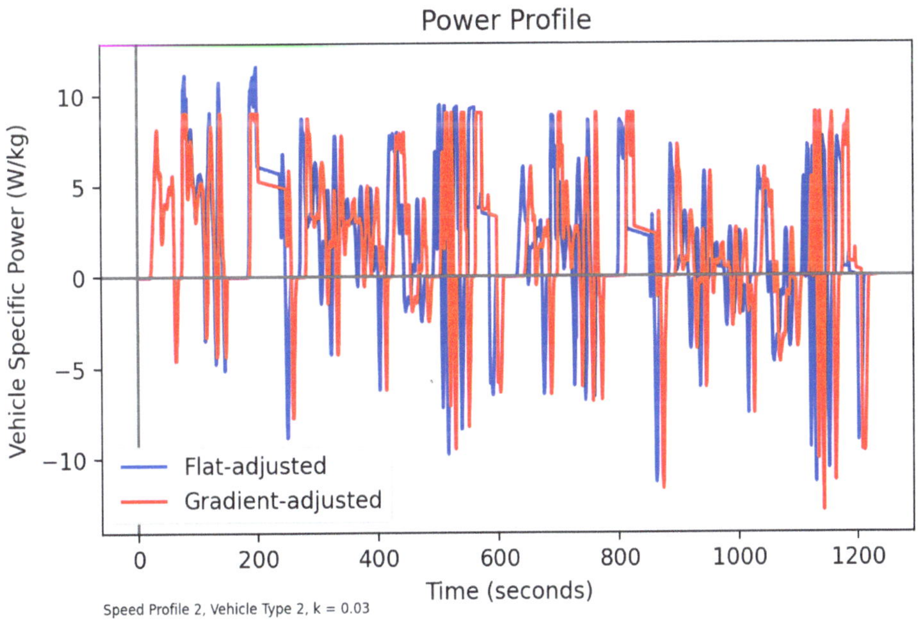

Figure 11. Comparison of VSP profiles following Flat ($k = 0$) vs. Gradient ($k = 0.03$) adjusted speed profiles, both evaluated on $k = 0.03$ terrain (Both adjusted from WMTC Reference Cycle 2; Vehicle Type 2).

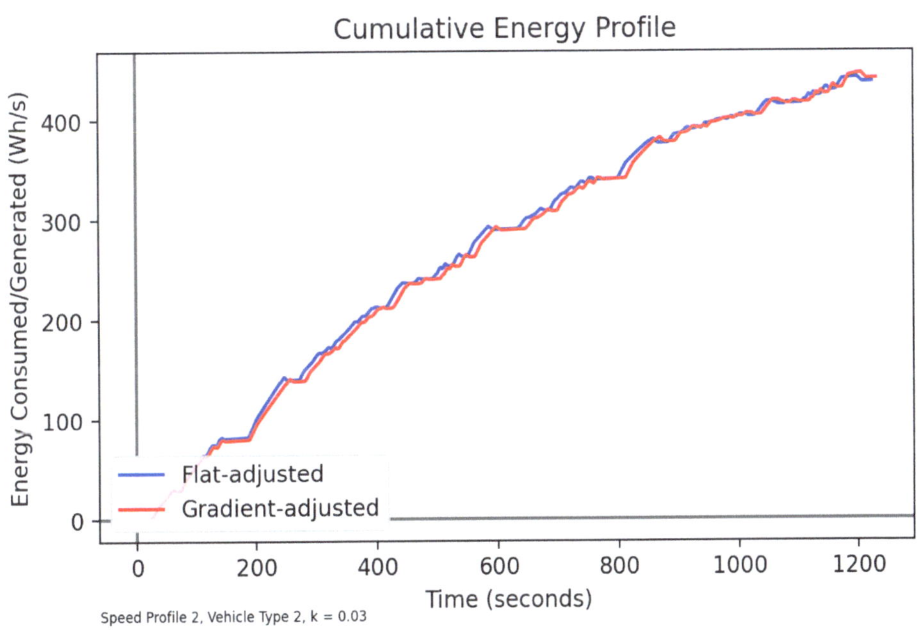

Figure 12. Comparison of cumulative energy profiles for Flat ($k = 0$) vs. Gradient ($k = 0.03$) adjusted speed profiles, both evaluated on $k = 0.03$ terrain (Both adjusted from WMTC Reference Cycle 2; Vehicle Type 2).

Our objective in studying energy impacts is, rather than evaluating the efficiency of different speed profiles, to explore the *sensitivity* of energy consumed, and especially to what extent bespoke speed profiles might be needed for undulating terrain. In Table 7, three speed profile assumptions are compared: the original reference speed profile ('Reference'), an adjusted profile computed assuming a flat terrain ('Flat adj speed'), and an adjusted profile that is bespoke to the assumption on gradient through k ('Grad adj speed'). This is repeated for three different assumptions on the parameter k in (19) and (20), and for three different vehicle types (which differ in their maximum VSP and energy consumption parameters, as described in Section 2.1). Relative to the reference profile, the differences between the flat and gradient profiles are extremely small, in the order of 1%. The direction of the differences is not the same across all vehicle types; for Vehicle Type 3, the flat profiles give a lower total energy consumption than the bespoke profiles, but for Vehicle Types 1 and 2 this is reversed. However, as noted earlier above, we do not place any significance on the direction of these differences, since we are not aiming to propose energy-optimal profiles in any sense. More important is that the *scale* of the differences at the level of the complete route section are small. Thus, depending on the objective of the study, it may be justifiable to assume the profiles developed for a flat profile on undulating terrain, even though this gives rise to some violations of the maximum VSP. On the other hand, if detailed dynamic control measures are to be evaluated, then a bespoke speed profile for any undulating terrain is likely to be needed. Table 8 confirms these impacts on regenerative energy.

Table 7. Total energy consumption (Wh) under alternative speed profiles, gradient assumptions, and vehicle types (symmetric hill, WMTC Reference Cycle 2).

	Vehicle Type 1			Vehicle Type 2			Vehicle Type 3		
k	0.01	0.03	0.05	0.01	0.03	0.05	0.01	0.03	0.05
Reference	463.9	462.0	457.1	448.2	445.5	439.1	591.5	594.9	597.7
Flat adj speed	457.5	452.4	449.4	443.5	437.5	432.8	591.2	593.6	596.6
Grad adj speed	458.59	455.5	454.2	444.5	440.8	438.2	590.8	593.0	593.9

Table 8. Percentage of regenerative energy under alternative speed profiles, gradient assumptions, and vehicle types (symmetric hill, WMTC Reference Cycle 2).

	Vehicle Type 1			Vehicle Type 2			Vehicle Type 3		
k	0.01	0.03	0.05	0.01	0.03	0.05	0.01	0.03	0.05
Reference	11.65	12.49	14.42	11.58	12.43	14.39	14.39	15.25	17.31
Flat adj speed	10.63	11.47	13.48	10.52	11.37	13.40	14.26	15.13	17.19
Grad adj speed	10.64	11.59	13.45	10.53	11.48	13.34	14.30	15.13	17.08

A similar approach was made now with a symmetric valley, as opposed to a symmetric hill, which corresponds to $k < 0$ in (19) and (20). Combining the results on total energy consumption in Tables 7 and 9, it can be seen that the average variation between the reference, flat adjusted speed, and gradient adjusted speed is less than -1% between $k = -0.01$ and $k = 0.01$, up to -3% for $k = -0.03$ and $k = 0.03$. The difference is reduced to -2% between $k = -0.05$ and $k = 0.05$.

When analyzing only the impact of slope profile on regeneration (Tables 8 and 10), it can be observed that the combination of speed and the type of slope profile can lead to increasing differences as the slope increases. For instance, the average variation for reference, flat adjusted speed, and gradient adjusted speed is around -3% between $k = -0.01$ and $k = 0.01$, -5% comparing $k = -0.03$ and $k = 0.03$ and -10% for $k = -0.05$ and $k = 0.05$. This just indicates the importance of the slope magnitude and its impact when combined with a speed schedule, even considering that the driving cycle is adjusted to have speed and acceleration values adequate to real-world driving.

Table 9. Total energy consumption (Wh) under alternative speed profiles, gradient assumptions, and vehicle types (symmetric valley, WMTC Reference Cycle 2).

	Vehicle Type 1			Vehicle Type 2			Vehicle Type 3		
k	−0.01	−0.03	−0.05	−0.01	−0.03	−0.05	−0.01	−0.03	−0.05
Reference	458.9	446.4	447.1	443.5	430.5	429.6	586.7	579.6	588.9
Flat adj speed	458.6	439.5	443.0	444.7	425.1	427.7	585.5	578.3	587.7
Grad adj speed	460.2	441.1	443.1	446.3	426.9	427.7	584.9	576.8	583.8

Table 10. Percentage of regenerative energy under alternative speed profiles, gradient assumptions, and vehicle types (symmetric valley, WMTC Reference Cycle 2).

	Vehicle Type 1			Vehicle Type 2			Vehicle Type 3		
k	−0.01	−0.03	−0.05	−0.01	−0.03	−0.05	−0.01	−0.03	−0.05
Reference	11.37	11.92	13.08	11.28	11.86	13.03	13.98	14.45	15.56
Flat adj speed	10.26	10.97	12.15	10.15	10.87	12.05	13.87	14.36	15.47
Grad adj speed	10.18	10.92	12.10	10.07	10.82	11.99	13.83	14.24	15.36

5. Case Study Application

Finally, we apply the method to a real-life road gradient profile, as measured in the city of Lisbon. This is an interesting case due to the undulating terrain, and because the data used were part of the study that calibrated the power and energy relationships, as described in Section 2.1. A stretch of road was considered of a similar length to that considered in WMTC Reference Cycle 2 (see Section 2.2), for which the road profile is illustrated in Figure 13. If the WMTC Reference Cycle 2 speeds are exactly followed, then the portion of the gradient up to the red line is used. Again considering Vehicle Type 2 (as defined in Table 7) and the WMTC reference speeds (Figure 1), the resulting VSP profile (Figure 14) again shows violations of the maximum VSP, suggesting that the reference speed profile is not suited to this vehicle type and terrain. Applying the methodology described in Section 3, a modified speed profile is produced (Figure 15). It is noticeable that, as would be anticipated, many of the speed modifications are associated with steeper uphill sections of the profile; for example, significant modifications are apparent towards the end of the cycle, corresponding to the final uphill section of Figure 13. A comparison of the VSP profiles corresponding to the reference and modified speeds is given in Figure 16, and confirms that the method is successful in constraining VSP to a feasible range, with the impacts on energy consumption depicted in Figure 17.

As a final experiment, we performed an analysis of the road slope in a forward and reversed direction, with the summary statistics for the original direction in Tables 11 and 12, and for the reverse direction in Tables 13 and 14. The adjusted results indicate a small difference of −0.8% in total travel time under the reverse topography compared with the original. However, at the same time, the total energy is reduced by 4.9%, regenerated energy decreases by 4.6%, while the percentage of time in a negative energy state increases by 6.5% for the reversed slope compared with the original topography. This suggests that the original road slope combined with the real-world cycle, after adjustment, is more energy demanding, but also produces higher energy from regeneration, although there are fewer opportunities to regenerate. Such a comparison demonstrates the complexity of real-world driving energy consumption on an L-category EV.

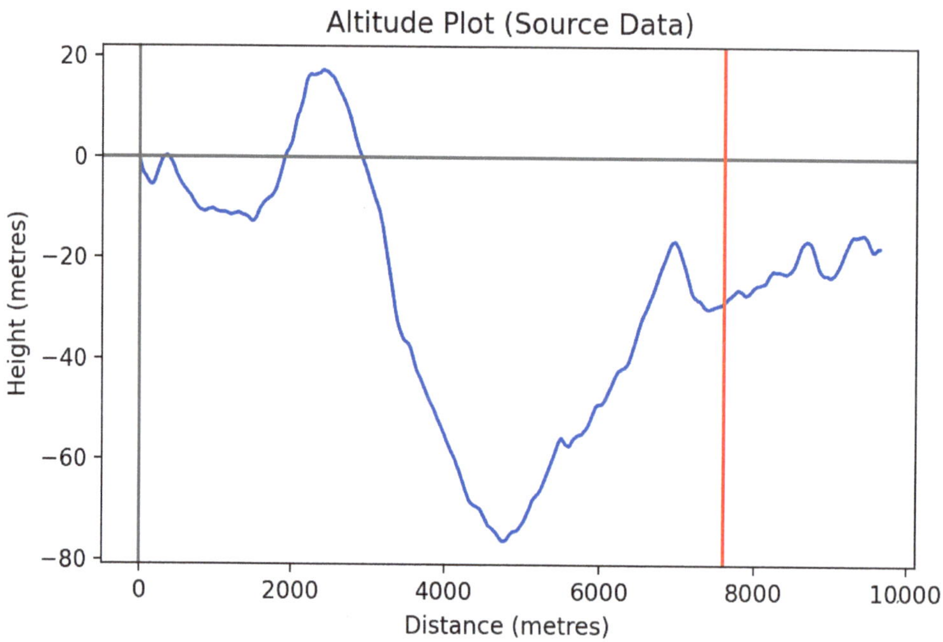

Figure 13. Studied road profile for Lisbon case study.

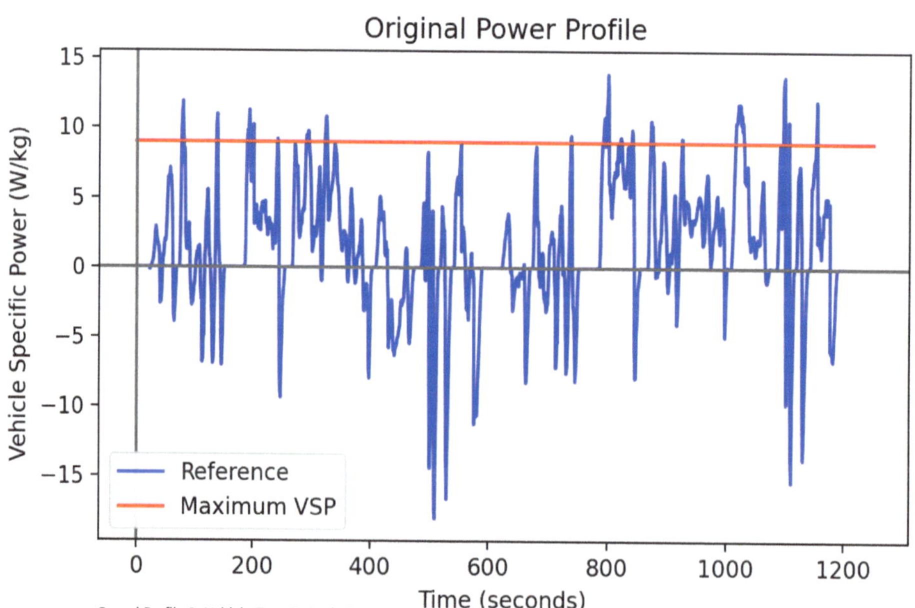

Figure 14. VSP profile obtained from applying WMTC Speed Profile 2 to the Lisbon case.

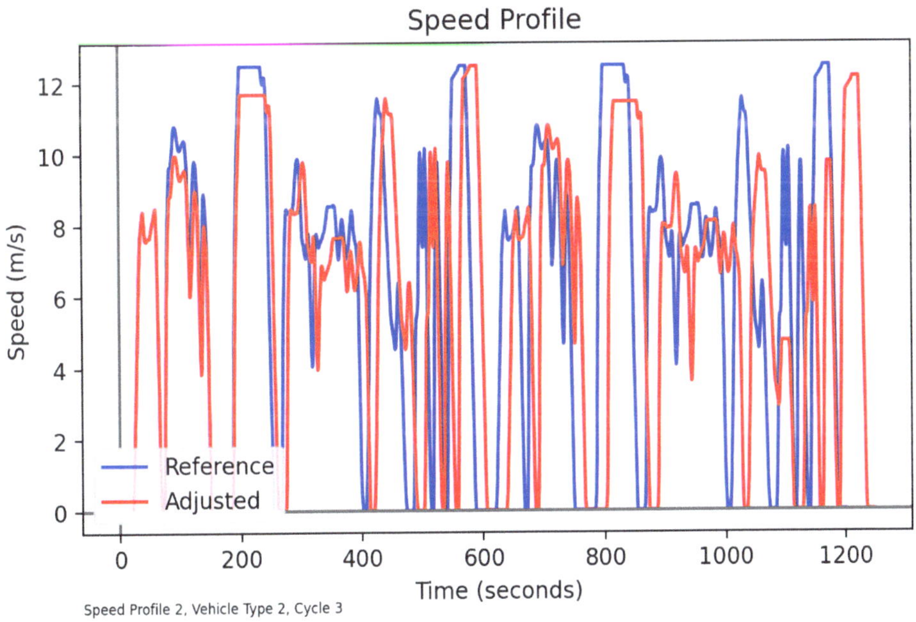

Figure 15. Reference and adjusted speed profiles for the Lisbon case.

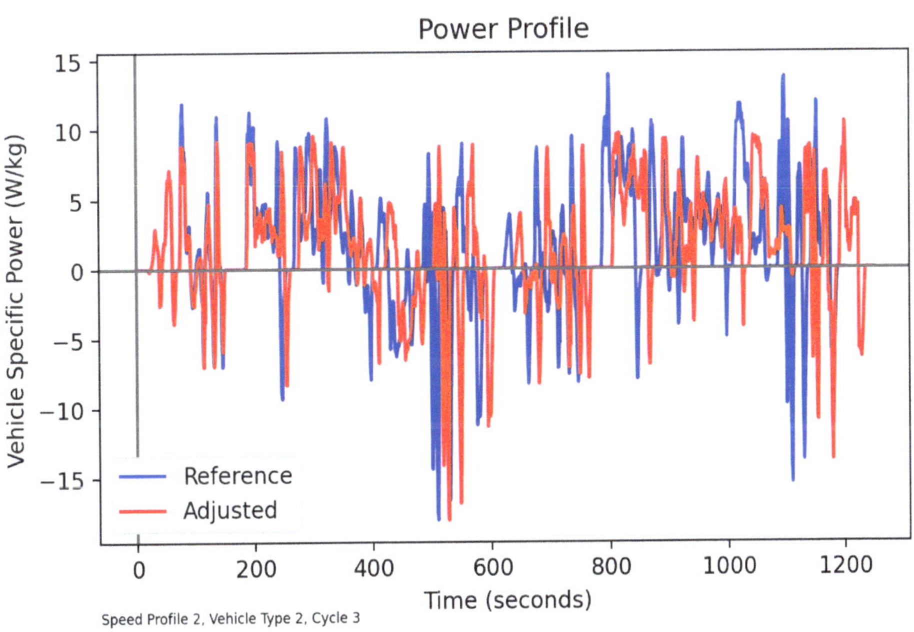

Figure 16. VSP under reference and adjusted speed profiles for the Lisbon case.

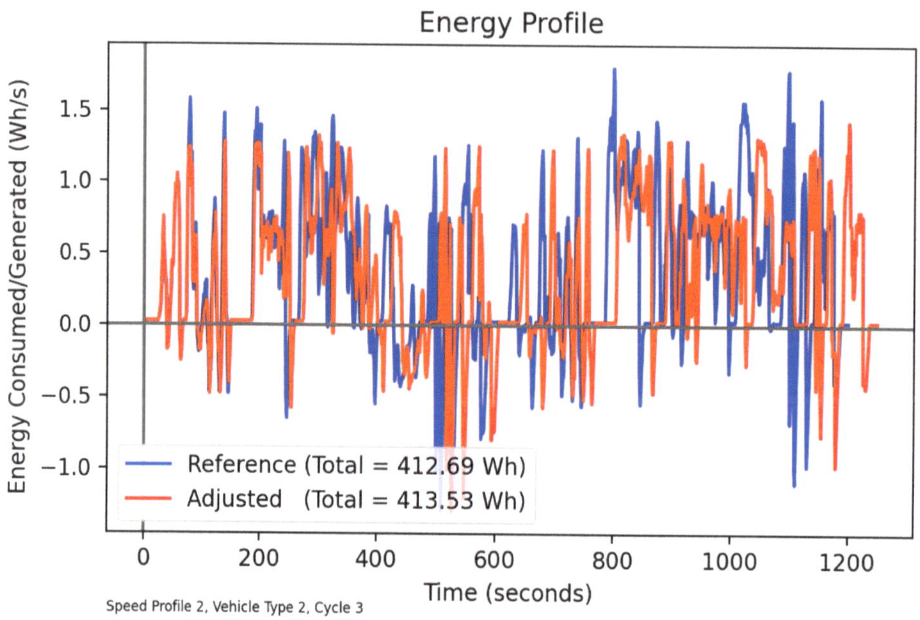

Figure 17. Energy consumption under reference and adjusted speed profiles for the Lisbon case.

Table 11. Kinetic summary statistics under reference and adjusted speed profiles for the Lisbon case.

Profile	Speed (m/s)		Positive Acceleration (m/s^2)		Negative Acceleration (m/s^2)		Total Idling Time (s)	Total Travel Time (s)
	Mean	Max	Mean	Max	Mean	Max		
Reference	6.33	12.50	0.46	1.64	−0.55	−1.74	227.0	1201.00
Adjusted	6.08	12.50	0.44	1.64	−0.55	−1.73	237.0	1250.89

Table 12. Energy-related summary statistics under reference and adjusted speed profiles for the Lisbon case.

Profile	Total Energy (Wh)	Regenerated Energy (Wh)	% Time in Negative Energy State	% Time in Positive Energy State
Reference	428.67	75.32	21.82	78.10
Adjusted	427.78	71.46	21.44	78.48

Table 13. Kinetic summary statistics under reference and adjusted speed profiles for the Lisbon case with reversed road profile.

Profile	Speed (m/s)		Positive Acceleration (m/s^2)		Negative Acceleration (m/s^2)		Total Idling Time (s)	Total Travel Time (s)
	Mean	Max	Mean	Max	Mean	Max		
Reference	6.33	12.50	0.46	1.64	−0.55	−1.74	227.0	1201.00
Adjusted	6.13	11.72	0.43	1.63	−0.53	−1.73	238.0	1240.42

Table 14. Energy-related summary statistics under reference and adjusted speed profiles for the Lisbon case with reversed road profile.

Profile	Total Energy (Wh)	Regenerated Energy (Wh)	% Time in Negative Energy State	% Time in Positive Energy State
Reference	411.95	74.19	24.40	75.52
Adjusted	406.59	68.16	22.84	77.08

6. Conclusions

In the context of an expected emergence of L-category electric vehicles in the urban context, a generic methodology has been developed to adjust reference speed profiles to be applicable to the performance characteristics of these specific vehicles, while also accounting for different terrains. The methodology comprises three main elements: bounding maximum instantaneous power by dynamically limiting acceleration, given the current speed and gradient; bounding decelerations to avoid physically impossible speeds that may otherwise arise due to previous adjustments to speeds; and inserting cruising periods to ensure that the modified profile covers the same route length, albeit over a longer period of time.

The methodology has been applied, firstly, to derive speed profiles for electric quadricycles on flat terrain, given reference speed profiles for motorcycles. The results show that adjustments made to cope with the power limitations of L-EV do not introduce significant differences in energy consumption, suggesting that the certification does not require an extensive modification. Secondly, it was explored to what extent bespoke speed profiles are necessary for undulating terrain, as opposed to adopting the speed profiles developed for flat terrain in all cases. The first set of tests considered a fictitious road profile that was easily adjustable, namely a symmetric hill *versus* a symmetric valley. In this case, it was found that the power limitations, as well as the combination of a fixed driving profile and different magnitudes of slope and shapes of slope, lead to differences of up to 5% in energy use and of up to 10% in regenerated energy. The second set of tests analyzed the impact of topography on real-world driving cycles and slope (by considering a real-world profile driven in both directions), and qualitatively confirmed the results obtained for the fictitious profile tests.

The developed speed profiles are useful in their own right for studies that wish to assess the impacts of different measures and policies on electric quadricycles, with the advantage of considering regenerative capabilities and road grade, which are crucial in the characterization of energy performance in real-world conditions. The developed method may be readily extended to other L-category electric vehicles, given the relevant input information on vehicle specific power relations and maximum VSP.

Author Contributions: D.W.: conceptualization; formal analysis; methodology; software; writing—original draft. P.B.: formal analysis; writing—review & editing. G.D.: funding acquisition; formal analysis; writing—review & editing. J.G.: data curation. H.C.: funding acquisition; writing—review & editing. All authors have read and agreed to the published version of the manuscript.

Funding: The University of Leeds authors acknowledge the support of EU H2020-EU.3.4 funding, under grant agreement ID 769,926 for the project ELVITEN (elviten-project.eu). The IST-ID authors acknowledge the support of Fundação para a Ciência e Tecnologia for IN+ Strategic Project UID/EEA/50009/2019 and for contract CEECIND/02589/2017.

Institutional Review Board Statement: Not applicable.

Informed Consent Statement: Not applicable.

Data Availability Statement: This study did not report any new data. The reference cycles that were adapted can be found in reference [46]. All other parameter values used are reported in the main body of the current paper.

Conflicts of Interest: The authors declare no conflict of interest.

References

1. Fulton, L.M. Three Revolutions in Urban Passenger Travel. *Joule* **2018**, *2*, 575–578. [CrossRef]
2. Lee, R.; Brown, S. Evaluating the role of behavior and social class in electric vehicle adoption and charging demands. *iScience* **2021**, *24*, 102914. [CrossRef] [PubMed]
3. Tushar, W.; Yuen, C.; Saha, T.; Chattopadhyay, D.; Nizami, S.; Hanif, S.; Alam, J.E.; Poor, H.V. Roles of Retailers in the Peer-to-Peer Electricity Market: A Single Retailer Perspective. *iScience* **2021**, *24*, 103278. [CrossRef]
4. Fachrizal, R.; Shepero, M.; Van der Meer, D.; Munkhammar, J.; Widén, J. Smart charging of electric vehicles considering photovoltaic power production and electricity consumption: A review. *eTransportation* **2020**, *4*, 100056. [CrossRef]
5. Transport & Environment. How Clean Are Electric Cars? 2020. Available online: https://www.transportenvironment.org/sites/te/files/downloads/T%26E%E2%80%99s%20EV%20life%20cycle%20analysis%20LCA_0.pdf (accessed on 3 February 2022).
6. IEA. Global EV Outlook 2020: Entering the Decade of Electric Drive. 2020. Available online: https://www.iea.org/reports/global-ev-outlook-2020 (accessed on 3 February 2022).
7. Zhang, H.; Zhang, J. An overview of modification strategies to improve $LiNi_{0.8}Co_{0.1}Mn_{0.1}O_2$ (NCM811) cathode performance for automotive lithium-ion batteries. *eTransportation* **2021**, *7*, 100105. [CrossRef]
8. Santucci, M.; Pieve, M.; Pierini, M. Electric L-category vehicles for smart urban mobility. *Transp. Res. Procedia* **2016**, *14*, 3651–3660. [CrossRef]
9. EU: Vehicle Categories. 2022. Available online: https://ec.europa.eu/growth/sectors/automotive-industry/vehicle-categories_en (accessed on 3 February 2022).
10. Transportpolicy.net. EU: Vehicle Definitions. 2022. Available online: https://www.transportpolicy.net/standard/eu-vehicle-definitions/ (accessed on 3 February 2022).
11. UNECE. Special Resolution No. 1 Concerning the Common Definitions of Vehicle Categories, Masses and Dimensions (S.R.1). World Forum for Harmonization of Vehicle Regulations, Report no. TRANS/WP.29/1045. 2005. Available online: https://unece.org/classification-and-definition-vehicles (accessed on 3 February 2022).
12. Ewert, A.; Brost, M.; Eisenmann, C.; Stieler, S. Small and Light Electric Vehicles: An Analysis of Feasible Transport Impacts and Opportunities for Improved Urban Land Use. *Sustainability* **2020**, *12*, 8098. [CrossRef]
13. Ministry of Road Transport and Highways. Gazette of India: Extraordinary, Part II Section 3, Sub-Section (i). 2014. Available online: https://morth.nic.in/sites/default/files/notifications_document/Notification_No_GSR_99E_dated_1922014%C2%A0.pdf (accessed on 3 February 2022).
14. Korean Transportation Safety Authority. Korean Safety Regulations on Micro Mobility Related to Passive Safety. Informal Document GRSP-65-23. 2019. Available online: https://unece.org/DAM/trans/doc/2018/wp29grsp/GRSP-65-23e.pdf (accessed on 3 February 2022).
15. National Highway Traffic Safety Administration. Federal Motor Vehicle Safety Standards: Low-Speed Vehicles. US Federal Register, Document Number 06-3590. 2006. Available online: https://www.federalregister.gov/documents/2006/04/19/06-3590/federal-motor-vehicle-safety-standards-low-speed-vehicles (accessed on 3 February 2022).
16. Pavlovic, A.; Fragassa, C. General considerations on regulations and safety requirements for quadricycles. *Int. J. Qual. Res.* **2015**, *9*, 657–674.
17. EEA. *The First and Last Mile*; European Environment Agency: Copenhagen, Denmark, 2019.
18. Martiskainen, M.; Sovacool, B.K.; Lacey-Barnacle, M.; Hopkins, D.; Jenkins, K.E.H.; Simcock, N.; Mattioli, G.; Bouzarovski, S. New dimensions of vulnerability to energy and transport poverty. *Joule* **2021**, *5*, 3–7. [CrossRef]
19. Sachs, C.; Burandt, S.; Mandelj, S.; Mutter, R. Assessing the market of light electric vehicles as a potential application for electric in-wheel drives. In Proceedings of the IEEE 6th International Electric Drives Production Conference (EDPC), Nuremberg, Germany, 30 November–1 December 2016; pp. 280–285.
20. Renault. Renault Twizy E-tech Electric. 2021. Available online: https://www.renault.pt/veiculos-eletricos/twizy.html (accessed on 3 February 2022).
21. Citröen. My Ami Orange. 2021. Available online: https://www.citroen.pt/ami (accessed on 3 February 2022).
22. EU. Sustainable and Smart Mobility Strategy—Putting European Transport on Track for the Future. Communication from the Commission to the European Parliament, The Council, The European Economic and Social Committee and the Committee of the Regions. 2020. Available online: https://eur-lex.europa.eu/legal-content/EN/TXT/HTML/?uri=CELEX:52020DC0789&from=EN (accessed on 3 February 2022).
23. Alves, J.; Baptista, P.C.; Gonçalves, G.A.; Duarte, G.O. Indirect methodologies to estimate energy use in vehicles: Application to battery electric vehicles. *Energy Convers. Manag.* **2016**, *124*, 116–129. [CrossRef]
24. Fiori, C.; Ahn, K.; Rakha, H.A. Power-based electric vehicle energy consumption model: Model development and validation. *Appl. Energy* **2016**, *168*, 257–268. [CrossRef]
25. Luin, B.; Petelin, S.; Al-Mansour, F. Microsimulation of electric vehicle energy consumption. *Energy* **2019**, *174*, 24–32. [CrossRef]
26. Xie, Y.; Li, Y.; Zhao, Z.; Dong, H.; Wang, S.; Liu, J.; Guan, J.; Duan, X. Microsimulation of electric vehicle energy consumption and driving range. *Appl. Energy* **2020**, *267*, 115081. [CrossRef]

27. Zhang, J.; Wang, Z.; Liu, P.; Zhang, Z.; Li, X.; Qu, C. Driving cycles construction for electric vehicles considering road environment: A case study in Beijing. *Appl. Energy* **2019**, *253*, 113514. [CrossRef]
28. Chiang, H.L.; Huang, P.H.; Lai, Y.M.; Lee, T.Y. Comparison of the regulated air pollutant emission characteristics of real-world driving cycle and ECE cycle for motorcycles. *Atmos. Environ.* **2014**, *87*, 1–9. [CrossRef]
29. Tutuianu, M.; Bonnel, P.; Ciuffo, B.; Haniu, T.; Ichikawa, N.; Marotta, A.; Pavlovic, J.; Steven, H. Development of the World-wide harmonized Light duty Test Cycle (WLTC) and a possible pathway for its introduction in the European legislation. *Transp. Res. Part D Transp. Environ.* **2015**, *40*, 61–75. [CrossRef]
30. Giechaskiel, B.; Zardini, A.; Martini, G. Particle emission measurements from L-category vehicles. *SAE Int. J. Engines* **2015**, *8*, 2322–2337. [CrossRef]
31. Burrows, A. EU Regulation on the Approval of L-Category Vehicles. 2013. Available online: https://www.zemo.org.uk/assets/presentations/EU%20Regulation%20on%20the%20Approval%20of%20L-Category%20Vehicles.pdf (accessed on 3 February 2022).
32. RESOLVE. Range of Electric Solutions for L-Category Vehicles. 2018. Available online: http://www.resolve-project.eu/ (accessed on 3 February 2022).
33. Kamble, S.H.; Mathew, T.V.; Sharma, G.K. Development of real-world driving cycle: Case study of Pune, India. *Transp. Res. Part D Transp. Environ.* **2009**, *14*, 132–140. [CrossRef]
34. Al-Wreikat, Y.; Serrano, C.; Sodré, J.R. Driving behaviour and trip condition effects on the energy consumption of an electric vehicle under real-world driving. *Appl. Energy* **2021**, *297*, 117096.
35. Kühne, R. Electric buses—An energy efficient urban transportation means. *Energy* **2010**, *35*, 4510–4513. [CrossRef]
36. Schücking, M.; Jochem, P. Two-stage stochastic program optimizing the cost of electric vehicles in commercial fleets. *Appl. Energy* **2021**, *293*, 116649. [CrossRef]
37. Welzel, F.; Klinck, C.-F.; Pohlmann, Y.; Bednarczyk, M. Grid and user-optimized planning of charging processes of an electric vehicle fleet using a quantitative optimization model. *Appl. Energy* **2021**, *290*, 116717. [CrossRef]
38. Jiménez-Palacios, J.L. Understanding and Quantifying Motor Vehicle Emissions with Vehicle Specific Power and TILDAS Remote-Sensing. Ph.D. Thesis, Massachusetts Institute of Technology, Massachusetts, MA, USA, 1999.
39. Frey, H.C.; Rouphail, N.M.; Zhai, H.; Farias, T.L.; Gonçalves, G.A. Comparing real-world fuel consumption for diesel- and hydrogen-fueled transit buses and implication for emissions. *Transp. Res. Part D* **2007**, *12*, 281–291. [CrossRef]
40. Frey, H.; Unal, A.; Rouphail, N.; Colyar, J. On-road measurement of vehicle tailpipe emissions using a portable instrument. *J. Air Waste Manag. Assoc.* **2003**, *53*, 992–1002. [CrossRef]
41. Rhys-Tyler, G.; Bell, M. Toward reconciling instantaneous roadside measurements of LDV exhaust emission with type approval driving cycles. *Environ. Sci. Technol.* **2012**, *46*, 10532–10538. [CrossRef]
42. Duarte, G.O.; Gonçalves, G.A.; Baptista, P.C.; Farias, T.L. Establishing bonds between vehicle certification data and real-world vehicle fuel consumption—A vehicle specific power approach. *Energy Convers. Manag.* **2015**, *92*, 251–265. [CrossRef]
43. Varella, R.; Duarte, G.; Baptista, P.; Sousa, L.; Villafuerte, P.M. *Comparison of Data Analysis Methods for European Real Driving Emissions Regulation*; SAE Technical Paper 2017-01-0997; SAE International: Warrendale, PA, USA, 2017. [CrossRef]
44. Aprillia, H.; Yang, H.-T.; Huang, C.-M. Optimal Decomposition and Reconstruction of Discrete Wavelet Transformation for Short-Term Load Forecasting. *Energies* **2019**, *12*, 4654. [CrossRef]
45. Shi, P.; Yang, W.; Sheng, M.; Wang, M. An Enhanced Empirical Wavelet Transform for Features Extraction from Wind Turbine Condition Monitoring Signals. *Energies* **2017**, *10*, 972. [CrossRef]
46. EU. Reglamento Delegado (UE) No. 134/2014 de la Comisión. Consolidated Text: Commission Delegated Regulation (EU) No. 134/2014 of 16 December 2013 Supplementing Regulation (EU) No. 168/2013 of the European Parliament and of the Council with Regard to Environmental and Propulsion Unit Performance Requirements and Amending Annex V Thereof. 2013. Available online: http://data.europa.eu/eli/reg_del/2014/134/2018-03-20 (accessed on 3 February 2022).

energies

MDPI

Article

Rockburst Intensity Level Prediction Method Based on FA-SSA-PNN Model

Gang Xu [1,2], Kegang Li [1,2,*], Mingliang Li [1,2], Qingci Qin [1,2] and Rui Yue [1,2]

1 School of Land and Resources Engineering, Kunming University of Science and Technology, Kunming 650093, China; 20202201005@stu.kust.edu.cn (G.X.); 20182101079@kust.edu.cn (M.L.); qqc1457186548@126.com (Q.Q.); yr@stu.kust.edu.cn (R.Y.)

2 Yunnan Key Laboratory of Sino-German Blue Mining and Utilization of Special Underground Space, Kunming 650093, China

* Correspondence: 20070001@kust.edu.cn

Abstract: To accurately and reliably predict the occurrence of rockburst disasters, a rockburst intensity level prediction model based on FA-SSA-PNN is proposed. Crding to the internal and external factors of rockburst occurrence, six rockburst influencing factors (σ_θ, σ_t, σ_c, σ_c/σ_t, σ_θ/σ_c, W_{et}) were selected to build a rockburst intensity level prediction index system. Seventy-five sets of typical rockburst case data at home and abroad were collected, the original data were preprocessed based on factor analysis (FA), and the comprehensive rockburst prediction indexes, CPI_1, CPI_2, and CPI_3, obtained after dimensionality reduction, were used as the input features of the SSA-PNN model. Sixty sets of rockburst case data were extracted as the training set, and the remaining 15 sets of rockburst case data were used as the test set. After the model training was completed, the model prediction results were analysed and evaluated. The research results show that the proposed rockburst intensity level prediction method based on the FA-SSA-PNN model has the advantages of high prediction accuracy and fast convergence, which can accurately and reliably predict the rockburst intensity level in a short period of time and can be used as a new method for rockburst intensity level prediction, providing better guidance for rockburst prediction problems in deep rock projects.

Keywords: rock mechanics; factor analysis; sparrow search algorithm; probabilistic neural network; rockburst intensity level prediction

1. Introduction

Rockburst is a deep underground rock construction process of hard and brittle surrounding rock due to excavation, mining, or other external disturbances, triggered by the rapid and violent release of the elastic properties gathered in the rock and leads to the production of surrounding rock fragments by bursting, rapid ejection, or throwing the dynamic destabilization phenomenon, which are sudden, random, and extremely hazardous geological hazards [1–3]. In recent years, with the reduction of shallow mineral resources, more and more underground rock works are moving deeper at an unprecedented rate, and the rockburst hazards problem is becoming increasingly prominent. These hazards have been a pressing problem in deep underground rock engineering, often causing huge losses to construction personnel, equipment, and buildings, which in turn seriously affects the construction process, so it is particularly important to accurately predict the occurrence of rockburst hazards. Accurate and reliable prediction of rockburst hazards effectively avoids and controls rockbursts, and rockburst prediction has become a hot spot for research in the field of deep underground rock engineering [4].

In order to accurately predict the intensity level of rockburst, many experts and scholars at home and abroad have carried out exploratory research on rockburst prediction methods, which can be classified into three categories: The first category is the acoustic emission technique [5], microseismic observation technique [6], and other methods of rockburst prediction based on field measurements; the second category is a single-factor prediction method, where the discrimination of rockburst intensity levels varies slightly with the criterion, such as Hoek criterion [7], N-Jhelum criterion [8], Erlang Mountain criterion [9], and Lujiayou criterion [10], etc. With the continuous research on the problem of rockburst prediction, a large number of scholars have gradually realized the complexity of the mechanism of rockburst and the many factors that induce rockburst [11–13], but it is difficult to accurately predict rockburst using only single-factor prediction methods. At present, non-linear theory uses more than just the third category of rockburst prediction methods, that is, the multi-factor integrated prediction method and the multi-factor integrated prediction method, to integrate rock mechanics parameters and a variety of rockburst criterion to achieve rockburst intensity level prediction. The multi-factor integrated prediction method, according to the different non-linear theory, is divided into two subcategories. The former is mainly based on mathematical methods to predict the rockburst intensity level, which are representative of ideal point method [14], cloud model theory [15,16], fuzzy comprehensive evaluation method [17–19], uncertainty measurement theory [20], gray system theory [21], the TOPSIS method [22], discriminant by distance method [23], and the extenics theory [3]. The latter is mainly based on intelligent algorithms to predict rockburst intensity levels, such as self-organizing feature mapping neural networks [24], machine learning [25,26], deep neural networks [27], generalized regression neural networks [28], and sarticle swarm optimization [29].

All of the above rockburst prediction methods have achieved some success, enriching the theory of rockburst prediction. However, there are still some shortcomings, for example, when ignoring the impact of correlation between rockburst prediction indicators on the prediction results, due to the diversity of factors affecting the occurrence of rockburst, there is a certain correlation between rockburst prediction indicators, which will not only lead to double calculation of indicators, increasing the workload in the prediction process, but will directly affect the accuracy and reliability of the prediction results. Therefore, the elimination of correlation between indicators is the key to accurate prediction of rockbursts, and research on the elimination of correlation between rockburst prediction indicators is necessary.

Factor analysis (FA) [30] is the extension and development of principal component analysis, which regroups the information of the original variables to find out the common factors affecting the variables and can make the factor variables more interpretable and give high naming clarity by rotation. At present, the application of factor analysis method in rockburst intensity level prediction is relatively small in order to eliminate the correlation between rockburst prediction indicators. This paper uses the factor analysis method to extract the characteristics of rockburst prediction indicators, using the original rockburst prediction indicators with minimal loss of information, with comprehensive rockburst indicators, as much as possible to reflect the original rockburst prediction indicators information, which is a good solution to the problem of overlapping information indicators.

Probabilistic neural networks (PNNs) were proposed by Dr. D. F. Specht in 1988 and can implement the functions of nonlinear learning algorithms using linear learning algorithms, with the advantages of simple structure, good expansion performance, fast convergence, and high fault tolerance [31]. However, when the probabilistic neural network is used, the problem of selecting the optimal smoothing factor is somewhat subjective and tedious. Therefore, this paper uses the sparrow search algorithm (SSA) to select the optimal smoothing factor, which has the advantages of being rapid and efficient when optimizing for a single objective and good merit-seeking ability and solves the problem of selecting the optimal smoothing factor very well.

Combining the above research, this paper selects 75 groups of typical rockburst case data, combines factor analysis method, sparrow search algorithm, and probabilistic neural network, and establishes a rockburst intensity level prediction method based on the FA-SSA-PNN model. The method has the advantages of simple logic, easy implementation, strong generalization ability of the model, high prediction accuracy, fast convergence, and applicability to small samples, which can be used as a new method for rockburst intensity level prediction. The present research results provide an important basis for predicting the rockburst intensity level in advance and provide preparation time for rockburst disaster prevention and control, and the method in this paper can also provide a reference for other geological hazard prediction problems similar to rockburst disasters.

2. Methods

2.1. Factor Analysis (FA)

Factor analysis (FA), is a multivariate statistical analysis method that combines multiple variables (or samples) with intricate relationships into a smaller number of factors. The calculation process is as follows:

With n rockburst predictor variables $X_1, X_2, \ldots, X_n$, n rockburst prediction variables can be represented by m factors $F_1, F_2, \ldots, F_m$ and the product of an $A_{n\times m}$ order factor loading matrix plus a special factor $\varepsilon = (\varepsilon_1, \varepsilon_2, \ldots, \varepsilon_n)$ $(n \geq m)$, while the established mathematical model of factor analysis is: $X_n = A_{n\times m}F_m + \varepsilon_n$, i.e., Equation (1):

$$\begin{bmatrix} X_1 \\ X_2 \\ \vdots \\ X_n \end{bmatrix} = \begin{bmatrix} a_{11} & a_{12} & \cdots & a_{1m} \\ a_{21} & a_{22} & \cdots & a_{2m} \\ \vdots & \vdots & \ddots & \vdots \\ a_{n1} & a_{n2} & \cdots & a_{nm} \end{bmatrix} \begin{bmatrix} F_1 \\ F_2 \\ \vdots \\ F_m \end{bmatrix} + \begin{bmatrix} \varepsilon_1 \\ \varepsilon_2 \\ \vdots \\ \varepsilon_n \end{bmatrix}, \quad (1)$$

$F_1, F_2, \ldots,$ and F_m are m independent common factors. The matrix $A_{n\times m}$ is called the factor loading matrix. a_{nm} denotes the weight of the nth variable on the mth factor variable, which reflects the importance of the common factor on the variable, and is important for explaining the common factor. The special factors in the model have a small effect relative to the main factor F_m in Equation (1) and are neglected in the study. The steps of the factor analysis are as follows:

The raw data need to be normalized before factor analysis to eliminate the effect of order of magnitude on the calculated results. The normalized treatment in this paper is specified as follows:

$$\begin{cases} X_{ij}^* = \left(X_{ij}^* - X_{\min}\right)/(X_{\max} - X_{\min}) \\ X_{ij}^* = \left(X_{\max} - X_{ij}^*\right)/(X_{\max} - X_{\min}) \end{cases}, \quad (2)$$

Equation: $X_{\max}$ is the maximum value in the sample; $X_{\min}$ is the smallest value in the sample; X_{ij}^* is the normalized data value of indicator X_{ij} in the sample, $i = 1, 2, \ldots, n$; $j = 1, 2, \ldots, m$.

After the normalized process, the factor loading matrix is calculated from the eigenvalues λ_i of the sample correlation matrix R, which is symmetric,

$$R = \begin{bmatrix} 1 & r_{21} & r_{31} & \cdots & r_{p1} \\ r_{21} & 1 & r_{21} & \cdots & r_{p2} \\ r_{31} & r_{32} & 1 & \vdots & r_{p3} \\ \vdots & \vdots & \vdots & \ddots & \vdots \\ r_{q1} & r_{q2} & r_{q3} & \cdots & 1 \end{bmatrix}, \quad (3)$$

Equation: r_{qp} is the correlation coefficient of the variable F_m with F_n.

$$r_{qp} = \frac{\mathrm{cov}(F_q F_p)}{\sqrt{D(F_q)}\sqrt{D}(F_p)}, \quad (4)$$

Equation: cov(F_q, F_p) is the covariance of F_q and F_p; $D(F_q)$ and $D(F_p)$ are the variances of F_q and F_p, respectively. The raw sample data have been normalized so that there is:

$$\left.\begin{array}{l} D(F_q) = D(F_p) = 1 \\ b_{ir} = a_{ir}\cos\theta + a_{ir}\sin\theta \\ r_{qp} = a_{qp} \end{array}\right\}. \quad (5)$$

From the above equation, a_{qp} can be regarded as the correlation coefficient between F_q and F_p, which also exactly indicates the degree of linear combination of X_m and F_m. λ_i (i = 1, 2, ..., m) are all the characteristic roots of the sample correlation matrix R, finding all the characteristic root λ_i (i = 1, 2, ..., m) of the sample correlation matrix R and the corresponding normalized eigenvectors U_i (i = 1, 2, ..., m),

$$\lambda_1 \geq \lambda_2 \geq \cdots \lambda_m \geq 0, \quad (6)$$

The eigenvectors are used to form the eigenvector matrix $U = (U_1, U_2, \ldots, U_m)$, and the load matrix is derived from the characteristic root and eigenvectors A,

$$A = (\sqrt{\lambda_1}U_1, \sqrt{\lambda_2}U_2, \cdots, \sqrt{\lambda_m}U_m, \quad (7)$$

The first k column vectors of the loading matrix A are used as the factor loading matrix, and the cumulative contribution of the factors is required,

$$\frac{\sum\limits_{i=1}^{k} \lambda_i}{\sum\limits_{i=1}^{m} \lambda_i} \geq 85\%. \quad (8)$$

The maximum variance method was chosen to perform the common factor rotation, and the new loading matrix B was obtained by left multiplying the loading matrix A with the orthogonal matrix Γ,

$$[\Gamma] = \begin{bmatrix} \cos\theta & -\sin\theta \\ \sin\theta & \cos\theta \end{bmatrix}. \quad (9)$$

Let the elements of the ith row and jth column of B be b_{ij}, then we have:

$$\left.\begin{array}{l} b_{ir} = a_{ir}\cos\theta + a_{ir}\sin\theta \\ b_{ir} = a_{ir}\cos\theta + a_{ir}\sin\theta \end{array}\right\}, \quad (10)$$

where θ is the orthogonal rotation angle, i = (1, 2, ..., p); r, j = (1, 2, ..., n). After this transformation, the aim is to polarize the loading matrix and spread the factor contributions as much as possible, i.e., the $b^2{}_{ir}$, $b^2{}_{ig}$ 2 sets of data are required to be spread as much as possible, the degree of dispersion is expressed by the variance of the samples (V_1, V_2, ..., V_n), and the total variance is required to be maximum, that is, the request:

$$G = V_1 + V_2 + \cdots + V_n - \max, \quad (11)$$

When the number of common factors is more than 2 (i.e., when m > 2), the orthogonal matrix Γ can generally only be obtained iteratively, and then a total of $C_m^2 = m(m-1)/2$ rotations are required, after which a second round of C_m^2 pairwise rotations can be performed, and so on until the variance converges to a certain limit and the rotation is stopped. The steps of factor analysis in this paper are shown in Figure 1.

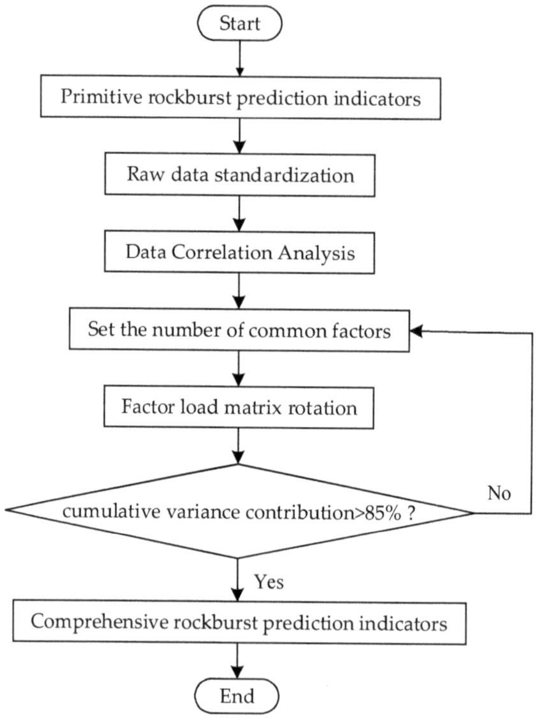

Figure 1. Factor analysis flow chart.

2.2. Sparrow Search Algorithm (SSA)

The Sparrow Search Algorithm [32] (SSA) is a new swarm intelligence optimization algorithm proposed in 2020 to simulate sparrow foraging and anti-predation behaviors, which divides the sparrows in the population into a discoverer, a follower, and a certain ratio of scouts. The finder is responsible for finding food, the follower joins the finder for foraging, and the scout is responsible for scouting the surrounding environment and alerting to the danger in time.

The sparrow as a discoverer has good adaptability. It will preferentially obtain food and provide foraging orientation for followers during foraging, and the location update of the discoverer in each iteration is described as in Equation (12):

$$\left.\begin{array}{l} x_{ij}^{t}\exp(\frac{-i}{\alpha\cdot iter_{\max}}), R_2 < ST \\ x_{ij}^{t}\exp(\frac{-i}{\alpha\cdot iter_{\max}}), R_2 < ST \end{array}\right\} = x_{ij}^{t+1}, \quad (12)$$

where t is the current number of iterations; L is an all-1 matrix of $1 \times d$; $iter_{\max}$ is the maximum number of iterations; Q and α are random numbers; x_{ij}^{t} denotes the position of the ith sparrow in the jth dimension at iteration t; R_2 and ST are the warning value and safety threshold. When $R_2 < ST$, it means there is no danger in the surrounding environment and the foraging range can continue to be expanded, and when $R_2 \geq ST$, it means the scout detects danger in the surrounding area and sends an alert to the population, telling the population to move to a safe area as soon as possible.

The updated description of the follower's location is as follows:

$$\left.\begin{array}{l} Q\exp(\frac{x_{wj}^{t}-x_{ij}^{t}}{i^2}), i > \frac{N}{2} \\ x_{pj}^{t+1} + \left|x_{ij}^{t} - x_{pj}^{t+1}\right| A^{+}L, other \end{array}\right\} = x_{ij}^{t+1}, \quad (13)$$

where A is a d-column matrix with random element values of 1 or -1 and $A^+ = A^T(AA^T)^{-1}$; x_{pj} is the optimal position of the discoverer; x_{wj} is the worst position in the game.

The scout generates the initial position randomly, and its subsequent position is updated as in Equation (14):

$$\left.\begin{array}{l} x_{ij}^t + \beta\left|x_{ij}^t - x_{bj}^t\right|, f_i > f_g \\ x_{ij}^t + K\frac{\left|x_{ij}^t - x_{WJ}^t\right|}{(f_i - f_w) + \varepsilon}, f_i = f_g \end{array}\right\} = x_{ij}^{t+1}, \tag{14}$$

where x_{bj} indicates the global best position; β denotes the iteration step size; f_i indicates the current level of adaptation; f_g and f_w denote the global optimal and worst adaptation degrees; K is a random number between $[-1, 1]$; ε is the smallest constant that prevents the denominator from going to zero.

After completing one round of iterations, as described above, the position of the population will be changed in some way, and each iteration will make the population change in the direction of better fitness, and finally the optimal fitness will be obtained.

2.3. Probabilistic Neural Network (PNN)

The Probabilistic Neural Network ((PNN)) [33,34] is a feed forward neural network based on radial basis neural network, which uses the Parzen window function to calculate the conditional probability density function of the samples to be recognized and then completes the classification and recognition of patterns by Bayes classification criterion, and its topology is shown in Figure 2.

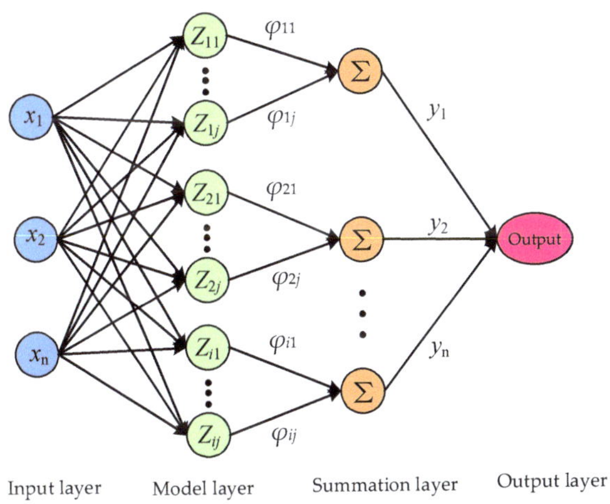

Figure 2. Topology of probabilistic neural network.

The role of the input layer is to receive the training sample data and multiply the values x of the input sample parameters with the weighting coefficients w_i to obtain the scalar product Z_i input to the model layer, as shown in Equation (15):

$$Z_i = xw_i, \tag{15}$$

The pattern layer is used to calculate the matching relationship between the input vector and each pattern and return a scalar value. The vector Z is input to the pattern layer, and the input and output relationship of the jth neuron of the ith class of patterns in the pattern layer is:

$$\varphi_{ij} = \frac{1}{(2\pi)^{\frac{1}{2}}\sigma^d} exp\left[-(Z - Z_{ij})(Z - Z_{ij})\right]^{T/\sigma^2}, \tag{16}$$

where φ_{ij} is the output value of the jth neuron of the ith class of patterns in the pattern layer; σ is the smoothing factor; d is the dimensionality of the sample space data; and Z_{ij} is the jth center of the ith class of samples.

The main role of the summation layer is linear summation and weighted averaging. The summation layer takes the outputs of neurons belonging to the same class in the pattern layer and makes a weighted average,

$$v_i = \sum_{j-1}^{L} \phi_{ij} / L, \tag{17}$$

where v_i is the output of class i, (i = 1, 2, . . . , n); n is the total number of training sample patterns; L denotes the number of neurons in class i.

The last layer is the output layer and the Bayesian classification rule is applied to the output of the summation layer, and the neuron with the maximum posterior probability density is found to have an output of 1 among all the output layer neurons, and the rest of the neurons have an output of 0,

$$y = \text{argmax}(f_i), \tag{18}$$

where y indicates the output of the output layer.

3. Dataset Preparations

3.1. Selection of Rockburst Prediction Indicators

The rockburst mechanism is complex and has significant randomness and suddenness. The selection of indicators is the key to accurately predicting the rockburst. The selection of predictive indicators should meet the following conditions: (1) less influenced by external factors, so the actual measured values of indicators are easy to obtain; (2) has a good representative, so it can accurately reflect the main characteristics of the occurrence of rockbursts; (3) capable of reflecting comprehensive information on rockburst characteristics. This paper is based on a large number of rockburst case study analyses to determine the rockburst prediction evaluation indicators.

From the geological structure of the occurrence of rockbursts, rockbursts usually occur in the deeper buried underground works and higher structural stress in the rock mass. From the structural surface of the rock, rockburst often occurs near the hard structural surface, and the more irregular the structural surface, the more likely to occur rockburst. The maximum tangential stress in the surrounding rock can reflect the above factors well, so the maximum tangential stress in the surrounding rock (σ_θ) is selected as the rockburst prediction evaluation indicators.

The occurrence of rockburst section form of the surrounding rock is mainly tensile damage, and rockburst usually occurs in the structural integrity and hard rock. The hardness of the rock is usually expressed in terms of uniaxial compressive strength. Through reading a large amount of literature, we found that the actual rockburst case of uniaxial tensile strength and uniaxial compressive strength is more documented, and most of the rock projects need to obtain these two mechanical properties, so the uniaxial tensile strength (σ_t) and uniaxial compressive strength (σ_c) are rockburst prediction evaluation indicators.

From an energy point of view, rockburst is the rapid release of energy gathered in high-energy reservoirs. Under the same stress conditions, the elastic energy index, the performance of rock aggregation, and the release of energy is positively correlated, so the rock elastic energy index (W_{et}) is selected as the rockburst prediction indicators. A number of rockburst cases have shown that the occurrence of rockbursts is closely related to the brittleness of the rock, and the brittleness coefficient of the rock is often used as a rockburst criterion. The stress coefficient is also commonly used as a rockburst criterion; therefore, the brittleness index (σ_{c}/σ_{t}) and the stress coefficient ($\sigma_{\theta}/\sigma_{c}$) are rockburst prediction evaluation indicators.

Comprehensive analysis of the above, according to the causes and characteristics of the occurrence of rockburst, six rockburst impact factors (σ_{θ}, σ_{t}, σ_{c}, σ_{c}/σ_{t}, $\sigma_{\theta}/\sigma_{c}$, W_{et}) were selected as the rockburst prediction indicators in this paper.

3.2. Sample Library of Rockburst Case Data

Rockburst is currently a common geological hazard in many underground rock projects at home and abroad, and many engineering rockburst cases have been well documented. In this paper, through literature research [35–38], based on the rockburst prediction evaluation indicators selected by the study, 75 groups of typical rockburst cases at home and abroad were selected, and some of the raw data are shown in Table 1, and the rockburst intensity level was divided into four levels, of which the actual distribution of rockburst levels is shown in Figure 3.

Table 1. Data of some domestic and international rockburst cases [35–38].

Serial Number	Primitive Rockburst Prediction Indicators						Comprehensive Rockburst Prediction Indicators			Actual Level
	σ_{θ}	σ_{c}	σ_{t}	$\sigma_{\theta}/\sigma_{c}$	σ_{c}/σ_{t}	W_{et}	CPI_1	CPI_2	CPI_3	
1	18.8	178	5.7	0.11	31.23	7.4	0.549	0.456	0.964	I
2	96.41	18.32	0.38	0.19	47.93	1.87	0.411	0.657	0.933	I
3	15.2	53.8	5.56	0.283	9.68	1.92	0.562	0.314	1.001	I
...	...	...	...	...	...	...	...	...	...	...
61	48	120	1.5	0.4	80	5.8	0.606	0.998	0.746	III
62	48.75	180	8.3	0.27	21.69	5	0.634	0.320	0.768	III
63	105	115	1.5	0.55	76.67	5.7	0.538	0.895	0.486	III
64	33.94	117.48	4.23	0.29	27.77	2.37	0.644	0.497	0.892	II
65	14.96	115	5	0.1	23	5.7	0.498	0.403	1.059	I
66	157.3	91.23	6.92	0.58	13.18	6.27	0.311	0.088	0.317	IV
67	91.43	157.63	11.96	0.58	13.18	6.27	0.559	0.108	0.397	IV
68	13.9	124	4.22	0.112	29.4	2.04	0.667	0.538	1.086	I
69	38.2	71.4	3.4	0.53	21	3.6	0.539	0.423	0.718	III
70	39.4	69.2	2.7	0.57	25.6	3.8	0.537	0.478	0.686	III
71	52	175	7	0.3	25	5.2	0.615	0.368	0.744	III
72	105	304.21	20.9	0.35	14.56	10.57	0.639	−0.094	0.331	IV
73	35.82	127.93	4.43	0.28	28.9	3.67	0.608	0.485	0.872	II
74	69.8	198	22.4	0.35	8.84	4.68	0.763	−0.062	0.570	II
75	55.4	176	7.3	0.31	24.11	9.3	0.452	0.290	0.683	III

Note: Groups 1 to 60 are the training sets; Groups 61 to 75 are the test sets.

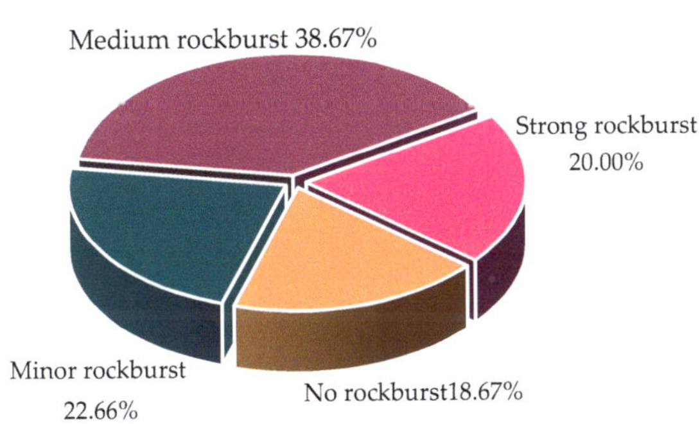

Figure 3. Rockburst actual intensity level distribution.

The number of rockburst case data collected in the least number of I samples, 14; the number of II samples is 17; the number of III samples is the most, 29; the number of IV samples is 15; the ratio of various types of samples is 1.4:1.7:2.9:1.5; there is a certain imbalance in the characteristics of various types of samples. However, the ratio of the maximum sample size to the minimum sample size is only slightly greater than 2. The imbalance problem of rockburst samples is small. Figure 4 shows the violin diagram of rockburst prediction evaluation indicators, whose horizontal coordinates indicate different rockburst levels, and vertical coordinates are rockburst prediction evaluation indicators. The violin chart is a combination of a box chart and a nuclear density chart, which gives a good indication of the shape of the distribution of the data. The white dot in the middle of the box line box indicates the median, the middle box line box indicates the interquartile range, the thin line extending from it represents the 95% confidence interval, and the outer shape is the nuclear density estimate.

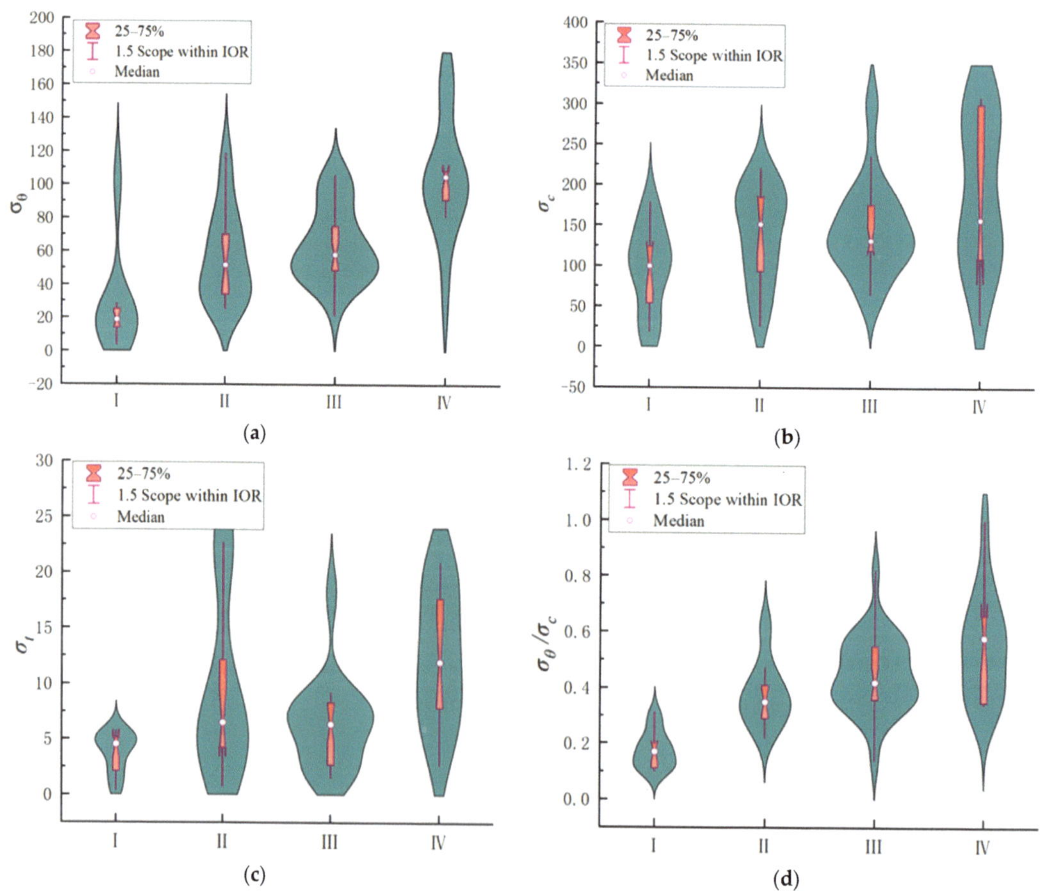

Figure 4. *Cont.*

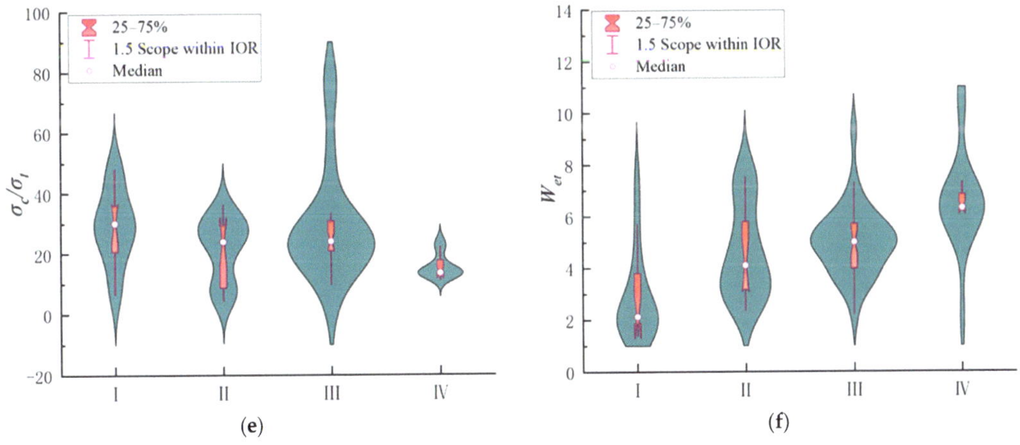

Figure 4. Violin diagram of rockburst prediction indicators: (**a**) σ_θ distribution; (**b**) σ_c distribution; (**c**) σ_t distribution; (**d**) σ_θ/σ_c distribution; (**e**) σ_c/σ_t distribution; (**f**) W_{et} distribution.

4. Implementation Process of FA-SSA-PNN Model

4.1. Model Construction Steps

The 75 groups of rockburst case data collected show that there is variability in the dimensionality, which in turn leads to a decrease in the accuracy of the rockburst prediction model. In order to eliminate the impact of the difference in the dimensionality between the indicators and improve the accuracy of the rockburst prediction model, it is necessary to reduce the original rockburst prediction data, the dimensionality of the resulting comprehensive rockburst prediction data into the rockburst prediction model, and the prediction results of the model for analysis and discussion.

In this paper, Matlab software to program the calculation of the neural network algorithm to establish the FA-SSA-PNN rockburst prediction model process is shown in Figure 5, and the main steps are as follows:

Step 1: Analysis of the impact of rockburst factors; the selection of rockburst prediction indicators.

Step 2: Collect rockburst case data according to the selected rockburst prediction indicators.

Step 3: Use factor analysis to reduce the dimensionality of the collected rockburst case data to obtain the comprehensive rockburst prediction index CPI_1, CPI_2, CPI_3.

Step 4: Partition the data set of the rockburst case data after dimensionality reduction processing; extract 80% of the overall rockburst prediction data samples as the training sets and 20% of the overall samples as the test sets.

Step 5: Imported the training samples into the SSA-PNN model and use the training for model training and updating parameters.

Step 6. After the training is completed, input the test samples to the model to test the network performance, get the rockburst intensity level prediction results, and calculate the accuracy of its rockburst intensity level prediction.

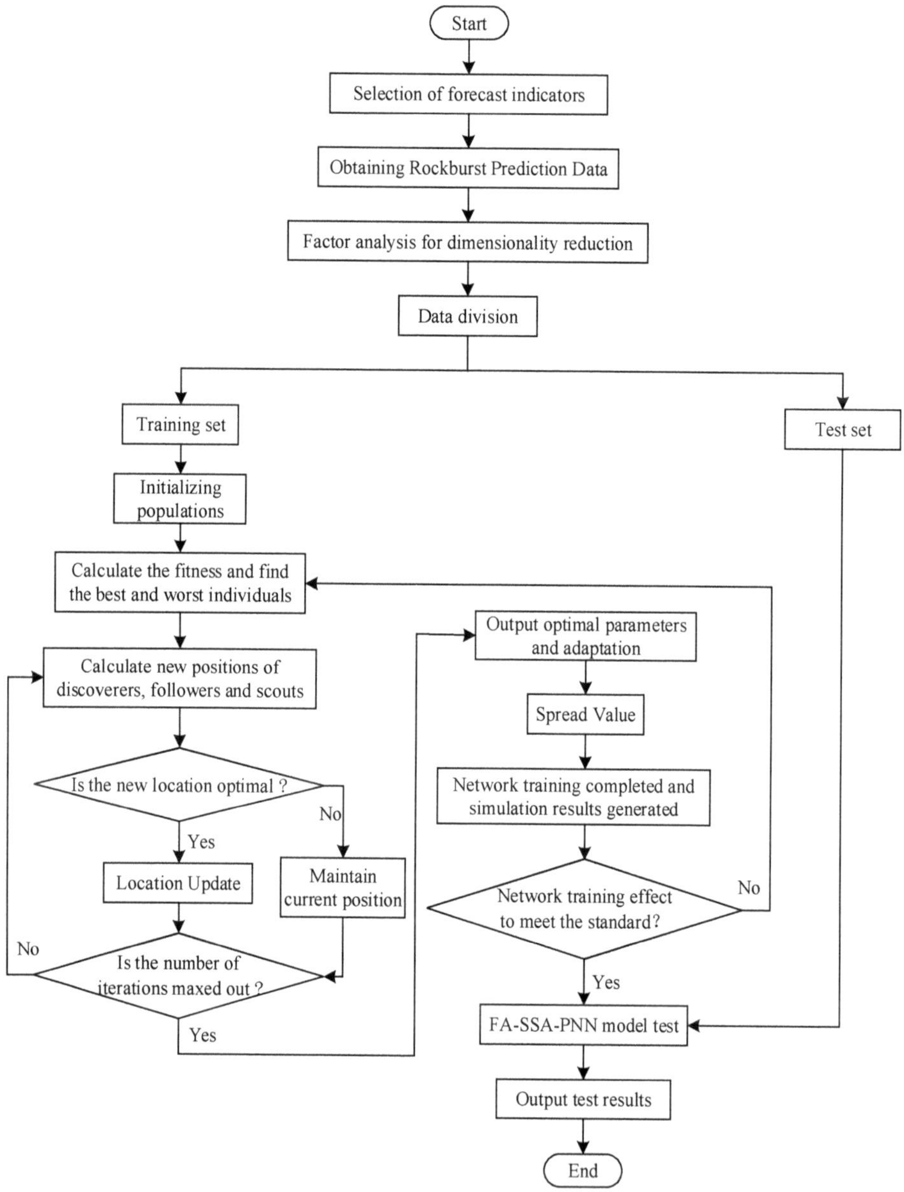

Figure 5. Flow chart of the rockburst prediction model of FA-SSA-PNN.

4.2. Test of Applicability of Factor Analysis

The rockburst cases at home and abroad were collected and organized, 75 groups of typical rockburst cases were selected as the sample data of the FA-SSA-PNN rockburst prediction model, the KMO test and Bartlett's spherical test were used to test the applicability of factor analysis on the sample data, and the test results and applicability test criteria are shown in Tables 2 and 3. It can be seen from Tables 2 and 3 that it is feasible to conduct factor analysis on the selected rockburst case data.

Table 2. Results of factor analysis method applicability test.

Kaiser-Meyer-Olkin test	KMO value	0.641
Bartlett spherical test	chi-squared test value Sig	187.075 0.000

Table 3. Applicability test criteria of factor analysis method.

Test Method	Range of Values	Factor Analysis Applicability
Kaiser-Meyer-Olkin test	>0.9	Perfect suitable
	0.8~0.9	Great suitable
	0.7~0.8	Relatively suitable
	0.6~0.7	Suitable
	0.5~0.6	Barely suitable
	<0.5	Not suitable
Bartlett spherical test	sig $\leq$ 0.01	Suitable

4.3. Data Processing

The absolute value of the correlation coefficient r reflects the degree of linear correlation between the two rockburst prediction evaluation indicators. When $|r| < 0.3$, it means that the correlation between the two rockburst prediction evaluation indicators is extremely weak and can be regarded as uncorrelated; when $0.3 < |r| < 0.5$, the two rockburst prediction evaluation indicators are low correlated; when $0.5 < |r| < 0.8$, the two rockburst prediction evaluation indicators are significantly correlated; when $0.8 < |r| < 1$, the two rockburst prediction evaluation indicators are extremely correlated. Correlation analysis of rockburst prediction evaluation indicators and the correlation coefficient between predictors is shown in Table 4. The absolute values of the correlation coefficients between σ_θ and σ_c, σ_θ and σ_t, σ_θ and σ_θ/σ_c, σ_c and σ_t, σ_c and W_{et}, and σ_t and σ_c/σ_t were all greater than 0.5, indicating that the rockburst prediction evaluation indicators were significantly correlated with each other and the sample data were suitable for factor analysis.

Table 4. Correlation coefficients among prediction evaluation indicators.

Indicators	σ_θ	σ_c	σ_t	σ_θ/σ_c	σ_c/σ_t	W_{et}
σ_θ	1.00	0.411	0.449	0.410	−0.114	0.541
σ_c	0.411	1.00	0.677	−0.089	−0.153	0.643
σ_t	0.449	0.677	1.00	0.142	−0.583	0.588
σ_θ/σ_c	0.410	−0.089	0.142	1.00	−0.220	0.276
σ_c/σ_t	−0.114	−0.153	−0.583	−0.220	1.00	−0.174
W_{et}	0.541	0.643	0.588	0.240	−0.174	1.00

Factor analysis was used to reduce the dimensionality of the standardized 75 sets of rockburst data, and Mardia gave the correspondence between the original number of variables and the number of principal factors after dimensionality reduction in Table 5. In this paper, 6 rockburst prediction evaluation indicators were selected as the original number of variables, so the number of principal factors after factor analysis was set to 3. Table 6 shows the total variance interpretation of the rockburst prediction evaluation indicators, and we can see that the eigen values of the first three factor variables are all greater than 1 and the cumulative contribution of the first three principal factors is 85.538% > 85%, indicating that the first three principal factors retain 85.538% of the information carried by the original variables, so the extraction of the first three principal factors as influencing factors is consistent with the previous setting.

Table 5. Relationship between the number of original variables and the number of main factors.

Number of original variables	5	7	8	9	11
Number of principal factors	2	3	4	5	6

Table 6. Total variance explained.

Principal Factor	Load Sum of Squares			Sum of Squared Rotating Loads		
	Eigen Value	Variance Contribution	Cumulative Variance Contribution	Eigen Value	Variance Contribution	Cumulative Variance Contribution
F_1	2.897	48.282%	48.282%	2.410	40.160%	40.160%
F_2	1.186	19.769%	68.051%	1.367	22.785%	62.945%
F_3	1.049	17.486%	85.538%	1.356	22.593%	85.538%

The changes in factor loadings before and after rotation are shown in Table 7. Combining the positive and negative correlations and the composite rate, it can be seen that the principal factor F_1 is significantly positively correlated with the rockburst prediction evaluation indicators σ_θ, σ_c, σ_t, σ_θ/σ_c, indicating that the principal factor F_1 concentrates on the maximum tangential stress, compressive strength, compressive strength, and the influence of the stress coefficient on the prediction results of rockburst. The principal factor F_2 is only positively correlated with the indicator σ_c/σ_t, indicating that the main factor F_2 combines the information of the indicators of the brittleness index, which can be referred to as the brittleness factor. The main factor F_3 is positively correlated with the indicator W_{et} only and can be referred to as the energy factor.

Table 7. Changes in factor loadings before and after rotation.

Indicators	Factor Loading before Rotation			Factor Loadings after Rotation		
	F_1	F_2	F_3	F_1	F_2	F_3
σ_θ	0.874	−0.150	−0.315	0.907	−0.130	−0.173
σ_c	0.823	−0.100	0.281	0833	−0.072	0.260
σ_t	0.769	−0.512	0.126	0.704	−0.622	0.570
σ_θ/σ_c	0.712	0.278	0.423	0.628	0.460	−0.606
σ_c/σ_t	0.344	0.875	0.123	−0.061	0.965	−0.118
W_{et}	0.489	−0.221	0.813	−0.029	−0.158	0.934

Table 8 shows the factor score coefficient matrix. The factor analysis reallocated the weights of the impact of rockburst prediction evaluation indicators on the principal factor and reduced the impact of poorly correlated rockburst prediction evaluation indicators on the principal factor, resulting in a functional expression between the principal factors Y_1, Y_2, Y_3 and the six rockburst prediction evaluation indicators, as follows (x_i^* is the standardized data value of x_i).

$$\left.\begin{aligned} Y_1 &= 0.243x_1^* + 0.444x_2^* + 0.212x_3^* - 0.166x_4^* + 0.185x_5^* + 0.363x_6^* \\ Y_2 &= 0.227x_1^* + 0.061x_2^* - 0.376x_3^* - 0.050x_4^* + 0.793x_5^* + 0.136x_6^* \\ Y_3 &= 0.409x_1^* - 0.266x_2^* - 0.103x_3^* - 0.736x_4^* + 0.006x_5^* + 0.095x_6^* \end{aligned}\right\}, \quad (19)$$

Table 8. Factor score coefficient matrix.

Indicators	Factor Score Coefficients		
	F_1	F_2	F_3
σ_θ	0.243	0.227	0.409
σ_c	0.444	0.061	−0.266
σ_t	0.212	−0.376	−0.103
σ_θ/σ_c	−0.166	−0.050	0.736
σ_c/σ_t	0.185	0.793	0.006
W_{et}	0.363	0.136	0.095

Standardized data are substituted into Equations (17)–(19) to obtain partial principal factor data (Table 1). The principal factor retains most of the information in the original data, so the three principal factors are comprehensive rockburst prediction evaluation indicators CPI_1, CPI_2, CPI_3.

4.4. Datasets Segmentation

The sample data of rockburst after factor analysis (see Table 1) were divided into datasets, and 20% of the 75 sets of rockburst case data were taken as the test set, while 80% of the remaining data were used as the training set of the neural network model. After the division, there were 60 sets of sample data in the training set, and the training set was used to train the neural network model and update the parameters. There were 15 sets of sample data in the test set, and the test set was used to evaluate the generalization ability of the model and test the real prediction accuracy of the model.

4.5. Model Parameter Setting and Implementation

The traditional PNN model uses the original rockburst prediction evaluation indicators (σ_θ, σ_t, σ_c, σ_c/σ_t, σ_θ/σ_c, W_{et}) as the input vectors of the model. The FA-SAA-PNN rockburst prediction model developed in this paper used factor analysis to preprocess the original rockburst prediction evaluation indicators, and the comprehensive rockburst prediction indicators CPI_1, CPI_2, CPI_3 obtained after factor analysis were used as the prediction input vectors of the model. The selection of the smoothing factor is the key to the performance of PNN networks, and when the value of the smoothing factor is too small, it tends to cause the network to be overfitted and in essence a nearest neighbor classifier; when the value of the smoothing factor is too large, the details cannot be fully distinguished so close to a linear classifier [39]. This paper makes use of the good global search ability of the SSA algorithm to optimize the smooth factor of PNN neural network. The algorithm has the advantages of being rapid and efficient when optimizing for a single objective, as well as good merit-seeking ability, which solves the problem of selecting the optimal smoothing factor and improves the accuracy of the prediction model.

At present, there is no uniform standard for rockburst intensity grading, and scholars have recognized the rockburst intensity level in four classes, respectively: no rockburst (I), minor rockburst (II), medium rockburst (III), and strong rockburst (IV). This paper uses the PNN network model output vector set to 1×4 line vector, the i class in the line vector of the i neuron output value of 1, and the rest of the neuron output value of 0, such as the output vector is (0, 0, 1, 0), which means that the prediction model predicts the sample data as a medium rockburst (III).

The main parameters of the FA-SSA-PNN model are shown in Table 9, and the rockburst prediction model is programmed and calculated in this paper using Matlab software version 2018b, and the code implementation is based on M language.

Table 9. Main parameters of FA-SSA-PNN model.

Serial Number	Parameters	Parameter Values
1	Number of neurons in the input layer	3
2	Number of neurons in the pattern layer	60
3	Number of neurons in summation layer	4
4	Number of neurons in the output layer	4
5	Mode layer activation function	Gauss function
6	Optimization parameters	Spread Value
7	Number of populations of SSA	100
8	Maximum number of iterations of SSA	20
9	Proportion of discoverers	70%
10	Scout's ratio	20%
11	Early warning values	0.6

5. Model Performance Evaluation and Comparison

To verify the merit of the FA-SSA-PNN rockburst prediction model, test samples were input into the FA-SSA-PNN model, FA-PNN model, PNN model, RF model, SVM model, and ANN model [39], and the prediction results of each model are shown in Figure 6. To comprehensively evaluate the classification performance of each model, F_1 value (the summed average of precision and recall), macro-average F_1 value (the arithmetic mean of F_1 for each category), and accuracy rate are used as the evaluation indicators of the models in this paper. F_1 value and macro-average F_1 value reflect the classification performance of the models for different rockburst intensity levels, and accuracy rate reflects the overall classification performance of the models.

The evaluation indicators for the six rockburst prediction models are shown in Table 10, and a comparison of the models shows that:

(1) The FA-SSA-PNN model does not improve the F_1 value of the primary rockburst compared to the PNN model; the F_1 value for Level 2 rockburst is increased by 50% (from 50% to 100%); the F_1 value for Level 3 rockburst is increased by 25.6% (from 66.7% to 92.3%); and the F_1 value for Level 4 rockburst is increased by 20% (from 80% to 100%).

(2) Compared with the original PNN model, the macro-average F_1 value reflecting the classification performance of the model for different rockburst intensity level increased by 18.9% (from 69.2% to 88.1%) after the introduction of FA dimensionality reduction, and the macro-average F_1 value improved but remained low, and then, after the optimization of the PNN neural network by the SSA algorithm, the macro-average F_1 value increased by another 5% (from 88.1% to 93.1%), and the macro-average F_1 values of the FA-SSA-PNN model were significantly higher than those of the other five rockburst prediction models.

(3) The accuracy of the FA-PNN model after the introduction of FA improved by 13.3% (from 66.7% to 80%) compared with the original PNN model, and then, after the optimization of the PNN neural network by the SSA algorithm, the accuracy of the model improved by another 13.3% (from 80% to 93.3%), and the prediction accuracy of the FA-SSA-PNNN model was significantly higher than that of the other models, verifying the advantages and disadvantages of the FA-SSA-PNN rockburst intensity level prediction model.

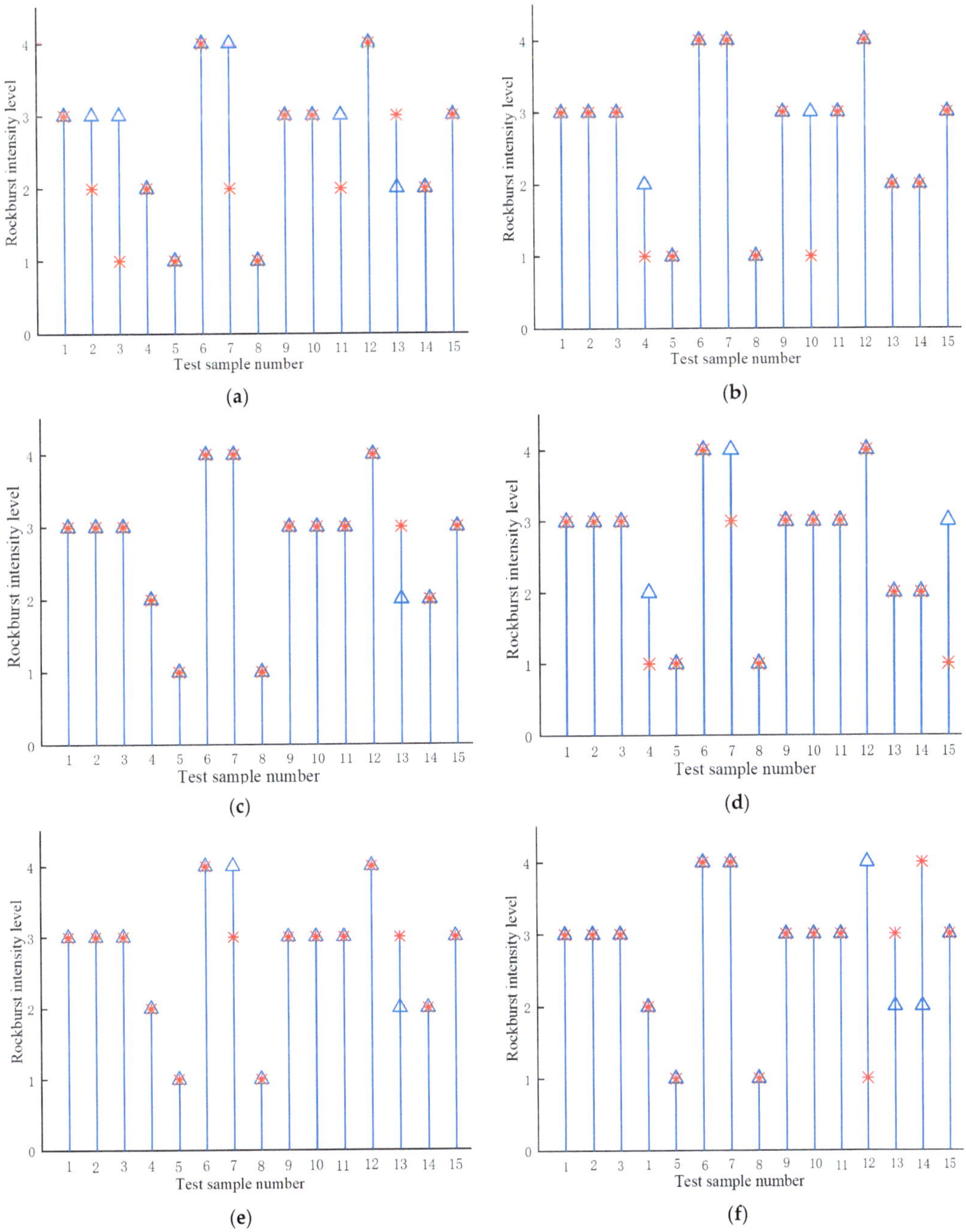

Figure 6. Rockburst prediction model test results: (a) PNN model; (b) FA-PNN model; (c) FA-SSA-PNN model; (d) ANN model; (e) SVM model; (f) RF model.

Table 10. Model evaluation indicators.

Evaluation Indicators	Intensity Level	PNN	FA-PNN	SSA-FA-PNN	ANN	SVM	RF
Accuracy rate	I	0.667	0.500	0.667	0.500	1.000	0.667
	II	0.400	1.000	1.000	1.000	1.000	1.000
	III	0.800	1.000	1.000	0.857	0.778	0.875
	IV	1.000	1.000	1.000	1.000	1.000	0.667
Recall Rate	I	1.000	1.000	1.000	1.000	1.000	1.000
	II	0.667	0.667	1.000	0.667	0.667	0.333
	III	0.571	0.857	0.857	0.857	1.000	1.000
	IV	0.667	1.000	1.00	0.800	0.667	0.667
F_1 value	I	0.800	0.667	0.80	0.667	1.000	0.800
	II	0.500	0.800	1.00	0.800	0.800	0.500
	III	0.667	0.923	0.923	0.857	0.875	0.933
	IV	0.80	1.000	1.00	0.667	0.800	0.667
Macro average F_1 value	-	0.692	0.881	0.931	0.781	0.86.9	0.725
Accuracy	-	0.667	0.800	0.933	0.800	0.800	0.867

6. Conclusions

As more and more underground rock projects move deeper at an unprecedented rate, the geological environment in which the rock masses are embedded is more complex, and the problem of rockburst hazards is becoming increasingly prominent. In this paper, based on 75 sets of typical rockburst case data collected, a rockburst intensity level prediction model based on FA-SSA-PNN is established, and F_1 value, macro-averaged F_1 value, and accuracy rate are introduced as the evaluation indexes of rockburst prediction model classification performance. This study proposes a new method for predicting the intensity level of rockbursts, which provides better guidance for the problem of predicting rockbursts in deep underground rock projects and can provide a reference for other geological hazard prediction problems similar to rockburst hazards, with the following main conclusions:

(1) The maximum tangentialstress of surrounding rock (σ_θ), uniaxial tensile strength (σ_t), uniaxial compressive strength (σ_c), brittleness index (σ_c/σ_t), stress coefficient (σ_θ/σ_c), and elastic energy index (W_{et}) of surrounding rock are selected to form a rockburst prediction index system. The characteristic information of the original rockburst prediction indexes was compressed and extracted by the factor analysis method, and three comprehensive rockburst prediction indexes, CPI_1,CPI_2, and CPI_3, were obtained. The introduction of factor analysis into the rockburst intensity level prediction eliminates the correlation between indicators and solves the problem of overlapping information of indicators, so that the comprehensive prediction index of rockburst after dimensionality reduction has a broader mathematical expression of Gaussian function in the PNN model.

(2) Fifteen sets of rockburst case data were sampled as test data, and the prediction results of the FA-PNN model were analyzed and compared with those of the original PNN model. It was found that the macro-average F_1 value and accuracy of the FA-PNN model were improved, with the macro-average F_1 value reaching 88.1% (from 69.6% to 88.1%) and the accuracy rate reaching 80% (from 66.7% to 80%).

(3) The SSA algorithm was used to select the smoothing factors in PNN to avoid the subjectivity and contingency of the existence of artificial preset smoothing factors. The comparison between the prediction results of FA-SSA-PNN rockburst prediction model and those of FA-PNN rockburst prediction model shows that, after the introduction of SSA algorithm, the accuracy of FA-SSA-PNN rockburst prediction model significantly improved, reaching 93.3% (increased from 80% to 93.3%), and the macro-average F_1 value is 93.1% (increased from 88.1% to 93.1%). Moreover, the SSA algorithm has good optimization ability and can complete the optimization of

smoothing factors in a few seconds. It greatly reduces the operation time of the model and improves the prediction efficiency of the model.

(4) The prediction results of the FA-SSA-PNN model were compared and analyzed with those of the FA-PNN model, PNN model, RF model, SVM model, and ANN model, and the results showed that the macro-averaged F_1 values and the prediction accuracy of the FA-SSA-PNN model were significantly higher than those of the other five models, which verified the feasibility and effectiveness of the FA-SSA-PNN rockburst prediction model.

The complexity of the rockburst mechanism and the many factors that induce rockburst, such as the traditional rockburst prediction methods, have not been able to make accurate and efficient predictions of the rockburst intensity level. Therefore, it will become more and more important to propose new methods for predicting rockburst intensity levels.

Author Contributions: Conceptualization, G.X. and K.L.; methodology, G.X.; software, G.X.; validation, Q.Q. and M.L.; formal analysis, G.X.; resources, R.Y.; data curation, G.X. and M.L.; writing—original draft preparation, G.X.; writing—review and editing, K.L. and Q.Q., supervision M.L. and R.Y.; visualization, Q.Q. and M.L.; supervision, R.Y.; project administration, Q.Q.; funding acquisition, K.L. All authors have read and agreed to the published version of the manuscript.

Funding: This research was supported by the National Natural Science Foundation of China (Grant No. 51934003) and the Yunnan major scientific and technological special project (Grant No. 202102AG050024).

Data Availability Statement: No applicable.

Conflicts of Interest: The authors declare no conflict of interest.

References

1. Feng, X.; Xiao, Y.; Feng, G.; Yao, Z.; Chen, B.; Yang, C.; Su, G. Study on the development process of rockbursts. *Chin. J. Rock Mech. Eng.* **2019**, *38*, 649–673. [CrossRef]
2. He, S.; Song, D.; Li, Z.; He, X.; Chen, J.; Li, D.; Tian, X. Precursor of Spatio-temporal Evolution Law of MS and AE Activities for Rock Burst Warning in Steeply Inclined and Extremely Thick Coal Seams Under Caving Mining Conditions. *Rock Mech. Rock Eng.* **2019**, *52*, 2415–2435. [CrossRef]
3. Xue, Y.; Li, Z.; Li, S.; Qiu, D.; Tao, Y.; Wang, L.; Yang, W.; Zhang, K. Prediction of rock burst in underground caverns based on rough set and extensible comprehensive evaluation. *Bull. Eng. Geol. Environ.* **2019**, *78*, 417–429. [CrossRef]
4. Li, X.; Gong, F.; Wang, S.; Li, D.; Tao, M.; Zhou, J.; Huang, L.; Ma, C.; Du, K.; Feng, F. Coupled static-dynamic loading mechanical mechanism and dynamic criterion of rockburst in deep hard rock mines. *Chin. J. Rock Mech. Eng.* **2019**, *38*, 708–723. [CrossRef]
5. Lin, M.; Gao, C.; Xia, Y.; Zhang, D.; Liu, X.; Liang, X. Rock burst initiation and precursors in a model specimen based on acoustic emission and infrared monitoring. *Arab. J. Geosci.* **2022**, *15*, 333. [CrossRef]
6. Ren, J.; Zhang, W.; Wu, Z.; Li, J.; Shen, Y. Microseismic Signals in Heading Face of Tengdong Coal Mine and Their Application for Rock Burst Monitoring. *Shock. Vib.* **2021**, *2021*, 6650446. [CrossRef]
7. Shirani Faradonbeh, R.; Shaffiee Haghshenas, S.; Taheri, A.; Mikaei, R. Application of self-organizing map and fuzzy c-mean techniques for rockburst clustering in deep underground projects. *Neural Comput. Appl.* **2019**, *32*, 8545–8559. [CrossRef]
8. Ma, C.; Chen, W.; Tan, X.; Tian, H.; Yang, J.; Yu, J. Novel rockburst criterion based on the TBM tunnel construction of the Neelum–Jhelum (NJ) hydroelectric project in Pakistan. *Tunn. Undergr. Space Technol.* **2018**, *81*, 391–402. [CrossRef]
9. Xu, L.; Wang, L. Study on the laws of rockburst and its forecasting in the tunnel of Erlang Mountain road. *Chin. J. Geotech. Eng.* **1999**, *21*, 569–572.
10. Lu, J. Study on mechanism of rockburst in a headrace tunnel. In *1st National Congress of Chinese Society for Numerical Calculation and Model Experiment of Rock Mechanics*; Southwest Jiaotong University Press: Chengdu, China, 1986; pp. 210–214.
11. Jiang, L.; Kong, P.; Zhang, P.; Shu, J.; Wang, Q.; Chen, L.; Wu, Q. Dynamic Analysis of the Rock Burst Potential of a Longwall Panel Intersecting with a Fault. *Rock Mech. Rock Eng.* **2019**, *53*, 1737–1754. [CrossRef]
12. Yang, Z.; Liu, C.; Zhu, H.; Xie, F.; Dou, L.; Chen, J. Mechanism of rock burst caused by fracture of key strata during irregular working face mining and its prevention methods. *Int. J. Min. Sci. Technol.* **2019**, *29*, 889–897. [CrossRef]
13. Ma, T.; Tang, C.; Tang, S.; Kuang, L.; Yu, Q.; Kong, D.; Zhu, X. Rockburst mechanism and prediction based on microseismic monitoring. *Int. J. Rock Mech. Min. Sci.* **2018**, *110*, 177–188. [CrossRef]
14. Xu, C.; Liu, X.; Wang, E.; Zhang, Y.; Wang, S. Rockburst prediction and classification based on the idealpoint method of information theory. *Tunn. Undergr. Space Technol.* **2018**, *81*, 382–390. [CrossRef]
15. Guo, J.; Zhang, W.; Zhao, Y. A multidimensional cloud model for rockburst prediction. *Chin. J. Rock Mech. Eng.* **2018**, *37*, 1199–1206. [CrossRef]

16. Lin, Y.; Zhou, K.; Li, J. Application of cloud model in rock burst prediction and performance comparison with three machine learnings algorithms. *IEEE Access* **2018**, *30*, 958–968. [CrossRef]
17. Wang, X.; Li, S.; Xu, Z.; Xue, Y.; Hu, J.; Li, Z.; Zhang, B. An interval fuzzy comprehensive assessment method for rock burst in underground caverns and its engineering application. *Bull. Eng. Geol. Environ.* **2019**, *78*, 5161–5176. [CrossRef]
18. Adoko, A.; Gokceoglu, C.; Wu, L.; Zuo, Q. Knowledge-based and data-driven fuzzy modeling for rockburst prediction. *Int. J. Rock Mech. Min. Sci.* **2013**, *61*, 86–95. [CrossRef]
19. Wang, C.; Wu, A.; Lu, H.; Bao, T.; Liu, X. Predicting rockburst tendency based on fuzzy matter–element model. *Int. J. Rock Mech. Min. Sci.* **2015**, *75*, 224–232. [CrossRef]
20. Shi, X.; Zhou, J.; Dong, L.; Hu, H.; Wang, H.; Chen, S. Application of unascertained measurement model to prediction of classification of rock burst Intensity. *Chin. J. Rock Mech. Eng.* **2010**, *29*, 2720–2726.
21. Li, Z.; Xue, Y.; Li, S.; Qiu, D.; Zhang, L.; Zhou, Y.; Zhou, B. Rock burst risk assessment in deep-buried underground caverns: A novel analysis method. *Arab. J. Geosci.* **2020**, *13*, 388. [CrossRef]
22. Peng, T.; Deng, H. Comprehensive evaluation on water resource carrying capacity in karst areas using cloud model with combination weighting method: A case study of Guiyang, southwest China. *Environ. Sci. Pollut. Res.* **2020**, *27*, 37057–37073. [CrossRef] [PubMed]
23. Gong, F.; Li, X. Distance discrimination method for rockburst occurrence and intensity classification prediction and its application. *Chin. J. Rock Mech. Eng.* **2007**, *26*, 1012–1018.
24. Yang, X.; Pei, Y.; Cheng, H.; Hou, X.; Lu, J. Prediction method of rockburst intensity grade based on SOFM neural network model. *Chin. J. Rock Mech. Eng.* **2021**, *40*, 2708–2715. [CrossRef]
25. Li, M.; Li, K.; Qin, Q.; Wu, S.; Liu, Y.; Liu, B. Discussion and selection of machine learning algorithm model for rockburst intensity grade prediction. *Chin. J. Rock Mech. Eng.* **2021**, *40*, 2806–2816. [CrossRef]
26. Tang, Z.; Xu, Q. Rockburst prediction based on nine machine learning algorithms. *Chin. J. Rock Mech. Eng.* **2020**, *39*, 773–781. [CrossRef]
27. Tian, R.; Meng, H.; Chen, S.; Wang, C.; Zhang, F. Prediction of intensity classification of rockburst based on deep neural network. *J. Chin. Coal. Soc.* **2020**, *45*, 191–201. [CrossRef]
28. Jia, Y.; Lu, Q.; Shang, Y. Rockburst prediction using particle swarm optimization and generalized regression neural network. *Chin. J. Rock Mech. Eng.* **2013**, *32*, 343–348.
29. Xue, Y.; Bai, C.; Qiu, D.; Kong, F.; Li, Z. Predicting rockburst with database using particle swarm optimization and extreme learning machine. *Tunn. Undergr. Space Technol.* **2020**, *98*, 103287. [CrossRef]
30. Zhou, J.; He, Y.; Huang, R.; Ju, N. Weights of slope stability evaluation indexes based on factor analysis method. *J. Southwest Jiaotong Univ.* **2015**, *50*, 325–330.
31. Wu, S.; Zhang, C.; Cheng, Z. Prediction of intensity classification of rockburst based on PCA-PNN principle. *J. Chin. Coal Soc.* **2019**, *44*, 2767–2776. [CrossRef]
32. Chen, Y.; Liu, Z.; Xiao, C.; Zhao, X.; Li, K.; Pang, L.; Shi, Y.; Li, S. Overlapping Peak Analysis of Soil Heavy Mettal X-Ray Fluorescence Spectra Based on Sparrow Search algorithm. *Spectrosc. Spectr. Anal.* **2021**, *41*, 2175–2180.
33. Cheng, B. Research on Transformer Fault Diagnosis Based on the Combination of Rough Set-Probabilistic Neural Network. Master's Thesis, Guangxi University, Nanning, China, 2008.
34. Du, Z.; Zhao, J.; Li, H.; Zhang, X. Afauit diagnosis method of a plunger pump based on SA-EMD-PNN. *J. Shock Vib.* **2019**, *38*, 145–152. [CrossRef]
35. Wang, Y.; Li, W.; Li, Q.; Xu, Y.; Tan, G. Method of fuzzy comprehensive evaluations for rockburst prediction. *Chin. J. Rock Mech. Eng.* **1998**, *17*, 493–501.
36. Zhou, J.; Li, X.; Shi, X. Long-term prediction model of rockburst in underground openings using heuristic algorithms and support vector machines. *Saf. Sci.* **2011**, *50*, 629–644. [CrossRef]
37. Dong, L.; Li, X.; Peng, K. Prediction of rockburst classifi-cation using Random Forest. *Trans. Nonferrous Met. Soc. China* **2013**, *23*, 472–477. [CrossRef]
38. Afraei, S.; Shahriar, K.; Madani, S. Developing intelligent classification models for rock burst prediction after recognizing significant predictor variables, Section 1: Literature review and data preprocessing procedure. *Tunn. Undergr. Space Technol.* **2019**, *83*, 324–353. [CrossRef]
39. Wang, J.; Zhou, Z.; Li, K.; Wang, H.; Fu, Z.; Li, X. Evaluation model for the risk grade of rock burst based on the R-type factor analysis and a probabilistic neural network. *J. Shock Vib.* **2019**, *38*, 192–203. [CrossRef]

entropy

MDPI

Article

A New Perspective on Cooking Stove Loss Coefficient Assessment by Means of the Second Law Analysis

Lomena Mulenda Augustin [1,2], Sumuna Temo Vertomene [1], Ndaye Nkanka Bernard [2], Amsini Sadiki [3,4,5,*] and Mbuyi Katshiatshia Haddy [1]

1 Centre de Recherche en Energies Renouvelables, Faculté Polytechnique, Université de Kinshasa, Avenue de l'Université N° 01, Commune de Lemba, BP 127 Kinshasa, Democratic Republic of the Congo; lomena.mulenda@ista.ac.cd or augustinlomena@gmail.com (L.M.A.); temo.sumuna@unikin.ac.cd (S.T.V.); haddy.mbuyi@unikin.ac.cd (M.K.H.)

2 Centre d'Etudes et de Recherches sur les Energies Renouvelables Kitsisa-Khonde (CERERK), ISTA-Kinshasa, Avenue Aérodrome N° 3930, Commune de Barumbu, BP 6593 Kinshasa, Democratic Republic of the Congo; ndaye.nkanka@ista.ac.cd

3 Institute for Energy and Power Plant Technology, Technische Universität Darmstadt, 64287 Darmstadt, Germany

4 Laboratoire de Modélisation Mécanique, Energétique et Matériaux, ISTA-Kinshasa, Avenue Aérodrome N° 3930, Commune de Barumbu, BP 6593 Kinshasa, Democratic Republic of the Congo

5 Institute for Reactive Flows and Diagnostics (RSM), Technische Universität Darmstadt, 64287 Darmstadt, Germany

* Correspondence: sadiki@ekt.tu-darmstadt.de; Tel.: +49-6151-16-28914/+49-6071-39-2212

Abstract: The chimney effect taking place in biomass cooking stoves results from a conversion process between thermal and mechanical energy. The efficiency of this conversion is assessed with the stove loss coefficient. The derivation of this quantity in cooking stove modelling is still uncertain. Following fluid mechanics, this loss coefficient refers to an overall pressure drop through stove geometry by performing an energy balance according to the first law of thermodynamics. From this approach, heat-transfer processes are quite ignored yet they are important sources of irreversibilities. The present work takes a fresh look at stove loss coefficient assessment relying on the second law of thermodynamics. The purpose in this paper is to identify the influence of operating firepower level on flow dynamics in biomass natural convection-driven cooking stoves. To achieve that, a simplified analytical model of the entropy-generation rate in the flow field is developed. To validate the model, experiments are conducted first on a woodburning stove without cooking pot to better isolate physical processes governing the intrinsic behaviour of the stove. Then, for the practical case of a stove operating with a cooking pot in place, data from published literature have served for validation. In particular, mass-flow rate and flue gas temperature at different firepower levels have been monitored. It turns out that losses due to viscous dissipations are negligible compared to the global process dissipation. Exergy analysis reveals that the loss coefficient should rather be regarded from now as the availability to generate flow work primarily associated with the heat-transfer Carnot factor. In addition, the energy flux applied as flow work has to be considered as pure exergy that is lost through consecutive energy-transfer components comprising the convective heat transfer to the cooking pot. Finally, this paper reports a satisfactory agreement that emerged between the exergy Carnot factor and the experimental loss coefficient at different fuel-burning rates.

Keywords: woodburning cooking stove; fuel-burning rate; buoyancy; loss coefficient; entropy-generation rate; Carnot factor; exergy

1. Introduction

The combustion-induced driven flow is the phenomenon that occurs in traditional cooking stoves widely used in the developing world, and mainly in rural areas, where the major portion of the population uses biomass fuel as the primary source of energy [1–3].

Inefficient stoves are important sources of emissions of pollutants hazardous to human health, such as carbon monoxide (CO), particulate matters (PM) and polycyclic aromatic hydrocarbons (PAHs) [4–7]. The Clean Cooking Alliance reports that every year four million people die from illnesses associated with smoke from cooking activities and, at the same time, burning woodfuels contributes to about 2% of global CO_2 emissions [8]. Thus, the challenge for clean cooking designers is to create user-friendly appliances that can maintain high overall efficiency and reduce harmful emissions to levels low enough to ensure health, environment and climate co-benefits [7].

Given its importance, the conception of clean cooking stoves is now increasingly deserving of the attention of researchers [9–11]. Starting in the 1980s, early modelling efforts have been initiated to design more efficient cooking stoves. Since then, two types of models have emerged from researchers. The primary type is a zonal model in which conservation of mass, momentum and energy are applied to different zones within the stove [9]. In the second type of model, Computational Fluid Dynamics (CFD) is used to represent detailed informations inside the computational domain, such as regions of high soot and CO concentrations [12–14]. While remarkable progress has been made in the modelling of cooking stoves, many questions remain unanswered.

Figure 1 presents significant features of a natural draft stove burning wood fuel including a horizontal combustion chamber, and under the cooking pot there is an insulated short chimney, inside which takes place a buoyant flow of hot gases [11]. Buoyancy occurring in the stove results from a conversion process between two forms of energy (internal and mechanical) conserving the overall energy according to the first law of thermodynamics. The efficiency of this conversion is globally assessed by the stove loss coefficient. The derivation of this quantity in cooking stove modelling is still uncertain. How this stove flow loss coefficient varies when fire takes place under different woodburning rates remains a challenging concern.

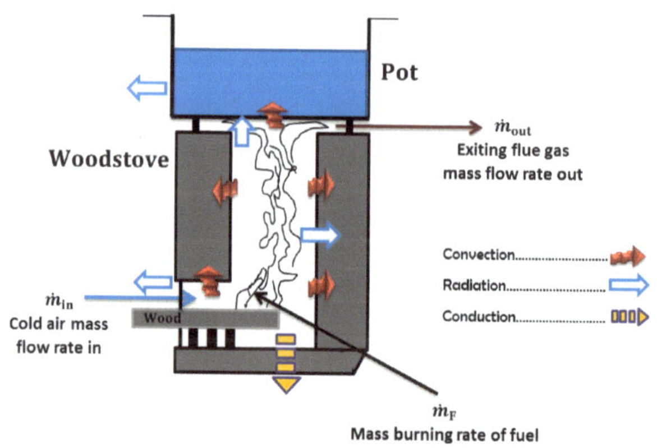

Figure 1. Schematic of a chimney woodstove cross-section with different heat-transfer modes [15].

To better understand the importance of the issue, it is worth mentioning in this study that in woodstove modelling, great attention has always been focused on the influence of design parameters such as geometry or insulation materials [16–19], whereas little prior consideration has been devoted to the operating firepower level impacting aerodynamics and chemistry in the stove. The firepower–performance dependency becomes a key issue whose interest has been growing only in the last 10 years [20]. The works of Agenbroad et al. [21–23] represent an important benchmark in identifying the influence of operating firepower on woodburning stove behaviour. Moving beyond empirical/observational approaches, these authors developed and validated experimentally on steady state assumption an analytical and simplified stove flow theory that predicts mass-flow rate and exhaust-gas temperature from stove design and operating firepower.

Despite the Agenbroad simplified stove flow model, and other previous investigations, there is still a limited understanding of variations in the stove flow loss coefficient. Published contributions explicitly addressing access to the stove loss coefficient (also named the discharge coefficient) are very scarce. Table 1 summarizes some of the few papers dealing with the loss coefficient issue in stove modelling.

Table 1. Loss coefficient C in small-scale biomass cooking stove modelling.

Reference Authors	System Configuration	Highlights of the Study
MacCarthy [9]	Open cooking fire, Shielded cooking fire.	The study referred to various correlations in literature. Fluid flow constants and equations have been collected, deduced from a general balance of forces. However any specific value of C has been reported.
Agenbroad [21–23]	Stove without pot and Stove with pot.	Analytical stove flow modelling considered by default C remaining constant for stove operations: (a) C= 0.5 for stove without pot. (b) C = 0.35 for stove with pot. However, variable C depending on operating firepower level was experimentally observed. In theory, model accounted contributions for both losses due to viscous effects and losses due to distributed heat addition. In the assumption of a more realistic linear density profile, model suggested to replace C by the product $C = C_{viscous} \cdot C_{heat}$. Reduction of available chimney effect results in $C_{heat} = \frac{\sqrt{2}}{2} \approx 0.707$. CFD-based loss coefficient predicted stove behaviour using pressure drop with comparison to validation results [21]. Effects for reacting flow are unknown.
Kshirsagar [24]	Stove with pot.	Model treated C as a variable which in itself depends upon other variables, i.e., inlet area and geometrical variation. Model predicted C in the range of 0.195–0.38. Effects for reacting flow are unknown.
Zube [15]	Stove with pot.	Model adapted for calculation experimental C values formerly determined in [21,23]. Model discussed heat-transfer efficiency of the three different HT modes. Theoretical calculations in MathCAD/Excel established some correlations between C, firepower, convection and combustion efficiency, pot gap adjustment, pot skirt adjustment, skirt height, etc.
Parajuli [25]	Two-pot enclosed mud cookstoves.	Mass-flow rate calculated on the pressure difference incorporating geometric loss coefficients to determine C. Thermal effects are not taken into account.

Following a common fluid mechanics approach, this loss coefficient is associated with an overall pressure drop through stove geometry. Thus, as a conclusion of the literature review in Table 1, losses in the flow field are supposed to be systematically characterized by empirical friction factors and single-valued head loss coefficients of different conduit components such as sudden contraction at inlet, friction loss in elbow length, loss due to friction in the pot gap zone, etc. This way of proceeding continues to appear in the eyes of many as the single rule to predict stove flow behaviour [11,26].

However, from a thermodynamics point of view, in addition to fluid friction, a real process can present other kinds of irreversibilities associated with heat-transfer mechanisms, mixing, chemical reactions, etc., and all resulting in the loss of process efficiency [27–29] related to the entropy-generation rate. Therefore, an energy devaluation (energy quality loss) manifests in a destruction of available work commonly known as exergy. Losses in the internal flow field of a technical application like energy conversion in a woodburning stove can from now be assumed to be losses of flow exergy. The notion on the quality of energy and its change during thermodynamic processes is today well addressed in many contributions, to name just a few, [30–36].

In recent years, the second law of thermodynamics in analyzing energy conversion in power-generating units permits a fresh look to evaluate some key features of the flow dynamics, heat transfer and chemical reactions through various systems. Many works dealing with entropy-generation analysis in flows involving heat transfer in natural convection processes can be found in the literature, e.g., [27,37]. These works concluded that the second law analysis plays a vital role in determining the frictional and heat-transfer losses [38]. Recently, [39–42] and Schmandt [43,44] also analyzed the basic principles of entropy and its role in the momentum and heat transfer. However, these authors made an attempt to understand the physics beyond convective heat-transfer processes in an original way by introducing some alternative non-dimensional parameters that allow to also assess qualitative aspects during the energy-transfer processes.

To the best of our knowledge, no scientific paper has addressed woodburning stove engineering from the angle of the second law analysis. In researching this, not a single paper contains reference to words like entropy or exergy. Even when Agenbroad et al. [21,22] mentioned reduction of chimney effect due to a non-ideal heat addition profile assumption, the entropy or exergy concept was not dealt with in their model (see Table 1). Thus, it is not surprising to see that researchers in the cooking stove community focused exclusively on viscous and frictional losses when addressing the issue of stove flow resistance. The evaluation is being performed as if losses were occurring in an "isothermal cold flow field", whereas heat-transfer processes are identified to be in turn important sources of irreversibilities [45].

The present contribution presents a new outlook on the stove flow loss coefficient assessment based on an exergetic analysis of the flowing fluid and making particular use of qualitative assessment numbers in energy-transfer processes. The study derives a simplified analytical model that permits to evaluate the entropy-generation rate due to steady-flow combustion and viscous dissipation in a natural draft shielded fire stove burning wood fuel. The effect of varying woodburning rate (or firepower) on the entropy-generation rate is assessed in a steady-state assumption. To validate this model, experiments have been conducted first using a G3300 cooking stove model without a cooking pot in place to better isolate the physical processes governing the intrinsic behaviour of the stove. Then, we referred to published literature [15,23,46] to confirm the practical case of a stove operating with a cooking piece in place.

The idea in carrying out this study is motivated by the observation, in the literature, of often erroneous results or simply the lack of information concerning the loss coefficient of cooking stoves. The loss coefficient is likely a determinant parameter in stove engineering. Without a comprehensive assessment method of this parameter, it will remain challenging to evaluate proper stove-operating behaviour. This work attempts to reconsider flow and heat transfer issues through a holistic approach.

This paper is organised as follows. The theoretical background along with the derivation of the loss coefficient is provided in Section 2. Since validation data shall be generated, the experimental setup together with the materials and methods utilized is introduced in Section 3. The obtained results and related discussions are presented in Section 4. The last Section 5 is devoted to conclusions.

2. Theoretical Formulation

This section provides the theoretical background in terms of thermodynamics and entropy-generation analysis. Thereby, the efficiency of the conversion process between thermal and mechanical energy is derived, and the loss coefficient is consistently provided.

2.1. Thermodynamics of Steady-Flow Combustion

2.1.1. Conservation of Mass and Energy

Air flow along the chimney is due to buoyancy forces that drive hot gases upward. The mass balance for the combustion chamber in Figure 1 yields:

$$\dot{m}_{\text{in}} + \dot{m}_{\text{F}} = \dot{m}_{\text{out}} \tag{1}$$

where $\dot{m}_{in}$ is the cold air mass-flow rate entering the combustion chamber, $\dot{m}_F$ is the mass-burning rate of the fuel and $\dot{m}_{out}$ is the exiting flue gas mass-flow rate.

Complete combustion of wood on a per-mole-of-fuel basis can be described by a generalized one-step overall reaction:

$$C_xH_yO_z + \left(x + \frac{y}{4} - \frac{z}{2}\right)(O_2 + 3.76\ N_2) \rightarrow x\ CO_2 + \frac{y}{2}\ H_2O + 3.76\left(x + \frac{y}{4} - \frac{z}{2}\right)N_2 \tag{2}$$

The first law of thermodynamics for a steady-flow combustion requires that the rate of heat transfer per mole of wood burnt balances the difference between the enthalpies of the reactant and product streams.

The expression for the total molar enthalpy is given as:

$$Q_{\text{m}} = x\ \bar{h}^{\circ}_{\text{f},CO_2} + \frac{y}{2}\ \bar{h}^{\circ}_{\text{f},H_2O} - \bar{h}^{\circ}_{\text{f},C_xH_yO_z} \tag{3}$$

The quantities $\bar{h}^{\circ}_{\text{f},i}$ in Equation (3) individually represent the enthalpy of formation of the compound i. Note that the $\bar{h}^{\circ}_{\text{f},i}$ for O_2 and N_2 are zero as they are all elementary substances.

The rate of total heat of combustion is linked to the low heating value (LHV) of the wood species as:

$$\dot{Q} = a \cdot Q_{\text{m}} = a \cdot M \cdot LHV \tag{4}$$

where a represents the molar rate of wood consumption and M its molecular weight.

The total enthalpy in a compact form reduces to:

$$\dot{Q} = \dot{m}_{\text{F}} \cdot LHV \tag{5}$$

Part of this thermal energy takes on the form of flow energy in the flue gas ($\dot{Q}_{\text{flue}}$), which is responsible for the buoyant flow through the stove and past the surface of the pot (Figure 1). An in-depth analysis in references such as [11,15] shows that the rest of the energy identifies principally components of radiative heat transfer from char bed to the pot bottom and/or to the surrounding environment ($\dot{Q}_{\text{char radiation}}$), heat loss from the stove through insulation ($\dot{Q}_{\text{heat loss}}$), radiative heat loss through the feed opening ($\dot{Q}_{\text{door loss}}$) and the heat loss due to evaporation and sensible heat of fuel moisture ($\dot{Q}_{\text{fuel moisture}}$).

Applying the energy balance to the entire cooking stove:

$$\dot{Q} = \dot{Q}_{\text{flue}} + \dot{Q}_{\text{char radiation}} + \dot{Q}_{\text{heat loss}} + \dot{Q}_{\text{door loss}} + \dot{Q}_{\text{fuel moisture}} \tag{6}$$

For simplicity, in the flame zone, air, wood volatiles and combustion gases can all be modeled as a single ideal gas. So, the enthalpy of reaction distributed over the air mass-flow rate $\dot{m}_{\text{A}}$ crossing the stove can be written:

$$\dot{Q}_{\text{flue}} = \dot{m}_{\text{A}} \cdot \bar{c}_{\text{p}} \cdot (T_{\text{H}} - T_0) \tag{7}$$

where $\bar{c}_{\text{p}}$ is a mean value of specific heat capacity at constant pressure, T_0 is the temperature of the ambient and T_{H} is the exit flue gas temperature. This means that $\bar{c}_{\text{p}}$ is determined:

$$\bar{c}_{\text{p}} = \sum_{k=1}^{N} c_{\text{p}k} Y_k \tag{8}$$

where c_{pk} are the specific heat capacities at constant pressure of species k and Y_k is their respective mass fractions, for $k = 1$ to N, and N is the number of species in the reacting mixture. As combustion of wood proceeds in excess air conditions, the thermophysical properties of air dominate and the mass heat capacity of the mixture is very close to that of air. The value of $\bar{c}_p$ for air changes only from 1000 to 1200 $\text{J}\cdot\text{kg}^{-1}\cdot\text{K}^{-1}$ when temperature varies from 300 to 1500 K, so that $\bar{c}_p$ can be considered with a good approximation to be constant [47]. As discussed in [21], comparing model-predicted mass-flow rate dependence with firepower using a polynomial in place of a constant specific heat capacity, the resulting error is relatively small and justifiable to neglect its variation. The next section derives the temperature equation in the combustion chamber.

2.1.2. Temperature Equation

One needs a consistent temperature profile within the stove prior to addressing the entropy-generation rate equation in Section 2.2. To achieve this, some additional simplifying assumptions are needed:

- Geometry of the domain is assimilated to a vertical cylindrical chimney.
- Flow is considered to be laminar, uniform (steady) and one dimensional axisymmetric.
- Heat addition proceeds gradually along the height of the chimney, see [21,22].
- Thermodynamic properties of the flue gases are the same as those of air.
- Radiative heat transfer of the flue gases is negligible, less than 1% of the flame radiation heat balance on the energy balance of the entire stove, as reported by [11,15].

The temperature equation for the incompressible 1-D reacting flow in steady state can be found in e.g., [47] as:

$$\rho V \frac{dT}{dx} = -\frac{k}{\bar{c}_p}\frac{d^2T}{dx^2} + \frac{\dot{Q}_{\text{flue}}}{\bar{c}_p h_c} \quad (9)$$

In this transport equation, ρ is the air density, V the velocity, k the air thermal conductivity, h_c the height of the combustion chamber and $\frac{\dot{Q}_{\text{flue}}}{h_c}$ the volume unit enthalpy of reaction added to the flue gas.

The general form of the solution to the differential equation is given as:

$$T(x) = C_1 e^{-\frac{\rho \bar{c}_p V}{k}x} + \frac{\dot{Q}_{\text{flue}}}{\rho \bar{c}_p V h_c}x + C_2 \quad (10)$$

The quantity $\frac{\rho \bar{c}_p V x}{k}$ (or $\frac{Vx}{\alpha}$ with α the thermal diffusivity) in Equation (9) can be associated with the Peclet number Pe along the flow, which is defined as the ratio of the bulk flow heat transfer by convection to the heat transfer by conduction. Scaling analysis can show that for all the parameters under study $Pe \geqslant 400$. Therefore, heat conduction as well as the first term in Equation (9) can be safely neglected.

Inserting the boundary condition $T(0) = T_0$ at the entrance, the temperature profile results in the same form as in [21] when assuming a uniform heat addition with rising flame:

$$T(x) = T_0 + \frac{\dot{Q}_{\text{flue}}}{\dot{m}_A \bar{c}_p h_c}x \quad (11)$$

and its gradient:

$$\frac{dT(x)}{dx} = \frac{\dot{Q}_{\text{flue}}}{\dot{m}_A \bar{c}_p h_c} \quad (12)$$

The heat released from combustion entrains air circulation due to the density difference between cold air and hot flue gases. The basics of fluid dynamics applied to a cooking stove is introduced next.

2.1.3. Energy of The Flowing Fluid

The rate of the total energy θ contained in the flowing fluid takes the form:

$$\theta = \dot{m}_{\rm A}\left[Pv + \left(u + e_{\rm k} + e_{\rm p}\right)\right] \quad (13)$$

where P is the fluid pressure, v the specific volume and u the specific internal energy, while $e_{\rm k}$ and $e_{\rm p}$ express the mass kinetic and the mass potential energy of the flow, respectively. The additional form of energy, the flow energy Pv, represents the energy needed to admit and evacuate the flow in the control volume [30].

Equation (13) can be rewritten taking into account the energy of the flow via the specific enthalpy, $h = u + Pv$, as:

$$\theta = \dot{m}_{\rm A}\left(h + \frac{V^2}{2} + gx\right) \quad (14)$$

From the integral form of Bernoulli's equation, the fluid flow through the combustion chamber is determined by a momentum balance of airflow due to buoyancy and pressure losses through friction, bends, expansions and contractions in the flow path [9] as:

$$\frac{\rho_{\rm H} V_{\rm H}^2}{2} = gh_{\rm c}(\rho_0 - \rho_{\rm H}) - \sum_{l=1}^{N} \rho_{\rm H} \frac{V_{\rm H}^2}{2}\left(f_l \frac{x_l}{D_{{\rm h},l}} + K_l\right) \quad (15)$$

where g is the gravity constant, $V_{\rm H}$ the hot gas velocity, $h_{\rm c}$ the chimney height and ρ_0 and $\rho_{\rm H}$, respectively, the ambient and the hot-gas density.

In subsonic combustion, as the flame speeds are small compared to the sound speed, pressure can be considered constant [47]. Thus, considering the ideal gas equation, the density change through the flame front can be directly related to the temperature change as:

$$\frac{\rho_0}{\rho_{\rm H}} = \frac{T_{\rm H}}{T_0} \quad (16)$$

Using the continuity equation, the mass-flow rate provided by the buoyantly driven flow in the common form of the chimney effect is then given by Equation (17):

$$\dot{m}_{\rm A} = CA\left(\frac{P}{RT_{\rm H}}\right)\sqrt{2gh_{\rm c}\left(\frac{T_{\rm H} - T_0}{T_0}\right)} \quad (17)$$

where C is the loss coefficient, T_0 the temperature of the ambient and $T_{\rm H}$ the hot gas temperature and A the flow cross-section area. P stands for the ambient pressure and R the perfect gas constant.

The loss coefficient is introduced to account for all inefficiencies in the chimney effect $(0 \leq C \leq 1)$. In a practical sense:

$$C = \frac{\dot{m}_{\rm A\ actual}}{\dot{m}_{\rm A\ theoretical}} \quad (18)$$

It would theoretically be 1 for an ideal cookstove with neither loss during the heat addition to kinetic energy conversion nor viscous dissipations. Indeed, it is debatable whether this could be possible; otherwise, what could be this limit value? The loss coefficient issue will be discussed in very wide and detailed terms in Section 4.

2.1.4. Second Law and Exergy Balance of the Flowing Fluid

The entropy-generation rate arising along the flow stream crossing the boundaries of a control volume can be determined on a mass-flow rate basis as a net entropy balance taking into account the in- and the outflowing entropies as well as that transferred by energy flows in the form of heat into and out of the system.

For a steady flow, single stream, the entropy-generation rate is given by:

$$\dot{S}_{\text{gen}} = \dot{m}_{\text{A}}(s_{\text{out}} - s_{\text{in}}) - \sum_{k=1}^{N} \frac{\dot{Q}_k}{T_k} \geqslant 0 \tag{19}$$

where s_{in} and s_{out} are, respectively, the entropy per unit mass of flow entering the system and that of the flow exiting the system, and $\dot{Q}_k$ is the heat transferred through the boundary at temperature T_k at location k.

The cooking stove combustion chamber acts like a producer of entropy as the heat transferred $\dot{Q}_{\text{flue}}$ responsible for the buoyant flow discharges in it at a given temperature. Exergy is the precious part of this thermal energy which can be used by work until it is part of the internal energy of the ambient.

The exergy balance during the steady-flow combustion as sketched in Figure 2 takes into account:

1. The rate of exergy flow by heat transfer to the flue gases $\dot{X}_{\text{heat}}$ that can be determined defining a Carnot factor η_{c_1} which determines the quality of the heat depending on its temperature:

$$\dot{X}_{\text{heat}} = \eta_{c_1} \cdot \dot{Q}_{\text{flue}} \tag{20}$$

According to the linear temperature profile adopted in Section 2.1.2, flame can be modeled as a heat reservoir along stove chimney height that supplies heat indefinitely at temperatures gradually raising from T_0 to T_{H}. So care is taken to determine a mean Carnot efficiency by integration:

$$\eta_{c_1} = \frac{1}{T_{\text{H}} - T_0} \int_{T=T_0}^{T_{\text{H}}} \left(1 - \frac{T_0}{T}\right) dT \tag{21}$$

and

$$\eta_{c_1} = 1 - \frac{T_0}{T_{\text{H}} - T_0} \cdot \ln \frac{T_{\text{H}}}{T_0} \tag{22}$$

The Carnot efficiency represents the fraction of the energy transferred from the heat source that can be converted to work, see Cengel [30].
The rate of exergy flow by heat transfer becomes:

$$\dot{X}_{\text{heat}} = \left(1 - \frac{T_0}{T_{\text{H}} - T_0} \cdot \ln \frac{T_H}{T_0}\right) \cdot \dot{Q}_{\text{flue}} \tag{23}$$

Otherwise, considering Equations (7) and (23), the difference $\left[(1 - \eta_{c_1}) \cdot \dot{Q}_{\text{flue}}\right]$ determines the rate at which exergy destruction due to heat transfer takes place:

$$\dot{X}_{\text{destroyed}_{(\text{heat})}} = \left(\frac{T_0}{T_{\text{H}} - T_0} \ln \frac{T_{\text{H}}}{T0}\right) \cdot \dot{m}_{\text{A}} \bar{c}_{\text{p}} (T_{\text{H}} - T_0) \tag{24}$$

$$\dot{X}_{\text{destroyed}_{(\text{heat})}} = \dot{m}_{\text{A}} \bar{c}_{\text{p}} \cdot T_0 \ln \frac{T_{\text{H}}}{T_0} \tag{25}$$

2. The rate of exergy change of the flow stream (exergy of the flowing fluid) across the combustion chamber that is written as:

$$\dot{X}_{\text{mass,in}} - \dot{X}_{\text{mass,out}} = \dot{m}_{\text{A}} \cdot \left[(h(T_0) - h(T_{\text{H}})) - T_0 \underbrace{\cdot (s(T_0) - s(T_{\text{H}}))}_{\text{Net entropy transfer by heat and mass}} + \frac{V_{\text{H}}^2}{2} + gh_{\text{c}} \right] \tag{26}$$

where $\dot{X}_{\text{mass,in}}$ is the rate of exergy transferred by mass when the mass in the amount of $\dot{m}_{\text{A}}$ enters the control volume and $\dot{X}_{\text{mass,out}}$ is the rate of exergy evacuated by mass when the mass in the same amount leaves the control volume [30]. Mass flow into the system is accompanied by enthalpy $h(T_0)$ and entropy $s(T_0)$, and out of the system by

$h(T_H)$ and $s(T_H)$, respectively. In Equation (26), $\frac{V_H^2}{2}$ and gh_c are, respectively, exergy change associated with the mass kinetic and the mass potential energy of the flow.

3. The rate of exergy loss (or exergy destruction) of useful work by any other mechanisms at a location k (i.e., interaction of hot flue gases with the surface of the pot or with inner stove chimney surface) that is given:

$$\dot{X}_{\text{destroyed},\,k} = T_0 \cdot \dot{S}_{\text{gen},\,k} \tag{27}$$

directly proportional to the rate of entropy generation $\dot{S}_{\text{gen},k}$ in this form of relation known as the Gouy–Stodola theorem [31].

Finally, neglecting exergy of kinetic energy and exergy of potential energy (compared to other terms in Equation (26)), the exergy balance of the flowing fluid is summarized as:

$$\left(1 - \frac{T_0}{T_H - T_0} \cdot \ln\frac{T_H}{T_0}\right) \cdot \dot{Q}_{\text{flue}} + \dot{X}_{\text{mass, in}} - \dot{X}_{\text{mass, out}} - \dot{X}_{\text{destroyed},\,k} = 0 \tag{28}$$

Otherwise, the energy flux applied as flow work $\Delta\dot{X}_{\text{flow}}$ can be written:

$$\Delta\dot{X}_{\text{flow}} = \dot{X}_{\text{mass,out}} - \dot{X}_{\text{mass,in}} = \eta_{C_1} \cdot \dot{Q}_{\text{flue}} - \dot{X}_{\text{destroyed},\,k} \tag{29}$$

Equation (29) is an important result in the analysis of convective heat transfer. This relation states that the energy flux applied as flow work is pure exergy which is lost in consecutive dissipation processes.

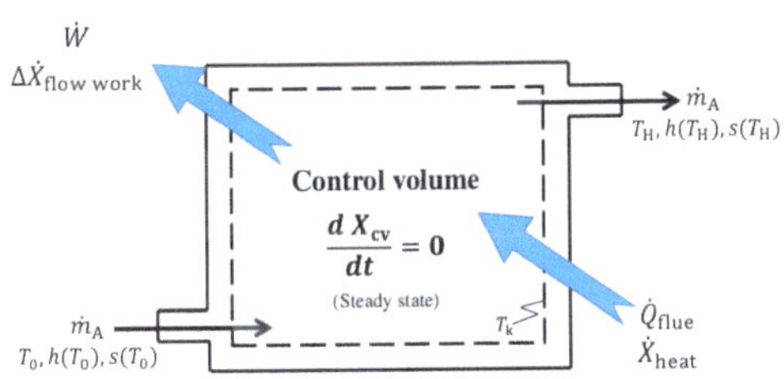

Figure 2. The rate of exergy change within the control volume $\dot{X}_{cv}$ is equal to the rate of net exergy transfer through the control volume boundary by heat $\dot{X}_{\text{heat}}$, work $\dot{W}$ and mass flow minus the rate of exergy destruction within the boundaries of the control volume. Note that in a steady state $\dot{X}_{cv}$ is zero.

Figure 3 shows the thermodynamic system equivalent to the simplified cooking stove model.

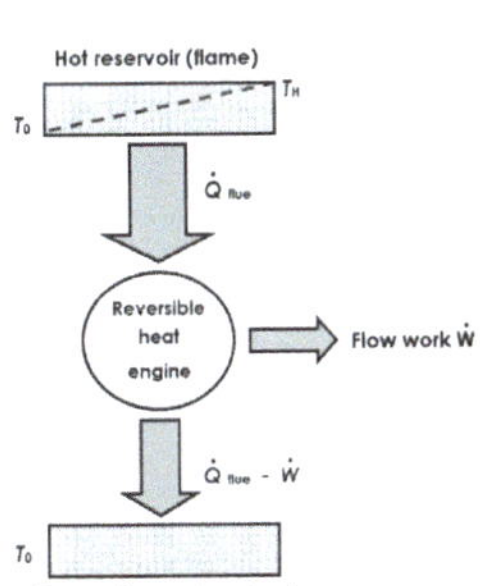

Figure 3. Open thermodynamic system in steady state equivalent to a reversible heat engine that operates between hot reservoir (flame) and atmospheric temperatures. The engine releases its work output into flow work, and rejects heat to the pot and to the surrounding environment.

2.1.5. Alternative Assessment Numbers in Energy Transfer Processes

As mentioned before, the quantity of energy is conserved, but its quality deteriorates during energy-transfer processes according to Equation (29). When thermodynamic considerations are added to the analysis and interpretation of convective heat-transfer situations, it turns out that the heat-transfer coefficient h or the Nusselt number Nu are no more precise parameters to access qualitative aspects of the energy-transfer processes, see [42]. In [39,48] are introduced some alternative non-dimensional parameters for a comprehensive characterization of the energy devaluation chain consecutive to the unit-transfer operations.

Energy Devaluation in Heat-Transfer Processes

The *energy-devaluation number* N_i for an energy-transfer operation i indicates how much of the *entropic potential* is used:

$$N_i = \frac{T_0 \cdot \dot{S}_{\text{gen},i}}{\dot{E}} \tag{30}$$

where $\dot{S}_{\text{gen},i}$ the entropy-generation rate reported to the unit-transfer operation i. This entropy generation is seen in the context of the devaluations of the energy-transfer rate $\dot{E}$ that happened prior to the-transfer operation i and that will happen afterwards.

Losses Due to Dissipation of Mechanical Work

Specifically, in a convective heat-transfer process, flow work rate is needed to maintain the flow into which heat transfer occurs. So a second coefficient is needed which is defined as the *exergy destruction number* N^E, indicating the loss of exergy in the flow field:

$$N^E = \frac{T_0 \cdot \dot{S}_{\text{gen},D}}{\dot{E}} \tag{31}$$

where $\dot{S}_{\text{gen},D}$ is the entropy dissipation rate due to dissipation of mechanical energy and $\dot{E}$ the kinetic energy.

Take care that in N^E it is the kinetic energy of the fluid flow which is used as a reference quantity, whereas N_i refers to the quantity of energy or heat transferred. However, it is not the kinetic energy that is devaluated but the energy that enters the system as flow work [39].

Overall Exergy Devaluation in Heat-Transfer Processes

For an overall assessment of a convective heat-transfer process, Herwig [39] refers subtly to the sum of exergy losses (in the temperature and in the flow field) to the exergy transferred in the process, which is the product $\eta_{c_2} \cdot \dot{Q}_{\text{c-p}}$. The *overall exergy loss number* reads:

$$\widehat{N}^E = \frac{T_0 \cdot \left(\dot{S}_{\text{gen}_{(\text{heat})}} + \dot{S}_{\text{gen}_{(\text{viscous})}}\right)}{\eta_{c_2} \cdot \dot{Q}_{\text{c-p}}} \tag{32}$$

In the equation above, the term η_{c_2} is the Carnot factor for the consecutive convective exergy-transfer process different from η_{c_1} which previously defined the exergy part of the energy transferred after combustion to the flue gases. The physical meaning of the Carnot factor η_{c_2} will be resumed in Section 2.2.2.

The measure for the quality of energy and its potential degradation in energy-transfer processes is entropy. In the next section we developed a simplified analytical expression of the total entropy-generation rate in the flowing fluid due to heat-transfer processes and viscous dissipations.

2.2. *Entropy Generation Rate: Analytical Solution*

2.2.1. Stove Operating without Cooking Pot

The infinitesimal change of the rate of entropy generation is:

$$d\dot{S}_{\text{gen}} = \dot{m}_{\text{A}} ds - \frac{\delta \dot{Q}_{\text{c-w}}}{T_{\text{g}}} \tag{33}$$

where $\dot{Q}_{c-w}$ is the rate of convective heat transferred from gases to the inner combustion chamber walls at temperature T_{g}.

In a steady-state regime for an insulated stove and considering the radiative heat transfer of the flame to be negligible [11,15], Equation (33) can be simplified to:

$$d\dot{S}_{\text{gen}} \approx \dot{m}_{\text{A}} ds \tag{34}$$

By virtue of the principle of conservation of energy, the infinitesimal mass sensible enthalpy increase dh in the flue gases and according to Equation (7) denotes:

$$dh = \bar{c}_{\text{p}} dT \tag{35}$$

The Gibbs–Duhem relation corresponding to the energetic fundamental relation is given:

$$Tds = dh - vdp \tag{36}$$

where v stands for the specific volume and p the pressure.

Rearranging Equations (34) and (35) in (33) gives:

$$d\dot{S}_{\text{gen}} = \dot{m}_{\text{A}} \cdot \left(\frac{\bar{c}_{\text{p}} dT}{T} - \frac{vdp}{T} \right) \tag{37}$$

and expressing the specific volume $v = \rho^{-1}$

$$d\dot{S}_{\text{gen}} = \dot{m}_{\text{A}} \cdot \left(\frac{\bar{c}_{\text{p}} dT}{T} - \frac{1}{\rho} \frac{dp}{T} \right) \tag{38}$$

Let us now introduce derivations with respect to x (the spatial coordinate):

$$\frac{d\dot{S}_{\text{gen}}}{dx} = \dot{m}_{\text{A}} \cdot \left(\frac{\bar{c}_{\text{p}}}{T(x)} \frac{dT(x)}{dx} - \frac{1}{\rho T(x)} \frac{dp}{dx} \right) \tag{39}$$

In a rearranged form, integration along the height of the combustion chamber gives:

$$\dot{S}_{\text{gen}} = \dot{m}_{\text{A}} \bar{c}_{\text{p}} \cdot \int_{x=0}^{h_{\text{c}}} \frac{1}{T(x)} \frac{dT(x)}{dx} dx - \dot{m}_{\text{A}} \cdot \int_{x=0}^{h_{\text{c}}} \frac{1}{\rho T(x)} \frac{dp}{dx} dx \tag{40}$$

In the second term of the right-hand side of the expression above, the pressure drop $-\frac{dp}{dx}$ evaluated on a finite distance can be related to the dynamic pressure as:

$$-\frac{\Delta p}{\Delta x} = K\rho \frac{V^2}{2x} \tag{41}$$

taking into account the definition of the chimney effect, see Equation (16):

$$-\frac{\Delta p}{\Delta x} = gK\rho \frac{T(x) - T_0}{T_0} \tag{42}$$

Note that the single K-value represents a total heat loss through the stove related to viscous dissipations in the fluid flow, see Equation (15).

Therefore, entropy-generation rate becomes explicitly related to mass-flow rate and temperature:

$$\dot{S}_{\text{gen}} = \dot{m}_{\text{A}}\bar{c}_{\text{p}} \cdot \int_{x=0}^{h_c} \frac{1}{T(x)} \frac{dT(x)}{dx} dx + \dot{m}_{\text{A}} \cdot \int_{x=0}^{h_c} \frac{gK(T(x) - T_0)}{T(x)T_0} dx \tag{43}$$

$$\dot{S}_{\text{gen}} = \dot{m}_{\text{A}}\bar{c}_{\text{p}} \cdot \int_{x=0}^{h_c} \frac{1}{T(x)} \frac{dT(x)}{dx} dx + \dot{m}_{\text{A}} \cdot \int_{x=0}^{h_c} gK \cdot \left(\frac{1}{T_0} - \frac{1}{T(x)}\right) \cdot dx \tag{44}$$

Substituting in Equation (44) the temperature profile $T(x)$ from Equation (11) and its derivative with respect to x from Equation (12):

$$\dot{S}_{\text{gen}} = \dot{m}_{\text{A}}\bar{c}_{\text{p}} \cdot \int_{x=0}^{h_c} \frac{\dot{Q}_{\text{flue}}}{\dot{m}_{\text{A}} h_{\text{c}}(T_0 + \frac{\dot{Q}_{\text{flue}}}{\dot{m}_{\text{A}}\bar{c}_{\text{p}}h_{\text{c}}} x)} dx + \dot{m}_{\text{A}} \cdot \int_{x=0}^{h_c} \frac{gK}{T_0} dx - \dot{m}_{\text{A}} \cdot \int_{x=0}^{h_c} gK \cdot \frac{1}{T_0 + \frac{(T_{\text{H}} - T_0)}{h_c} x} dx \tag{45}$$

Likewise, considering the expression of the heat addition $\dot{Q}_{flue}$ in Equation (7):

$$\dot{S}_{\text{gen}} = \dot{m}_{\text{A}}\bar{c}_{\text{p}} \cdot \int_{x=0}^{h_c} \frac{T_{\text{H}} - T_0}{T_0 h_{\text{c}} + (T_{\text{H}} - T_0)x} dx + \dot{m}_{\text{A}} \cdot \int_{x=0}^{h_c} \frac{gK}{T_0} dx - \dot{m}_{\text{A}} \cdot \int_{x=0}^{h_c} gK \cdot \frac{1}{T_0 + \frac{(T_{\text{H}} - T_0)}{h_c} x} dx \tag{46}$$

Finally, the entropy generation in the flow stream that results from the heat transfer and frictional pressure drop processes is:

$$\dot{S}_{\text{gen}} = \underbrace{\dot{m}_{\text{A}}\bar{c}_{\text{p}} \cdot \ln \frac{T_{\text{H}}}{T_0}}_{\dot{S}_{\text{gen(heat)}}} + \underbrace{K\dot{m}_{\text{A}} g h_{\text{c}} \cdot \left(\frac{1}{T_0} - \frac{1}{T_{\text{H}} - T_0} \cdot \ln \frac{T_{\text{H}}}{T_0}\right)}_{\dot{S}_{\text{gen(viscous and frictional pressure drop)}}} \tag{47}$$

The first term on the right-hand side represents the entropy-generation rate due to heat transfer. Note that this term multiplied by T_0 matches the expression of the exergy-destruction rate due to heat transfer in Equation (25). The second term represents the contributions due to viscous processes.

$$\dot{S}_{\text{gen}} = \dot{m}_{\text{A}}\bar{c}_{\text{p}} \cdot \left[\ln \frac{T_{\text{H}}}{T_0} + K \frac{g h_{\text{c}}}{\bar{c}_{\text{p}} T_0} \cdot \left(1 - \frac{T_0}{T_{\text{H}} - T_0} \cdot \ln \frac{T_{\text{H}}}{T_0}\right)\right] \tag{48}$$

An entropy-generation number N_s introduced by Bejan [49] can be defined in a dimensionless form as:

$$N_s = \frac{\dot{S}_{\text{gen}}}{\dot{m}_{\text{A}}\bar{c}_{\text{p}}} = \ln \frac{T_{\text{H}}}{T_0} + K \frac{g h_{\text{c}}}{\bar{c}_{\text{p}} T_0} \cdot \left(1 - \frac{T_0}{T_{\text{H}} - T_0} \cdot \ln \frac{T_{\text{H}}}{T_0}\right) \tag{49}$$

Considering the expression of Carnot efficiency in Equation (21):

$$N_s = \ln \frac{T_{\text{H}}}{T_0} + \eta_{\text{c}_1} \cdot \frac{g h_{\text{c}}}{\bar{c}_{\text{p}} T_0} \cdot K \tag{50}$$

It appears that the entropy-generation rate due to viscous dissipations is directly related to the exergy Carnot factor η_{c_1}. The dimensionless quantity $\frac{g h_{\text{c}}}{\bar{c}_{\text{p}} T_0}$ is also known as the Gebhart number, accounting for the viscous dissipation of thermal energy in natural convection processes [37]. The single K-value represents a total heat-loss coefficient associated with the conduit components. Thus, the Carnot factor plays the role of a weighting parameter for the potential to generate entropy by frictional pressure-drop effects. The higher the flue gas temperature, the higher the exergy flow rate, and the more viscous dissipations are accounted for.

2.2.2. Stove Operating with Cooking Pot

Let us consider now the practical case of the stove operating with a cooking pot containing, let us say, a given quantity of water. Hot flue gases interact with the outer surface of the pot. Therefore, this convective heat transfer contributes to the destruction of exergy flow.

The rate of thermal energy transfer to pot $\dot{Q}_{c-p}$ is affected by the convective coefficient h changing with the mass-flow rate, the pot exposed surface area $A_{\rm p}$ and the difference between the gases temperature $T_{\rm g}$ and the averaged pot surface temperature $T_{\rm p}$:

$$\dot{Q}_{\text{c-p}} = h \cdot A_{\rm p} \cdot (T_{\rm g} - T_{\rm p}) \tag{51}$$

The convective coefficient is related to the Nusselt number:

$$h = \frac{Nu \cdot k}{D} \tag{52}$$

where k is the thermal conductivity of the flue gas and D is the value of the chimney diameter.

The thermal energy transfer to pot can be rewritten:

$$\dot{Q}_{c-p} = \frac{Nu \cdot k \cdot A_{\rm p} \cdot (T_{\rm g} - T_{\rm p})}{D} \tag{53}$$

Zube [15] proposed an average Nu for a fully developed free jet impinging on a flat plate:

$$Nu = 0.565 \cdot Pr^{0.5} \cdot Re^{0.5} \tag{54}$$

A constant Prandtl number Pr of air is assumed to be around 0.7 at 1 atm in the range of temperature between 300 K and 1500 K. The Reynolds number Re is written:

$$Re \equiv \frac{\rho V D}{\mu(T)} \tag{55}$$

where ρ is the air density, $\mu(T)$ the air dynamic viscosity function of temperature, V the velocity of the fluid and D the diameter of the cylindrical stove chimney. The dependence of the stove flow Reynolds number on stove operation can be determined in the function of the mass-flow rate and the cross-section area A of the stove chimney as shown:

$$Re \equiv \frac{\rho V D A}{\mu(T) A} = \frac{2\,\dot{m}_{\rm A}}{\mu(T) \cdot \sqrt{A\pi}} \tag{56}$$

The convective heat transfer sketched in Figure 4 reveals a "temperature gap" between the flame and the exposed surface of the pot. The heat-transfer interaction $\dot{Q}_{\text{c-p}}$ across this space remains undiminished [31].

The pot surface temperature $T_{\rm p}$ can be determined by defining the overall heat-transfer coefficient U between the three media sketched in Figure 4:

$$U = \frac{\dot{Q}_{\text{c-p}}}{A_{\rm p} \cdot (T_{\rm g} - T_{\rm g})} \tag{57}$$

The electrical analogy of resistance gives means to calculate U:

$$U = \frac{1}{\frac{1}{h} + \frac{L}{k_{\text{metal}}} + \frac{1}{h_{\text{water}}}} \tag{58}$$

where h is the convective heat-transfer coefficient in the "temperature gap" between hot gases and external cooking pot surface, k_{metal} is the thermal conductivity of the metal (often aluminium) and L is its thickness and h_{water} is the free convection heat-transfer coefficient between internal pot surface and water. The order of magnitude of the heat-transfer coefficient is around 200 $W{\cdot}m^{-2}{\cdot}K^{-1}$ for h, $\frac{1}{160{,}000}$ $W{\cdot}m^{-2}{\cdot}K^{-1}$ for $\frac{L}{k_{metal}}$ and around 5000 $W{\cdot}m^{-2}{\cdot}K^{-1}$ for h_{water} [50]. It can be seen that conduction resistance through the metal pot and internal surface convective resistance to water are negligible. The overall heat-transfer coefficient U becomes almost equal to the convective heat-transfer coefficient in the temperature gap h; therefore, $T_p \approx T_{water}$.

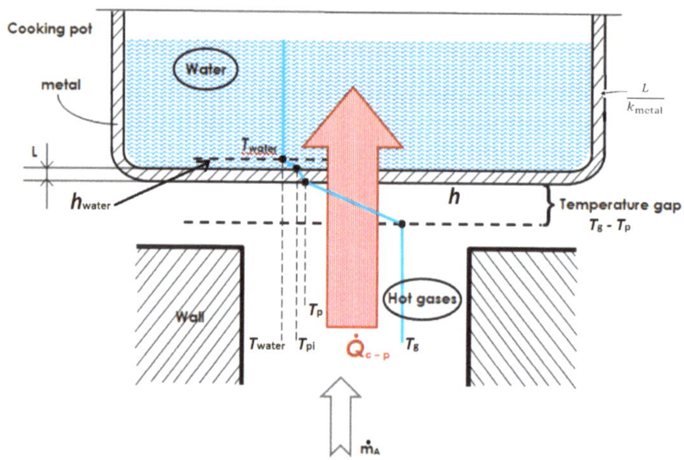

Figure 4. Hot gases at temperature T_g convect a certain amount of energy $\dot{Q}_{c\text{-}p}$ to the external pot surface at temperature T_p. Then the heat is conducted through the metal (pot) of small thickness and finally convected from the internal surface of the pot at temperature T_{pi} into water at temperature T_{water}.

Thus, the entropy-generation rate in this temperature gap can be written:

$$\dot{S}_{gen_{(c\text{-}p)}} = \frac{\dot{Q}_{c\text{-}p}}{T_p} - \frac{\dot{Q}_{c\text{-}p}}{T_g} = \dot{Q}_{c\text{-}p} \cdot \frac{(T_g - T_p)}{T_g T_p} \tag{59}$$

The loss of exergy as a result of this irreversibility is:

$$\dot{X}_{destroyed_{(c\text{-}p)}} = T_0 \cdot \dot{S}_{gen_{(c\text{-}p)}} = \frac{T_0}{T_p}\left(1 - \frac{T_p}{T_g}\right) \cdot \dot{Q}_{c\text{-}p} = \eta_{c_2} \cdot \dot{Q}_{c\text{-}p} \tag{60}$$

Caution must be taken when analyzing consecutive energy-transfer operations. The very question largely discussed in [48] is how to put energy-transfer assessment in the right perspective. Section 4 will address the way to deal with alternative non-dimensional parameters in order to assess adequately the overall energy devaluation in the present application.

3. Experimental Setup, Materials and Methods

In our experimental part, tests have been conducted solely for the basic case of stove without cooking pot. Given the complexity, validation for the practical case with pot in place refers to calculations and data in [15,22,46].

The two properties that most predominantly characterize the flow are mass-flow rate and temperature. The task is to assess the entropy-generation rate and associated quantities by means of the mass-flow rate and exhaust-gas temperature measured when the cooking stove is tested at different operating firepower levels.

In practice, differing firepower level is achieved by varying by hand the fuel-feed rate, the mass of fuel in the combustion chamber and fuel spacing [20]. Furthermore, though the actual fire is an intrinsically transient phenomenon, the stove will be considered to operate under steady state conditions by averaging its temporal behaviour. The next paragraphs present successively the stove, the fuel properties and the experimental protocol.

3.1. The Stove

Environfit G3300 in Figure 5 is a stick burning wood fuel cookstove, developed on the basis of the rocket elbow principle by Envirofit International, Inc. (Fort Collins, CO, USA) (http://www.envirofit.org/ (accessed on 8 March 2020)). Many papers related to this model have been published by researchers at the Colorado State University [15,22,23,51]. Table 2 gives the G3300 geometrical parameters.

Figure 5. G3300 envirofit cookstove model.

Table 2. G3300 stove model geometrical parameters.

Parameter	Value	Unit
Chimney diameter	100	mm
Chimney height	220	mm
Outer stove diameter	230	mm
Air entrance area	160×100	mm^2
Entrance area/Chimney area ratio	2.04	–

3.2. Wood Properties and Preparation

The wood used for this experiment is Entandrophragma Cylindricum (Sapele), a tropical species widely found in many African regions and commonly known as red wood because of its reddish tint. Table 3 gives the elemental analysis of Sapele from the literature report [52].

Table 3. Elemental analysis of Sapele.

Element	Percentage (%)
Carbon	54.6
Hydrogen	4.7
Oxygen	40.7
Sulfur	0
Nitrogen	0.01

The wood is moderately heavy, with a density of 560–750 kg·m^{-3} at 12% moisture content. Chemical analysis from [53] shows that Sapele wood is slightly alkaline (pH = 8) and the net-heat of combustion of Sapele in the air-dry state (8% relative humidity) was measured to be 17.1 MJ·kg^{-1} and one can infer for an oven-dried sample a low heating value of 18.8 MJ·kg^{-1} [54]. To improve repeatability, Sapele wood in all our tests was used oven dried. Wood cribs of square and rectangular cross-sections were prepared: $2.5 \times 2.5 \times 33$ cm (stick) and $1.3 \times 2.5 \times 33$ cm ($\frac{1}{2}$ stick). Typical sample stick and $\frac{1}{2}$ stick (half) weighing, respectively, 96 and 48 g presented specific area ratios of 240 and 320 m^{-1}, the calculation being limited to only the tip of long pieces of wood inserted a small distance (≈2.5 cm) into the fire.

3.3. Testing Protocol

Experiments have been performed using the Environfit G3300 wood cookstove under a hood. Each test consisted of 15 min data sampling periods over which firepower is held as constant as possible. Temporal recording of fuel mass reduction, bulk flow temperature and O_2 concentration in the exhaust gas sample was also simultaneously measured. Temporal averaging was then applied to the 15 min data samples giving the values used for different firepower sample points. The sampling periods were counted from the time when the firepower reached approximately steady state behaviour excluding start-up and shut-down periods [20]. Data averaging was performed using GNUPLOT version 5.2 patchlevel 2.

3.3.1. Measuring Fuel-Mass-Burning Rate and Firepower

The mass-burning rate of fuel was calculated as in [46] by keeping the stove on a sufficiently robust balance. Time intervals for every 0.01 kg fuel reduction were noted down for every batch of wood burned, while an experienced operator tries to maintain flame intensity as constant as possible. The average mass-burning rate of fuel was determined as the ratio of 0.01 kg to the average time for a set of readings t_{avg}:

$$\dot{m}_F = \frac{0.01}{t_{avg}} \quad (61)$$

Operating firepower was calculated using expression Equation (5). Table 4 presents batch load characteristics, the stack giving wood cribs composition, the entrance area allowed by the stacking of wood cribs, the inlet area ratio (IAR) defined as the ratio of the cross-sectional area unoccupied by the wood at the feed door to the total entrance area [46] and the averaged fuel-mass-burning rate that resulted following pseudo-steady state firepower levels.

Table 4. Fuel batch loads.

# test	Stack Wood	Entrance Area (m^2)	IAR	Fuel Burning Rate ($kg{\cdot}s^{-1}$)
1	$\frac{1}{2}+\frac{1}{2}$ sticks	0.0153	0.95625	0.027×10^{-3}
2	2 sticks	0.01475	0.921875	0.037
3	$1+\frac{1}{2}$ sticks	0.015	0.9375	0.050
4	$1+\frac{1}{2}$ sticks	0.015	0.9375	0.059
5	$\frac{1}{2}+\frac{1}{2}$ sticks	0.0153	0.95625	0.085
6	$\frac{1}{2}+\frac{1}{2}+\frac{1}{2}$ sticks	0.015	0.9375	0.128
7	2 sticks bis	0.01475	0.921875	0.149
8	2 sticks	0.01475	0.921875	0.170
9	4 sticks	0.0135	0.84375	0.219
10	3 sticks	0.014	0.875	0.229
11	4 sticks	0.0135	0.84375	0.309
12	$2+\frac{1}{2}$ sticks	0.0144	0.9	0.314

3.3.2. Measuring $\%O_2$ and Calculating Mass-Flow Rate

The O_2 concentration in the exhaust gas sample was determined via the syngas analyser GASBOARD-3100P of Cubic-Ruiyi Instrument based on ECD, where a fraction of the total flow is drawn by a suction pump through a sample line to the real-time (0.5 Hz) sensor.

The air mass-flow rate is calculated with the method in [21,22] using the stack exhaust volumetric $\%O_2$ instead of flow meters. This simplified exhaust $\%O_2$ approach is worth being exposed here again.

Combustion of wood in an excess of air can be described by the generalized form one-step reaction:

$$aC_xH_yO_z + b(O_2 + 3.76N_2) \rightarrow dCO_2 + eH_2O + fO_2 + b(3.76N_2) \tag{62}$$

where a, b, d, e and f correspond to the stoichiometric coefficients expressed in molar rate units.

Stack exhaust $\%O_2$ is related to the overall one-step reaction summarized in Equation (61) as shown in Equation (62). The molar rate of oxygen (b) can be calculated as shown in Equation (63), (b) the function of the molar rate of fuel consumption (a), $\%O_2$ concentration and the elemental composition of the fuel ($x = 4.6$, $y = 4.7$, $z = 2.5$).

$$\%O_2 = \frac{f \cdot 100}{d + e + f + 3.76b} \tag{63}$$

$$b = \frac{a(z - 2x - \frac{y}{2})}{\%O_2 \cdot 4.76 - 100} \times (50 - \frac{\%O_2}{2} - \%O_2 \cdot ax) \tag{64}$$

At the end, the mass-flow rate of air is obtained by multiplying (b) by the molecular weight, as shown in Equation (64).

$$\dot{m}_A = b\frac{\text{kgmol}}{\text{s}} \cdot (32 + 3.76 \cdot 28)\frac{\text{kg}}{\text{kgmol}} \tag{65}$$

3.3.3. Measuring Temperature

Bulk flow temperature is measured using a K-type thermocouple placed at the approximate center of the chimney about 1 cm above the chimney exit.

4. Results: Validation and Discussion of the Second Law Approach

4.1. Stove without Pot

The thermal properties of air described by quantities including specific enthalpy and entropy at different temperatures are given in thermodynamic tables, for example, in [55]. Table 5 outlines the experimental values of time-averaged exiting gas temperature (column 4) and air mass-flow rate (column 6) at different fuel burning rates for the G3300 stove operating without cooking pot. The corresponding firepower levels were obtained according to Equation (5). Oxygen percentage values enter air mass-flow rate calculations following Equations (62)–(64). Experimental mass-flow rates and exit gas temperatures permit calculation of the stove flow Reynolds number using Equation (56). Values of *Re* vary in the interval between 438 and 1358. These low Reynolds numbers verify the initial hypothesis that the flow encountered is laminar.

Table 5. Fuel burning rate, Firepower, Bulk flow temperature, Air mass-flow rate and entropy-generation rate for a G3300 woodburning stove without cooking pot.

			Experimental Results			Table Values	
#Test	$\dot{m}_F$ (kg·s^{-1})	*Firepower* (kW)	T_H (K)	O_2 (%)	$\dot{m}_A$ (kg·s^{-1})	$s(T_H) - s(T_0)$ (kJ·kg^{-1}·K^{-1})	$\dot{S}_{gen_1}$ (kJ·K^{-1}·s^{-1})
1	0.027×10^{-3}	0.5	421	19.5	1.08×10^{-3}	0.346	0.37×10^{-3}
2	0.034	0.63	439	19.35	1.26	0.353	0.44
3	0.050	0.94	527	18.78	0.60	0.558	0.89
4	0.059	1.1	543	18.54	2.33	0.608	1.42
5	0.085	1.6	484	17.85	2.21	0.460	1.02
6	0.128	2.4	684	16.36	2.58	0.871	2.25
7	0.149	2.8	694	17.53	3.64	0.888	3.23
8	0.170	3.2	800	17.43	4.08	1.059	4.32
9	0.219	4.1	818	14.66	3.55	1.084	3.85
10	0.229	4.3	873	14.66	3.52	1.158	4.08
11	0.309	5.8	995	11.38	3.75	1.292	4.84
12	0.314	5.9	994	12.64	4.21	1.291	5.44

Air mass-flow rates along with specific entropy changes (column 7) were used as inputs to calculate the total entropy-generation rate $\dot{S}_{gen}$ appearing in the last column of Table 5. From this simplified model, it appears that for all the parameters under study, the contribution of the entropy-generation rate due to viscous dissipation in Equation (48) is less than 0.1% compared to the total entropy-generation rate. Following Gebhart [37] the dimensionless number $\frac{gh_c}{\bar{c}_p T_0}$ in Equation (48) shows that the effects of viscous dissipation in natural convection is appreciable indeed when the induced kinetic energy becomes appreciable compared to the amount of heat transferred. This occurs when either the equivalent body force is large (g) or when the convection region (h_c) is extensive, but this is not the case here. This quantity remains of the order of 10^{-5} and therefore the effects of viscous dissipations can be safely ignored; this was the same as assumed in [38] citing earlier works devoted to the role of irreversibility distribution ratio on the total entropy generation. That said, it can be concluded that the major source of irreversibilities in fluid flow through a stove is in the heat-transfer process. In contrast, entropy generation due to viscous dissipation and fluid friction is negligible. Hence, for the rest of discussion, entropy due to viscous dissipation will not be shown as a separate component.

Figure 6 depicts the evolution of the dimensionless entropy-generation number N_s with respect to the exhaust-gas temperature. Considering Equation (49), the model-predicted $\left(\ln \frac{T_H}{T_0}\right)$ sample points obtained from experiments at different operating fire-power levels agree well with thermodynamic table values of the specific entropy balance into and out of the combustion chamber reported to the specific heat as $\left(\frac{s(T_H)-s(T_0)}{\bar{c}_p}\right)$.

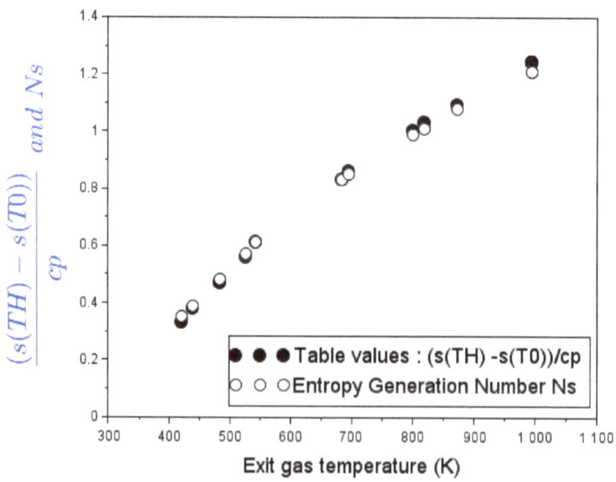

Figure 6. Specific-entropy-to-specific-heat ratio on the basis of measures on the G3300 stove operating without cooking pot and model-predicted dimensionless entropy number N_s .

Table 6 gives the rate of the total energy variation of the flowing fluid $\Delta\theta$ in column 4, which is almost equivalent to the rate of sensible enthalpy increase obtained using Equations (7) and (13). Column 5 gives the rate of exergy destroyed $\dot{X}_{destroyed_{(heat)}}$ calculated by means of Equation (25). Note that the difference between the two corresponds approximately to the rate of the flow exergy balance $\Delta\dot{X}_{flow}$ taking place according to Equation (29).

Now let us examine the link between the two energy conversion determinants, namely the loss coefficient C and the Carnot factor η_{C_1}. Equation (17) permits to equate experimental values of C for the G3300 woodburning stove operating without pot. On the other hand, Equation (22) completes Table 6 with values of the Carnot factor η_{C_1} obtained by integration along the chimney height. It is important to note that the ratio of flow exergy rate $\Delta\dot{X}_{flow}$ to total energy of the flowing fluid $\Delta\theta$ leads to the same Carnot factor results. Thus, Figure 7 plots the loss coefficient C and the Carnot factor η_{C_1} sample points obtained at different exhaust-gas temperatures. This figure shows that η_{C_1} and C trend overlay.

Table 6. The rate of exergy by heat transfer and associated parameters for the G3300 stove operating without cooking pot.

Firepower (kW)	T_H (K)	$\dot{m}_A$ (kg·s^{-1})	$\Delta\theta$ (kW)	$\dot{X}_{destr(heat)}$ (kW)	$\Delta\dot{X}_{flow}$ (kW)	η_{C_1}
0.5	421	1.08×10^{-3}	0.133	0.111	0.022	0.16
0.63	439	1.26	0.179	0.133	0.032	0.18
0.94	527	1.60	0.372	0.266	0.096	0.26
1.1	543	2.33	0.598	0.422	0.165	0.28
1.6	484	2.21	0.416	0.303	0.093	0.22
2.4	684	2.58	1.029	0.669	0.369	0.36
2.8	694	3.64	1.490	0.964	0.542	0.36
3.2	800	4.08	2.140	1.287	0.885	0.41
4.1	818	3.55	1.932	1.147	0.814	0.42
4.3	873	3.52	2.130	1.215	0.943	0.44
5.8	995	3.75	2.783	1.443	1.349	0.48
5.9	994	4.21	3.120	1.619	1.511	0.48

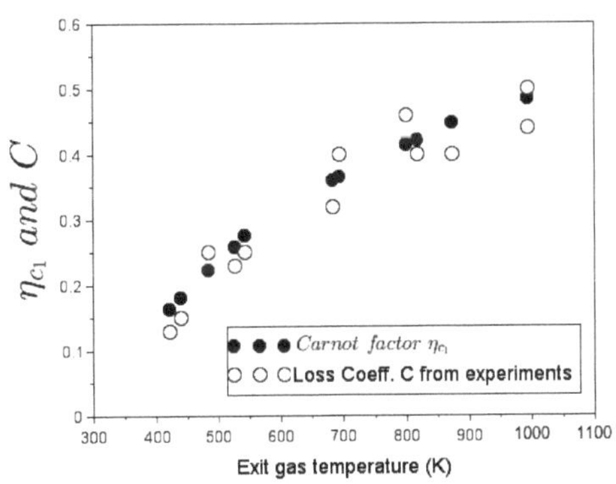

Figure 7. The Carnot factor η_{C_1} and the loss coefficient C in function of the flue gas temperature for the G3300 stove without pot.

It appears in Figure 7 that the loss coefficient coincides remarkably with the Carnot factor that is identified as a measure of the quality of heat transfer. By definition, the Carnot factor η_{C_1} defines the exergy part of the energy transferred. Thus, according to values in Table 6, when the cooking stove operates at low firepower levels (e.g., firepower = 0.5 kW and exit gas temperature = 421 K), a big amount of heat of combustion (≈85%) is not available in the form of useful (potential) work, in the sense that it does not participate in generating fluid motion, whereas at higher firepower levels (e.g., 5.9 kW and 994 K), almost half of the primary energy contributes to flow work and only half of the energy content is degraded.

In the basic case of the cookstove running without a cooking piece, the energy that enters the system as flow work is pure exergy that subsequently devaluates by dissipation processes according to Equation (29). The second case of the stove equipped with a cooking pot will show how flue gases and pot interaction degrade this flow work potential.

4.2. Stove with Cooking Pot

Contrary to the preceding case, in a stove operating with a pot in place, two energies are subjected to degradation in the convective heat-transfer process (the transferred thermal energy and the needed flow work). Table 7 illustrates the calculation results for mass-flow rate $\dot{m}_{\mathrm{A}}$, hot gas temperature impinging the bottom surface of the pot T_{g} (column 3) for the one-door rocket stove at different firepower levels in Zube [15].

The exit gas temperature T_{exit} (column 4) can be determined with the function of the rate of heat transfer to pot $\dot{Q}_{\text{c-p}}$ in a rearrangement of the first law of thermodynamics:

$$T_{\mathrm{exit}} = T_{\mathrm{g}} - \frac{\dot{Q}_{\text{c-p}}}{\dot{m}_{\mathrm{A}}\bar{c}_{\mathrm{p}}} \tag{66}$$

The hot gas temperature falls down from the exposed pot bottom surface to the exit port, as upward flows transport *exergy* to the pot. Then, the final water temperature inside the pot for each test can be also obtained in this way:

$$T_{\mathrm{water}} = \frac{\dot{Q} \cdot t_{\mathrm{test}} \cdot \eta_{\mathrm{th}}}{m_{\mathrm{water}} \cdot \bar{c}_{\mathrm{p_{water}}}} + T_0 \tag{67}$$

where $\dot{Q}$ is the operating firepower, t_{test} is the time duration of the test, η_{th} is the stove overall thermal efficiency, $\bar{c}_{\mathrm{p_{water}}}$ is the isobaric mass-specific approximate heat capacity of water between 20 °C and 100 °C given 4.180 kJ·kg $^{-1}$·K $^{-1}$ and T_0 is the temperature of the water at the beginning of the test.

Table 7 also presents convective heat-transfer parameters, namely the Reynolds number Re, the Nusselt number Nu and the convective heat-transfer coefficient h. However, Nu and h do not cover qualitative aspects of energy-transfer processes. As mentioned in Section 2.1.5, alternative assessment parameters are required to indicate how energy is used [39,41,42]. To achieve that, the stove system was divided into two energy-transfer unit components:

- The first component concerns adding thermal energy from combustion to flue gases. This unit operation is assessed by the energy-devaluation number noted N_h, introduced in Equation (30).
- The second component concerns transferring heat from flue gases to the pot. This unit is assessed by exergy destruction number N^E in Equation (31).

Then, the energy utilization for the entire process is assessed by the overall exergy destruction number $\hat{N}^E$, see Equation (32). Table 8 presents a schematic of these alternative energy assessment parameters applied to the stove domain and calculation methods for the two consecutive energy-transfer components.

Table 7. Mass-Flow Rate, Temperature and Their Effects on Heat Transfer—4 in Elbow with pot at different operating firepower.

FP (kW)	$\dot{m}_A$ (kg·s^{-1})	T_g (at Pot Bottom) (K)	[1] T_{exit} (K)	V (m·s^{-1})	Re	Nu	h (kW·m^{-2}·K^{-1})	$\dot{Q}_{c\text{-}p}$ (kW)
1.5	3.08 × 10^{-3}	619	540	0.72	1352	17.2	8.3 × 10^{-3}	0.245
2	3.00	705	598	0.78	1176	16.2	8.4	0.322
2.5	2.87	792	651	0.83	1027	15.0	8.6	0.405
3	2.67	879	698	0.88	911	14.2	8.6	0.482
3.5	2.45	965	744	0.85	752	12.9	8.4	0.542
4	1.51	1052	722	0.57	438	9.8	8.8	0.499

[1] Adapted from Zube [15].

Table 8. Energy-devaluation number N_h, Exergy destruction number N^E, Overall exergy destruction number $\widehat{N}^E$ and Entropy indirect calculation methods for the two consecutive energy-transfer components of the cooking stove sketched in Figure 8.

	Energy Transfer Component	$\dot{S}_{\text{gen},i}$	N_i
N_h energy devaluation number	adding thermal energy to flue gases	$\dot{m}_A \cdot (s(T_{\text{exit}}) - s(T_0))$	$\frac{T_0 \cdot \dot{m}_A \cdot (s(T_{\text{exit}}) - s(T_0))}{\dot{m}_A \cdot (h(T_{\text{exit}}) - h(T_0))}$
N^E exergy destruction number	transferring heat from flue to pot (conv. heat transf)	$\dot{m}_A \cdot (s(T_{\text{exit}}) - s(T_g))$	$\frac{T_0 \cdot \dot{m}_A \cdot (s(T_{\text{exit}}) - s(T_g))}{\dot{E}}$
$\widehat{N}^E$ **Overall exergy loss number**	**transferring exergy to the process**	$\dot{m}_A \cdot (s(T_{\text{exit}}) - s(T_g))$	$\frac{T_0 \cdot \dot{m}_A \cdot (s(T_{\text{exit}}) - s(T_g))}{\eta_{c_2} \cdot \dot{Q}_{c-p}}$

Figure 8 represents a value diagram sketching the degradation of energy consecutive to heat-transfer processes through the stove.

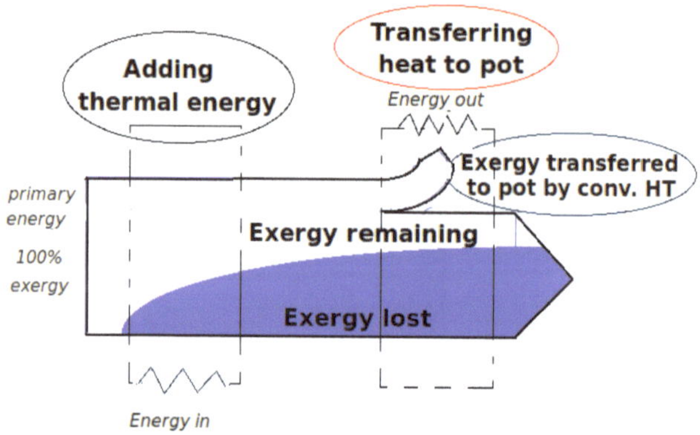

Figure 8. The value diagram of exergy destruction (loss) in a natural convection-driven woodburning stove operating with a pot.

Let us now show in practice, and step by step, the way to determine the alternative assessment numbers related to the devaluation chain of the two consecutive unit-transfer operations in the 4 in Elbow rocket stove of [15] taken as reference.

By writing an energy balance in this system, Table 9 presents, respectively, the rate of sensible enthalpy increase in the flue gases and the rate of exergy loss due to heat transfer. Then their difference gives the starting rate of exergy flow accompanied (in the last column) by the Carnot factor η_{c_1} that assesses this flow work potential. To look closely, values of exergy flow rate in Table 9 are near to the rate of thermal energy transfer to pot $\dot{Q}_{\text{c-p}}$ proposed by [15] and included in Table 7.

It is worth noting that for a stove with a pot in place, difficulties may arise in terms of assuming a priori a proper temperature profile as the flame interacts with the pot surface. Therefore the exergetic Carnot factor of the heat transfer to the flue gases can be accessed indirectly:

$$\eta_{c_1} = 1 - N_h = 1 - \frac{T_0 \cdot \dot{m}_A \cdot (s(T_{\text{exit}}) - s(T_0))}{\dot{m}_A \cdot (h(T_{\text{exit}}) - h(T_0))} = 1 - \frac{T_0 \cdot (s(T_{\text{exit}}) - s(T_0))}{(h(T_{\text{exit}}) - h(T_0))} \tag{68}$$

The Carnot factor η_{c_1} is subjected to devaluation dictated by the overall exergy loss number $\widehat{N}^E$, a concept largely developed in references like [39,44,48]. To evaluate $\widehat{N}^E$, one needs to determine first the second exergetic Carnot factor η_{c_2} for the subsequent convective heat transfer to pot. Thus, according to Equation (59), η_{c_2} is given :

$$\eta_{c_2} = \frac{T_0}{T_p}\left(1 - \frac{T_p}{T_g}\right) \tag{69}$$

So following Equation (32), the overall exergy loss number $\widehat{N}^E$ reads:

$$\widehat{N}^E = \frac{T_0 \cdot \dot{m}_A \cdot (s(T_{exit}) - s(T_0))}{\eta_{c_2} \cdot \dot{Q}_{c\text{-}p}} \tag{70}$$

This energy-transfer quality assessment number $\widehat{N}^E$ can be interpreted as the ratio of the rate by which exergy in the flowing fluid is lost to the rate by which exergy is transferred from a convective heat-transfer process to the pot. Table 10 presents the resulting overall exergy loss number $\widehat{N}^E$ for the 4 in Elbow woodburning cooking stove operating with pot.

Table 9. Sensible enthalpy increase, loss of exergy due to heat transfer to flue gases, exergy flow due to convective heat transfer to pot for a 4 in Elbow with pot at different operating firepowers.

Firepower (kW)	T_{exit} (K)	Sensible Enthalpy Gained by Flue Gases $\dot{m}_A \cdot (h(T_{exit}) - h(T_0))$ (kW)	Loss of Exergy Due to Heat Transfer to Flue Gases $T_0 \cdot \dot{m}_A \cdot (s(T_{exit}) - s(T_0))$ (kW)	Exergy Flow Consecutive to Heat Addition $\dot{X}_{heat}$ (kW)	η_{c_1}
1.5	540	0.780	0.530	0.250	0.32
2	598	0.944	0.606	0.338	0.36
2.5	651	1.065	0.650	0.415	0.39
3	698	1.127	0.660	0.467	0.42
3.5	744	1.154	0.650	0.504	0.44
4	722	0.675	0.387	0.288	0.43

Table 10. Exit gas temperature, Exergy losses along the bottom surface of the pot to the exit after impinging, Exergy transferred to pot and Overall exergy destruction number $\widehat{N}^E$ in a 4 in Elbow with pot at different operating firepower levels.

Firepower (kW)	T_{exit} (K)	$T_0 \cdot \dot{m}_A \times (s(T_{exit}) - s(T_g))$ (kW)	$\eta_{c_2} \cdot \dot{Q}_{c\text{-}p}$ (kW)	$\widehat{N}^E$
1.5	540	0.123	0.117	1.05
2	598	0.144	0.156	0.92
2.5	651	0.163	0.205	0.80
3	698	0.177	0.251	0.71
3.5	744	0.185	0.286	0.65
4	722	0.165	0.266	0.62

Finally, the *global devaluated exergy factor* from the two consecutive unit operations is by definition:

$$\eta_E = \eta_{c_1} \cdot \widehat{N}^E \tag{71}$$

Table 11 summarizes numerical results of all qualitative assessment numbers for the two energy-transfer components of the 4 in Elbow cooking stove in application.

Table 11. Exergetic Carnot factors and Energy-devaluation numbers in a 4 in Elbow with pot at different operating firepowers and exit-gas temperatures.

Firepower (kW)	T_{exit} (K)	N_h	η_{c_1}	η_{c_2}	$\widehat{N}^E$	η_E
1.5	540	0.68	0.32	0.48	1.05	0.34
2	598	0.64	0.36	0.49	0.92	0.33
2.5	651	0.61	0.39	0.51	0.80	0.31
3	698	0.58	0.42	0.52	0.71	0.30
3.5	744	0.56	0.44	0.53	0.65	0.29
4	722	0.57	0.43	0.53	0.62	0.27

Figure 9 depicts values of the devaluated Carnot factor η_E compared to the loss coefficient of the 4 in Elbow with pot. Note that raw data adopted for the adjustment of loss coefficient in Zube [15] have been collected from the experimental works of Agenbroad et al. [21–23].

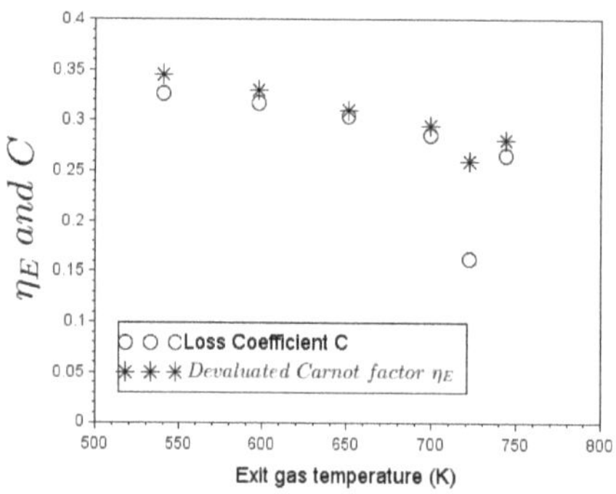

Figure 9. Values of devaluated exergy Carnot factor parameter compared to loss coefficient.

A satisfactory agreement emerges between the devaluated exergy Carnot factor η_E and the loss coefficient C at different exiting gas temperatures in Figure 9. The same decreasing trend of the loss coefficient when firepower increases in a woodburning stove equipped with a pot can be observed referring to experimental raw data in [23,46]. The form of the overall exergy destruction number $\widehat{N}^E$ in Equation (70) becomes very instructive in terms of explaining the decline of the overall exergy Carnot factor or the stove loss coefficient.

To take values in Table 11 as an example, the overall exergy destruction number $\widehat{N}^E$ for the first test is roughly equal to unity. This means that the rate by which exergy is being transferred to the pot is nearly the same as the rate by which exergy would be lost within the flowing fluid. However, for the rest of the tests, as firepower increases, the drop of $\widehat{N}^E$ means that the *entropic potential* of convective heat transfer becomes much higher than that generated in the working flow; therefore, the availability to set fluid in motion decreases. This concept of entropic potential in energy-transfer operations is largely developed in [48].

Futhermore, it appears that exergy destruction effects become important when convective heat-transfer potential increases. The devaluated exergy factor for the overall process η_E in Equation (71) is indeed a product of two competing terms: η_{c_1} and $\widehat{N}^E$. When flue gas temperature increases, the Carnot factor η_{c_1} increases as well, but in contrast the overall exergy destruction number $\widehat{N}^E$ decreases.

5. Conclusions

The second law of thermodynamics analysis was performed to assess the loss coefficient in buoyantly-driven biomass cooking stoves. Accordingly, a simplified mathematical model of the entropy-generation rate in the flow field was developed. To validate the model, experiments were conducted first on a G3300 woodburning cookstove operating without pot to better isolate physical processes governing the basic behaviour of the stove. For the practical case of a stove operating with the cooking pot in place, data from published literature have served for validation. In particular, mass-flow rate and flue gas temperature at different firepower levels have been monitored.

For the parameters under study, it turned out that the entropy generation due to fluid friction is negligible compared to the global dissipation process. Therefore, heat-transfer processes are revealed to be the main source of irreversibilities in the flowing fluid. Energy flux applied as flow work in cooking stove is pure exergy which is lost in consecutive dissipative processes. Furthermore, analysis shows:

- In the stove without pot: Experimental values of the stove loss coefficient at different exhaust-gas temperatures coincide with the heat Carnot factor. Thus, the energy transfer in the cookstove becomes thermodynamically assimilable to a reversible engine that releases its work output into buoyant flow work. This thesis leads to a novel definition of the loss coefficient as a measure of exergy flow.
- In the stove with a cooking piece (pot) in place: As upward hot gases transfer exergy to pot, both the transferred thermal energy and the needed flow work degrade. Alternative heat-transfer parameters such as exergy Carnot factor and energy-devaluation numbers were introduced to account for the destruction of exergy in the overall process. A clear relationship emerged between devaluated exergy Carnot factor and experimental values of the loss coefficient at different flue gas temperatures.

The second law analysis somewhat changes the paradigm in stove engineering by bringing quite a different perspective to the traditional concept of the so-called *loss coefficient*. From now, this *flow loss coefficient* can rather be regarded as the *availability* of internal energy to generate (buoyant) flow work through the stove. Therefore, the magnitude of this *reversible work* depends upon operating conditions and consecutive energy-transfer processes undergone following the stove operating at high or low firepower levels. Minimizing entropy generation with a view to optimizing energy-transfer processes in biomass cooking stoves remains a potential application for future works.

Author Contributions: Conceptualization, L.M.A.; methodology, L.M.A. and M.K.H.; software, L.M.A.; validation, L.M.A., S.T.V., M.K.H. and A.S.; formal analysis, A.S., S.T.V. and M.K.H.; investigation, L.M.A, M.K.H., A.S. and N.N.B.; resources, N.N.B., M.K.H. and A.S.; data curation, M.K.H.; writing—original draft preparation, L.M.A.; writing—review and editing,A.S., L.M.A. and M.K.H.; visualization, A.S., S.T.V. and N.N.B.; supervision, S.T.V., M.K.H., A.S. and N.N.B.; project administration, M.K.H. and A.S.; funding acquisition, A.S. All authors have read and agreed to the published version of the manuscript.

Funding: This research was funded by the German Research Foundation DFG, through the project SA 836/15-1. The APC was funded by the Open Access Publishing Fund of the Technical University of Darmstadt.

Data Availability Statement: Not applicable.

Acknowledgments: All our gratitude to the Forest Investment Program Coordination Unit (UC-PIF) of the Democratic Republic of the Congo, Ministry of Environment and Sustainable Development (MEDD), for lab testing equipments funding. The corresponding author gratefully acknowledges fruitful discussions with Sadiki's research group within the CRC/Transregio 150 "Turbulent, chemically reactive, multi-phase flows near wall" program founded by DFG. The authors gratefully acknowledge the financial support by the Open Access Publishing Fund of Technische Universität Darmstadt.

Conflicts of Interest: The authors declare no conflict of interest.

Abbreviations

The following abbreviations are used in this manuscript:

MDPI	Multidisciplinary Digital Publishing Institute
DOAJ	Directory of open access journals
TLA	Three letter acronym
LD	Linear dichroism

References

1. WHO (World Health Organisation). WHO Guidelines for Indoor Air Quality: Household Fuel Combustion. Available online: https://www.who.int/publications/i/item/9789241548885 (accessed on 23 May 2021).
2. WHO (World Health Organisation). Burning opportunities. Available online: https://www.afro.who.int/publications/burning-opportunity-clean-household-energy-health-sustainable-development-and (accessed on 25 May 2019).
3. *ISO 19867-1:2018*; Clean Cookstoves and Clean Cooking Solutions—Harmonized Laboratory Test Protocols—Part 1: Standard Test Sequence for Emissions and Performance, Safety and Durability. ISO: Geneva, Switzerland, 2018.
4. Clark, M.L.; Peel, J.L.; Balakrishnan, K.; Breysse, P.N.; Chillrud, S.N.; Naeher, L.P.; Rodes, C.E.; Vette, A.F.; Balbus, J.M. Health and household air pollution from solid fuel use: The need for improved exposure assessment. *Environ. Health Perspect.* **2013**, *121*, 1120–1128. [CrossRef] [PubMed]
5. Bilsback, K.R.; Dahlke, J.; Fedak, K.M.; Good, N.; Hecobian, A.; Herckes, P.; L'Orange, C.; Mehaffy, J.; Sullivan, A.; Tryner, J.; et al. A Laboratory Assessment of 120 Air Pollutant Emissions from Biomass and Fossil Fuel Cookstoves. *Environ. Sci. Technol.* **2019**, *53*, 7114–7125. [CrossRef]
6. Bond, T; Venkataraman, C.; Masera, O. Global atmospheric impacts of residential fuels. *Energy Sustain. Dev.* **2004**, *8*, 20–32. [CrossRef]
7. Amegah, A.K.; Jaakkola, J.K. Household air pollution and the sustainable development goals. *Bull. World Health Organ.* **2016**, *94*, 215–221. [CrossRef] [PubMed]
8. CCA (Clean Cooking Alliance). The Value of Clean Cooking. Available online: https://cleancooking.org/the-value-of-clean-cooking/ (accessed on 14 May 2022).
9. MacCarthy, N.A. A Zonal Model to Aid in the Design of Household Biomass Cookstoves. Master's Thesis, Iowa State University, Ames, IA, USA, 2013.
10. Sedighi, M.; Salarian, H. A comprehensive review of technical aspects of biomass cookstoves. Renewable and Sustainable. *Renew. Sustain. Energy Rev.* **2017**, *70*, 656–665. [CrossRef]
11. Kshirsagar, M.P. Kalamkar, V.R. A mathematical tool for predicting thermal performance of natural draft biomass cookstoves and identification of a new operational parameter. *Energy* **2015**, *93*, 188–201. [CrossRef]
12. Deshmukh, G.K; Gupta, R. Numerical Simulation of Wood-Volatiles and Air combustion in Differentially Heated Diffuser Tube under Free Convection. *Int. J. Eng. Trends Technol.* **2013**, *4*, 7–8.
13. Mahmoud, K.; Dimitris, A. Wood stove combustion modeling and simulation: Technical review and recommendations. *Int. Commun. Heat Mass Transf.* **2021**, *127*, 105423.
14. Brewster, J.J. *CFD Design of Improved Cookstoves*; University of Strathclyde: Glasgow, UK, 2008.
15. Zube, D.J. Heat Transfer Efficiency of Biomass Cookstoves. Master's Thesis, Colorado State University, Fort Collins, CO, USA, 2010.
16. Baldwin, S. *Biomass Stoves: Engineering Design Development and Dissemination*, 1st ed.; Vita Publications: Arlington, TX, USA, 1988.
17. De Lepeleire, G.; Christiaens, M. Heat transfer and cooking woodstove modelling. *Indian Acad. Sci.* **1983**, *6*, 35–45. [CrossRef]
18. Bryden, M.; Still, D.; Scott, P.; Hoffa, G.; Ogle, D.; Bailis, R.; Goyer, K. *Design Principles for Wood Burning Cook Stoves*; Aprovecho Research Center: Cottage Grove, OR, USA; Shell Foundation: London, UK; U.S. Environmental Protection Agency: Washington, DC, USA, 2006.
19. Kshirsagar, M.P.; Kalamkar, V.R. User-centric approach for the design and sizing of natural convection biomass cookstoves for lower emissions. *Energy* **2016**, *115*, 1202–1215. [CrossRef]
20. Bilsback, K.R.; Eilenberg, S.R.; Good, N.; Heck, L.; Johnson, M.; Kodros, J.K.; Lipsky, E.M.; L'Orange, C.; Pierce, J.R.; Robinson, A.L.; et al. The Firepower Sweep Test: A novel approach to cookstove laboratory testing. *Indoor Air* **2018**, *28*, 936–949. [CrossRef] [PubMed]
21. Agenbroad, J. A Simplified Model for Understanding Natural Driven Flow in Biomass Cooking Stoves. Master's Thesis, Colorado State University, Fort Collins, CO, USA, 2010.
22. Agenbroad, J.; DeFoort, M.; Kirkpatrick, A.; Kreutzer, C. A simplified model for understanding natural convection driven biomass cooking stoves—Part 1: Setup and baseline validation. *Energy Sustain. Dev. Elsevier* **2011**, *15*, 160–168. [CrossRef]
23. Agenbroad, J.; DeFoort, M.; Kirkpatrick, A.; Kreutzer, C. A simplified model for understanding natural convection driven biomass cooking stoves—Part 2: With cook piece operation and the dimensionless form. *Energy Sustain. Dev.* **2011**, *15*, 169–175. [CrossRef]
24. Kshirsagar, M.P.; Kalamkar, V.R. A comprehensive review on biomass cookstoves and a systematic approach for modern cookstove design. *Energy* **2014**, *30*, 580–603. [CrossRef]
25. Parajuli, A.; Agrawal, S.; Tharu, J.K.; Kamat, A.K.; Jha, A.K.; DarlamiBilsback, H.B. A simplified model for understanding the performance of two-pot enclosed mud cookstoves. *Clean Energy* **2019**, *3*, 288–306. [CrossRef]
26. Prapas, J.; Baumgardner, M.E.; Marchese, A.J.; Willson, B.; DeFoort, M. Influence of chimneys on combustion characteristics of buoyantly driven biomass stoves. *Energy Sustain. Dev.* **2014**, *23*, 286–293. [CrossRef]
27. Oliveski, R.D.C.; Macagnan, M.H.; Coppeti, J.B. Entropy generation and natural convection in rectangular Cavities. *Appl. Therm. Eng.* **2009**, *29*, 1417–1425. [CrossRef]
28. Agrebi, S.; Dreßler, L.; Nicolai, H.; Ries, F.; Nishad, K.; Sadiki, A. Analysis of Local Exergy Losses in Combustion Systems Using a Hybrid Filtered Eulerian Stochastic Field Coupled with Detailed Chemistry Tabulation: Cases of Flames D and E. *Energies* **2021**, *14*, 6315. [CrossRef]

29. Dressler, L.; Nicolai, H.; Agrebi, S.; Ries, F.; Sadiki, A. Computation of Entropy Production in Stratified Flames Based on Chemistry Tabulation and an Eulerian Transported Probability Density Function Approach. *Entropy* **2022**, *24*, 615. [CrossRef]
30. Cengel, Y.; Boles, M. *Thermodynamics: An Engineering Approach*, 9th ed.; McGraw-Hill Education: New York, NY, USA, 2015; pp. 329–484, ISBN 978-0-07-339817-4.
31. Bejan, A. *Advanced Engineering Thermodynamics*, 4th ed.; John Wiley and Sons Inc.: Hoboken, NJ, USA, 2016.
32. Bejan, A. *Entropy Generation Minimization—The Method of Thermodynamic Optimization of Finite-Size Systems and Finite-Time Operations*, 1st ed.; CRC Press LLC: New York, NY, USA, 1996.
33. Keenan, J.H. *Thermodynamics*; Wiley: New York, NY, USA, 1941.
34. Ries, F.; Li, Y.; Nishad, K.; Janicka, J.; Sadiki, A. Entropy Generation Analysis and Thermodynamic Optimization of Jet Impingement Cooling Using Large Eddy Simulation. *Entropy* **2019**, *21*, 129. [CrossRef]
35. Sciacovelli, A.; Verda, V.; Sciubba, E. Entropy generation analysis as a design tool—A review. *Renew. Sustain. Energy Rev.* **2015**, *43*, 1167–1181. [CrossRef]
36. Li, Y.; Rico Cortes, L.F.; Hamel, H.; Nishad, K.; Biondo, L.; Ries, F. Prediction of Heat Transfer and Fluid Flow Effects on Entropy Generation in a Monolithic Catalytic Converter Using Large-Eddy Simulation. *Entropy* **2022**, *24*, 602. [CrossRef] [PubMed]
37. Gebhart, B. Effects of viscous dissipation in natural convection. *J. Fluid Mech.* **1962**, *14*, 225–233. [CrossRef]
38. Biswal P.; Basak, T. Entropy generation vs energy efficiency for natural convection based energy flow in enclosures and various applications: A review. *Renew. Sustain. Energy Rev. Elsevier* **2017**, *80*, 1412–1457. [CrossRef]
39. Herwig, H.; Redecker, C. Heat transfer and Entropy: Peer reviewed, Chapter 6. In *Heat Transfer Studies and Applications*; Salim, N.K., Ed.; IntechOpen: London, UK, 2015; pp. 150–169. Available online: https://www.intechopen.com/chapters/48786 (accessed on 23 March 2021).
40. Herwig, H. The Role of Entropy Generation in Momentum and Heat Transfer. *J. Heat Transf.* **2012**, *134*, 031003. [CrossRef]
41. Herwig, H.; Schmandt, B. How to Determine Losses in a Flow Field: A Paradigm Shift. *Entropy* **2014**, *16*, 2959–2989. [CrossRef]
42. Herwig, H. What exactly is the Nusselt Number in Convective Heat Transfer Problems and are There Alternatives? *Entropy* **2016**, *18*, 198. [CrossRef]
43. Schmandt, B.; Herwig, H. Internal Flow Losses: A Fresh Look at Old Concepts towards the Second Law Analysis. *J. Fluids Eng.* **2011**, *133*, 051201. [CrossRef]
44. Schmandt, B.; Herwig, H. A Standard Method to Determine Loss Coefficients of Conduit Components Based on the Second Law of Thermodynamics. In Proceedings of the ASME 2012 10th International Conference on Nanochannels, Microchannels, and Minichannels ICNMM2012, Rio Grande, Puerto Rico, 8–12 July 2012.
45. Datta, P. Entropy Generation in a Confined Laminar Diffusion Flame. *Combust. Sci. Technol.* **2000**, *159*, 39–56. [CrossRef]
46. Pande, R.R.; Kshirsagar, M.P.; Kalamkar, V.R. Experimental and CFD analysis to study the effect of inlet area ratio in a natural draft biomass cookstove. *Environ. Dev. Sustain.* **2018**, *22*, 1897–1911. [CrossRef]
47. Poinsot, T.; Veynante, D. *Theoretical and Numerical Combustion*, 3rd ed.; T. Poinsot: Paris, France, 2016.
48. Wenterodt, T.; Herwig, H. The Entropic Potential Concept: A New Way to Look at Energy Transfer Operations. *Entropy* **2014**, *16*, 2071–2084. [CrossRef]
49. Bejan, A. General criterion for rating heat-exchanger performance. *Int. J. Heat Mass Transf.* **1978**, *21*, 655–658. [CrossRef]
50. Lienhard, J., IV; Lienhard, V.J. *A Heat Transfer Textbook*, 5th ed.; Dover Publications: Mineola, NY, USA, 2019; pp. 154–196.
51. Dischino, K. Methods for Particulate Matter Emissions Reduction in Wood Burning Cookstoves. Master's Thesis, Colorado State University, Fort Collins, CO, USA, 2015.
52. Onuegbu, V.I.; Alabi, I. Characterisation of Sawdust from Entandrophragma Cylindricum (Sapele), Diospyruscrassiflora (Ebony), Triplochitonscleroxylon (Obeche) and Chlorophor a Excelsa (Iroko) Woods. *Int. J. Sci. Eng. Res.* **2018**, *9*, 1569–1576.
53. Günther, B.; Gebauer, K.; Barkowski, R.; Rosenthal, M.; Bues, C.-T. Calorific value of selected wood species and wood products. *Eur. J. Wood Wood Prod.* **2012**, *70*, 755–757. [CrossRef]
54. Sawerysyn, J.P. La combustion du bois et ses impacts sur l'environnement. *Air Pur* **2012**, *81*, 7–16.
55. Wark, K. *Thermodynamics*, 4th ed.; McGraw-Hill: New York, NY, USA, 1983.

energies

MDPI

Article

Shannon Entropy in Stochastic Analysis of Some Mems

Marcin Kamiński [1,*] and Alberto Corigliano [2]

[1] Department of Structural Mechanics, Łódź University of Technology, 90-924 Łódź, Poland
[2] Department of Civil and Environmental Engineering, Politecnico di Milano, 20133 Milano, Italy; alberto.corigliano@polimi.it
* Correspondence: marcin.kaminski@p.lodz.pl

Abstract: This work is focused on the numerical determination of Shannon probabilistic entropy for MEMS devices exhibiting some uncertainty in their structural response. This entropy is a universal measure of statistical or stochastic disorder in static deformation or dynamic vibrations of engineering systems and is available for both continuous and discrete distributions functions of structural parameters. An interval algorithm using Monte Carlo simulation and polynomial structural response recovery has been implemented to demonstrate an uncertainty propagation of the forced vibrations in some small MEMS devices. A computational example includes stochastic nonlinear vibrations described by the Duffing equation calibrated for some micro-resonators, whose damping is adopted as a Gaussian, uniformly and triangularly distributed input uncertainty source.

Keywords: Shannon entropy; MEMS vibrations; Duffing equation; random damping

1. Introduction

It is well known that the probabilistic entropy of the given random variable or process represents an averaged level of uncertainty in its numerical or experimental realizations. It is also a mathematical extension of its thermodynamic origin, invented by Boltzmann, and now it may serve as a universal measure of disorder in the given engineering system. The first method of its calculation was proposed by Claude Shannon [1,2], although the term "entropy" was introduced by Johann von Neumann; this has been used multiple times to analyse disorder in various engineering systems [3]. The original definition by Shannon includes a summation of probabilities of various admissible states of the given experiment multiplied with their logarithms; such an entropy was determined from the very beginning using various estimation methods [4]. Various further extensions and modifications have been worked out independently and presented by Renyi [5,6], Tsallis [7–10], and Kolmogorov and Sinai [11,12]. They are available now for both discrete and continuous probability distributions of the stochastic response for a given engineering system, and are also applicable in some non-engineering studies [13].

Despite the calculation method, maximum entropy is equivalent to minimum knowledge about some phenomenon and vice versa. The majority of research related to probabilistic engineering mechanics recalls most frequently the maximum entropy principle [14,15], which states that the probability distribution giving the best representation of the current knowledge state about a given engineering system is the one having the largest probabilistic entropy.

The main numerical difficulty in probabilistic entropy calculation is the necessity of the final probability density function availability. This makes many of the stochastic computer techniques like Bayesian methods [16], Karhunen-Loeve, polynomial chaos, or stochastic perturbation technique [17] simply inapplicable; this is the reason why the traditional Monte Carlo simulation scheme [18,19] may be preferred in this context. Some studies in the area of applications of probabilistic structural dynamics have been documented in the literature [20]. An interesting alternative could be the determination of probabilistic distance (divergence) in between two probability densities, which can represent the general input signal and the output response. It can be useful in civil engineering to model limit functions, being an inherent part of reliability assessment. Such a probabilistic distance may link extreme structural effort of the given nature with its corresponding capacity (in the context of deformations and/or stresses), so it could be an alternative measure of the structural safety or may be employed for sensitivity in engineering design [21]. A huge variety of relative entropy models is available now in the literature: it is necessary to recall the Kullback–Leibler approach [22] (exhibiting a well-known lack of symmetry), Jensen–Shannon, Hellinger [23], Chernoff [24], Mahalanobis [25], or Bhattacharyya [26] probabilistic differences.

The purpose of this study is the determination of probabilistic entropy fluctuations in some micro-resonator-forced vibration problems with uncertainty described by the Duffing equation; its understanding seems to be necessary for further, more general, numerical analyses of the disorder and uncertainty propagation in nonlinear vibrations of various MEMS devices [27–29]. Let us note that entropy propagation in stochastic structural dynamics has been studied before for some elastic beams with cross-sectional uncertainty [30]. The need for randomness analysis in the case of these devices is motivated by a lot of different reasons: (i) material and geometrical imperfections, (ii) environmental actions, measurements, and vibrations (as well as their sources), (iii) statistical scattering in the shape, response and material characteristics of their structural components, (iv) coupling in-between different physical fields, phenomena and properties (thermo-electro-magneto-mechanical) and their time delay in various exploitation conditions [31]. The damping coefficient of these micro-devices is chosen as the input uncertainty source because this parameter decisively affects its dynamic response, and its probability distribution is modelled using alternatively uniform, triangular as well as Gaussian distributions; all these distributions are based on the same upper and lower bounds for the uncertainty interval. The deterministic vibrations spectrum is obtained numerically using the Runge–Kutta–Fehlberg algorithm [32] implemented in the computer algebra system MAPLE, and this is the basis for the polynomial Least Squares Method [33] approximation of the micro-resonator displacements concerning the damping coefficient. Further parts programmed in the same system include Monte Carlo sampling (MCS) of the damping parameter and the same sampling of the dynamic response and final Shannon entropy calculation using a formula adjacent to the interval representation of the statistical data. Due to the highly nonlinear character of the physical phenomenon and of the corresponding dynamic equilibrium equation, numerical simulation is the only uncertainty analysis method; analytical formulas derived and widely used for linear stochastic systems cannot be simply used in this case. The approach proposed in this work may be used for more advanced MEMS analyses using commercial Finite Element Method codes like COMSOL or ABAQUS, for computer simulation of inelastic deformation in solids, for disorder propagation analysis in various computational fluid mechanics problems.

2. Governing Equations of the Problem

Problem 1. *Determine Shannon entropy $H(y(c;t))$ for a dynamical system having a single degree of freedom and governed by the following differential equation:*

$$m\ddot{y}(t) + c\dot{y}(t) + k_1 y(t) + k_2 y^2(t) + k_3 y^3(t) = F\sin(\omega t) \tag{1}$$

where m denotes the mass of the vibrating structure, c stands, obviously, for the damping coefficient, discrete values k_1, k_2 and k_3 are first, second and third-order stiffness coefficients related to various physical fields and sources, F and ω are the amplitude and frequency of the modulation signal. External forcing has been chosen as perfectly periodic at the initial stage to verify how stationary signals affect the structural response of the MEMS. It also enables for some verification of the output signals and is frequently chosen in various theoretical and numerical studies. Initial *equation of motion is solved with traditional initial conditions equivalent to static equilibrium in the undeformed configuration:*

$$y(t=0) = 0,\ \dot{y}(t=0) = 0, \tag{2}$$

Shannon entropy is to be determined from its classical definition at any discrete-time moment τ. It yields [3,34]:

$$H(y(c,\tau)) = -\sum_{i=1}^{n} p_i(y(c,\tau))\ \log(p_i(y(c,\tau))), \tag{3}$$

where n denotes the total number of subintervals enabling discretization of the entire probability domain of the given structural response. Let us note that this formula could be generalized towards the continuous PDFs as

$$\hat{H}(y(c,\tau)) = -\int_{-\infty}^{+\infty} p_y(x)\ \log(p_y(x))\, dx. \tag{4}$$

This formula would be applicable if only any reliable numerical method of determination of the density $p_y(x)$ would be available. The majority of such an approach would be insensitivity to the output PDF partition during post-processing of the Monte Carlo simulation results. Interval analysis for estimation of the Shannon entropy [35,36] is further employed, so that: let us consider a continuous random variable b with its probability density function $g_b(x)$ discretized by a set of subintervals of constant length δ. The mean value theorem leads to the following representation of $p_b(x)$ in each ith subinterval:

$$p_b(x_i) = \frac{1}{\delta}\int_{i\delta}^{(i+1)\delta} p_b(x)\, dx. \tag{5}$$

Let us introduce further the quantized random variable $\hat{x}$ defined as

$$\hat{x} \equiv x_i \ for\ i\delta \leq \hat{x} \leq (i+1)\delta, \tag{6}$$

with probability equal to

$$p_i = \int_{i\delta}^{(i+1)\delta} p_b(x)\, dx = \delta\, p_b(x_i). \tag{7}$$

According to the Shannon definition, probabilistic entropy corresponding to a partition of $p_b(x)$ into n equal subsets equals [37,38]

$$\begin{aligned} H(\hat{x}) &= -\sum_{i=1}^{n} p_i \log(p_i) = -\sum_{i=1}^{n} \delta p_b(x_i) \log(\delta p_b(x_i)) = \\ &= -\sum_{i=1}^{n} \delta p_b(x_i) \log(p_b(x_i)) - \log(\delta). \end{aligned} \tag{8}$$

Finally, calculation of probabilistic moments of the structural response at any discrete-time moment τ undergoes, using classical statistical estimators (as in [39]),

$$E[y(c,\tau)] = \frac{1}{M}\sum_{j=1}^{M} y_j(c,\tau), Var(y(c,\tau)) = \frac{1}{M-1}\sum_{j=1}^{M} \{y_j(c,\tau) - E[y_j(c,\tau)]\}^2. \tag{9}$$

M denotes in these relations the total number of random trials, while the coefficient of variation of the structural response is obtained by a ratio of standard deviation to the corresponding expectation. Further determination of probabilistic entropy for the given MEMS device proceeds directly thanks to the determination and partition of the probability distribution function of its structural response. This in turn proceeds via the iterative solution of the following equation:

$$m\ddot{y}_{(j)}(t) + c_{(j)}\dot{y}_{(j)}(t) + k_1 y_{(j)}(t) + k_2 y_{(j)}^2(t) + k_3 y_{(j)}^3(t) = F\sin(\omega t), \quad (10)$$

Since the damping coefficient is taken as the Gaussian input variable with expectation and standard deviation equal to $E[c]$ and $\sigma(c)$, respectively, its discrete realizations are obtained via an equidistant partition of the interval $[E[c] - 3\sigma(c), E[c] + 3\sigma(c)]$. These discrete solutions substituted to the Least Squares Method fitting procedure enable one to determine the structural response as a polynomial of its damping, and the order of this polynomial is optimized using minimization of the fitting variance and RMS error as well as maximization of the correlation factor of the approximating polynomial to the fitting dataset. Having determined the response polynomial, one may apply the classical MCS strategy to make a histogram of this response, whose further processing results in Shannon entropy value. Such an algorithm enables the application of non-Gaussian parameters or even a couple of various uncertainty sources at the same time.

3. Mems Device Description

The so-called L-shaped micro-resonator is the engineering device under consideration here. It consists of two parallel electrodes (the driving one and the sensing one; see Figure 1). External acceleration includes in this resonator some longitudinal force and it is modelled as a slender beam having a constant cross-section, inertia moment, and Young's modulus. It is axially constrained at both ends and it oscillates in bending mode due to a certain electrostatic actuation. Some relaxation at one end of such a micro-resonator is modelled by introducing an equivalent axial spring. It is well known that L-shaped micro-resonators exhibit a coupled non-linear response, which consists of electrostatic non-linearity resulting from the parallel plates actuation as well as mechanical non-linearity in the axial stretching. The equation of motion of such a resonant beam including both effects may be directly derived from the Hamilton principle [40] and it is relevant to the equivalent single degree of freedom (d.o.f.) dynamic model. Needless to say, such a micro-device is very sensitive to the operation temperature range [40], but this effect has been postponed here due to the brevity of presentation and discussion.

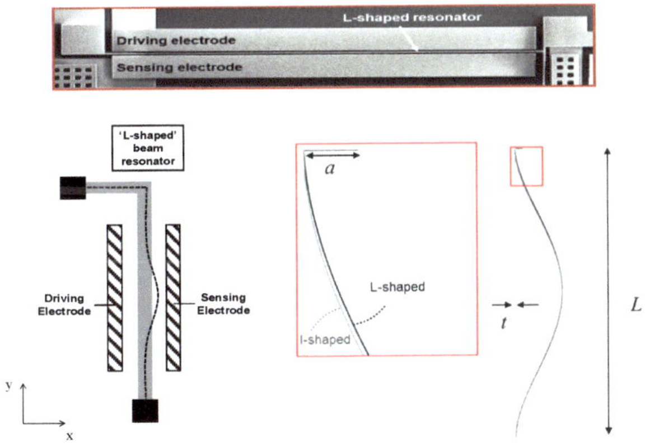

Figure 1. L-shaped micro-resonator.

Numerical analysis reported here has been completed with experimental input data adopted after [41]. Mechanical and electrical contributions k_m and k_e to the overall stiffness of the micro-resonator are represented by the following stiffness coefficients k_i, i = 1, 2, 3, which in turn are taken as

$$k_1 = (k_{m1} - k_{e1}) = (0.829 - 0.068) = 0.761 \text{ N/m}, \quad (11)$$

$$k_2 = 0 \text{ N/m}^2,$$

$$k_3 = (k_{m3L} - k_{e3}) = (1.45 \times 10^{11} - 2.2 \times 10^{10}) = 12.3 \times 10^{10} \text{ N/m}^3.$$

The second order stiffness k_2 simply vanishes for the perfect micro-resonators, where the gaps in between the resonator and driving, as well as sensing electrode are equal to each other. In this specific case, Equation (10) becomes the Duffing equation, whose further numerical solution enables stochastic response analysis. Some other physical phenomena appearing in the micro-resonators together with their mathematical consequences have been reported in detail in [40]. The effective mass of the micro-resonator was calculated in [41] from its length L = 400 μm, width t = 1.2 μm, out of the plane thickness w = 15 μm, the silicon mass density $\rho = 2330 \frac{\text{kg}}{\text{m}^3}$. The value $m = 0.3965 \times M = 6.65 \times 10^{-12}$ kg is obtained, using the formula that gives the equivalent mass m as a fraction of the total beam mass $M = 16.78 \times 10^{-12}$ kg. The mean value of the damping coefficient has been proposed using the formula $c = \frac{1}{Q}\sqrt{km} \quad \left[\frac{\text{Nsec}}{\text{m}}\right]$, where k includes all the stiffnesses introduced in Equation (1); the expected value of damping parameter c has been taken as equal to $E[c] = [0.00394] \times 10^{-6}$ [Nsec/m] and the coefficient of variation of this physical parameter is taken further from the interval $\alpha(c) \in [0.00, 0.20]$. The harmonic external force representing electrostatic actuation is introduced as

$$F\sin(\omega t) = \eta\, v_a(t), \quad (12)$$

where

$$\eta = \alpha\, V_p \frac{\varepsilon_0 w L}{d^2}, \quad \alpha = 0.523, \quad V_p = 2 \div 9 \ [V], \quad \varepsilon_0 = 8.8541878176 \times 10^{-12} \ [F/m].$$

$$w = 15\,\mu m, \quad L = 400\,\mu m, \quad d = 2.1\,\mu m, \quad v_a(t) = v_a \sin(\omega t), \quad v_a = 5 \times 10^{-3} \div 1 \times 10^{-1} \ [V] \quad (13)$$

In the above relations, α denotes the coefficient related to the mechanical behavior of the resonator, V_p is the bias voltage, ε_0 is the absolute vacuum permittivity constant, d means the gap between the oscillating beam and the electrode, $v_a(t)$ is the actuation voltage, usually modulated at the mechanical frequency of the oscillating beam ω. Finally, the external force has the following multiplier: $F = 56.7 \times 10^{-10}$ $[N]$, while a frequency ω has been adopted as 10^3.

4. Numerical Simulation and Discussion

The entire numerical simulation has been programmed and performed in the computer algebra system MAPLE 2019.2. This computational environment has been preselected because it offers a Runge–Kutta–Fehlberg numerical solution of the Duffing equation, the Least Squares Method polynomial approximations for the given discrete datasets, and also random number generators, statistical estimation of the resulting probabilistic characteristics, together with histogram creation. Some additional procedures enabling statistical optimization of approximation order, Shannon entropy estimation have been additionally implemented into a single MAPLE script. The computer system MAPLE offers also interoperability with other computer simulation systems (both academic and commercial), building and usage of the user's own Windows applications, as well as macro-procedures supervising cyclic usage of other mathematical and engineering software.

It is of paramount importance when the Monte Carlo simulation is to be carried out. It should be mentioned that neither the chosen computer system nor other commercial mathematical packages directly enable computations of probabilistic entropy in the context of differential equations solution.

A flowchart of numerical analysis programmed in the system MAPLE is schematically shown in Figure 2 below. The following notation has been adopted: (i) M means the total number of random samples in the Monte Carlo scheme, (ii) N denotes the number of discrete time moments in dynamic analysis, while (iii) P corresponds to the order of polynomial approximation, whose optimal choice is P′.

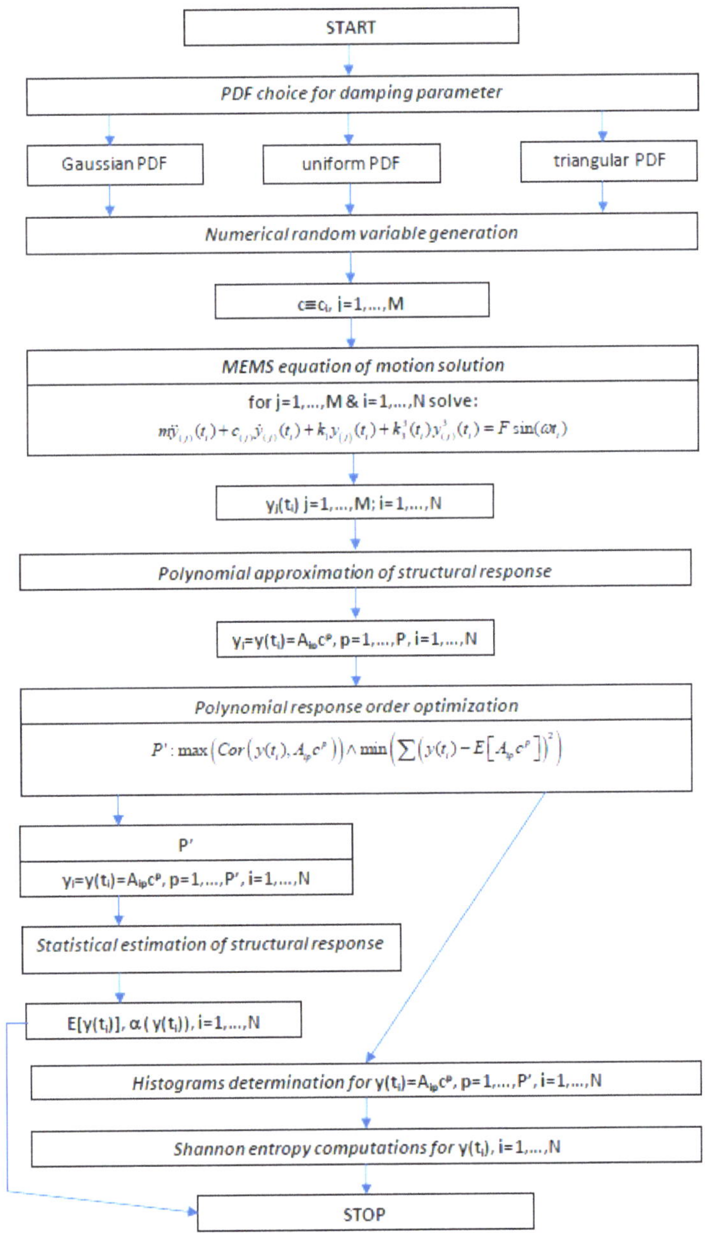

Figure 2. Flowchart of probabilistic entropy numerical simulation.

One of the most important parts in this methodology is numerical recovery of the polynomial response functions and statistical optimization of its order. They are carried out both for several input discrete damping coefficients and the corresponding structural outputs. The main goal is to find the polynomial, whose order guarantees that the polynomial is the closest to the datasets linking the damping and structural response. Such a procedure is completed at each discrete time moment for the given domain of structural vibrations. Various order fitting polynomials are numerically determined using the Least Squares Method approach (consecutively from 2 to 9) and then, the variance as well as the correlation factor are computed for all these orders. Optimal polynomial minimizes the variance (also the mean square error) and maximizes the correlation factor at the same time. It can be concluded from the results contained in Table 1 that the 4th order polynomial basis has been found as the most efficient. It is confirmed with the results shown in this table that optimal 4th order approximation minimizes the squares sum in the LSM procedure, while the differences in this parameter obtained for various orders may have a few orders of magnitude.

Table 1. Statistical optimization results for the polynomial response order.

Polynomial Order	Correlation	RMS Error	Squares Sum	Fitting Variance
2	0.944648	6.50235×10^{-12}	4.65126×10^{-22}	4.65228×10^{-23}
3	0.999390	7.22270×10^{-13}	5.76870×10^{-24}	5.86760×10^{-25}
4	0.999571	6.09214×10^{-13}	4.11380×10^{-24}	2.90649×10^{-24}
5	−0.594698	7.08217×10^{-10}	5.51726×10^{-18}	2.45674×10^{-19}
6	−0.924625	8.02584×10^{-9}	7.08554×10^{-16}	4.98402×10^{-16}
7	0.936636	8.11866×10^{-7}	7.25038×10^{-12}	3.53456×10^{-12}
8	−0.942430	8.61679×10^{-5}	8.16740×10^{-8}	1.25495×10^{-6}
9	−0.937340	6.38454×10^{-4}	4.48386×10^{-6}	6.55977×10^{-7}

Having an optimal polynomial basis for the structural response, we studied the uncertain behavior of the micro-resonator in terms of its first two probabilistic moments computed for the input coefficient of variation of damping equal to $\alpha(c) = 0.10$. Expected values and coefficients of variations have been presented in the left and the right graph of Figure 3 and they are included here for a comparison with further Shannon entropies. It is seen that initial expectations show some periodicity, but the most important conclusion is that the resulting coefficient of variation exhibits an enormously large value at a certain discrete time of the vibration process. The value larger than 1.0 was never observed in classical solid mechanics with uncertainty and follows the coupled electro-mechanical character of this phenomenon. It may be of special importance while analyzing the reliability of such a microdevice, where traditionally second order moments play a decisive role.

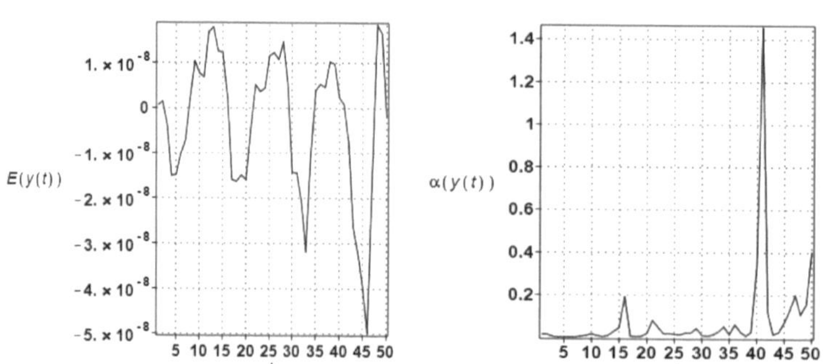

Figure 3. The expected values of the MEMS device vibration (**left**) and its coefficient of variation (**right**).

After determination of the first two moments of the structural response, it is necessary to make an initial numerical study concerning the numerical convergence of Shannon entropy (and its possible sensitivity) while increasing the partitioning density of the structural response histogram. A Gaussian random variable has been preselected for this study, and it has an expected value equal to 10, and also a few different standard deviations taken in turn as $\sigma = [0.05, 0.10, 0.15, 0.20, 0.25, 0.30]$. The results corresponding to the increasing number of subintervals n have been contrasted in Figure 4. It is seen that the entropy under consideration is less sensitive to this partitioning for smaller standard deviations of the given uncertainty source (especially when $\sigma = 0.05$ here). Huge statistical dispersion (close to $\sigma = 0.20$ and higher) needs may cause some numerical discrepancies for a smaller n close to 10 and demands each time determination of the stability region starting usually from $n = 25$.

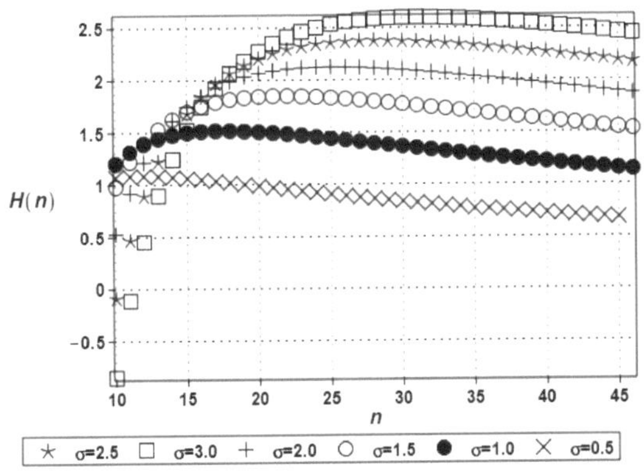

Figure 4. Convergence rate for the interval entropy computations.

A determination of the satisfactory partition of both input and output uncertainty guarantees efficient estimation of probabilistic entropy, which can be evaluated when the coefficient of variation reaches its extreme value. Shannon entropy is expected to be more efficient in uncertainty evaluation while periodic or quasi-periodic vibrations are analyzed, because of the huge increase of the coefficient of variation, especially when the expected value tends to 0 (cf. Figure 3, right diagram). Application of this technique to multi degrees of freedom models for MEMS is possible by replacing Equation (10) with its matrix counterpart and by determination of the optimal polynomial responses in a local manner—for each degree of freedom separately.

Three different probability distributions having the same variability ranges have been tested to model statistical scattering of structural damping in this MEMS system, namely, in turn—Gaussian, uniform, and also the triangular one. The total number of random trials in the MCS algorithm has been set as 200,000 and probabilistic entropy has been computed in any second for a time period $t \in [0.0, 50.0\ \mathrm{s}]$. The range of Gaussian distribution has been restricted following the three-sigma rule. The results of numerical simulation have been collected in Figures 5, 7, and 9 in the form of probabilistic Shannon entropies time variations, and also in the form of their logarithms, cf., Figures 6, 8, and 10; the right additional column has been added to better expose time fluctuations second by a second. These data have been contrasted for four different input coefficients of variation of structural damping α (c) = [0.05, 0.10, 0.15, 0.20] to see an impact of the input statistical scattering of this damping.

Figures 5–10 show clearly that probabilistic entropies (and their logarithms) have the same patterns for all three different probability distributions and that they all reach extreme values at the very end of the simulated dynamic process. A comparison of Figure 3 (right) with Figure 5, Figure 7, and Figure 9 exhibits that the pattern of time fluctuations of the resulting CoVs and these adjacent to probabilistic entropies are almost the same. Generally, the variability ranges of all entropies for three different PDFs seem to be very similar to each other. However, as one could expect after the maximum entropy principle, the largest values of Shannon entropy are obtained when the damping coefficient has Gaussian distribution, then—in turn—for the uniform and triangular PDFs. It is also seen that the resulting probabilistic entropy is proportional to the input coefficient of variation of structural damping, i.e., they increase with each other.

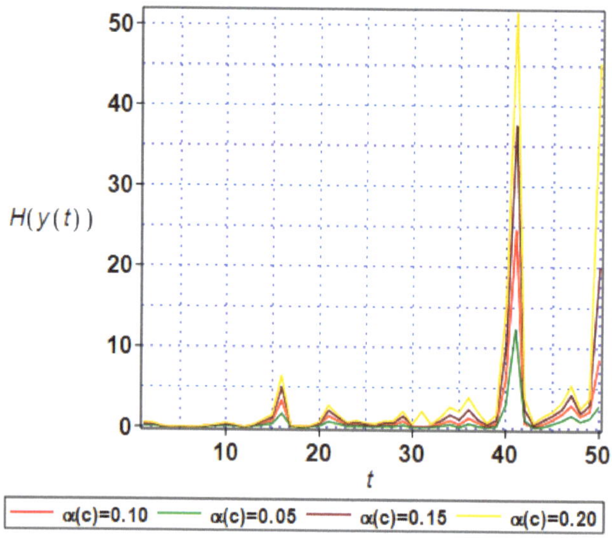

Figure 5. Shannon entropies time fluctuations for Gaussian structural damping.

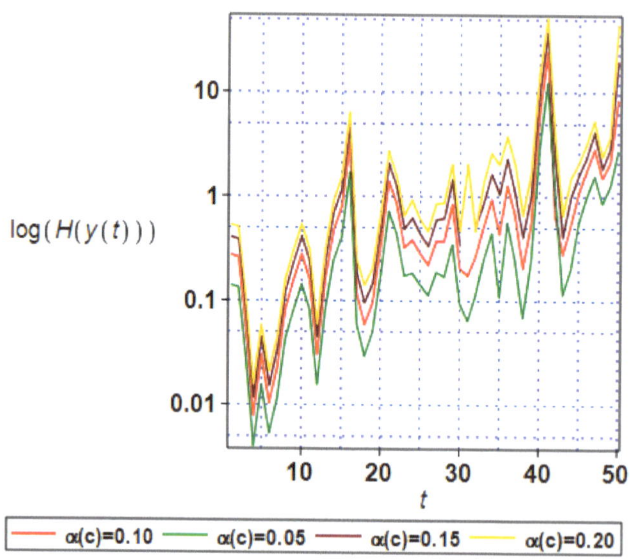

Figure 6. Shannon entropies logarithm time fluctuations for Gaussian structural damping.

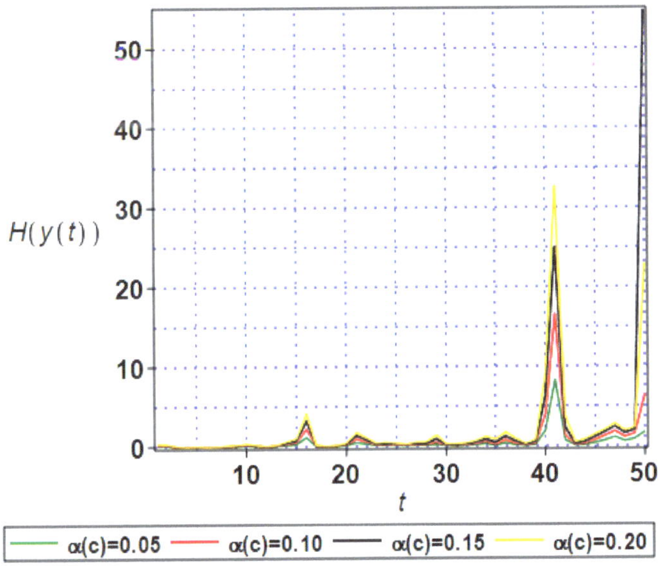

Figure 7. Shannon entropies time fluctuations for uniformly distributed structural damping.

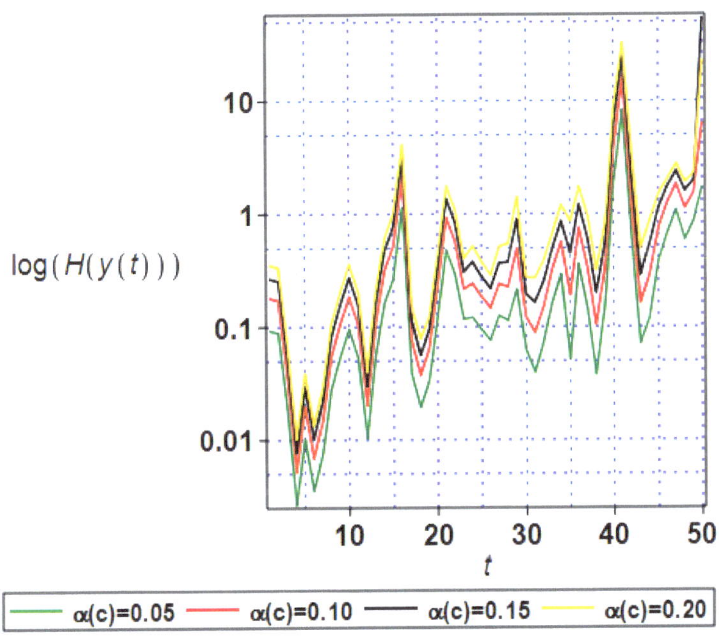

Figure 8. Shannon entropies logarithm time fluctuations for uniformly distributed structural damping.

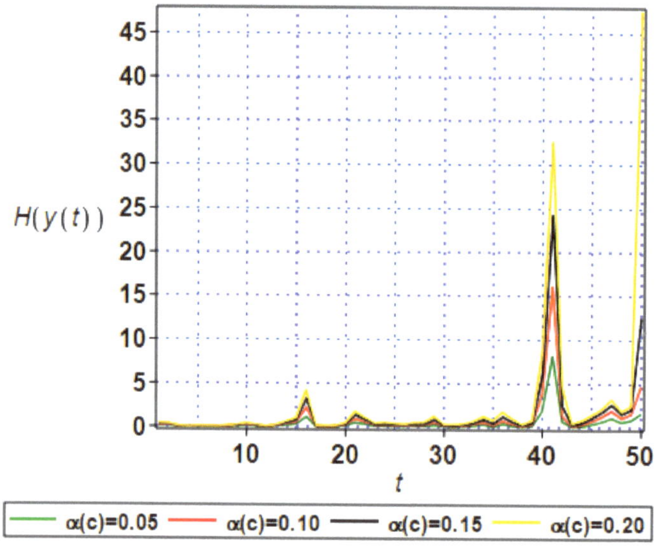

Figure 9. Shannon entropies time fluctuations for triangularly distributed structural damping.

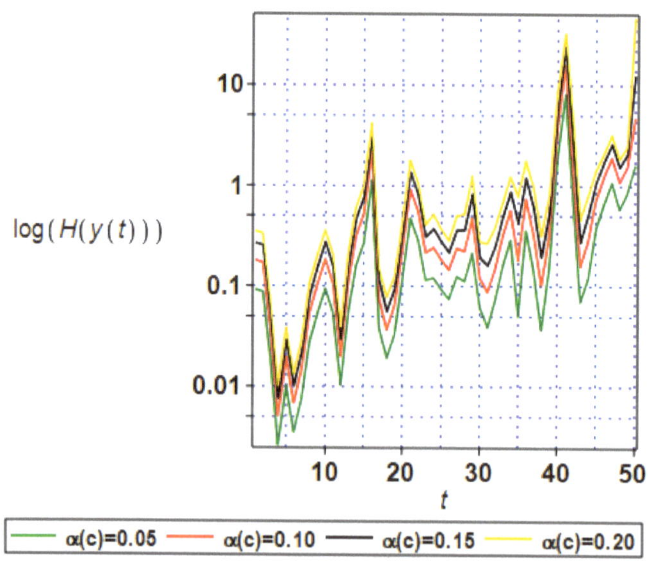

Figure 10. Shannon entropies logarithm time fluctuations for triangularly distributed structural damping.

A very important aspect is that computational time relevant to determination probabilistic entropy is almost the same, while for the increasing number of random samples above 10^6, it is even shorter than statistical estimation of the first two probabilistic moments of the structural response. If one could estimate the first four probabilistic characteristics (including also skewness and kurtosis [41]), then the entropy calculus brings definitely more information about uncertainty propagation with definitely smaller computer effort. This is seen for the single-degree-of-freedom system, whereas the Stochastic Finite Element Method analysis of large scale systems would show huge disproportion of computer power consumption. Therefore, a proper calculation and interpretation of probabilistic entropy may bring essential time savings while using the Monte Carlo simulation approach.

Although time variations of these entropies exhibit some remarkably high and rapid fluctuations, the mean values of probabilistic entropy increase systematically in each case study. It means that uncertainty (disorder) in each dynamical system subjected to harmonic excitation increases in the presence of random damping. This is, in turn, quite consistent with engineering practice and shows that this system becomes less and less predictable together with an increase of vibrations time. Finally, one may observe that this vibration undergoes some critical points, where the resulting CoV reaches huge extreme values, while the uncertainty level for the remaining part throughout this vibration process is a few times smaller. This especially happens while the mean value of the displacement is very close to the initial value (see Figure 4).

5. Concluding Remarks

A simulation-based approach for the determination of probabilistic entropy in some MEMS devices, whose dynamic response follows the Duffing equation, has been demonstrated in this work. A majority of the proposed approach has no upper bounds on the input uncertainty level and no limitation for the number of various design random parameters. Interval representation of the given probability density function applied together with the Monte Carlo simulation scheme of the structural response enables for reliable and relatively fast computation of probabilistic entropy propagation in many engineering systems including MEMS devices. As was demonstrated, it does not demand a large effort in the discretization of the resulting probability distribution—one can model non-Gaussian distribution of any physical, material, or geometrical parameters. It can be done both using some analytical solutions provided with the computer algebra programs like MAPLE as well as thanks to its common application with some commercial FEM systems (like COMSOL, ANSYS, or ABAQUS). It would be very instructive to compare the mechanical model presented and discussed above with a full multi-physics simulation of the electro-magneto-mechanical system vibrations to discover similarities and also possible divergence. An influence of the temperature field on the MEMS signal analysis and vibrations would be very important from the technological point of view [40], where uncertainty seems to be quite natural.

Probabilistic Shannon entropy exhibits almost the same extreme values and the corresponding time moments as the output coefficient of variation of the dynamical system response. It has been demonstrated that both approaches based on probabilistic characteristics of the structural responses and their probabilistic entropy exhibit huge values, many times larger than in traditional civil engineering structures, where such MEMS devices are applicable. Such a huge output uncertainty needs special modelling attention and excludes application of less precise stochastic methods like lower order expansions; in this context, the proposed entropy-based structural analysis has no limitations. That is why this methodology seems to be very promising in further uncertainty quantification in elasto-dynamics; civil engineering applications of this apparatus can be especially beneficial in seismic uncertainty analysis [42]. However, numerical Shannon entropy determination may be sensitive to the PDF partitioning of the structural response, so some initial sensitivity analysis would be required. Further computational experiments should include relative probabilistic entropies analysis, where an additional divergence (like Kullback–Leibler, for instance) in between a forced vibration and the resulting extreme structural response, would be considered. A possible alternative for the Monte Carlo simulation-based determination of entropy could be the Probability Transformation Method proposed by Falsone [43].

Author Contributions: Conceptualization, M.K. and A.C.; methodology, M.K.; software, A.C.; validation, M.K. and A.C.; formal analysis, M.K.; investigation, M.K. and A.C.; resources, A.C.; data curation, A.C.; writing—original draft preparation, M.K.; writing—review and editing, M.K. and A.C.; visualization, M.K.; supervision, A.C.; project administration, M.K.; funding acquisition, M.K.; literature overview, M.K. and A.C. All authors have read and agreed to the published version of the manuscript.

Funding: This paper was written in the framework of the research grant OPUS no 2021/41/B/ST8/02432, entitled "Probabilistic entropy in engineering computations", and is sponsored by the National Science Center in Poland.

Institutional Review Board Statement: This does not apply in case of this manuscript.

Informed Consent Statement: This does not apply in case of this manuscript.

Data Availability Statement: This does not apply in case of this manuscript.

Acknowledgments: This does not apply in case of this manuscript.

Conflicts of Interest: Both authors declare no conflict of interest.

References

1. Shannon, C.E. A mathematical theory of communication. *Part I Bell Sys. Tech. J.* **1948**, *27*, 379–423. [CrossRef]
2. Shannon, C.E. A mathematical theory of communication. *Part II Bell Sys. Tech. J.* **1948**, *27*, 623–656. [CrossRef]
3. Kamiński, M. On Shannon entropy computations in selected plasticity problems. *Int. J. Num. Meth. Engrg.* **2021**, *122*, 5128–5143. [CrossRef]
4. Beirlant, J.; Dudewicz, E.J.; Gyoerfi, L.; van der Meulen, E.C. Nonparametric entropy estimation: An overview. *Int. J. Math. Stat. Sci.* **1997**, *6*, 17–39.
5. Renyi, A. On measures of entropy and information. In Proceedings of the Fourth Berkeley Symposium on Mathematical Statistics and Probability, Berkeley, CA, USA, 20 June–30 July 1960; pp. 547–561.
6. Renyi, A. On the dimension and entropy of probability distributions. *Acta Math. Acad. Sci. Hung.* **1959**, *10*, 193–215. [CrossRef]
7. Abe, S.; Rajagopal, A.K. Revisiting disorder and Tsallis statistics. *Science* **2003**, *300*, 249–251. [CrossRef]
8. Kamiński, M. Tsallis entropy in dual homogenization of random composites using the stochastic finite element method. *Int. J. Num. Meth. Engrg.* **2018**, *113*, 834–857. [CrossRef]
9. Tsallis, C. *Introduction to Non-extensive Statistical Mechanics: Approaching a Complex World*; Springer: New York, NY, USA, 2009.
10. Tsallis, C. Possible generalization of Boltzmann-Gibbs statistics. *J. Stat. Phys.* **1988**, *52*, 479–487. [CrossRef]
11. Cornfeld, I.P.; Fomin, S.F.; Sinai, Y.A. *Ergodic Theory*; Springer: Berlin/Heidelberg, Germany, 1981.
12. Latora, V.; Baranger, M. Kolmogorov-Sinai entropy rate versus physical entropy. *Phys. Rev. Letters* **1999**, *82*, 520–523. [CrossRef]
13. Leydesdorff, L. The production of probabilistic entropy in structure/action contingency relations. *J. Soc. Evol. Syst.* **1995**, *18*, 339–356. [CrossRef]
14. Li, H.S. Maximum entropy method for probabilistic bearing strength prediction of pin joints in composite laminates. *Compos. Struct.* **2013**, *106*, 626–634. [CrossRef]
15. Shi, X.; Teixeira, A.P.; Zhang, J.; Soares, C.G. Structural reliability analysis based on probabilistic response modelling using the maximum entropy method. *Eng. Struct.* **2014**, *70*, 106–116. [CrossRef]
16. Erickson, G.J.; Smith, C.R. *Maximum-Entropy and Bayesian Methods in Science and Engineering*; Kluwer: Dordrecht, The Netherlands; Boston, MA, USA; London, UK, 1988.
17. Kamiński, M. *The Stochastic Perturbation Method for Computational Mechanics*; Wiley: Chichester, UK, 2013.
18. Cruz, M.E.; Patera, A.T. A parallel Monte-Carlo finite element procedure for the analysis of multicomponent media. *Int. J. Num. Meth. Engrg.* **1995**, *38*, 1087–1121. [CrossRef]
19. Hurtado, J.E.; Barbat, A.H. Monte-Carlo techniques in computational stochastic mechanics. *Arch. Comput. Meth. Engrg.* **1998**, *5*, 3–30. [CrossRef]
20. Liu, Y.K. *Probabilistic Theory of Structural Dynamics*; McGraw-Hill: New York, NY, USA, 1967.
21. Liu, H.; Chen, W.; Sudjianto, A. Relative entropy based method for probabilistic sensitivity analysis in engineering design. *ASME J. Mech. Des.* **2006**, *128*, 326–336. [CrossRef]
22. Kullback, S.; Leibler, R.A. On information and sufficiency. *Ann. Math. Statist.* **1951**, *22*, 79–86. [CrossRef]
23. Hellinger, E. Neue begruendung der theorie quadratischer formen von unendlichenvielen veraenderlichen. *J. Fűr Die Reine Und Angew. Math.* **1909**, *136*, 210–271. [CrossRef]
24. Chernoff, H. A measure of asymptotic efficiency for tests of a hypothesis based on the sum of observations. *Ann. Math. Stat.* **1952**, *23*, 493–507. [CrossRef]
25. Mahalanobis, P. On the generalized distance in statistics. *Proc. Natl. Inst. Sci. India* **1936**, *2*, 49–55.
26. Bhattacharyya, A. On a measure of divergence between two statistical populations defined by their probability distributions. *Bull. Calcutta Math. Soc.* **1943**, *35*, 99–109.
27. Corigliano, A.; Ardito, R.; Comi, C.; Frangi, A.; Ghisi, A.; Mariani, S. *Mechanics of Microsystems*; John Wiley & Sons: Hoboken, NJ, USA, 2018; ISBN 978-1-119-05383-5.
28. Lucas, V.; Golinval, J.C.; Paquay, S.; Nguyen, V.D.; Noels, L.; Wu, L. A stochastic computational multiscale approach; Application to MEMS resonators. *Comput. Methods Appl. Mech. Eng.* **2015**, *294*, 141–167. [CrossRef]
29. Sung, S.H.; Park, J.W.; Nagayama, T.; Jung, H.J. A multi-scale sensing and diagnosis system combining accelerometers and gyroscopes for bridge health monitoring. *Smart Mat. Struct.* **2014**, *23*, 015005. [CrossRef]

30. Batou, A.; Ritto, T.G.; Sampaio, R. Entropy propagation analysis in stochastic structural dynamics: Application to a beam with uncertain cross sectional area. *Comput. Mech.* **2014**, *54*, 591–601. [CrossRef]
31. Wang, R.M.; Zhang, H.M.; Zhang, Y.N. Bifurcation and vibration resonance in the time delay Duffing system with fractional internal and external damping. *Meccanica* **2022**, *57*, 999–1015. [CrossRef]
32. Forsythe, G.E.; Malcolm, M.A.; Moler, C.B. *Computer Methods for Mathematical Computations*; Prentice Hall: Hoboken, NJ, USA, 1977.
33. Björck, A. *Numerical Methods for Least Squares Problems*; SIAM: Philadelphia, PA, USA, 1996.
34. Cover, T.M.; Thomas, J.A. *Elements of Information Theory*, 2nd ed.; Wiley-Interscience: New York, NY, USA, 2006.
35. Al-Omari, A.I. Estimation of entropy using random sampling. *J. Comput. Appl. Math.* **2014**, *261*, 95–102. [CrossRef]
36. Xu, Z.; Shi, Y.; Zhao, Q.; Li, W.; Liu, K. Extreme interval entropy based on symbolic analysis and a self-adaptive method. *Entropy* **2019**, *21*, 238. [CrossRef]
37. Shao, Y.; Hahn, M.G. Limit theorems for the logarithm of sample spacings. *Stat. Prob. Lett.* **1995**, *24*, 121–132. [CrossRef]
38. Vasicek, O. A test for normality based on sample entropy. *J. Roy. Stat. Soc.* **1976**, *38*, 54–59. [CrossRef]
39. Feller, W. *An Introduction to Probability Theory and Applications*; Wiley: New York, NY, USA, 1965.
40. Comi, C.; Zega, V.; Corigliano, A. Non-linear mechanics in resonant inertial micro sensors. *Int. J. Non-Linear Mech.* **2020**, *120*, 103386. [CrossRef]
41. Kamiński, M.; Corigliano, A. Numerical solution of the Duffing equation with random coefficients. *Meccanica* **2015**, *50*, 834–857. [CrossRef]
42. Zembaty, Z. Non-stationary random vibrations of a shear beam under high frequency seismic effects. *Soil Dyn. Earthq. Engrg.* **2007**, *27*, 1000–1011. [CrossRef]
43. Falsone, G.; Settineri, D. Explicit solutions for the response probability density function of nonlinear transformations of static random inputs. *Probabilistic Eng. Mech.* **2013**, *33*, 79–85. [CrossRef]

energies

MDPI

Article

McCARD Criticality Benchmark Analyses with Various Evaluated Nuclear Data Libraries

Ho Jin Park [1], Mohammad Alosaimi [2], Seong-Ah Yang [1], Heejeong Jeong [3] and Sung Hoon Choi [3,*]

1 Department of Nuclear Engineering, Kyung Hee University, Yongin 17104, Korea
2 King Abdullah City for Atomic and Renewable Energy, Riyadh 12244, Saudi Arabia
3 Korea Atomic Energy Research Institute, Daejeon 34057, Korea
* Correspondence: cshoon@kaeri.re.kr; Tel.: +82-42-868-8642

Abstract: International Criticality Safety Benchmark Evaluation Project (ICSBEP) criticality analyses were conducted using the McCARD Monte Carlo code for 85 selected benchmark problems with 7 evaluated nuclear data libraries (ENDLs): ENDF/B-VII.1, ENDF/B-VIII.0, JENDL-4.0, JENDL-5.0, JEFF-3.3, TENDL-2021, and CENDL-3.2. Regarding the analyses, it was confirmed that the k_{eff} results are sensitive to the ENDL. It is noted that the new-version ENDLs show better performance in the fast benchmark cases, while on the other hand, there are no significant differences in k_{eff} among the different ENDLs in the thermal benchmark cases. The sensitivity of the k_{eff} results depending on the ENDL may impact nuclear core design parameters such as the shutdown margin, critical boron concentration, and power defects. This study and k_{eff} results will be a good reference in the development of new types of nuclear cores or new design codes.

Keywords: Monte Carlo; McCARD; criticality analysis; ICSBEP; ENDF/B-VIII.0; ENDF/B-VII.1; JENDL-4.0; JENDL-5.0; TENDL-2021; CENDL-3.2; JEFF-3.3

1. Introduction

In various nuclear engineering applications, atomic and nuclear data are widely used as important and critical inputs to solve particle transport balance equations. Many research institutes have provided the nuclear data as evaluated nuclear data libraries (ENDLs) in a traditional ENDF-6 (evaluated nuclear data file) format, which are processed from measurements, compilations, and evaluations. The ENDF-6 format includes general information, resonance parameter data, reaction cross section, angular distribution, and their covariance data. The Cross Section Evaluation Working Group, organized by the United States (i.e., Brookhaven, Oak Ridge, and Argonne national Laboratories) and international nuclear societies, has released ENDF/B ENDLs. Among the versions in this series, ENDF/B-VII.1 [1] is widely used in particle transport simulation codes for nuclear reactor physics and core design analysis. An up-to-date version ENDF/B-VIII.0 [2], was released in February 2018. This version includes new evaluation data of the six nuclides (i.e., ^{1}H, ^{16}O, ^{56}Fe, ^{235}U, ^{238}U, ^{239}Pu) from the CIELO (Collaborative International Evaluation Library Organization) project. At the time of release, the neutron-reaction evaluation data for 557 materials in ENDF/B-VIII.0 were totally new or partially updated, including improved thermal neutron scattering data. Meanwhile, the Japan Atomic Energy Research Institute and Japanese Nuclear Data Committee (JNDC) have been continuously providing a series of Japanese Evaluated Nuclear Data Libraries (JENDLs), including JENDL-4.0 [3] released in May 2010 and an up-to-date version JENDL-5.0 [4] released in December 2021. In JENDL-5.0, the number of neutron sub-libraries was increased from 406 to 795 and the energy region was extended from 20 MeV to 200 MeV. Otherwise, the OECD (Organization for Economic Cooperation and Development)/NEA (Nuclear Energy Agency) Data Bank has coordinated the Joint Evaluated Fission and Fusion (JEFF) ENDL development for the last 35 years. Released in November 2017 JEFF-3.3 [5] provided 562 evaluations for neutron reactions. In another example, the Paul Scherrer Institute (PSI) and International Atomic Energy Agency nuclear data section developed a series of TENDL ENDLs. TENDL [6] provides the outputs of the TALYS code [7] to analyze and predict nuclear reactions. The latest version, TENDL-2021, provides 2813 evaluations for neutron reactions while ENDF/B-VIII.0 has 557 isotopic data files. And as a final example, the China Nuclear Data Center has released a series of Chinese general purpose Evaluated Nuclear Data Library (CENDL). CENDL-3.2 [8] is the latest release of CENDL, which has ENDF-6 formatted neutron reactions for 272 isotopes.

As stated above, a variety of ENDLs have been released and continuously updated by their providers around the world for use in various nuclear physics research and applications. To validate the newly developed ENDLs, the integral testing work has been performed using various benchmark problems. The International Criticality Safety Benchmark Evaluation Project (ICSBEP) [9] is one of the representative integral testing programs from critical experiments. The ICSBEP criticality analysis problems were classified by various fuel types and system spectrums. The ICSBEP handbook provides an overview of experiments, benchmark specifications, and some results for sample calculations by KENO-V in the SCALE code package [10], MCNP [11], and ONEDANT/TWODANT in the DANTSYS code package [12]. There are many studies and results for the ICSBEP benchmark problems with various ENDLs [13–16].

Recently, the Korea Atomic Energy Research Institute (KAERI) and King Abdullah City for Atomic and Renewable Energy (K.A.CARE) established the KAERI-K.A.CARE joint R&D center at KAERI to continue effective and close cooperation for the establishment of the National Nuclear Laboratory in Saudi Arabia. This center has carried out various joint R&D programs, an example of which is a project called "Application of a Monte-Carlo Neutron/Photon Transport Simulation Code for Advanced Shielding Design of Nuclear Reactors". The main goal of this project is to train K.A.CARE engineers in a nuclear core shielding design analysis and to validate the McCARD [17] Monte Carlo (MC) code to be used for the advanced shielding design and analyses of new-type reactors. To validate the capability of the McCARD code for criticality analyses, KAERI and K.A.CARE engineers performed criticality analyses with the McCARD code and the up-to-date ENDLs.

In this study, seven ENDLs—ENDF/B-VII.1, ENDF/B-VIII.0, JENDL-4.0, JENDL-5.0, JEFF-3.3, TENDL-2021, and CENDL-3.2—were tested and examined by performing McCARD criticality analyses for selected ICSBEP benchmark problems. Section 2 briefly describes the configuration of the selected ICSBEP problems for criticality analysis and explains how to generate the continuous energy cross section from the raw ENDLs. Section 3 presents the results of the ICSBEP criticality analyses calculated by the McCARD MC code with the various ENDLs. The results are provided by categorizing the fuel fissile isotopes, fuel form, and system spectrum. A summary and conclusions are given in Section 4.

2. Evaluated Nuclear Data Libraries and ICSBEP Benchmarks for Criticality Analyses

2.1. Evaluated Nuclear Data Libraries

Various up-to-date evaluated nuclear data libraries are first prepared for the integral testing work via MC criticality analyses. First of all, the most up-to-date NJOY code [18] and its user inputs for all nuclides at three temperature points (300 K, 600 K, and 900 K) were prepared to process the raw ENDLs and to generate MC continuous energy cross section libraries in ACE format. Figure 1 shows the general flow chart of the ACE-formatted continuous-energy (CE) nuclear data library generation in the NJOY code. Neutron CE cross sections for each isotope are generated by the flow of the RECONR, BROADR, UNRESR, PURR, and ACER modules in NJOY, whereas the thermal scattering cross sections are generated by the RECONR, BROADR, LEAPR, THERMR, and ACER modules. The RECONR module reconstructs point-wise cross sections from ENDF resonance parameters and interpolation schemes, which are then processed into Doppler-broadens and thins point-wise cross sections by the BROADR module. The UNRESR and PURR modules generate effective self-shielded point-wise cross sections and probability tables in unresolved energy regions. For the thermal scattering cross section generation, LEAPR calculates the thermal scattering law while THERMR produces cross sections and energy-to-matrices for free or bound scattering in the thermal energy range. Lastly, the ACER module prepares libraries in ACE format for a CE MC code (e.g., MCNP, McCARD, RMC).

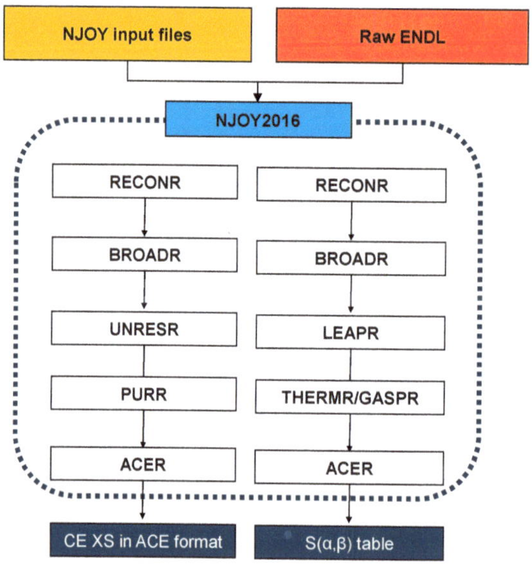

Figure 1. A flowchart of Monte Carlo CE library generation in NJOY code.

Table 1 summarizes the newly generated Monte Carlo CE cross section libraries. In this study, only the five most often used thermal scattering cross sections (i.e., H in H_2O, D in D_2O, Be metal, Be in BeO, and C in graphite) were generated for all ENDLs. As shown in Table 1, there is no thermal scattering cross section data in CENDL-3.2 and only one thermal scattering cross section data in TENDL-2021. Accordingly, the lack of thermal scattering cross section data was substituted by ENDF/B-VIII.0. In general, the ENDF/B cross section library has been used all around the world in various research and fields, and among the ENDF/B versions, ENDF/B-VIII.0 is the latest version. According to this, we used the thermal scattering cross section data for the lack of other ENDL thermal scattering cross section data.

Table 1. A summary of the generated Monte Carlo CE cross section libraries.

ENDL	Years of Release	Number of Generated CE Libraries/Total Number	
		Neutron	Thermal Scattering
ENDF/B-VII.1	2011	393/423	5/21
ENDF/B-VIII.0	2018	544/557	5/34
JENDL-4.0	2011	405/406	5/16
JENDL-5.0	2021	233/795	5/37
JEFF-3.3	2018	558/562	5/20
TENDL-2021 *	2021	630/2813	1/1
CENDL-3.2	2020	270/272	None

* CE neutron reaction library was taken from the official website.

2.2. Selected International Criticality Benchmark Problems

To perform the integral testing work for criticality capability, 85 benchmark problems were selected from the ICSBEP handbook [9]. The 85 ICSBEP benchmarks were selected from the well-known relevant experiments (i.e., godiva, jezebel, flattop) or the problems that have the results by MCNP with various ENDLs. In general, they boil down to three criteria: fuel fissile isotope, fuel form, and system spectrum. Fuel fissile isotopes can be categorized into high-enriched uranium (HEU), low-enriched uranium (LEU), plutonium (PU), ^{233}U (U233), and mixed composition (MIX). Fuel forms are defined as metal (MET), compound (COMP), and solution (SOL), and system spectrum is classified as fast (FAST) and thermal (THERMAL). The ICSBEP handbook provides the identification (ID) for each benchmark problem as a combination of the fuel isotope, fuel form, and spectrum type.

Table 2 lists the 85 selected ICSBEP benchmark problems, providing benchmark IDs, categories, reference k_{eff}, and short IDs for the sake of convenient reference. The McCARD inputs for each ICSBEP benchmark problem were prepared. All the McCARD calculations were performed by employing 10,000 neutron particles per cycle with 1000 active cycles and 50 inactive cycles. The initial neutron sources were uniformly distributed in the system boundary for MC eigenvalue calculations. Figures 2 and 3 show the neutron energy spectra for five fast benchmarks (i.e., Jezebel, Jezebl-240, Godiva, Flattop-25, and Jezebel-233) and six thermal benchmarks (i.e., LCT001c1, LCT002c1, LCT006c1, ORNL-1, PNL-3, and ORNL-11), respectively. In the fast benchmarks, the energy spectra are similar to the energy distribution of neutrons from fission reactions. In the thermal benchmarks, the neutron energy spectra are attributed to neutron moderation or slowing-down. As shown in Table 2, the thermal scattering law (TSL) sub-library for light water was only used in this ICSBEP benchmark analyses.

Table 2. A list of selected International Criticality Safety Benchmark Problems.

No.	Short Name	Handbook ID	Category	Benchmark k_{eff}	Uncertainty of k_{eff}	TSL *
1	Jezebel	PU-MET-FAST-001	Pu Fast	1.0000	0.00200	-
2	Jezebel-240	PU-MET-FAST-002	Pu Fast	1.0000	0.00200	-
3	PMF-020	PU-MET-FAST-020	Pu Fast	0.9993	0.00170	-
4	PMF-022	PU-MET-FAST-022	Pu Fast	1.0000	0.00210	-
5	PMF-005	PU-MET-FAST-005	Pu Fast	1.0000	0.00130	-
6	PMF-006	PU-MET-FAST-006	Pu Fast	1.0000	0.00300	-
7	PMF-010	PU-MET-FAST-010	Pu Fast	1.0000	0.00180	-
8	PMF-011	PU-MET-FAST-011	Pu Fast	1.0000	0.00100	Light water
9	Godiva	HEU-MET-FAST-001	HEU Fast	1.0000	0.00100	-
10	Flattop-25	HEU-MET-FAST-028	HEU Fast	1.0000	0.00300	-
11	HMF-002 c2	HEU-MET-FAST-002 c2	HEU Fast	1.0000	0.00300	-
12	HMF-002 c3	HEU-MET-FAST-002 c3	HEU Fast	1.0000	0.00300	-
13	HMF-002 c4	HEU-MET-FAST-002 c4	HEU Fast	1.0000	0.00300	-
14	HMF-002 c5	HEU-MET-FAST-002 c5	HEU Fast	1.0000	0.00300	-
15	HMF-002 c6	HEU-MET-FAST-002 c6	HEU Fast	1.0000	0.00300	-
16	HMF-004	HEU-MET-FAST-004	HEU Fast	1.0020	-	-
17	HMF-018	HEU-MET-FAST-018	HEU Fast	1.0000	0.00140	-
18	HMF-027	HEU-MET-FAST-027	HEU Fast	1.0000	0.00250	-
19	HMF-032 c1	HEU-MET-FAST-032 c1	HEU Fast	1.0000	0.00130	-
20	HMF-032 c2	HEU-MET-FAST-032 c2	HEU Fast	1.0000	0.00260	-
21	HMF-032 c3	HEU-MET-FAST-032 c3	HEU Fast	1.0000	0.00130	-
22	HMF-032 c4	HEU-MET-FAST-032 c4	HEU Fast	1.0000	0.00130	-
23	Jezebel-233	U233-MET-FAST-001	U233 Fast	1.0000	0.00100	-
24	U233-MF003	U233-MET-FAST-003	U233 Fast	1.0000	0.00100	
25	U233-MF004	U233-MET-FAST-004	U233 Fast	1.0000	0.00070	-
26	U233-MF005	U233-MET-FAST-005	U233 Fast	1.0000	0.00300	-
27	Flattop-23	U233-MET-FAST-006	U233 Fast	1.0000	0.00100	-
28	MMF-001	MIX-MET-FAST-001	MIX Fast	1.0000	0.00160	-
29	MMF-002 c1	MIX-MET-FAST-002 c1	MIX Fast	1.0000	0.00440	-
30	MMF-002 c2	MIX-MET-FAST-002 c2	MIX Fast	1.0000	0.00440	-
31	MMF-002 c3	MIX-MET-FAST-002 c3	MIX Fast	1.0000	0.00440	-
32	ORNL-1	HEU-SOL-THERM-013 c1	HEU Thermal	1.0012	0.00260	Light water

Table 2. *Cont.*

No.	Short Name	Handbook ID	Category	Benchmark k_{eff}	Uncertainty of k_{eff}	TSL *
33	ORNL-2	HEU-SOL-THERM-013 c2	HEU Thermal	1.0007	0.00360	Light water
34	ORNL-3	HEU-SOL-THERM-013 c3	HEU Thermal	1.0009	0.00360	Light water
35	ORNL-4	HEU-SOL-THERM-013 c4	HEU Thermal	1.0003	0.00360	Light water
36	ORNL-10	HEU-SOL-THERM-032	HEU Thermal	1.0015	0.00260	Light water
37	LCT-001 c1	LEU-COMP-THERM-001 c1	LEU Thermal	1.0000	0.00310	Light water
38	LCT-001 c2	LEU-COMP-THERM-001 c2	LEU Thermal	0.9998	0.00310	Light water
39	LCT-001 c3	LEU-COMP-THERM-001 c3	LEU Thermal	0.9998	0.00310	Light water
40	LCT-001 c4	LEU-COMP-THERM-001 c4	LEU Thermal	0.9998	0.00310	Light water
41	LCT-001 c5	LEU-COMP-THERM-001 c5	LEU Thermal	0.9998	0.00310	Light water
42	LCT-001 c6	LEU-COMP-THERM-001 c6	LEU Thermal	0.9998	0.00310	Light water
43	LCT-001 c7	LEU-COMP-THERM-001 c7	LEU Thermal	0.9998	0.00310	Light water
44	LCT-001 c8	LEU-COMP-THERM-001 c8	LEU Thermal	0.9998	0.00310	Light water
45	LCT-002 c1	LEU-COMP-THERM-002 c1	LEU Thermal	0.9997	0.00200	Light water
46	LCT-002 c2	LEU-COMP-THERM-002 c2	LEU Thermal	0.9997	0.00200	Light water
47	LCT-002 c3	LEU-COMP-THERM-002 c3	LEU Thermal	0.9997	0.00200	Light water
48	LCT-006 c1	LEU-COMP-THERM-006 c1	LEU Thermal	1.0000	0.00200	Light water
49	LCT-006 c2	LEU-COMP-THERM-006 c2	LEU Thermal	1.0000	0.00200	Light water
50	LCT-006 c3	LEU-COMP-THERM-006 c3	LEU Thermal	1.0000	0.00200	Light water
51	LCT-006 c4	LEU-COMP-THERM-006 c4	LEU Thermal	1.0000	0.00200	Light water
52	LCT-006 c5	LEU-COMP-THERM-006 c5	LEU Thermal	1.0000	0.00200	Light water
53	LCT-006 c6	LEU-COMP-THERM-006 c6	LEU Thermal	1.0000	0.00200	Light water
54	LCT-006 c7	LEU-COMP-THERM-006 c7	LEU Thermal	1.0000	0.00200	Light water
55	LCT-006 c8	LEU-COMP-THERM-006 c8	LEU Thermal	1.0000	0.00200	Light water
56	LCT-006 c9	LEU-COMP-THERM-006 c9	LEU Thermal	1.0000	0.00200	Light water
57	LCT-006 c10	LEU-COMP-THERM-006 c10	LEU Thermal	1.0000	0.00200	Light water
58	LCT-006 c11	LEU-COMP-THERM-006 c11	LEU Thermal	1.0000	0.00200	Light water
59	LCT-006 c12	LEU-COMP-THERM-006 c12	LEU Thermal	1.0000	0.00200	Light water
60	LCT-006 c13	LEU-COMP-THERM-006 c13	LEU Thermal	1.0000	0.00200	Light water
61	LCT-006 c14	LEU-COMP-THERM-006 c14	LEU Thermal	1.0000	0.00200	Light water
62	LCT-006 c15	LEU-COMP-THERM-006 c15	LEU Thermal	1.0000	0.00200	Light water
63	LCT-006 c16	LEU-COMP-THERM-006 c16	LEU Thermal	1.0000	0.00200	Light water
64	LCT-006 c17	LEU-COMP-THERM-006 c17	LEU Thermal	1.0000	0.00200	Light water
65	LCT-006 c18	LEU-COMP-THERM-006 c18	LEU Thermal	1.0000	0.00200	Light water
66	LCT-010 c9	LEU-COMP-THERM-010 c9	LEU Thermal	1.0000	0.00280	Light water
67	LCT-010 c11	LEU-COMP-THERM-010 c11	LEU Thermal	1.0000	0.00280	Light water
68	LCT-010 c12	LEU-COMP-THERM-010 c12	LEU Thermal	1.0000	0.00280	Light water
69	LCT-010 c14	LEU-COMP-THERM-010 c14	LEU Thermal	1.0000	0.00280	Light water
70	LCT-010 c15	LEU-COMP-THERM-010 c15	LEU Thermal	1.0000	0.00280	Light water
71	LCT-010 c16	LEU-COMP-THERM-010 c16	LEU Thermal	1.0000	0.00280	Light water
72	LCT-010 c17	LEU-COMP-THERM-010 c17	LEU Thermal	1.0000	0.00280	Light water
73	LCT-010 c18	LEU-COMP-THERM-010 c18	LEU Thermal	1.0000	0.00280	Light water
74	LCT-017 c13	LEU-COMP-THERM-017 c13	LEU Thermal	1.0000	0.00310	Light water
75	LCT-017 c15	LEU-COMP-THERM-017 c15	LEU Thermal	1.0000	0.00310	Light water
76	LCT-017 c18	LEU-COMP-THERM-017 c18	LEU Thermal	1.0000	0.00310	Light water
77	LCT-017 c21	LEU-COMP-THERM-017 c21	LEU Thermal	1.0000	0.00310	Light water
78	IPEN/MB-01	LEU-COMP-THERM-077	LEU Thermal	1.0003	0.00100	Light water
79	LMT-007 c1	LEU-MET-THERM-007 c1	LEU Thermal	0.9983	0.01140	Light water
80	LMT-007 c2	LEU-MET-THERM-007 c2	LEU Thermal	0.9976	0.00680	Light water
81	PNL-3	PU-SOL-THERM-011 c18-1	Pu Thermal	1.0000	0.00520	Light water
82	PNL-4	PU-SOL-THERM-011 c18-6	Pu Thermal	1.0000	0.00520	Light water
83	PNL-5	PU-SOL-THERM-011 c16-5	Pu Thermal	1.0000	0.00520	Light water
84	PST011c16-1	PU-SOL-THERM-011 c16-1	Pu Thermal	1.0000	0.00520	Light water
85	ORNL-11	U233-SOL-THERM-008	U233 Thermal	1.0006	0.00290	Light water

* TSL is a thermal scattering law sub-library.

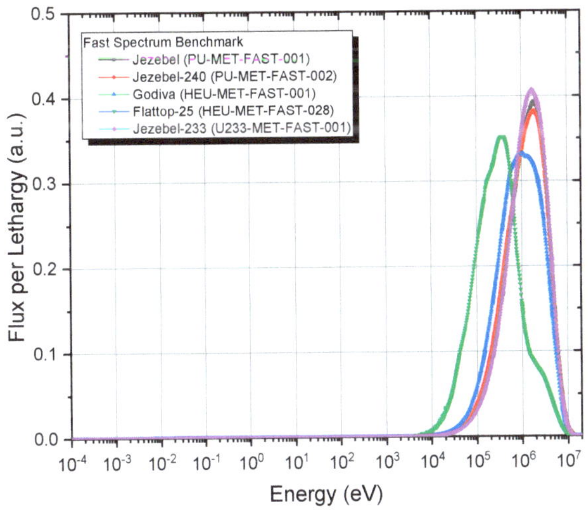

Figure 2. Spectra of example fast benchmark problems.

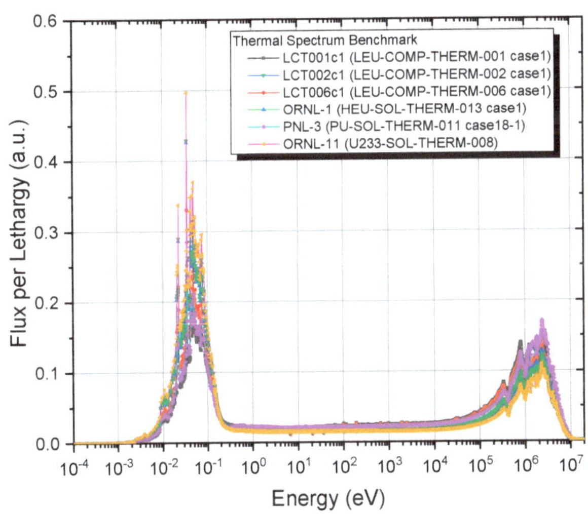

Figure 3. Spectra of example thermal benchmark problems.

3. ICSBEP Criticality Benchmark Analyses by McCARD

3.1. Fast Criticality Benchmarks

Table 3 shows the k_{eff} values calculated by McCARD with the seven ENDLs (ENDF/B-VII.1, ENDF/B-VIII.0, JENDL-4.0, JENDL-5.0, JEFF-3.3, TENDL-2021, and CENDl-3.2). Figure 4 plots the difference ($\Delta\rho_{cal}$) between the calculated and experimental k_{eff} for the 31 fast benchmark problems calculated by

$$\Delta\rho_{cal} = \frac{k^i_{cal} - k^i_{\exp}}{k^i_{cal} \cdot k^i_{\exp}} \cdot 10^5. \quad (1)$$

Table 3. k_{eff} values for the fast benchmarks with the different evaluated nuclear data libraries.

No.	Short Name	k_{eff} (McCARD) *						
		ENDF/B-VII.1	ENDF/B-VIII.0	JENDL-4.0	JENDL-5.0	JEFF-3.3	TENDL-2021	CENDL-3.2
1	Jezebel	1.00021	0.99995	0.99837	0.99925	0.99951	1.00004	1.00193
2	Jezebel-240	1.00041	1.00160	0.99849	0.99870	1.00151	1.00361	1.00271
3	PMF-020	0.99836	0.99697	0.99551	0.99826	0.99955	1.00023	0.99714
4	PMF-022	0.99870	0.99815	0.99694	0.99816	0.99800	0.99856	1.00034
5	PMF-005	1.00078	0.99944	1.00184	0.99890	1.00135	0.99663	1.00127
6	PMF-006	1.00140	1.00011	0.99906	1.00195	1.00366	1.00396	0.99946
7	PMF-010	0.99986	0.99799	0.99720	0.99972	1.00047	1.00095	0.99876
8	PMF-011	1.00041	1.00066	1.00194	1.00072	1.00108	1.00024	1.00222
9	Godiva	0.99991	1.00014	0.99760	0.99925	1.00005	1.00074	0.99974
10	Flattop-25	1.00299	1.00107	0.99812	1.00081	1.00433	1.00544	1.00184
11	HMF-002 c2	1.00248	1.00031	0.99762	1.00066	1.00381	1.00493	1.00073
12	HMF-002 c3	1.00063	0.99877	0.99593	0.99881	1.00206	1.00340	0.99918
13	HMF-002 c4	0.99990	0.99780	0.99497	0.99808	1.00118	1.00208	0.99782
14	HMF-002 c5	1.00021	0.99812	0.99520	0.99845	1.00162	1.00268	0.99882
15	HMF-002 c6	1.00160	0.99957	0.99677	0.99970	1.00294	1.00372	0.99983
16	HMF-004	1.00296	1.00189	1.00378	1.00223	1.00236	1.00139	1.00463
17	HMF-018	1.00006	1.00007	0.99764	0.99882	1.00032	1.00096	1.00044
18	HMF-027	1.00074	1.00046	1.00156	1.00323	1.00445	1.00134	1.00297
19	HMF-032 c1	1.00415	1.00183	0.99897	1.00233	1.00509	1.00567	1.00301
20	HMF-032 c2	1.00476	1.00246	0.99953	1.00296	1.00548	1.00652	1.00335
21	HMF-032 c3	1.00022	0.99821	0.99579	0.99862	1.00082	1.00148	0.99967
22	HMF-032 c4	1.00070	0.99973	0.99754	0.99973	1.00136	1.00220	1.00029
23	Jezebel-233	0.99974	1.00029	0.99906	0.99978	1.00084	1.00102	1.00112
24	U233-MF003	0.99917	0.99951	0.99861	1.00035	1.00144	1.00138	1.00038
25	U233-MF004	0.99835	0.99939	1.00012	0.99710	1.00000	0.99672	0.99966
26	U233-MF005	0.99578	0.99719	0.99590	0.99647	0.99709	0.99670	0.99555
27	Flattop-23	0.99888	0.99998	0.99849	1.00043	1.00352	1.00322	0.99900
28	MMF-001	0.99961	0.99949	0.99788	0.99895	0.99899	0.99934	0.99885
29	MMF-002 c1	1.00543	1.00375	1.00166	1.00421	1.00673	1.00714	1.00376
30	MMF-002 c2	1.00573	1.00393	1.00172	1.00419	1.00688	1.00731	1.00415
31	MMF-002 c3	1.00577	1.00447	1.00173	1.00391	1.00723	1.00861	1.00340

* The statistical uncertainties of the calculated k_{eff} are less than 10 pcm.

Here, $k_{\exp}$ and k_{cal} are the experimental and calculated k_{eff} for the i-th benchmark problem, respectively. The statistical uncertainties of the calculated k_{eff} are less than 10 pcm. Accordingly, error bars of the calculated k_{eff} are not marked in Figure 4 because they are relatively small compared to the uncertainties of the reference k_{eff} values. Overall, the values from JENDL-4.0 are lower than those from the other ENDLs whereas JEFF-3.3 and TENDL-2021 have higher values than the other ENDLs. For statistical analyses, root mean square (RMS) error and chi square (χ^2) can be utilized as indicators to confirm the differences between the experimental and calculated k_{eff}. Typically, RMS error and chi square values can be calculated by

$$RMS\ error(\%) = \sqrt{\frac{1}{N}\sum_{i=1}^{N}\left(k_{cal}^{i} - k_{\exp}^{i}\right)^2}, \tag{2}$$

$$\chi^2 = \frac{1}{N}\sum_{i=1}^{N}\left(\frac{k_{cal}^{i} - k_{\exp}^{i}}{\sigma_{\exp}^{i}}\right)^2. \tag{3}$$

where $\sigma_{\exp}$ is the uncertainty of $k_{\exp}$ provided from each benchmark document [9]. The number of benchmark problems is N.

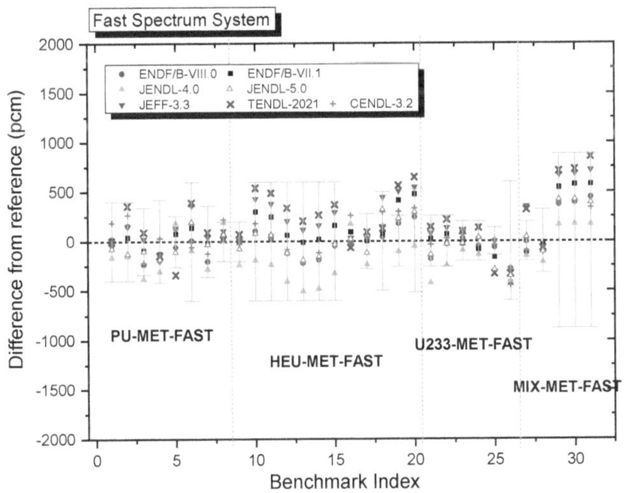

Figure 4. The difference between calculated and experimental k_{eff} by ENDL for the ICSBEP fast benchmarks.

Table 4 shows the RMS errors and chi square values for the 31 fast benchmark problems. It is observed that the new version ENDLs show better performance than the old versions in the fast benchmarks. The RMS error of ENDF/B-VII.1 is 244 pcm, whereas that of ENDF/B-VIII.0 is 179 pcm. The RMS error of JENDL-4.0 is 258 pcm compared to that of JENDL-5.0 at 199 pcm. It is noted that ENDF/B-VIII.0 has the smallest RMS error and chi square value among the ENDLs. In the 31 fast benchmarks, the average uncertainty of k_{exp} is about 220 pcm.

Table 4. RMS errors and chi square values of the 31 fast benchmarks for different evaluated nuclear data libraries.

ENDL	RMS Error (pcm)	χ^2
ENDF/B-VII.1	244	1.02
ENDF/B-VIII.0	179	0.72
JENDL-4.0	258	1.39
JENDL-5.0	199	1.06
JEFF-3.3	322	1.38
TENDL-2021	374	1.79
CENDL-3.2	211	0.97

3.2. Thermal Criticality Benchmarks

Table 5 presents the k_{eff} values calculated by McCARD for the 54 thermal benchmark problems, and Figure 5 shows the difference between the calculated and experimental k_{eff}. In the LEU-COMP-THERMAL cases, CENDL-3.2 showed lower results than the other ENDLs, whereas JEFF3.3 and TENDL-2021 showed relatively higher results. Table 6 shows the RMS errors and chi square values for the thermal benchmark problems. When excluding the PU-SOL-THERMAL cases, there were no significant differences in k_{eff} among the different ENDLs in the thermal benchmark cases. In the PU-SOL-THERML cases, the difference in k_{eff} ranged from −1327 pcm to 2220 pcm. In all thermal cases, RMS errors ranged from 252 pcm to 512 pcm, while the chi square values were from 0.72 to 1.24. In the 54 thermal benchmarks, the average uncertainty of k_{exp} is about 297 pcm. However, for the thermal benchmarks excluding PU-SOL-THERMAL, RMS errors ranged from 180 pcm to 272 pcm and chi square values were from 0.63 to 0.96.

Table 5. k_{eff} values for the thermal benchmarks with the different evaluated nuclear data libraries.

No.	Short Name	k_{eff} (McCARD) *						
		ENDF/B-VII.1	ENDF/B-VIII.0	JENDL-4.0	JENDL-5.0	JEFF-3.3	TENDL-2021	CENDL-3.2
32	ORNL-1	0.99772	0.99785	0.99889	0.99761	0.99634	0.99680	0.99656
33	ORNL-2	0.99679	0.99712	0.99851	0.99723	0.99577	0.99652	0.99587
34	ORNL-3	0.99329	0.99370	0.99487	0.99357	0.99256	0.99320	0.99209
35	ORNL-4	0.99483	0.99535	0.99648	0.99565	0.99408	0.99490	0.99370
36	ORNL-10	0.99883	0.99844	0.99846	0.99780	0.99695	0.99793	0.99734
37	LCT-001 c1	1.00013	0.99996	1.00063	0.99996	0.99980	1.00102	0.99786
38	LCT-001 c2	0.99941	0.99914	0.99997	0.99872	0.99899	0.99988	0.99706
39	LCT-001 c3	0.99903	0.99850	0.99869	0.99877	0.99878	1.00013	0.99675
40	LCT-001 c4	0.99970	0.99931	1.00030	0.99924	1.00025	1.00067	0.99721
41	LCT-001 c5	0.99763	0.99715	0.99803	0.99715	0.99775	0.99870	0.99538
42	LCT-001 c6	0.99958	0.99926	0.99925	0.99938	0.99946	1.00044	0.99770
43	LCT-001 c7	0.99907	0.99857	0.99910	0.99896	0.99874	0.99977	0.99650
44	LCT-001 c8	0.99774	0.99749	0.99736	0.99792	0.99704	0.99929	0.99585
45	LCT-002 c1	0.99907	0.99767	0.99962	0.99911	0.99942	0.99919	0.99730
46	LCT-002 c2	1.00038	0.99899	1.00060	1.00059	1.00083	1.00021	0.99868
47	LCT-002 c3	0.99993	0.99857	1.00010	0.99968	1.00024	0.99962	0.99832
48	LCT-006 c1	1.00015	0.99960	1.00115	0.99929	1.00191	1.00216	0.99872
49	LCT-006 c2	1.00079	0.99995	1.00144	1.00016	1.00188	1.00286	0.99884
50	LCT-006 c3	1.00055	0.99971	1.00132	0.99983	1.00212	1.00274	0.99880
51	LCT-006 c4	1.00027	0.99980	1.00098	0.99988	1.00140	1.00203	0.99862
52	LCT-006 c5	1.00021	0.99954	1.00093	0.99944	1.00133	1.00213	0.99809
53	LCT-006 c6	1.00051	1.00001	1.00144	0.99991	1.00209	1.00245	0.99880
54	LCT-006 c7	1.00046	0.99983	1.00105	0.99976	1.00170	1.00204	0.99868
55	LCT-006 c8	1.00027	0.99987	1.00108	0.99966	1.00119	1.00247	0.99884
56	LCT-006 c9	1.00039	0.99999	1.00112	1.00014	1.00105	1.00190	0.99817
57	LCT-006 c10	1.00032	0.99999	1.00083	0.99979	1.00084	1.00109	0.99797
58	LCT-006 c11	1.00031	0.99994	1.00101	1.00001	1.00135	1.00186	0.99790
59	LCT-006 c12	0.99993	0.99981	1.00078	0.99961	1.00109	1.00169	0.99794
60	LCT-006 c13	1.00009	0.99946	1.00054	0.99969	1.00065	1.00104	0.99792
61	LCT-006 c14	1.00039	1.00019	1.00097	1.00015	1.00040	1.00093	0.99835
62	LCT-006 c15	1.00025	0.99988	1.00076	0.99994	1.00073	1.00089	0.99781
63	LCT-006 c16	1.00015	0.99987	1.00087	1.00006	1.00050	1.00098	0.99805
64	LCT-006 c17	1.00014	0.99966	1.00051	1.00027	1.00044	1.00055	0.99805
65	LCT-006 c18	1.00013	0.99962	1.00066	0.99970	1.00034	1.00121	0.99791
66	LCT-010 c9	1.00020	0.99921	1.00122	1.00043	1.00038	1.00034	0.99882
67	LCT-010 c11	1.00091	0.99952	1.00121	1.00070	1.00077	1.00134	0.99951
68	LCT-010 c12	1.00012	0.99861	1.00050	1.00018	0.99991	0.99953	0.99840
69	LCT-010 c14	1.00189	1.00025	1.00228	1.00126	1.00366	1.00362	1.00060
70	LCT-010 c15	1.00267	1.00086	1.00286	1.00139	1.00391	1.00393	1.00131
71	LCT-010 c16	1.00306	1.00134	1.00347	1.00231	1.00469	1.00354	1.00161
72	LCT-010 c17	1.00255	1.00069	1.00293	1.00216	1.00405	1.00367	1.00168
73	LCT-010 c18	1.00249	1.00049	1.00306	1.00179	1.00392	1.00352	1.00109
74	LCT-017 c13	0.99924	0.99831	0.99855	0.99854	0.99890	0.99931	0.99700
75	LCT-017 c15	0.99790	0.99745	0.99782	0.99797	0.99908	0.99882	0.99695
76	LCT-017 c18	0.99938	0.99830	0.99917	0.99858	1.00001	1.00022	0.99819
77	LCT-017 c21	0.99890	0.99760	0.99881	0.99824	0.99942	0.99986	0.99732
78	IPEN/MB-01	1.00302	1.00220	1.00146	1.00289	1.00268	1.00289	1.00080
79	LMT-007 c1	1.00005	0.99768	0.99909	0.99775	1.00097	1.00032	0.99699
80	LMT-007 c2	0.99939	0.99776	0.99899	0.99884	1.00027	0.99945	0.99705
81	PNL-3	0.99371	0.98738	0.99426	0.98745	0.98857	0.98690	1.00470
82	PNL-4	0.99953	0.99315	1.00073	0.99197	0.99484	0.99278	1.01131
83	PNL-5	1.00569	0.99961	1.00799	0.99956	1.00128	0.99996	1.01956
84	PST011c16-1	1.00903	1.00312	1.01202	1.00320	1.00479	1.00275	1.02270
85	ORNL-11	1.00120	0.99968	0.99624	1.00281	1.00135	1.00007	0.99956

* Statistical uncertainties of the calculated k_{eff} are less than 10 pcm.

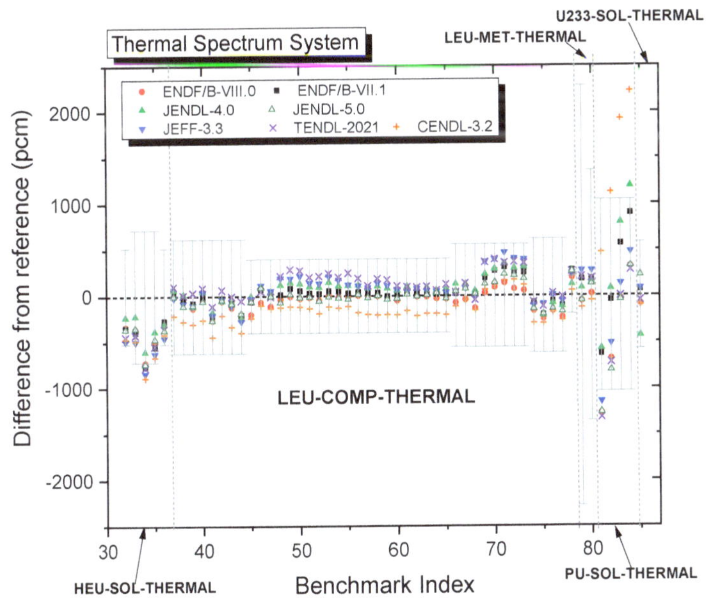

Figure 5. The difference between calculated and experimental k_{eff} by ENDL for the ICSBEP thermal benchmarks.

Table 6. RMS errors and chi square values of the 54 thermal benchmarks for different evaluated nuclear data libraries.

ENDL	All Cases		Cases Excluding PU-SOL-THERMAL		PU-SOL-THERMAL	
	RMS Error (pcm)	χ^2	RMS Error (pcm)	χ^2	RMS Error (pcm)	χ^2
ENDF/B-VII.1	252	0.76	195	0.71	620	1.19
ENDF/B-VIII.0	265	0.72	180	0.63	735	1.43
JENDL-4.0	279	0.77	190	0.68	777	1.48
JENDL-5.0	274	0.77	186	0.69	762	1.48
JEFF-3.3	308	0.96	257	0.93	674	1.31
TENDL-2021	311	0.99	241	0.95	760	1.48
CENDL-3.2	512	1.24	272	0.96	1619	3.05

Regarding these results, it can be observed that there are no significant differences in k_{eff} between the new and old version ENDLs. The RMS error of ENDF/B-VII.1 is 252 pcm whereas that of ENDF/B-VIII.0 is 265 pcm. Similarly, the RMS error of JENDL-4.0 is 279 pcm while that of JENDL 5.0 is 274 pcm. In the same manner as the fast benchmark cases, the JEFF-3.3 results are very similar to the TENDL-2021 results; the RMS errors of JEFF-3.3 and TENDL-2021 are 308 pcm and 311 pcm, respectively. In the PU-SOL-THERMAL cases, there is wide disparity in k_{eff} among the ENDLs as shown in Figure 5 and Table 6. It is worth mentioning that the difference in the thermal ^{239}Pu cross sections among the ENDLs affects the k_{eff} in the thermal spectrum system with fuels containing a significant fraction of plutonium.

3.3. Code-to-Code Comparison for ICSBEP Benchmarks

For code verification and validation, the McCARD results were compared to the MCNP results obtained from References [2,13] for the selected ICSBEP benchmark problems. Figure 6 shows the difference between k_{eff} values by the McCARD and MCNP calculations, and Table 7 summarizes the k_{eff} differences between the two codes for each benchmark category as RMS differences. The difference ($\Delta\rho_{MCNP}$) in k_{eff} between the McCARD and MCNP codes was calculated by

$$\Delta\rho_{MCNP} = \frac{k_{McCARD} - k_{MCNP}}{k_{McCARD} \cdot k_{MCNP}} \cdot 10^5. \quad (4)$$

where k_{McCARD} and k_{MCNP} are the k_{eff} by the McCARD and MCNP codes, respectively. In the fast benchmark cases, the RMS difference for ENDF/B-VII.1 was 26 pcm whereas those for ENDF/B-VIII.0 and JENDL-4.0 were 29 and 28 pcm, respectively. In the thermal benchmark cases, the RMS difference for ENDF/B-VII.1 was 53 pcm, while those for ENDF/B-VIII.0 and JENDL-4.0 were both 45 pcm. In the NJOY processing, the thermal scattering cross sections are sensitively affected by the thermal scattering law parameters, which are used in the LEAPR module. Accordingly, the difference between the thermal scattering cross sections used in the McCARD and MCNP calculations may have led to the increased RMS difference in the thermal benchmarks. In all benchmark cases, the RMS differences ranged from 40 pcm to 49 pcm. Considering that the statistical uncertainties of the MCNP results were less than 100 pcm, it was concluded that the k_{eff} results between McCARD and MCNP are in excellent agreement.

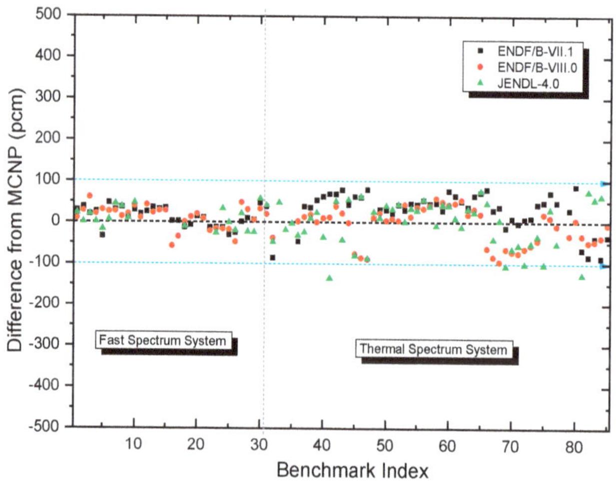

Figure 6. The difference between k_{eff} values by McCARD and MCNP calculations for the selected ICSBEP benchmark problems.

Table 7. The RMS difference between k_{eff} results by McCARD and MCNP calculations.

ENDL	RMS Difference in k_{eff} (pcm)		
	FAST	THERMAL	Total
ENDF/B-VII.1	26	53	45
ENDF/B-VIII.0	29	45	40
JENDL-4.0	28	45	49

4. Uncertainty Analyses of Criticality in ICSBEP Benchmarks

4.1. Uncertainty of k_{eff} Due to Uncertainty of Cross Sections

This section was prepared to provide an understanding of the difference in k_{eff} among the ENDLs with their cross section covariance data. In general, the mean of MC estimates on a criticality (i.e., k_{eff}) and its variance can be expressed by

$$\overline{k_{eff}} = \lim_{N\to\infty}\frac{1}{N}\sum_{i=1}^{N} k_{eff}^{i}, \tag{5}$$

$$\sigma^2\left[k_{eff}\right] = \lim_{N\to\infty}\frac{1}{N}\sum_{i=1}^{N}\left(k_{eff}^{i} - \overline{k_{eff}}\right)^2. \tag{6}$$

If one assumes that the total uncertainty on k_{eff} comes from statistical uncertainties of MC calculations and cross section uncertainties by their covariance data, Equation (6) can be rewritten as

$$\sigma^2\left[k_{eff}\right] = \lim_{N\to\infty}\frac{1}{N}\sum_{i=1}^{N}\left(k_{eff}^{i} - < k_{eff}^{i} > + < k_{eff}^{i} > - \overline{k_{eff}}\right)^2. \tag{7}$$

The angular bracket in $<k_{eff}>$ means the operator implying the expected value of a quantity on it. By the first-order Taylor expansion for $<k_{eff}>$ about the mean values of nuclear reaction cross section, $< k_{eff}^{i} > - \overline{k_{eff}}$ can be expressed by

$$< k_{eff}^{i} > - \overline{k_{eff}} \approx \sum_{i}\sum_{\alpha}\sum_{g}\left(\left(x_{\alpha,g}^{i}\right)_k - \overline{x_{\alpha,g}^{i}}\right)\left(\frac{\partial k_{eff}}{\partial x_{\alpha,g}^{i}}\right). \tag{8}$$

$x_{\alpha,g}^{i}$ is the α-type microscopic cross section of isotope i for energy group g. Substituting Equation (8) into Equation (7), one can obtain

$$\sigma^2\left(k_{eff}\right) = \sigma_S^2\left(k_{eff}\right) + \sigma_X^2\left(k_{eff}\right) \tag{9}$$

where

$$\sigma_S^2\left(k_{eff}\right) = \lim_{N\to\infty}\frac{1}{N}\sum_{i=1}^{N}\left(k_{eff}^{i} - < k_{eff}^{i} >\right)^2, \tag{10}$$

$$\begin{aligned}\sigma_X^2\left(k_{eff}\right) &= \lim_{N\to\infty}\frac{1}{N}\sum_{j=1}^{N}\left(< k_{eff}^{j} > - \overline{k_{eff}}\right)^2 \\ &= \sum_{i,\alpha,g}\sum_{i\prime,\alpha\prime,g\prime}\mathrm{cov}\left[x_{\alpha,g}^{i}, x_{\alpha\prime,g\prime}^{i\prime}\right]\left(\frac{\partial k_{eff}}{\partial x_{\alpha,g}^{i}}\right)\left(\frac{\partial k_{eff}}{\partial x_{\alpha\prime,g\prime}^{i\prime}}\right).\end{aligned} \tag{11}$$

$\sigma_S^2\left(k_{eff}\right)$ is the statistical contribution on the variance of k_{eff} whereas $\sigma_X^2\left(k_{eff}\right)$ is commonly known as the sandwich equation for S/U analyses. $\mathrm{cov}[x_{\alpha,g}^{i}, x_{\alpha\prime,g\prime}^{i\prime}]$ is the cross section covariance matrix from each ENDL. The sensitivity coefficients can be calculated by the MC perturbation technique. This S/U analysis capability was already implemented in the McCARD code [19].

To examine the uncertainty in k_{eff} due to the uncertainties of the cross sections, the benchmark problems, which have the largest difference in k_{eff} among ENDLs, were selected for each category. According to it, the uncertainty quantification in k_{eff} for Jezebel-240, Flattop-25, LCT-006c1, and PNL-5 were performed with the covariance data in each ENDL. ENDF/B-VII.1, ENDF/B-VIII.0, JENDL-4.0, JENDL-5.0, JEFF-3.3, and TENDL-2021 provide the covariance data for ν and cross section on the MF31 and MF33 sections in each ENDL, whereas there is no covariance data in the CENDL-3.2.

Table 8 shows the error of k_{eff} from reference and the uncertainty in k_{eff} due to the uncertainty of cross sections ($=\sigma_X^2\left(k_{eff}\right)$) for each ENDLs by the McCARD S/U calculations. The standard deviations of the errors among ENDLs are 194 pcm, 265 pcm, 121 pcm, and 358 pcm for Jezebel-240, Flattop-25, LCT-006 c1, PNL-5 benchmarks, respectively. Overall, it is noted that the errors of k_{eff} are less than the uncertainties of k_{eff} by the covariance data from each ENDL except the PNL-5 case with JENDL-4.0. Regarding the results, it was observed that the cross section data used in the four benchmarks have instability or uncertainty, and this led to the error of k_{eff} from the reference.

Table 8. Error of k_{eff} from reference and uncertainty of k_{eff} due to the uncertainty of cross sections for the four ICSBEP benchmarks.

ENDL	Jezebel-240		Flattop-25		LCT-006 c1		PNL-5	
	Error from Ref. *	Unc. from Cov. **	Error from Ref. *	Unc. from Cov. **	Error from Ref. *	Unc. from Cov. **	Error from Ref. *	Unc. from Cov. **
ENDF/B-VII.1	41	616	298	1456	15	904	566	624
ENDF/B-VIII.0	160	966	107	1134	−40	529	−39	1146
JENDL-4.0	−151	523	−188	848	115	919	793	510
JENDL-5.0	−130	684	81	879	−71	435	−44	1110
JEFF-3.3	151	906	431	1217	191	685	128	737
TENDL-2021	360	400	541	1222	216	683	−4	730
Standard Deviation (Error from Ref.)	194	-	265	-	121	-	358	-

* Error of k_{eff} from Reference (pcm). ** Uncertainty of k_{eff} due to the uncertainty of cross sections (pcm) = $\sigma_X^2\left(k_{eff}\right)$.

Meanwhile, O. Cabellos et al. presented the uncertainties of k_{eff} from the covariance data of various ENDLs by NDaST in the ICSBEP benchmark suite [20]. In the HEU category, the averaged uncertainties in k_{eff} due to the ^{235}U covariance data for ENDF/B-VIII.0, JENDL-3.3T4, ENDF/B-VII1, and JENDL-4.0 were 1012 pcm, 1190 pcm, 1345 pcm, and 679 pcm, whereas the averaged uncertainties of k_{eff} due to the ^{239}Pu covariance data in the PU-SOL-THERM category were 1157 pcm, 967 pcm, 608 pcm, and 687 pcm. It was noted that they were very similar to the uncertainties of the Flattop-25 in the HEU category and the PNL-3 in the PU-SOL-THERM category by the McCARD code.

4.2. Quantitative Analysis for Group-Wise Reactivity

This section shows the results of the quantitative analyses for the reactivity differences between ENDF/B-VII.1 and the other ENDL. In the quantitative analysis, the differences in absorption and fission cross sections between ENDF/B-VII.1 and the other ENDL can be expressed by the reactivity differences in the "pcm" unit for each energy group. The reactivity differences due to the difference of the absorption and fission cross section between ENDF/B-VII.1 and the other ENDL can be calculated by

$$\Delta\rho_{a,g}^{i} = \left[\frac{1}{(k_\infty)_{E71}} - \frac{\sum_{g',k'} N^{k'}\phi_{g'}\left(x_{a,g'}^{k'}\right)_{E71} - N^{i}\phi_{g}\left(\left(x_{a,g}^{i}\right)_{E71} - \left(x_{a,g}^{i}\right)_{OTR}\right)}{\sum_{g',k'} N^{k'}\phi_{g'}\left(\nu x_{f,g'}^{k'}\right)_{E71}}\right], \quad (12)$$

$$\Delta\rho_{f,g}^{i} = \left[\frac{1}{(k_\infty)_{E71}} - \frac{\sum_{g',k'} N^{k'}\phi_{g'}\left(x_{a,g'}^{k'}\right)_{E71}}{\sum_{g',k'} N^{k'}\phi_{g'}\left(\nu x_{f,g'}^{k'}\right)_{E71} - N^{i}\phi_{g}\left(\left(\nu x_{f,g}^{i}\right)_{E71} - \left(\nu x_{f,g}^{i}\right)_{OTR}\right)}\right], \quad (13)$$

where

$$(k_\infty)_{E71} = \frac{\sum\limits_{g\prime,k\prime} N^{k\prime}\phi_{g\prime}\left(\nu x^{k\prime}_{f,g\prime}\right)_{E71}}{\sum\limits_{g\prime,k\prime} N^{k\prime}\phi_{g\prime}\left(x^{k\prime}_{a,g\prime}\right)_{E71}}$$

$\left(x^i_{a,g}\right)_{E71}$ and $\left(x^i_{a,g}\right)_{OTR}$ means the ENDF/B-VII.1 and the other ENDL absorption cross section of isotope i for energy group g. $\left(\nu x^i_{f,g}\right)_{E71}$ and $\left(\nu x^i_{f,g}\right)_{OTR}$ are the product of the number of neutrons by a fission (ν) and the g-th group fission cross section of isotope i for ENDF/B-VII.1 and the other ENDL, respectively. The reactivity difference indicates the contribution of the difference in the cross section to the error in reactivity or criticality [21].

Figures 7 and 8 show the reactivity difference due to the difference of ^{239}Pu absorption and fission cross sections between ENDF/B-VII.1 and the other ENDLs for the PNL-5 benchmarks. The group-wise reactivity analyses due to the ^{239}Pu cross section changes were conducted out because ^{239}Pu is a major fuel isotope in the PNL-5 benchmark. The reactivity difference ($\Delta\rho_{E71}$) between ENDF/B-VII.1 and the other ENDL was calculated by

$$\Delta\rho_{E71} = \frac{k_{OTR} - k_{E71}}{k_{OTR} \cdot k_{E71}} \cdot 10^5. \quad (14)$$

k_{E71} and k_{OTR} are the k_{eff} by ENDF/B-VII.1 and the other ENDL. Table 9 presents the sum of group-wise reactivity differences due to the ^{239}Pu cross section changes. There are considerable reactivity differences due to the changes of ^{239}Pu absorption and fission cross sections at the thermal energy ranges (10^{-3}~1 eV). The individual group reactivity differences ranged from −1000 pcm to 900 pcm, but the group-wise reactivity differences due to absorption and fission cross section changes have the opposite sign. Therefore, the effects on the absorption and fission cross section changes were canceled out each other. It is observed that the sum of the reactivity changes by ^{239}Pu cross sections ranged from −259 pcm to 288 pcm. Meanwhile, the total reactivity difference ranged from −1353 pcm to 610 pcm because the leakage effects and the reactivity changes by the other nuclides (^{240}Pu, ^{1}H, ^{16}O, Fe, Ni, Cr) were considered in these total reactivity analyses. In the PNL-5 criticality analyses, the k_{eff} of ENDF/B-VIII.0, JENDL-5.0, JEFF-3.3, TENDL-2021 were less than ENDF/B-VII.1 whereas those of JENDL-4.0 and CENDL-3.2 were larger than ENDF/B-VII.1.

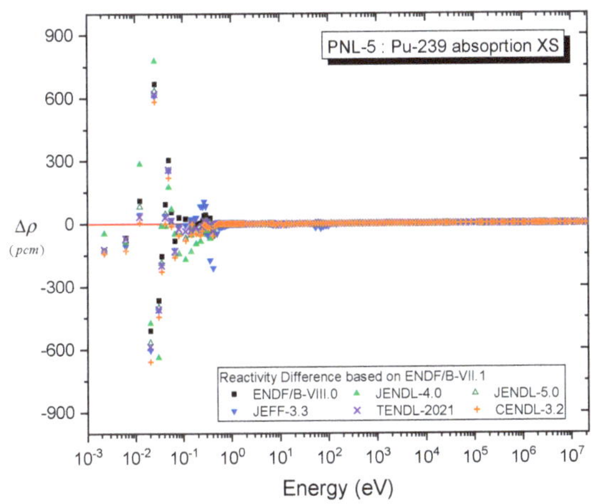

Figure 7. The reactivity difference due to the difference of ^{239}Pu absorption cross sections between ENDF/B-VII.1 and the other ENDL for PNL-5.

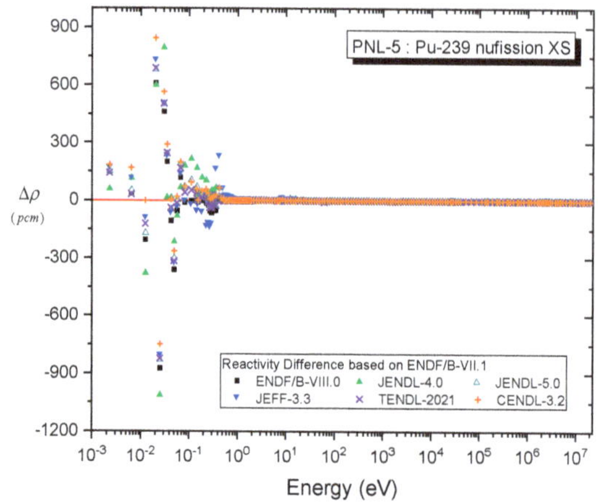

Figure 8. The reactivity difference due to the difference of ^{239}Pu fission cross sections between ENDF/B-VII.1 and the other ENDL for PNL-5.

Table 9. The sum of group-wise reactivity difference due to the difference of ^{239}Pu fission cross sections between ENDF/B-VII.1 and the other ENDLs for PNL-5.

ENDL	k_{eff}	$\Delta\rho_{E71}$ (pcm) *	Sum of Group-Wise Reactivity Differences (pcm)		
			^{239}Pu Absorption	^{239}Pu Fission	^{239}Pu Total
ENDF/B-VII.1	1.00569	-	-	-	-
ENDF/B-VIII.0	0.99961	−605	98	−266	−168
JENDL-4.0	1.00799	227	−1071	1163	92
JENDL-5.0	0.99956	−610	−873	816	−57
JEFF-3.3	1.00128	−438	−1084	796	−288
TENDL-2021	0.99996	−570	−910	701	−209
CENDL-3.2	1.01956	1353	−1642	1900	259

* $\Delta\rho_{E71}$ is the total reactivity difference between ENDF/B-VII.1 and the other ENDLs.

5. Conclusions

In this study, ICSBEP criticality analyses were conducted using the McCARD code for 85 selected benchmark problems with seven evaluated nuclear data libraries (ENDLs): ENDF/B-VII.1, ENDF/B-VIII.0, JENDL-4.0, JENDL-5.0, JEFF-3.3, TENDL-2021, and CENDL-3.2. To prepare some of the up-to-date ENDLs (i.e., ENDF/B-VIII.0, JENDL-5.0, JEFF-3.3, CENDL-3.2) for McCARD calculations, continuous energy nuclear data libraries in ACE format were generated by the NJOY code. Regarding the criticality analyses, it was noted that the k_{eff} results were sensitive to the ENDL. It is worth mentioning that the new version ENDLs showed better performance in the fast benchmark cases, while there were no significant differences in k_{eff} among the different ENDLs in the thermal benchmark cases. In all benchmark cases, the TENDL-2021 results were very similar to the JEFF-3.3 results because TENDL-2021 shared the raw nuclear data of the JEFF ENDL for ^{1,2,3}H, ^{3,4}He, ^{6,7}Li, ^{10,11}B, ^{7,9}Be, ^{12,13}C, ^{14,15}N, ^{16,17,18}O, ^{19}F, ^{232}Th, ^{233,235,238}U and ^{239}Pu isotopes.

The sensitivity of the k_{eff} results to the different ENDLs may impact certain nuclear core design parameters such as shutdown margin, critical boron concentration, and power defects. Consequently, nuclear core designers should consider this sensitivity to the ENDL as a margin of uncertainty. This study and k_{eff} results will be a good reference for the development of new types of nuclear cores or new design codes.

Author Contributions: Conceptualization, H.J.P.; methodology, H.J.P.; software, H.J.P. and S.H.C.; validation, H.J.P., M.A., S.H.C. and S.-A.Y.; formal analysis, H.J.P. and S.-A.Y.; investigation, H.J.P.; resources, H.J.P.; data curation, H.J.P.; writing—original draft preparation, H.J.P.; writing—review and editing, S.H.C.; visualization, H.J.P.; supervision, H.J.P.; project administration, S.H.C.; funding acquisition, H.J. and S.H.C. All authors have read and agreed to the published version of the manuscript.

Funding: This research was supported by KAERI and King Abdullah City for Atomic and Renewable Energy (K.A.CARE), Kingdom of Saudi Arabia, within the Joint Research and Development Center.

Conflicts of Interest: The authors declare no conflict of interest. The funders had no role in the design of the study; in the collection, analyses, or interpretation of data; in the writing of the manuscript, or in the decision to publish the results.

Abbreviations

ENDL	Evaluated Nuclear Data Library
JAERI	Japan Atomic Energy Research Institute
JENDL	Japanese Evaluated Nuclear Data Library
JEFF	Joint Evaluated Fission and Fusion
CENDL	Chinese general purpose Evaluated Nuclear Data Library
ICSBEP	International Criticality Safety Benchmark Problem
RMS	Root Mean Square
LEU	Low-Enriched Uranium
HEU	High-Enriched Uranium
MET	Metal
COMP	Compound
SOL	Solution
TSL	Thermal Scattering Law

References

1. Chadwick, M.B.; Herman, M.; Obložinský, P.; Dunn, M.; Danon, Y.; Kahler, A.; Smith, D.; Pritychenko, B.; Arbanas, G.; Arcilla, R.; et al. ENDF/B-VII.1 Nuclear Data for Science and Technology: Cross Sections, Covariances, Fission Product Yields and Decay Data. *Nucl. Data Sheets* **2011**, *112*, 2887–2996. [CrossRef]
2. Brown, D.A.; Chadwick, M.B.; Capote, R.; Kahler, A.C.; Trkov, A.; Herman, M.W.; Sonzogni, A.A.; Danon, Y.; Carlson, A.D.; Dunn, M.; et al. ENDF/B-VIII.0: The 8th Major Release of the Nuclear Reaction Data Library with CIELO-project Cross Sections, New Standards and Thermal Scattering Data. *Nucl. Data Sheets* **2018**, *148*, 1–142. [CrossRef]
3. Shibata, K.; Iwamoto, O.; Nakagawa, T.; Iwamoto, N.; Ichihara, A.; Kunieda, S.; Chiba, S.; Furutaka, K.; Otuka, N.; Ohsawa, T.; et al. JENDL-4.0: A New Library for Nuclear Science and Engineering. *J. Nucl. Sci. Technol.* **2011**, *48*, 1–30. [CrossRef]
4. Iwamoto, O.; Iwamoto, N.; Shibata, K.; Ichihara, A.; Kunieda, S.; Minato, F.; Nakayama, S. Status of JENDL. *EPJ Web Conf.* **2020**, *239*, 09002. [CrossRef]
5. Plompen, A.J.M.; Cabellos, O.; Jean, C.D.S.; Fleming, M.; Algora, A.; Angelone, M.; Archier, P.; Bauge, E.; Bersillon, O.; Blokhin, A.; et al. The Joint Evaluated Fission and Fusion Nuclear Data Library, JEFF-3.3. *Eur. Phys. J. A* **2020**, *56*, 181. [CrossRef]
6. Koning, A.J.; Rochman, D.; Sublet, J.-C.; Dzysiuk, N.; Fleming, M.; van der Marck, S. TENDL: Complete Nuclear Data Library for Innovative Nuclear Science and Technology. *Nucl. Data Sheets* **2019**, *155*, 1–55. [CrossRef]
7. Koning, A.J.; Duijvestijn, M.C.; Hilaire, S. TALYS-1.0. In Proceedings of the International Conference on Nuclear Data for Science and Technology, Nice, France, 22–27 April 2007; Volume 211.
8. Ge, Z.; Xu, R.; Wu, H.; Zhang, Y.; Chen, G.; Jin, Y.; Shu, N.; Chen, Y.; Tao, X.; Tian, Y., et al. CENDL-3.2: The New Version of Chinese General Purpose Evaluated Nuclear Data Library. *EPJ Web Conf.* **2020**, *239*, 09001. [CrossRef]
9. *International Handbook of Evaluated Criticality Safety Benchmark Experiments*; OECD Nuclear Energy Agency Report NEA/NSC/DOC(95)03; OECD Nuclear Energy Agency: Paris, France, 1998.
10. Greene, N.M.; Petrie, L.M.; Westfall, R.M.; Bucholz, J.A.; Hermann, O.W.; Fraley, S.K. *SCALE: A Modular Code System for Performing Standardized Computer Analyses for Licensing Evaluations*; ORNL/TM-2005/39; U.S. Nuclear Regulatory Commission: Washington, DC, USA, 2005.
11. MCNP. *User Manual–Code Version 6.2*; LA-UR-17-29981; Los Alamos National Security LLC: Los Alamos, MN, USA, 2017.
12. Alcouffe, R.E.; Baker, R.S.; Brinkley, F.W.; Marr, D.R.; O'Dell, R.D.; Walters, W.F. *DANTSYS: A Diffusion Accelerated Neutron Particle Transport Code System*; LA-12969-M; Los Alamos National Lab.: Los Alamos, NM, USA, 1995.
13. van der Marck, S.C. Benchmarking ENDF/B-VII.1, JENDL-4.0, and JEFF-3.1.1 with MCNP6. *Nucl. Data Sheets* **2012**, *113*, 2935–3005. [CrossRef]

14. Park, H.J.; Kang, H.; Lee, H.C.; Cho, J.Y. Comparison of ENDF/B-VIII.0 and ENDF/B-VII.1 in Criticality, Depletion Benchmark, and Uncertainty Analyses by McCARD. *Ann. Nucl. Energy* **2019**, *131*, 443–459. [CrossRef]
15. Zheng, L.; Huang, S.; Wang, K. Criticality Benchmarking of ENDF/B-VIII.0 and JEFF-3.3 Neutron Data Libraries with RMC code. *Nucl. Eng. Technol.* **2020**, *52*, 1917–1925. [CrossRef]
16. Kim, K.-S.; Wieselquist, W.A. Neutronic Characteristics of ENDF/B-VIII.0 Compared to ENDF/B-VII.1 for Light-Water Reactor Analysis. *J. Nucl. Eng.* **2021**, *2*, 318–335. [CrossRef]
17. Shim, H.J.; Han, B.-S.; Jung, J.-S.; Park, H.-J.; Kim, C.-H. McCARD: Monte Carlo Code for Advanced Reactor Design and Analysis. *Nucl. Eng. Technol.* **2012**, *44*, 161–176. [CrossRef]
18. MacFarlane, R.E.; Muir, D.W.; Boicourt, R.M.; Kahler, A.C., III; Conlin, J.L. *The NJOY Nuclear Data Processing System Version 2016*; LA-UR-17-20093; Los Alamos National Lab.: Los Alamos, NM, USA, 2016.
19. Park, H.J.; Shim, H.J.; Kim, C.H. Uncertainty Propagation in Monte Carlo Depletion Analysis. *Nucl. Sci. Eng.* **2011**, *167*, 196–208. [CrossRef]
20. Cabellos, O.; Dyrda, J.; Soppera, N. Checking, Processing and Verification of Nuclear Data Covariances. *EPJ Nucl. Sci. Technol.* **2018**, *4*, 39. [CrossRef]
21. Park, H.J.; Shim, H.J.; Joo, H.G.; Kim, C.H. Generation of Few-group Diffusion Theory Constants by Monte Carlo Code McCARD. *Nucl. Sci. Eng.* **2012**, *172*, 66–77. [CrossRef]

entropy

MDPI

Article

Investigation of a Negative Step Effect on Stilling Basin by Using CFD

Lei Jiang, Minjun Diao * and Chuan'ai Wang

State Key Laboratory of Hydraulics and Mountain River Development and Protection, Sichuan University, Chengdu 610065, China
* Correspondence: diaomingjun@scu.edu.cn

Abstract: The negative-step stilling basin is an efficient and safe energy dissipator for high-head, large-unit discharge high-dam projects. However, studies of the effects of the negative step on the hydraulic performance of a high-dam stilling basin have not been conclusive. In the present study, a 2D RANS-VOF numerical model was developed to simulate the flow field of a negative-step stilling basin. The numerical model was validated with a physical model and then used to simulate and test the performance of the negative-step stilling basin with different step heights and incident angles. The results showed that the flow pattern, the free-surface profile, the velocity profile, the characteristic lengths are strongly influenced by the step geometry. Increasing the height of the step will increase the relative flow depth and the reattachment length in the basin, but reduce the bottom velocity and the roller length. The incident angle has no significant influence on the flow pattern of the negative-step stilling basin, and increasing the incident angle of the step will reduce the bottom velocity and the reattachment length. Both the step height and the incident angle have no significant influence on the energy dissipation efficiency because of the high submergence conditions in this study.

Keywords: negative-step stilling basin; hydraulic performance; 2D RANS-VOF numerical model; flow pattern; free-surface profile; velocity profile; energy dissipation efficiency

1. Introduction

In recent years, more and more high-head, large-unit discharge high-dam hydropower projects have adopted hydraulic jump stilling basins. Compared with the ski-jump energy dissipator, hydraulic jump stilling basins are more adaptable to complex terrain, less likely to cause stronger flood discharge atomization, and have less impact on the environment. However, the slab at the bottom of high-dam stilling basins is easily damaged by floods. The most notable cases of slab damage are the stilling basin of Malpaso Dam, the Karnafuli Dam, and the Wu Qiang-Xi Hydroelectric Station [1].

Studies have shown that excessive bottom velocity is the most important cause of slab damage problems [2]. To reduce the bottom velocity, researchers have proposed applying the negative-step stilling basin (see Figure 1) to high-dam hydropower projects. Experiences of Xiangjiaba, Guandi, and other hydropower stations (as shown in Table 1) show that the negative-step stilling basin can effectively solve the slab damage problem [2,3]. However, hydraulic characteristics of hydraulic jumps at a negative step under a high-head, large-unit discharge condition have not been sufficiently investigated, which leads to difficulties for designers to accurately determine the step geometry parameters of the negative-step stilling basin of high dams.

Hydraulic jumps at a negative step and a negative-step stilling basin have been extensively studied and documented [5–9]. Moore and Morgan [10] showed four types of flow patterns at a negative step, that is A-jump, Wave-jump, B-jump, and minimum B-jump. Furthermore, they derived an expression for the sequent depth ratio, $Y = h_2/h_1$, in terms of the approaching Froude number, F_1, and the relative pressure head on the step, h_d/h_1. Hager and Bretz [6] showed that the shape of the negative step did not have a significant effect on the transition of the flow regime, but the relative step height had a significant effect on the sequent depth ratios and the relative energy dissipation. The relative length of the roller can be given as a function of F_1. Eroglu and Tastan [11] investigated the local energy losses at positive and negative steps in subcritical open channel flows. They found empirical equations for local energy losses depending on the Froude number and the relative step height. However, in the above studies, only low-head flows were considered.

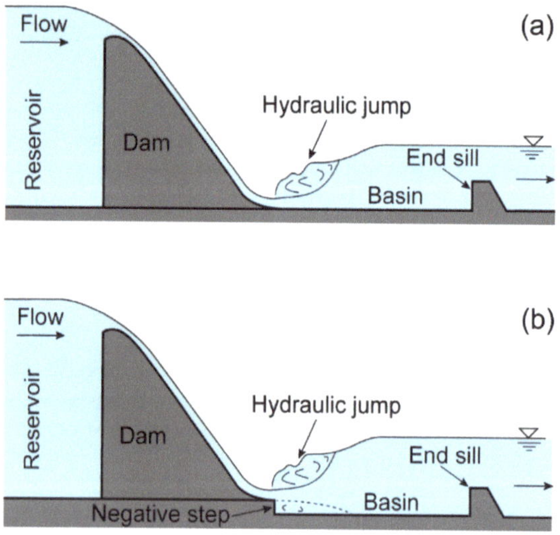

Figure 1. Schematic diagram of hydraulic jump stilling basins: (**a**) traditional stilling basin; (**b**) negative-step stilling basin.

Table 1. Representative high-head projects in China using the negative-step stilling basin [4].

No.	Name	Dam Height (m)	Design Discharge (m^3/s)	Step Height (m)
1	Xiangjiaba	162.0	41,200	9.0
2	Huangjinping	95.5	5650	-
3	Jin'anqiao	160.0	11,668	-
4	Guanyinyuan	159.0	16,900	7.5
5	Liyuan	155.0	11,361	15.8
6	Guandi	168.0	14,000	6.5
7	Tingzikou	110.0	34,500	8.0

In recent years, there has been an increasing number of studies on negative-step stilling basins due to the construction of many negative-step stilling basins in China. For example, Sun et al. [3] conducted a 1:90 model test with the Xiangjiaba Hydropower Station, and the results showed that the bottom velocity in the basin was significantly reduced after the use of the negative step, and the height of step had a significant impact on the bottom pressure and energy dissipation efficiency. Li [2] performed a model test study on the pressure and uplift force on the floor of the negative-step stilling basin. Jia et al. [12,13] and Huang et al. [4] investigated the pulsating pressure and wave characteristics of the negative-step stilling basin, respectively. Most of the previous research has been conducted by means of physical model tests for specific projects to analyze certain hydraulic characteristics (e.g., Bottom velocity, bottom pressure, pulsating pressure, and wave) of the negative-step stilling basin. These studies have increased the understanding of the negative-step stilling basin of high dams and successfully guided the design and operation of specific projects. However, because physical model experiments are expensive and time consuming, step geometry has rarely been used in previous studies. Therefore, there are no general conclusions about the influence of negative step geometry on the hydraulic performance of stilling basins.

Apart from the traditional experimental approach to studying hydraulic jumps, computational fluid dynamics (CFD) techniques are also useful tools with undoubtedly growing potential as computing power increases. A large number of studies on hydraulic structures in recent years have been conducted using CFD techniques. Although there are many turbulence models available, the Reynolds averaged Navier–Stokes (RANS) modeling is frequently applied to hydraulic structures including hydraulic jump stilling basins [14]. For example, Carvalho et al. [15] used RANS modeling with RNG *k*-ε closure and a two-dimensional volume-of-fluid (VOF) method to numerically model a stilling basin downstream of a spillway with a Froude number $F_1 = 6$. The computed velocity, pressure obtained by this method were in good agreement with the laboratory measurements. Valero et al. [14] applied the same method to simulate the flow field of a USBR Type III hydraulic jump stilling basin under the design and adverse conditions for stepped and smooth spillways. The fine grid convergence index (GCI) as suggested by Celik et al. [16] was used in their study to perform a mesh sensitivity analysis. Zhou and Wang [17] simulated the 3D flow field among a compound stilling basin using the commercial CFD package FLOW-3D. The simulation results of four turbulence models (standard *k*-ε, RNG *k*-ε, realizable *k*-ε, and large eddy simulation turbulence models) were compared, and the RNG *k*-ε turbulence model showed the most successful agreement with the experimental results. Macian-Perez et al. [18] investigated the performance of a USBR Type II stilling basin using a CFD numerical model. The sequent depth ratio, the roller and hydraulic jump lengths, and the energy dissipation efficiency were analyzed in their study. The previous research showed that CFD complements experimental studies and the inherent measurement limitations, providing additional insight into complex flows [14]. In particular, when properly applied, RANS-based one- and two-equation turbulence models have been shown to provide accurate averaged force, distribution, and velocity fields [14]. Therefore, CFD has become an effective tool to study the hydraulic performance of stilling basins.

In general, additional research is needed regarding the influence of the negative step geometry on the hydraulic performance of the stilling basins of high dams. In this study, 2D RANS-VOF numerical simulations were performed to achieve the research goal. The initial negative-step stilling basin was designed based on the Guandi Hydropower Station. Nine negative steps with different heights (d) and different incident angles (θ) were selected to study the effects of step geometry on the flow characteristics of the negative-step stilling basin. The flow patterns, velocity profiles, characteristic lengths, and energy dissipation efficiency of the negative-step stilling basin were analyzed.

2. Experiment Setup

An initial negative-step stilling basin (Run3) was designed according to the hydraulic conditions of the Guandi Hydropower Station in the Yalong River region, Sichuan Province, China. The physical model test was carried out at a geometrical scale of 1:80. The model, consisting of a spillway and a stilling basin, was assembled in a 50-cm-wide and 60-cm-deep channel. A flap gate was installed at the downstream end of the channel to adjust the tailwater depth. The spillway and the stilling basin were made of transparent Plexiglass with a thickness of 8 mm, allowing for flow observation. The prototype spillway has gates and piers. However, in order to generalize this study, no gates or piers were installed in the model spillway, which can make the downstream flow approximately two-dimensional. Figure 2 shows the device of the physical model test.

Figure 2. Photo of the physical model and equipment.

Both the mean free-surface longitudinal profile and the maximum forward average velocity profile were measured in the physical tests. The discharge was measured by means of a calibrated V-notch weir. The mean free-surface was measured using a point gauge. The origin of the (x, y) coordinate system was located at the beginning of the stilling basin. The bottom velocity (y = 2.0 cm) profiles were measured using a flowmeter. The following dimensionless quantities were used for a comparison between the experiments and numerical simulations:

$$X = \frac{x - x_1}{L_1} \tag{1}$$

$$Y = \frac{y - h_1}{h_t - h_1} \tag{2}$$

$$U_b = \frac{u_b}{u_1} \tag{3}$$

where x_1 is the jump toe position; L_1 is the roller length; h_1 is the flow depth at jump toe; u_b is the bottom velocity (y = 2.0 cm); u_1 is the mean flow velocity at the jump toe [18].

3. Numerical Simulations

3.1. Flow Equations

The 2D Reynolds-averaged continuity and Navier–Stokes equations (RANS) were used in the present study to simulate the flow field of the negative-step stilling basins. For an incompressible, Newtonian fluid flow, flow equations can be expressed as [19]:

$$\frac{\partial u_i}{\partial x_i} = 0 \tag{4}$$

$$\rho\left(\frac{\partial u_i}{\partial t} + u_j\frac{\partial u_i}{\partial x_j}\right) = -\frac{\partial p}{\partial x_i} + \mu\frac{\partial^2 u_i}{\partial x_j \partial x_j} + \frac{\partial}{\partial x_j}\left(-\rho\overline{u_i' u_j'}\right) + \rho K_i \tag{5}$$

where u_i is the velocity component in x_i direction; t is the time; ρ is the density; p is the pressure; μ is the dynamic viscosity; $-\rho\overline{u_i' u_j'}$ is the turbulence stresses; u_i' and u_j' are the horizontal and vertical velocity fluctuations, respectively; and K_i is the body forces.

The turbulence stress is derived from the following linear constitutive equation [20]:

$$-\rho\overline{u_i' u_j'} = \mu_t\left(\frac{\partial \overline{u}_i}{\partial x_j} + \frac{\partial \overline{u}_j}{\partial x_i}\right) - \frac{2}{3}\sigma_{ij}\rho k \tag{6}$$

where μ_t is the eddy viscosity; σ_{ij} is the Kronecker delta; and k is the turbulent kinetic energy, $k = \overline{u_i u_i}/2$.

In the present numerical simulations, to determine the turbulent viscosity μ_t in Equation (6), the RNG k-ε turbulence model, based on the Boussinesq eddy-viscosity assumption, was used. The RNG k-ε turbulence model usually provides better results when simulating swirling flows relative to the standard k-ε model [21,22]. The RNG k-ε turbulence model expresses the turbulent viscosity in terms of the turbulent kinetic energy k and the turbulent kinetic energy dissipation rate ε as follows [23]:

$$\mu_t = \rho C_\mu \frac{k^2}{\varepsilon} \tag{7}$$

where C_μ is the turbulence constant, $C_\mu = 0.09$. The corresponding k and ε equations are as follows:

$$\frac{\partial(\rho k)}{\partial t} + \frac{\partial(\rho k \overline{u}_j)}{\partial x_j} = \frac{\partial}{\partial x_j}\left(\Gamma_k \frac{\partial k}{\partial x_j}\right) + G_k - \rho\varepsilon \tag{8}$$

$$\frac{\partial(\rho\varepsilon)}{\partial t} + \frac{\partial(\rho\varepsilon \overline{u}_j)}{\partial x_j} = \frac{\partial}{\partial x_j}\left(\Gamma_\varepsilon \frac{\partial \varepsilon}{\partial x_j}\right) + C_{1\varepsilon}\frac{\varepsilon}{k}G_k - C_{2\varepsilon}\rho\frac{\varepsilon^2}{k} - R \tag{9}$$

where Γ_k and Γ_ε are the diffusivity terms, $\Gamma_k = (\mu + \mu_t/\sigma_k)$, $\Gamma_\varepsilon = (\mu + \mu_t/\sigma_\varepsilon)$. G_k is expressed as:

$$G_k = -\rho\overline{u_i' u_j'}\frac{\partial u_i}{\partial x_j} \tag{10}$$

In the RNG k-ε turbulence model, $C_{1\varepsilon} = 1.44$, $C_{2\varepsilon} = 1.92$, $\sigma_k = 1.0$, $\sigma_\varepsilon = 1.3$, and the coefficient R in Equation (9) is given by [24]:

$$R = \frac{\rho C_\mu \eta^3 (1 - \eta/\eta_0)}{1 + \beta\eta^3}\frac{\varepsilon}{k} \tag{11}$$

where η is the relative strain parameters, $\eta = Sk/\varepsilon$; S is the mean strain rate, $S = \sqrt{S_{ij}S_{ij}}$; $\eta_0 = 4.38$ and $\beta = 0.012$. Regarding the free surface modeling, the VOF method was adopted [25], which employs an additional variable called volume fraction (α) to capture the free surface at the interface between the air and the water. Based on the VOF method, $\alpha = 1$ denotes a region that is fully inundated with water, and $\alpha = 0$ corresponds to air. The free-surface volume fraction was specified to be $\alpha = 0.5$ [26]. The governing equations were numerically solved using Fluent computer code (ANSYS FLUENT Academic v.19.0).

3.2. Numerical Simulation Runs and Solution Domain

The negative-step stilling basin numerical simulation matrix is summarized in Table 2. In Table 2, d is the negative step height; θ is the step incident angle; Q is the flow discharge; q is the unit discharge; h_1, u_1, and F_1 are the mean flow depth, mean velocity, and Froude number at the toe section, respectively; and h_t is the tailwater depth. Stilling basins with six different negative step heights (d = 0 cm, 2.5 cm, 5 cm, 7.5 cm, 10 cm, 15 cm) and five different incident angles (θ = 0°, 5°, 10°, 15°, 20°) were investigated. It is important to note that the stilling basin matches the physical model dimensions. All CFD tests were performed under the design discharge Q = 0.09 m^3/s (discharge per unit width is q = 0.18 m^2/s), and the tailwater depth was 0.388 m.

Table 2. The simulation conditions.

Run	d (cm)	θ (°)	Q (m^3/s)	q (m^2/s)	h_1 (m)	u_1 (m/s)	F_1	h_t (m)
M1	0	0	0.09	0.18	0.033	5.386	9.4	0.388
M2	5	0	0.09	0.18	0.034	5.262	9.1	0.388
M3	10	0	0.09	0.18	0.037	5.091	8.6	0.388
M4	12.5	0	0.09	0.18	0.035	5.087	8.6	0.388
M5 (S1)	15	0	0.09	0.18	0.036	4.945	8.3	0.388
S2	15	5	0.09	0.18	0.038	4.782	7.9	0.388
S3	15	10	0.09	0.18	0.034	5.367	9.4	0.388
S4	15	15	0.09	0.18	0.034	5.301	9.2	0.388
S5	15	20	0.09	0.18	0.035	5.126	8.7	0.388

Figure 3 shows the schematic diagram of the calculation domain. As shown in Figure 3, the calculation domain consists of the reservoir region, spillway channel, stilling basin, and downstream region. The distance from the inlet boundary to the weir crest exceeded 10 times the height of the design heads and the outlet boundary was far enough from the end sill to ensure the formation of stable tailwater. Referring to the downstream stilling basin slab, the dam height reached 1.58 m. The length of the stilling basin was 2.125 m and an end sill of 0.25 m height was located at the end of the stilling basin. The length of the downstream region behind the height of the end sill was 2 m, which is enough to ensure that the downstream flow has sufficient distance to fully develop.

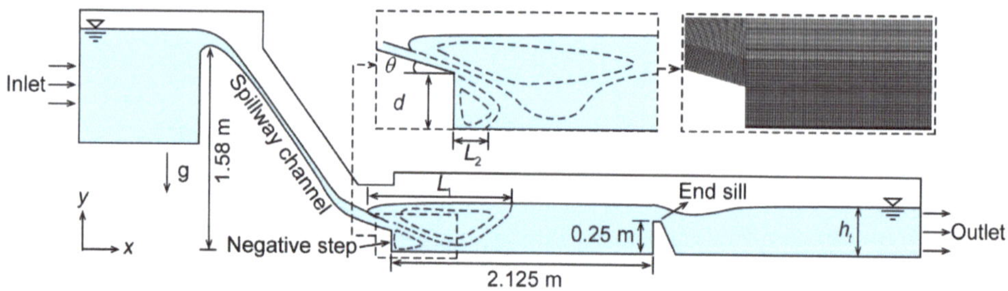

Figure 3. Schematic view of the calculation domain.

3.3. Numerical Settings

The explicit volume scheme was used to solve the volume fraction equation. The PISO (pressure implicit with splitting of operators) algorithm [27], coupled with a second-order backward scheme, was used to solve the momentum equation. The time term was discreted by a first-order implicit scheme [28]. A time step of $\Delta t = 0.0001$ s was set for time discretization, which gave 400,000 iterations for the simulation of each case. The CFD results of each case shows that the inlet and outlet flow discharges tend to be constant when the computation time $t > 30$ s, and a total simulation time of $T = 40$ s can guarantee that each simulation is converged.

3.4. Meshing and Boundary Conditions

Structured mesh systems were used for the entire solution domain. For both upstream and downstream parts, a coarse mesh in the flow direction was used to reduce the number of cells and computing time [29]. Ten boundary layers contracting with a ratio of 1.1 were placed at the bottom wall of the stilling basin, and the grid point of the first layer was placed in the log law region ($30 < y+ < 200 \sim 400$) [17]. A constant average velocity was used at the upstream inlet boundary for the water phase. The value of the inlet average velocity was calculated according to the design discharge (Table 2). The initial upstream water depth in the inlet boundary was given according to the test conditions. The outlet boundary was defined as a pressure outlet condition with a specified water height, which was realized through user defined functions (UDF). The upper boundary of the domain was defined as the pressure inlet condition. No-slip boundary conditions were imposed on all walls. Additionally, the standard wall function was used to compute the velocity profiles near solid boundaries [29].

3.5. Discretization Error Analysis

Discretization error analysis was carried out based on the grid convergence index (GCI) method [16]. The GCI method can be calculated as follows [16]. Here, Run M5 was used to estimate the discretization error. Three mesh sizes, $N_1 = 95{,}900$, $N_2 = 54{,}985$, and $N_3 = 34{,}015$, (fine mesh, medium mesh, and coarse mesh), were tested. The fine-grid convergence index can be computed as

$$GCI_{21}^{fine} = \frac{1.25E_{21}}{r_{21}^{P} - 1} \quad (12)$$

where $E_{21} = |(\phi_1 - \phi_2)/\phi_1|$ is the relative error between the medium and fine grids, where ϕ_1 and ϕ_2 are the fine- and medium-grid solutions for the velocity; $r_{21} = \lambda_2/\lambda_1$ is the grid refinement factor; λ_1 and λ_2 are the representative cell size of the fine- and medium-grid solutions, respectively; and P is the local order of accuracy. Table 3 illustrates the calculation procedure for the three selected grids. Figure 4 shows the calculation results of GCI. Computation of the GCI indicated that the discretization uncertainty in the vertical velocity profiles for the fine grid was within 11.3% (0.259 m/s), indicating good performance. Therefore, the fine-grid solution was then applied to the remaining numerical models.

Table 3. Calculations of the discretization error.

y (m)	ϕ_1	ϕ_2	ϕ_3	E_{21}	E_{32}	r_{21}	r_{32}	P	GCI_{21}^{fine}
0.017	2.742	2.799	2.804	0.021	0.002	1.3	1.3	9.45	0.7
0.035	2.667	2.744	2.890	0.029	0.053	1.3	1.3	2.42	10.8
0.052	2.292	2.427	2.764	0.059	0.139	1.3	1.3	3.48	11.3
0.069	1.824	1.947	2.367	0.067	0.215	1.3	1.3	4.69	6.3
0.087	1.415	1.476	1.919	0.043	0.300	1.3	1.3	7.59	1.2
0.104	1.054	1.108	1.416	0.051	0.277	1.3	1.3	6.62	1.5
0.122	0.767	0.788	0.835	0.027	0.059	1.3	1.3	3.14	2.0
0.139	0.477	0.513	0.266	0.075	0.481	1.3	1.3	7.36	0.8
0.156	0.233	0.291	0.170	0.248	0.416	1.3	1.3	2.82	6.6
0.174	0.067	0.156	0.190	1.337	0.219	1.3	1.3	3.66	6.9
0.191	0.196	0.193	0.294	0.015	0.525	1.3	1.3	13.51	0.0
0.208	0.341	0.301	0.411	0.115	0.363	1.3	1.3	3.91	2.7
0.226	0.472	0.459	0.521	0.027	0.136	1.3	1.3	6.05	0.4
0.243	0.590	0.561	0.613	0.048	0.092	1.3	1.3	2.30	4.3
0.261	0.681	0.674	0.679	0.011	0.008	1.3	1.3	1.34	2.2
0.278	0.739	0.739	0.723	0.000	0.022	1.3	1.3	24.00	0.0
0.295	0.785	0.796	0.752	0.014	0.055	1.3	1.3	5.35	0.4
0.313	0.807	0.824	0.769	0.022	0.067	1.3	1.3	4.43	1.0
0.330	0.804	0.834	0.779	0.038	0.067	1.3	1.3	2.33	4.5

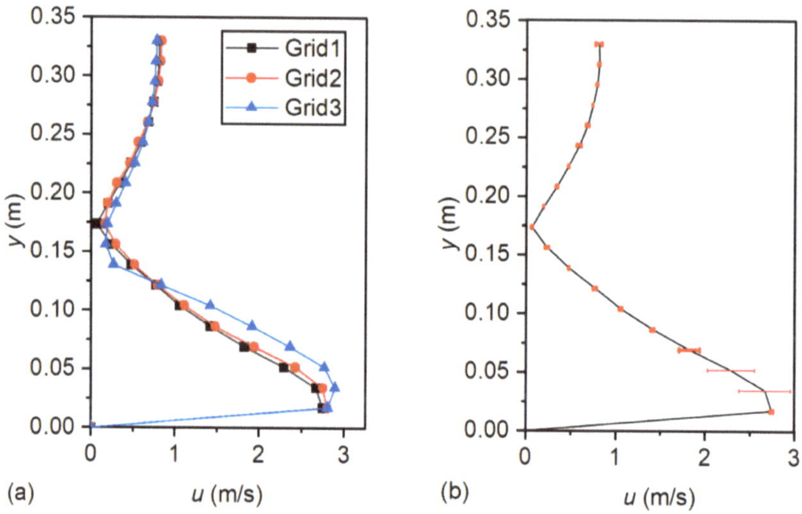

Figure 4. Vertical velocity profiles at x = 0.5 m (Run M3): (**a**) results of different meshes; and (**b**) fine-mesh solution, with discretization error bars.

4. Results

4.1. Validation of the Numerical Model

Run M3 was selected to perform the validation of the model test and numerical simulation. The mean free-surface profile and bottom velocity profile were selected to verify the accuracy of the numerical simulations. The results of the numerical simulation and model test were compared, as shown in Figure 5. According to Figure 5, the overall trend of the simulated mean free-surface profile and bottom velocity profile agreed well with the experimental results. To quantify the reliability of the numerical simulation, the root mean square error (RMSE) and the mean absolute percent error (MAPE) were used to evaluate the accuracy of the numerical results [26]. For the mean free-surface profile, the associated RMSE and MAPE were 0.048 m and 13.2%, respectively. For the bottom velocity profile, the associated RMSE and MAPE were 0.118 m/s and 9.5%, respectively. The computed results of the statistical indicators show that the numerical simulation had high accuracy.

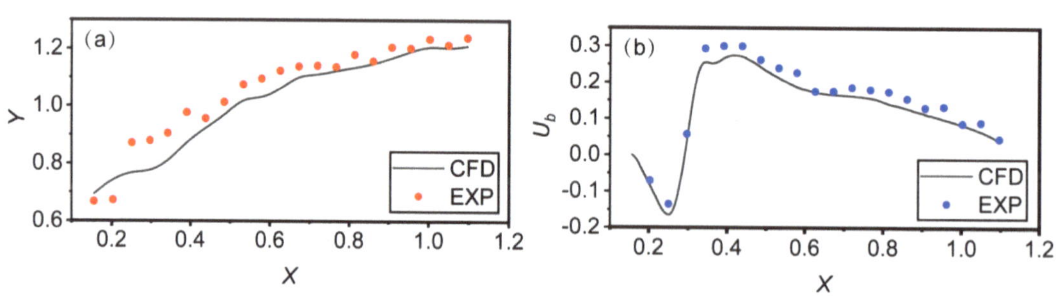

Figure 5. Validation of the numerical model: (**a**) mean free-surface longitudinal profile; and (**b**) bottom velocity profile (y = 2.0 cm).

4.2. Flow Pattern

The computed flow patterns in the negative-step stilling basins are shown in Figure 6. It can be seen that due to the high tailwater depth, the toe of the hydraulic jump moved upstream into the sloping channel, and the roller lay partly in the sloping, and partly in the horizontal channel portions [30,31]. This flow pattern was defined as the B-jump [31]. For the B-jump of the negative-step stilling basin, a reattachment region is created underneath the jet, and a roller region is created above the jet. The pressure in the reattachment region is low, causing the jet to deflect to the basin bottom and eventually attach to the wall to form a wall jet. It can also be seen from Figure 6 that the area of the reattachment region increased significantly with the increase in the step height. Moreover, Run M5 had a tendency to develop toward surface jump, which indicates that the height of the can has a large influence on the flow pattern. However, the flow pattern in the basin did not change much under different incident angles.

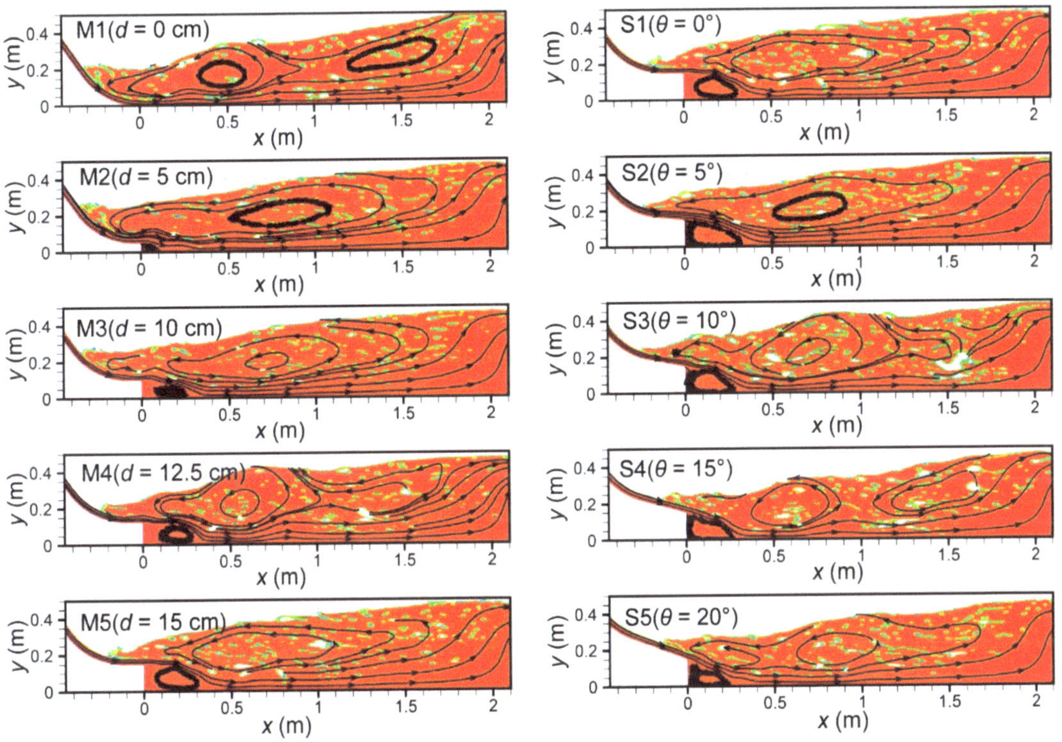

Figure 6. Flow pattern for the stilling basin with a negative step.

For the B-jump, people are usually very concerned about its jump toe position. The flowing submergence parameter E ($E = 1 - y_1/h_t$), as suggested by Hager, was used to describe the jump toe location [30]. Here, y_1 is the vertical distance between the basin floor and roller leading edge; h_t is the tailwater depth. The simulation results of the submergence parameter E in this study are given in Table 4. It can be seen that the value of E was in the range of 0.29 to 0.51 and E did not show a regular variation with the increase in the step height and incident angle. This is mainly due to the pulsating position of the jump toe, which has a great influence on the measurement accuracy.

Table 4. Submergence parameters under different cases.

Run	d (cm)	θ (°)	x_1 (m)	y_1 (m)	E
M1	0	0	−0.380	0.196	0.41
M2	5	0	−0.383	0.241	0.29
M3	10	0	−0.377	0.235	0.30
M4	12.5	0	−0.365	0.229	0.32
M5 (S1)	15	0	−0.219	0.164	0.51
S2	15	5	−0.248	0.182	0.46
S3	15	10	−0.227	0.191	0.43
S4	15	15	−0.149	0.191	0.43
S5	15	20	−0.153	0.206	0.39

The results of the dimensionless free-surface profile are shown in Figure 7. The numerical simulation results were compared with the bibliographic data from three different sources. It should be noted that the profiles of Bakhmeteff and Matzke [32] and Wang and Chanson [33] were originally obtained from research on CHJ. The profiles of Macián-Pérez [34] were obtained from the USBR II stilling basin. From the figure, it can be seen that the negative-step stilling basin had a larger flow depth compared to the CHJ and USBR II stilling basin. Comparing Run M1 and Run M5, it can be seen that the flow depth in the negative-step stilling basin will increase with the height of the step, and the location of the largest increase in the flow depth is at the beginning of the basin. Comparing Run S1 and Run S5, it can be seen that the flow depth in the negative-step stilling basin will decrease with the increase in the incident angle.

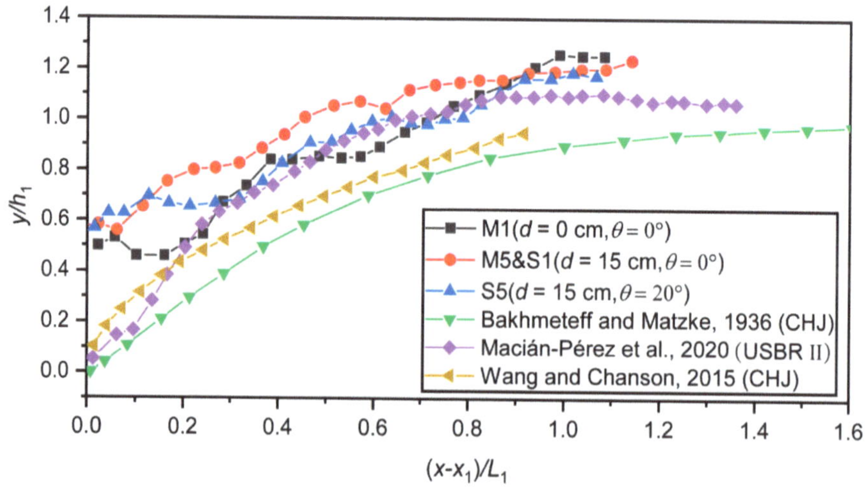

Figure 7. Dimensionless free-surface profile.

4.3. Velocity Profile

The velocity profile distributions reflect the structure of the flow in the stilling basin. In this section, the velocity profiles were obtained and averaged. The influence of the step height and incident angle on the velocity profiles was analyzed. Typical vertical velocity profiles of the negative-step stilling basin are shown in Figure 8 (Run M3). The results show that the vertical velocity profiles clearly reflect the structure of the B-jump in the negative-step stilling basin. The effect of the step on vertical flow velocity is concentrated in the reattachment region ($X < 0.4$ for Run M3). Within this range, the direction of the vertical velocity changes twice. A maximum forward velocity and a maximum backward velocity could be found at the jet core and the reverse flow region, respectively. Further downstream, the vertical velocity distribution law was gradually consistent with the typical submerged hydraulic jump vertical velocity profiles in the roller region, and a stable wall jet formed. At the end of the stilling basin, the vertical velocity distribution was similar to that of the open channel flow, with the maximum forward velocity appearing near the free surface.

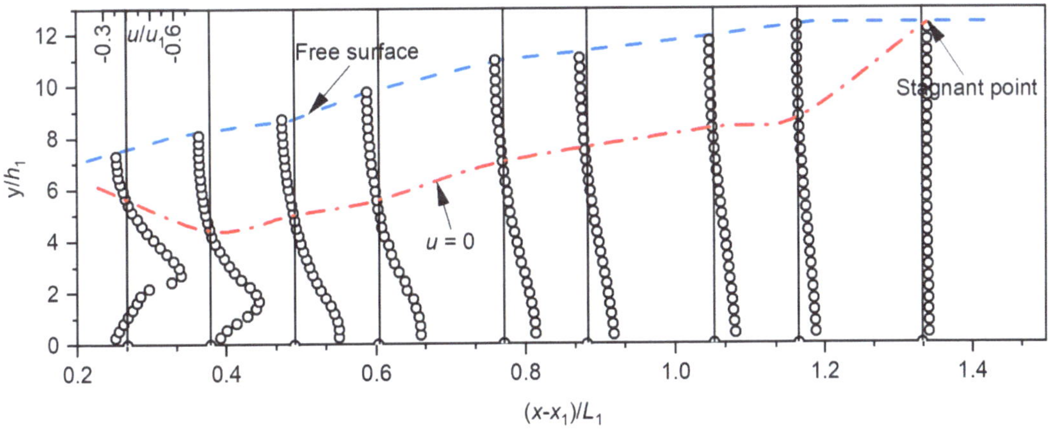

Figure 8. Typical vertical velocity profiles of the stilling basin with a negative step (Run M3).

The high-speed flow decreased rapidly after entering the stilling basin. The evolution of the section's maximum velocity reflects the decay process of the flow energy, and so represents a potential cause of cavitation damage [14]. Therefore, researchers tend to use the maximum velocity as an important indicator to describe the performance of a stilling basin. In the studied cases, the dimensionless maximum velocity decay is presented in Figure 9. Here, u_{max} corresponds to the maximum velocity located at a transversal section at a distance $(x - x_1)/L_1$. From Figure 9, we can see that although the step parameters varied from case to case, the maximum velocity basically decayed along the $u_{max}/u_1 = 1.19e^{-1.84(x-x_1)/L_1}$ ($R^2 = 0.93$) curve, and sudden changes in the maximum velocity decay were observed downstream of the negative step.

Bottom velocity (u_b) is also an important reference indicator for the design of the stilling basin. In the present study, the velocity at the position of $y = 2$ cm above the basin floor was taken as the critical bottom velocity. The effect of the step height on the bottom velocity distribution is shown in Figure 10. For traditional stilling basin (Run M1), u_b decreases along the x-axis direction, and the maximum bottom velocity (u_{bmax}) is generally found at the beginning of the basin. For the negative-step stilling basin (Run M2~M5), the bottom velocity profiles were characterized by a reverse flow range (reattachment region) downstream of the basin head, and a minimum bottom velocity (u_{bmin}) could be found in this reverse flow range. Subsequently, u_b increases rapidly along the flow direction to u_{bmax}. As the step height increases, u_{bmax} decreases gradually, and the location of the maximum bottom velocity (x_{bmax}) moves downstream gradually. The effect of the incident angle on the bottom velocity distribution is shown in Figure 11. It can be seen that with the increase in the incident angle, u_{bmax} gradually increases and its position moves upstream. Table 5 lists the extreme bottom velocity and positions of each case. Comparing M1 and M5, it can be seen that u_{bmax} decreased by 55.3% when $d/h_1 = 0$ increased from 0 to 4.12; comparing S1 and S5, it can be seen that u_{bmax} increased by 24.5% when θ increased from 0 to 20°.

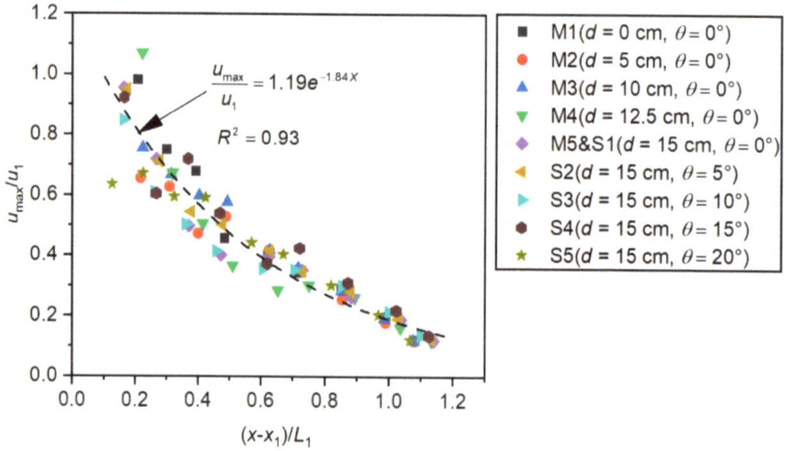

Figure 9. Maximum forward velocity decay along the *x*-axis direction.

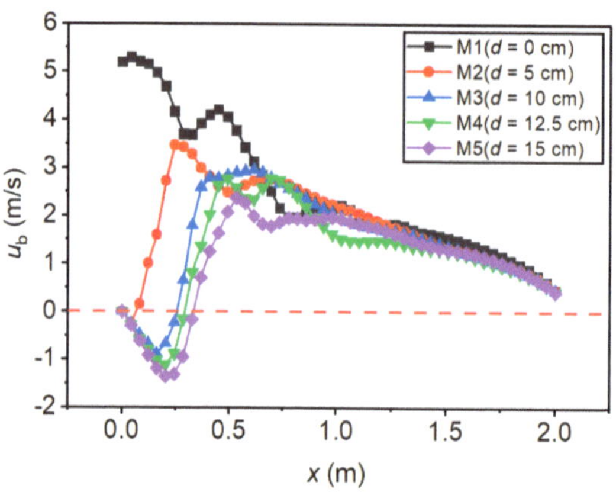

Figure 10. Bottom velocity profiles under different step heights.

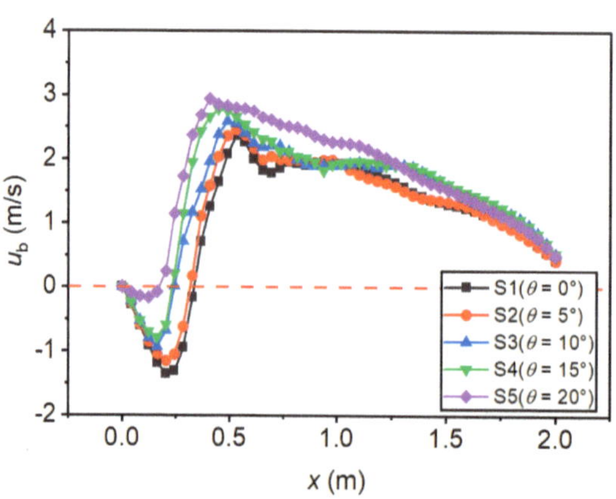

Figure 11. Bottom velocity profiles under different incident angles.

Table 5. The extreme bottom velocity and position.

Run	x_{bmin} (m)	u_{bmin} (m/s)	x_{bmax} (m)	u_{bmax} (m/s)
M1	-	-	0.041	5.293
M2	0.041	−0.239	0.245	3.473
M3	0.163	−0.892	0.612	2.954
M4	0.204	−1.110	0.694	2.784
M5	0.204	−1.356	0.531	2.365
S2	0.204	−1.148	0.531	2.448
S3	0.163	−0.932	0.490	2.588
S4	0.163	−0.781	0.449	2.785
S5	0.122	−0.169	0.408	2.944

4.4. Roller Length and Reattachment Length

The length characteristics of the hydraulic jump are usually described by the length of the surface roller or the length of the hydraulic jump. However, since the latter is not clearly defined and it is not possible to assess the length of sloping hydraulic jumps [31], the results of the length of the surface roller (L_1) are presented herein. The reattachment length (L_2) was defined as the flow distance from the source of the jet ($x = 0$) to the point of the jet and the bottom of the basin. According to [35], the mean streamlines method was used for the reattachment length. This method defines the reattachment point as the point where the mean streamline intersects with the bottom of the basin. The effects of step height on L_1 and L_2 are shown in Figure 12. From Figure 12, it can be seen that the effect of step height on L_1 and L_2 was significant. When d/h_1 increases, L_1/h_1 decreases gradually, while L_2/h_1 increases gradually. According to the computational results, the following regression equations can be obtained to estimate the roller length and reattachment length when $d/h_1 = 0 \sim 4.12$, $F_1 = 8.3 \sim 9.3$, $\theta = 0^\circ$:

$$\frac{L_1}{h_1} = -0.90\left(\frac{d}{h_1}\right)^2 + 1.20\frac{d}{h_1} + 64.70 \tag{13}$$

$$\frac{L_2}{h_1} = 2.42\frac{d}{h_1} + 0.21 \tag{14}$$

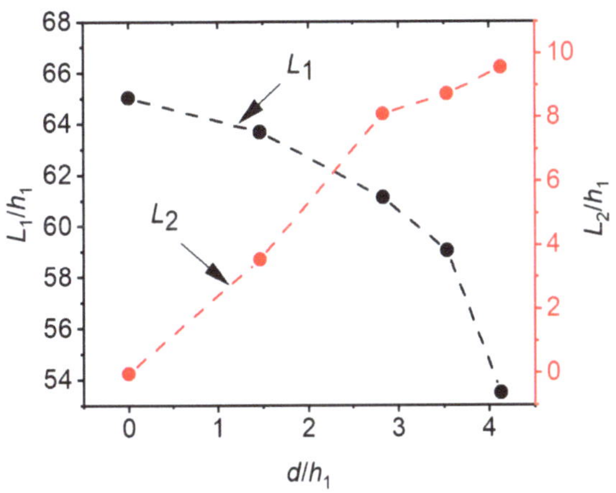

Figure 12. Effects of step height on the characteristic length.

The predictive capability of these equations is high ($R^2 = 0.98$ and $R^2 = 0.96$, respectively).

The effects of the incident angle on L_1 and L_2 are shown in Figure 13. From Figure 13, it can be seen that d/h_1 did not change regularly as θ increased, while it showed a linear decreasing trend. When θ increased from 0 to 20° L_1/h_1 only increased by 7.49%, but L_2/h_1 decreased by 23.7%, which indicates that the incident angle had no significant effect on the roller length, but it had a significant effect on the length of reattachment. Finally, according to the numerical calculation results, it is known that in this study, the ratio of the reattachment length to the surface roller length (L_2/L_1) was in the range of 0~0.18.

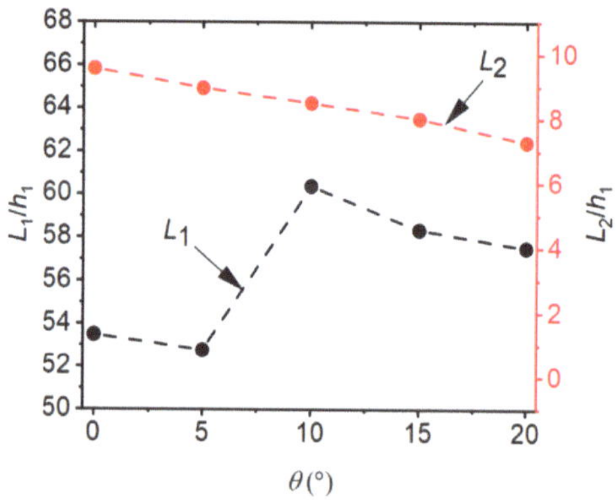

Figure 13. Effects of the incident angle on the characteristic length.

4.5. Energy Dissipation Efficiency

In this study, the energy dissipation efficiency (η) was computed in accordance with [36] as the difference between energy heads at the upstream section $x = -2.0$ m (E_0) and the downstream section $x = 2.1$ m (E_t), normalized to η

$$\eta = \frac{E_0 - E_t}{E_0} \tag{15}$$

Figures 14 and 15 show η as a function of the relative step height (d/h_1) and the incident angle (θ), respectively. As Figure 14 shows, the energy dissipation efficiency decreased with the relative step height, a finding consistent with Chen's [37] experiment. η decreased by 1.0% when d/h_1 increased from 0 to 4.12. As Figure 15 shows, the energy dissipation efficiency decreased as the incident angle increased, and the energy dissipation efficiency decreased by 1.8% as θ increased from 0 to 20°.

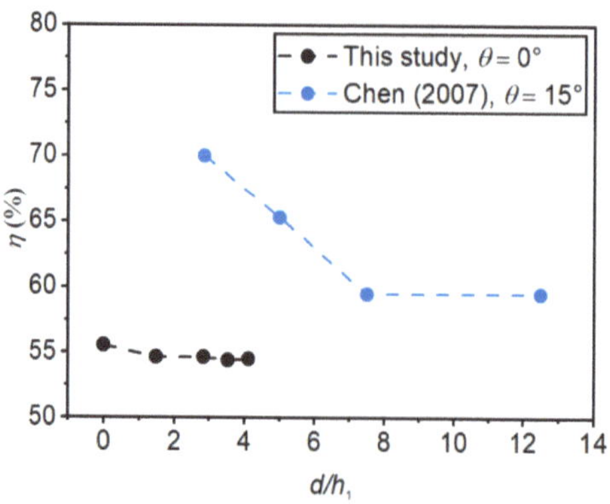

Figure 14. Effect of the step height on the energy dissipation rate.

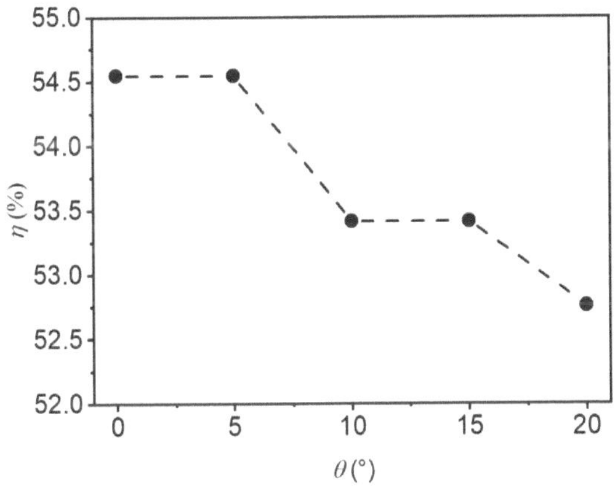

Figure 15. Effect of the incident angle on the energy dissipation rate.

5. Discussion

The present study provides a detailed description of the flow behavior of the negative-step stilling basin. The B-jump in the negative-step stilling basin is significantly influenced by the step height, while it is relatively unaffected by the incident angle. At a certain tailwater depth, a B-jump in the negative-step stilling basin will change to a surface jump when the step height increases to a critical step height (d_c) [38]. Zheng [39] proposed an empirical model of d_c:

$$d_c = \left(4.05\sqrt[3]{F_1^2} - \frac{h_2}{h_1}\right) \tag{16}$$

A reliable design for stilling basins must consider the hydraulic jump developed within the structure [18]. Thus, surface jumps should be avoided as much as possible. That is, when designing a negative-step stilling basin, one should ensure that d is less than d_c.

The present study showed that the relative flow depths (d/h_1) of the negative-step stilling basin were greater at the same location compared to the CHJ and USBR II stilling basins. This is due to the fact that this study had a larger unit discharge and tailwater depth compared with previous studies, which resulted in a larger flow depth in the basin. In general, since the water depth in the negative-step stilling basin is higher compared to the traditional stilling basins, higher sidewalls are required to ensure the safety of the project.

The present study shows that the velocity distribution downstream of the negative step is very complex. The results of the vertical velocity distribution show that a maximum forward velocity and a maximum backward velocity can be found downstream of the step. The backward velocity is generated by the backflow formed downstream of the step. The protection of the basin floor should be increased since previous studies have indicated that there is a significant velocity gradient in the backflow area, which is also the area with the greatest pulsing pressure [2,40]. The bottom velocity was also analyzed in this study. The bottom velocity is an important indicator for the design of the basin. In China, the bottom flow velocity is required to be less than 35 m/s. The results of this study confirm that the bottom velocity can be reduced significantly by increasing the step height, and increased significantly by increasing the incident angle. This is mainly because increasing the step height will increase the jet's movement distance in the water, greatly reducing the flow velocity of the main jet as it reaches the bottom. In contrast, by increasing the incident angle, the distance between the jet and the bottom can be shortened and the high velocity of the jet to the bottom can be maintained. This shows that the design of the negative-step stilling basin needs to consider the influence of the step height and the incident angle to achieve the most ideal design geometry.

The length of the roll reflects the concern about energy dissipation efficiency because rolling motion and associated turbulent shearing are the predominant energy dissipation mechanisms in hydraulic jumps. It may be further linked to the erosion length at the bottom, and thus to the protection range of the stilling basin. For an efficient stilling basin operation, it is advisable to use a longer roller length (thus having a larger range of high energy dissipation) [41]. According to the present findings, this may be achieved by adopting a higher step. The reattachment length is also an important characteristic length of the negative-step stilling basin, because it relates to the development of the flow downstream of the step and the size of the reattachment region [35]. The results of this study show that the reattachment length maintains a good linear relationship with both the step height and incident angle. This is important for predicting the reattachment length. However, due to the limited sample in this study, it failed to provide a good prediction equation.

The results show that the step height and the incident angle have little effect on the energy dissipation efficiency. This is mainly due to the high flow depth in the basin, which leads to a large proportion of potential energy and a small proportion of kinetic energy. Therefore, under the same upstream and downstream flow conditions, the calculated energy dissipation efficiency does not have much difference. This conclusion is more consistent with Hager's [5] study, which concluded that when $F_1 > 8$, the relative step height no longer has a significant effect on the energy dissipation efficiency of the negative-step hydraulic jumps. In contrast, Chen's [37] study showed that increasing the step height could significantly reduce the dissipation efficiency. This is mainly because the unit discharge and tailwater depth in Chen's study were relatively smaller than in this study, so the energy dissipation effect was more significantly affected by the step height.

6. Conclusions

In the present study, the influence of a negative step on the hydraulic performance of the stilling basin of high dams was investigated for the design conditions. Nine CFD models were conducted using 2D RANS equations with a VOF model for the free-surface tracking and the RNG k-ε turbulence model. The results show that:

(1) The influence of step height on the flow regime is significant and the influence of incident angle on the flow regime is insignificant. Increasing the step height will lead to an increase in the flow depth in the basin, so the negative-step stilling basin needs a higher sidewall than the traditional stilling basin. However, the step height should be less than the critical height to avoid the surface jump in the basin.

(2) Increasing the step height will significantly reduce the bottom flow velocity, and increasing the incident angle will significantly increase the critical bottom flow velocity. Therefore, the design of the negative-step stilling basin needs to consider the influence of the step height and the incident angle to achieve the most ideal design geometry.

(3) Using a higher step height can reduce the roller length of the jump in the negative-step stilling basin, thus shortening the length of the dissipation basin. The reattachment length maintains a good linear relationship with both the step height and incident angle.

(4) Within the scope of this study, changing the step height and incident angle did not have a strong effect on the total energy dissipation efficiency of the negative-step stilling basin.

Author Contributions: Conceptualization, L.J. and M.D.; Methodology, L.J.; Software, L.J.; Validation, L.J. and M.D.; Formal analysis, L.J. and M.D.; Investigation, L.J.; Resources, M.D.; Data curation, L.J.; Writing—original draft preparation, L.J.; Writing—review and editing, L.J. and M.D.; Visualization, L.J. and C.W.; Supervision, M.D.; Project administration, M.D. All authors have read and agreed to the published version of the manuscript.

Funding: This research received no external funding.

Data Availability Statement: Data available within the article.

Acknowledgments: We would like to thank the anonymous referees for their helpful comments and suggestions.

Conflicts of Interest: The authors declare no conflict of interest.

Abbreviations

d	Step height
θ	Incident angle
Q	Flow discharge
q	Unit discharge
h_1	Mean depth at the jump toe
u_1	Mean velocity at the jump toe
F_1	Froude number at the jump toe
h_t	Tailwater depth
GCI	Grid convergence index
x_1	Jump toe position
L_1	Roller length
h_1	Flow depth at jump toe
u_b	Bottom velocity
E	Submergence parameter
u_{max}	Maximum velocity
L_2	Reattachment length
η	Energy dissipation efficiency
d_c	Critical step height

References

1. Barjastehmaleki, S.; Fiorotto, V.; Caroni, E. Spillway Stilling Basins Lining Design via Taylor Hypothesis. *J. Hydraul. Eng.* **2016**, *142*, 04016010. [CrossRef]
2. Li, S. Research on Hydrodynamic Loads of Stilling Basin with Drop Sill. Ph.D. thesis, Tianjin University, Tianjin, China, 2012. (In Chinese)
3. Sun, S.; Liu, H.; Xia, Q.; Wang, X. Study on Stilling Basin with Step-down Floor for Energy Dissipation of Hydraulic Jump in High Dams. *J. Hydraul. Eng.* **2005**, *361*, 1188–1193. (In Chinese)
4. Huang, G.; Diao, M.; Jiang, L.; Wang, C.; Jia, W. Experimental Study on Wave Characteristics of Stilling Basin with a Negative Step. *Entropy* **2022**, *24*, 445. [CrossRef] [PubMed]
5. Hager, W.H. B-Jumps at Abrupt Channel Drops. *J. Hydraul. Eng.* **1985**, *111*, 861–866. [CrossRef]
6. Hager, W.H.; Bretz, N.V. Hydraulic Jumps at Positive and Negative Steps. *J. Hydraul. Res.* **1986**, *24*, 237–253. [CrossRef]
7. Ohtsu, I.; Yasuda, Y. Transition from Supercritical to Subcritical Flow at an Abrupt Drop. *J. Hydraul. Res.* **1991**, *29*, 309–328. [CrossRef]
8. Mossa, M. On the Oscillating Characteristics of Hydraulic Jumps. *J. Hydraul. Res.* **1999**, *37*, 541–558. [CrossRef]
9. Tokyay, N.D.; Altan-Sakarya, A.B. Local Energy Losses at Positive and Negative Steps in Subcritical Open Channel Flows. *Water SA* **2011**, *37*, 237–244. [CrossRef]
10. Moore, W.L.; Morgan, C.W. The Hydraulic Jump at an Abrupt Drop. *Trans. Am. Soc. Civ. Eng.* **1959**, *124*, 507–516. [CrossRef]
11. Eroğlu, N.; Taştan, K. Local Energy Losses for Wave-Type Flows at Abrupt Bottom Changes. *J. Irrig. Drain. Eng.* **2020**, *146*, 04020029. [CrossRef]
12. Jia, W.; Diao, M.; Jiang, L.; Huang, G. Time-Frequency Characteristics of Fluctuating Pressure on the Bottom of the Stilling Basin with Step-Down Floor Based on Hilbert–Huang Transform. *Math. Probl. Eng.* **2021**, *2021*, 7246488. [CrossRef]
13. Jia, W.; Diao, M.; Jiang, L.; Huang, G. Fluctuating Characteristics of the Stilling Basin with a Negative Step Based on Hilbert-Huang Transform. *Water* **2021**, *13*, 2673. [CrossRef]
14. Valero, D.; Bung, D.B.; Crookston, B.M. Energy Dissipation of a Type III Basin under Design and Adverse Conditions for Stepped and Smooth Spillways. *J. Hydraul. Eng.* **2018**, *144*, 04018036. [CrossRef]
15. Carvalho, R.F.; Lemos, C.M.; Ramos, C.M. Numerical Computation of the Flow in Hydraulic Jump Stilling Basins. *J. Hydraul. Res.* **2008**, *46*, 739–752. [CrossRef]
16. Celik, I.B.; Ghia, U.; Roache, P.J.; Freitas, C.J. Procedure for Estimation and Reporting of Uncertainty Due to Discretization in CFD Applications. *J. Fluids Eng.-Trans. ASME* **2008**, *130*, 078001.
17. Zhou, Z.; Wang, J. Numerical Modeling of 3D Flow Field among a Compound Stilling Basin. *Math. Probl. Eng.* **2019**, *2019*, 5934274. [CrossRef]
18. Macián-Pérez, J.F.; Vallés-Morán, F.G.; García-Bartual, R. Assessment of the Performance of a Modified USBR Type II Stilling Basin by a Validated CFD Model. *J. Irrig. Drain. Eng.* **2021**, *147*, 04021052. [CrossRef]

19. Kirkgoz, M.S.; Akoz, M.S.; Oner, A.A. Numerical Modeling of Flow over a Chute Spillway. *J. Hydraul. Res.* **2009**, *47*, 790–797. [CrossRef]
20. Akoz, M.S.; Gumus, V.; Kirkgoz, M.S. Numerical Simulation of Flow over a Semicylinder Weir. *J. Irrig. Drain. Eng.* **2014**, *140*, 04014016. [CrossRef]
21. Macián-Pérez, J.F.; Bayón, A.; García-Bartual, R.; Amparo, L.-J.P.; Vallés-Morán, F.J. Characterization of Structural Properties in High Reynolds Hydraulic Jump Based on CFD and Physical Modeling Approaches. *J. Hydraul. Eng.* **2020**, *146*, 04020079. [CrossRef]
22. Bombardelli, F.A.; Meireles, I.; Matos, J. Laboratory Measurements and Multi-Block Numerical Simulations of the Mean Flow and Turbulence in the Non-Aerated Skimming Flow Region of Steep Stepped Spillways. *Environ. Fluid Mech.* **2011**, *11*, 263–288. [CrossRef]
23. Launder, B.E.; Spalding, D.B. The Numerical Computation of Turbulent Flows. *Comput. Methods Appl. Mech. Eng.* **1974**, *3*, 269–289. [CrossRef]
24. Yakhot, V.; Orszag, S.A. Renormalization-Group Analysis of Turbulence. *Phys. Rev. Lett.* **1986**, *57*, 1722–1724. [CrossRef] [PubMed]
25. Hirt, C.W.; Nichols, B.D. Volume of Fluid (VOF) Method for the Dynamics of Free Boundaries. *J. Comput. Phys.* **1981**, *39*, 201–225. [CrossRef]
26. Helal, E.; Abdelhaleem, F.S.; Elshenawy, W.A. Numerical Assessment of the Performance of Bed Water Jets in Submerged Hydraulic Jumps. *J. Irrig. Drain. Eng.* **2020**, *146*, 04020014. [CrossRef]
27. Issa, R.I. Solution of the Implicitly Discretised Fluid Flow Equations by Operator-Splitting. *J. Comput. Phys.* **1986**, *62*, 40–65. [CrossRef]
28. Maranzoni, A.; Tomirotti, M. 3D CFD Analysis of the Performance of Oblique and Composite Side Weirs in Converging Channels. *J. Hydraul. Res.* **2021**, *59*, 586–604. [CrossRef]
29. Salehi, S.; Mostaani, A.; Azimi, A.H. Experimental and Numerical Investigations of Flow over and under Weir-Culverts with a Downstream Ramp. *J. Irrig. Drain. Eng.* **2021**, *147*, 04021029. [CrossRef]
30. Kawagoshi, N.; Hager, W.H. B-Jump in Sloping Channel, II. *J. Hydraul. Res.* **1990**, *28*, 461–480. [CrossRef]
31. Hager, W.H. B-Jump in Sloping Channel. *J. Hydraul. Res.* **1988**, *26*, 539–558. [CrossRef]
32. Bakhmeteff, B.A.; Matzke, A.E. The Hydraulic Jump in Terms of Dynamic Similarity. *Trans. Am. Soc. Civ. Eng.* **1936**, *101*, 630–647. [CrossRef]
33. Wang, H.; Chanson, H. Experimental Study of Turbulent Fluctuations in Hydraulic Jumps. *J. Hydraul. Eng.* **2015**, *141*, 04015010. [CrossRef]
34. Macián-Pérez, J.F.; Vallés-Morán, F.J.; Sánchez-Gómez, S.; De-Rossi-Estrada, M.; García-Bartual, R. Experimental Characterization of the Hydraulic Jump Profile and Velocity Distribution in a Stilling Basin Physical Model. *Water* **2020**, *12*, 1758. [CrossRef]
35. Nyantekyi-Kwakye, B.; Clark, S.P.; Tachie, M.F.; Malenchak, J.; Muluye, G. Flow Characteristics within the Recirculation Region of Three-Dimensional Turbulent Offset Jet. *J. Hydraul. Res.* **2015**, *53*, 230–242. [CrossRef]
36. Padulano, R.; Fecarotta, O.; Del Giudice, G.; Carravetta, A. Hydraulic Design of a USBR Type II Stilling Basin. *J. Irrig. Drain. Eng.* **2017**, *143*, 04017001. [CrossRef]
37. Chen, H. Experimental Study on the Hydraulic Characteristics of Hydraulic Jump Stilling Basin with a Drop Step. Master's Thesis, Kunming University of Technology, Kunming, China, 2007. (In Chinese).
38. Wang, Y.; Politano, M.; Weber, L. Spillway Jet Regime and Total Dissolved Gas Prediction with a Multiphase Flow Model. *J. Hydraul. Res.* **2019**, *57*, 26–38. [CrossRef]
39. Zheng, X. Hydraulic Design Study of the Step-Down Bottom Energy Dissipator. Master's Thesis, Kunming University of Technology, Kunming, China, 2010. (In Chinese).
40. Armenio, V.; Toscano, P.; Fiorotto, V. On the Effects of a Negative Step in Pressure Fluctuations at the Bottom of a Hydraulic Jump. *J. Hydraul. Res.* **2000**, *38*, 359–368. [CrossRef]
41. Bai, R.; Wang, H.; Tang, R.; Liu, S.; Xu, W. Roller Characteristics of Preaerated High-Froude-Number Hydraulic Jumps. *J. Hydraul. Eng.* **2021**, *147*, 04021008. [CrossRef]

energies

MDPI

Article

Numerical Study of the Impact of Fluid–Structure Interaction on Flow Noise over a Rectangular Cavity

Paweł Łojek *,†, Ireneusz Czajka † and Andrzej Gołaś †

Department of Power Systems and Environmental Protection Facilities, Faculty of Mechanical Engineering and Robotics, AGH University of Science and Technology, Mickiewicz 30 av, 30-059 Kraków, Poland
* Correspondence: lojek@agh.edu.pl
† These authors contributed equally to this work.

Abstract: Fluid–structure interactions (FSI) can significantly affect flow and the acoustic field generated by it. In this article, simulations of the flow over a rectangular cavity are conducted with and without taking FSI into account. The aim of this research is to conduct a numerical study of the flow over a cavity and to verify whether interactions between the flow and the elastic structure can significantly affect the flow itself or the acoustic pressure field. Four cases involving flexible walls with different material parameters and one reference case with rigid walls were analysed. The two-directional fluid–structure coupling between the flow and cavity walls was simulated. The simulations were performed with the volume and finite element methods using OpenFOAM software to solve the fluid field, CalculiX software to solve the displacement of the structure, and the preCICE library to couple the codes and computed fields. The acoustic analogy of Ffowcs-Williams and Hawkings and the libAcoustics library were used to calculate the sound pressure. The simulation results showed that FSI has a significant influence on sound pressure in terms of both pressure amplitudes and levels as well as in terms of noise frequency composition. There was a significant increase in the sound pressure compared to the case with rigid walls, especially for frequencies above 1 kHz. The frequencies at which this occurred are related to the natural frequencies of the cavity walls and the Rossiter frequencies. Overlap of these frequencies may lead to an increase in noise and structural vibrations, which was observed for one of the materials used. This study may provide insight into the flow noise generation mechanism when fluid–structure interactions are taken into account. The conclusions presented here can form a basis for further work on aerodynamic noise in the presence of thin-walled structures.

Keywords: aeroacoustics; fluid–structure interaction; duct noise

1. Introduction

Flow over a cavity is widely used as a model of flow over discontinuities, even ones with complex-shape. Historically, this research object arose from the problem of noise generated by aircraft weapon bays [1] and landing gear. A cavity is used as a model to describe flow phenomena and noise generation by car mirrors and door cavities [2], pantographs, recesses on the roofs of trains [3], and other vehicle discontinuities. Cavities are often used to analyse the flow in different parts of ventilation ducts. This was the goal of Radavich [4], who investigated the flow inside quarter-wave resonators, and of Lafon [5], who identified the tones in gate valves inside a duct.

The flow over a cavity has been widely investigated in the past via theoretical, numerical, and experimental techniques due to the complex phenomena involved, such as vortex shedding, free shear layer instability, and pressure oscillations. Rossiter [6], in his experimental study, was one of the first to attempt to derive the dependence describing the pressure fluctuations occurring in cavities. Under certain conditions, the flow past cavities starts to oscillate in a self-sustaining manner. Rockwell [7] describes three different types of self-sustaining oscillations of flow: fluid dynamic, fluid resonant, and fluid elastic. Most of the works dealing with cavity flow and noise focuses on fluid dynamic and fluid resonant oscillations, ignoring the effect of flexible cavity walls on the flow over the cavity and the aeroacoustic feedback it generates. There are only few studies on the third type of interaction, fluid elastic; these are mostly limited to unidirectional fluid–structure coupling, and assume that the vibrating structure has no influence on the flow. Yokohama [8] researched the impact of flow on the vibrations of a flexible cantilever beam attached to the upstream and downstream edges of a cavity. Thangamani [9] investigated the possible ways to harvest energy from flow by attaching a flexible piezoelectric beam at the downstream wall.

There are studies in which bidirectional fluid–structure coupling has been analysed. However, these are usually limited to the simplified case of a lid-driven cavity, which is a well known benchmark case for CFD. This model involves solving only the flow within the cavity, with the flow over the cavity replaced by appropriate boundary conditions. Khanafer [10] examined the effect of a heated flexible cavity bottom on heat transfer in such a system. Alsabery [11] researched a similar problem involving heat transfer in a lid-driven cavity with elastic walls and with a hot rotating cylinder in the middle of the cavity. Sun [12] analysed the vibrations of the lid-driven cavity walls themselves under flow in terms of the dependence of the rigidity and Reynolds number. Sabbar [13] investigated the flow over a cavity with a downstream flexible wall at a low Reynolds number and with a heat source at the bottom of the cavity. Most of the above works describing the fluid–structure interactions ocurring in cavities are modeled and solved numerically, and are limited only to the case of a lid-driven cavity. There is very little research focusing on the FSI effects ocurring in cavities placed in channels or open spaces. Moreover, the most common consideration of flexible walls is to increase heat transfer, while the research largely ignores their impact on the noise generated by the flow.

In the present work, we investigate fluid elastic oscillations and analyse their impact on the noise generated by the flow over a cavity. The length L to depth D ratio is used to describe the type of cavity. The flow over a cavity with an L/D ratio of 4 was analysed along with the vibrations of the cavity walls. The main assumption of this study was to model a cavity inside a ventilation duct. Hence, the thickness of elastic cavity walls was similar to the thickness of typical duct walls. We used four different sets of material parameters to model the different flexible walls of the cavity and one model with rigid walls as a reference model. The adopted materials were typical materials for making ventilation ducts. The flow velocities were typical of those found in the ducts. To the best of our knowledge, such research on the fluid–structure interactions cavity flow has not yet been performed. There are no numerical studies taking into account the influence of the structural vibrations generated by the flow on the aeroacoustic noise in the case of flow over a cavity.

To investigate this phenomenon, finite volume and finite element methods were used. We used the detached eddy simulation method described by Strelets [14] with the $k-\omega$ SST model developed by Menter [15] to compute the flow over the cavity. Typically, the large-eddy simulation model is used for hybrid flow acoustics simulations [16]. However, a model combining the LES and RANS models, namely, the DES model, is now used more often in this type of simulation [17,18] as well as in cavity noise problems [19]. We chose the $k-\omega$ SST DES model due to the fact that it combines the high accuracy of the LES model with the high computational speed of the RANS model. To model the fluid-induced vibrations, it was first necessary to solve the dynamics of the cavity walls. This was carried out using the classical finite element method. The bidirectional coupling between the flow and displacement fields was modeled. The values of forces exerted on the walls by the fluid and nodal displacement of the structure were exchanged. The preCICE library [20] was used to couple the fields and model the fluid–structure interaction.

The main objective of the study was to compute the acoustic pressure generated by the flow. The acoustic analogy of Lighthill [21] and its extension provided by Ffowcs-Williams and Hawkings [22] were used to achieve this. We used the hybrid CFD-CAA method to compute the noise based on the fluid flow simulation results. This approach combines two-way fluid–structure coupling with the use of acoustic analogies, and has been successfully used in the analysis of aeroacoustic and hydroacoustic noise [23] and in case of forced oscillations of a cylinder in flow [24]. The computational model itself was verified and compared with experimental results by Turek and Hron [25] in the study conducted by Chourdakis et al. [26]. The presented research is a continuation and summary of the preliminary analyzes presented in [27].

This study is important due to the possible consequences of taking into account flexible walls in the flow, especially when analysing noise in thin-walled systems. Additionally, this type of analysis can provide information on possible energy recovery of flows over cavities when flow-induced vibrations are taken into account. This type of work has already been carried out by Thangamani [9]; however, as mentioned earlier, the influence of the vibrating structure on the flow itself was ignored. Moreover, this problem can be developed towards possible damping of vibrations and noise in cavities by placing additional elastic vibrating structures in them, as shown in [28,29].

In addition, the findings of this study can be used in other industries where flows over cavities occur, primarily aviation and high-velocity rail transport. In addition, the observations made in this work can be used in analyses of the noise generated by flows in the presence of thin-walled systems more generally, not necessarily cavities/ducts.

The outline of this paper is as follows. In Section 2, the mathematical models of flow and structural vibrations are described along with the numerical methods used to solve them. In Section 3, the analysed model, simulation initial and boundary conditions, and grid independence study are presented. The results of simulations involving the acoustic pressure at the receiver, forces acting on the cavity, and flow fields are described and analysed in Section 4. Finally, the main findings and conclusions are provided in Section 5.

2. Mathematical Model and Numerical Methods

2.1. Flow Modeling

In this study, turbulent fluid motion was described by the $k-\omega$ SST DES (Detached Eddy Simulation) model [14]. This model solves RANS (Reynolds-Averaged Navier–Stokes) equations with the $k-\omega$ SST turbulence model in the boundary layer and LES (Large Eddy Simulation) in the separated region. The incompressible form of RANS equations were used for relatively low flow velocities, i.e., those for which the Mach number is much smaller than unity. The RANS continuity and momentum equations are provided by (1) and (2) [30]:

$$\frac{\partial \overline{v}_i}{\partial x_i} = 0 \tag{1}$$

$$\rho\frac{\partial \overline{v}_i}{\partial t} + \rho\overline{v}_j\frac{\partial \overline{v}_i}{\partial x_j} = \frac{\partial \overline{p}}{\partial x_i} + \frac{\partial}{\partial x_j}\left(\overline{\tau}_{ij} - \rho\overline{v_i'v_j'}\right) \tag{2}$$

where $v_i = \overline{v} + v_i'$ is the i-th component of the velocity, $i = x, y, z$ in the Cartesian coordinate system, $p = \overline{p} + p'$ is the pressure, ρ is the density, $\overline{\tau}_{ij}$ is the laminar viscous stress tensor, and $-\rho\overline{v_i'v_j'} = \tau_{ij}^R$ is the Reynolds stress tensor. In the above equation, a bar denotes an averaged component and prime denotes a fluctuating component, unless stated otherwise.

The Bousinessq hypothesis introduced the concept of eddy viscosity μ_t (3), which allows the Reynolds stress tensor to be modeled by means of turbulence models.

$$\tau_{ij}^R = 2\mu_t\left(S_{ij} - \frac{1}{3}\frac{\partial u_k}{\partial x_k}\delta_{ij}\right) - \frac{2}{3}k\delta_{ij} \tag{3}$$

In this work, we have chosen the $k-\omega$ SST model as suitable for aerodynamics [15]. In this model, two equations are solved: one for turbulent kinetic energy k (4), and one for the specific turbulent dissipation rate ω (5):

$$\frac{D\rho k}{Dt} = \tau_{ij}\frac{\partial u_i}{\partial x_j} - \beta^* \rho \omega k + \frac{\partial}{\partial x_j}\left[(\mu + \sigma_k \mu_t)\frac{\partial k}{\partial x_j}\right] \tag{4}$$

$$\frac{D\rho\omega}{Dt} = \frac{\gamma}{\nu_t}\tau_{ij}\frac{\partial u_i}{\partial x_j} - \beta\rho\omega^2 + \frac{\partial}{\partial x_j}\left[(\mu + \sigma_\omega \mu_t)\frac{\partial \omega}{\partial x_j}\right] + 2\rho(1 - F_1)\sigma_{\omega 2}\frac{1}{\omega}\frac{\partial k}{\partial x_j}\frac{\partial \omega}{\partial x_j} \tag{5}$$

where $\frac{D}{Dt} = \frac{\partial}{\partial t} + u_i \frac{\partial}{\partial x_i}$ is the material derivative, k is the turbulent kinetic energy, ω is the specific turbulent dissipation rate, $\beta^*, \sigma_k \sigma_\omega, \gamma, a_1$ are model constants, and F_1, F_2 are additional functions.

The model is described in greater detail in [15]. The turbulent eddy viscosity is then computed with

$$\mu_t = \frac{\rho a_1 k}{\max(a_1 \omega, S F_2)} \tag{6}$$

The LES continuity and momentum equations are provided by (7) and (8) [30]. In this kind of simulation, the flow is not decomposed by Reynolds averaging, as before, but by spatial filtering into the filtered $\overline{\phi}$ and sub-filtered ϕ' parts:

$$\frac{\partial \overline{v}_i}{\partial x_i} = 0 \tag{7}$$

$$\frac{\partial \overline{v}_i}{\partial t} + \frac{\partial}{\partial x_j}(\overline{v}_i \overline{v}_j) = -\frac{1}{\rho}\frac{\partial \overline{p}}{\partial x_i} + \frac{\partial}{\partial x_j}\left(\tau_{ij}^S + 2\nu \overline{S}_{ij}\right) \tag{8}$$

where $\tau_{ij}^S = \overline{u}_i \overline{u}_j - \overline{u_i u_j}$ is the subgrid-scale (SGS) stress tensor.

This SGS stress tensor has to be modeled, and in case of $k - \omega$ SST DES, the $k - \omega$ SST model is used as a subgrid-scale model. The switching between LES and RANS is based on the values of the maximum spatial step Δ and the length scale $\tilde{l}$, which are defined by Strelets [14]. If the length scale is less than $C_{DES}\Delta$, then the $k - \omega$ SST model is used; otherwise, the LES equations are solved for mesh cells with δ_x, δ_y, and δ_z dimensions:

$$\Delta = \max(\delta_x, \delta_y, \delta_z) \tag{9}$$

$$\tilde{l} = \min(l_{k-\omega}, C_{DES}\Delta) \tag{10}$$

OpenFOAM v.2012 was used to perform the fluid flow simulations. OpenFOAM is an open-source CFD software that uses the finite-volume method [31]. Because of its robustness, the PIMPLE algorithm [32], which combines SIMPLE (Semi-Implicit Method for Pressure Linked Equations) [33] and PISO (Pressure-Implicit with Splitting of Operators) [34], was used to solve the equations of the flow field. The SIMPLE algorithm was used to determine the initial conditions in the computational domain based on boundary conditions. To compute the gradient and divergence terms, the upwind and standard linear Gauss methods were used.

2.2. Structural Modeling

The motion of the structural part of the model, that is, the cavity walls, was described by the general momentum equation from structural analysis without damping terms [35]:

$$[M]\{\ddot{U}\} + [K]\{U\} = \{F\} \tag{11}$$

where $\{\ddot{U}\} = \{A\}$ is the global acceleration vector, $\{U\}$ is the global displacement vector, $[M]$ is the global mass matrix, $[K]$ is the global stiffness matrix, and $\{F\}$ is the global force vector.

The damping terms was not taken into account, as the simple dynamic model and the lack of additional coefficients were assumed. This equation is discretized in space by they finite element method and in time by finite differences. It is solved using the α-method, which is an extension of the Newmark method [35].

Equation (11) is discretized in space by the finite element method. The local (element) stiffness and mass matrices are provided by [35]

$$[K]_{e(iK)(jM)} = \int_{V_{0e}} \varphi_{i,L}\Sigma^{KLMN}(\theta)\varphi_{j,N}dV_e \tag{12}$$

$$[M]_{e(iK)(jM)} = \int_{V_{0e}} \rho_0\varphi_i\varphi_j dV_e \tag{13}$$

$$\{F\}_{e(iK)} = \int_{A_{0e}} \overline{T}^k_{(N)}\varphi_i dA_e + \int_{V_{0e}} \rho_0 f^k\varphi_i dV_e + \int_{V_{0e}} \left[\beta^{KL}(\theta)T - \gamma^{KL}\right]\varphi_{i,L}dV_e \tag{14}$$

where Σ^{KLMN} is the free energy function, $KLMN$ are the material coordinates, θ is the absolute temperature, ρ_0 is the mass density, $\overline{T}^K$ is the traction, f^k is the force per unit mass, β^{KL} is the thermal stress tensor per unit temperature, γ^{KL} is the residual stress tensor, and φ represents the shape functions.

After Equation (11) is discretized in time, it takes the form

$$[M]\{A\}_{n+1} + [K]\{U\}_{n+1} = \{F\}^{ext}_{n+1} \tag{15}$$

$$\{A\}_{n+1} = A_n + \{\Delta A\} \tag{16}$$

$$\{U\}_{n+1} = \{U\}_n + \Delta t\{V\}_{n+1} + \frac{1}{2}(\Delta t)^2[(1-2\beta)\{A\}_n + 2\beta\{A\}_{n+1}] \tag{17}$$

where the subscript n denotes the current and $n+1$ the next time step.

The algorithm is second order accurate and unconditionally stable for $\alpha \in \left[\frac{-1}{3}, 0\right]$ [35]. The parameter β is provided by

$$\beta = \frac{1}{4}(1-\alpha)^2 \tag{18}$$

In this study, we used an eight-node brick element, which is a general-purpose fully integrated brick element, sometimes called a linear brick element. The shape functions of this element are provided by [36]:

$$\varphi = \frac{1}{8}(1+\xi\xi_i)(1+\eta\eta_i)(1+\zeta\zeta_i) \quad i = 1,2,\ldots,8 \tag{19}$$

where ξ_i, η_i, ζ_i are the local coordinates of the th point. Very small thickness of the walls of the cavity and the ventilation duct encouraged the use of shell elements, however this turned out to be impossible due to the limitations of the preCICE library.

The structural modeling was performed using CalculiX open source finite element sofware [35].

2.3. Coupling of the Fields

As mentioned before, the coupling between the fluid flow field, computed with OpenFOAM and the solid displacement field computed with CalculiX was carried out using the preCICE coupling library. preCICE allows both explicit and implicit methods for solving interface equations and data mapping between different physical fields and computational meshes. In addition, it allows for communication between different codes, not only OpenFOAM and CalculiX [20].

In this study, we used the serial explicit coupling scheme. This scheme uses the nth timestep values $x_1^{(n)}$ of a solver S_1 as boundary values for the $n+1$ time step value of a solver S_2 and solution $x_2^{(n+1)}$, then uses the result for computation of $x_1^{(n+1)}$. Nearest-neighbor and nearest-projection methods were used for data mapping between fields. These were chosen due to the simplicity of their working principles and the fact that the size of the interface edges is similar. As mentioned before, bidirectional fluid–structure interaction was modeled. The displacement and force fields had to be exchanged between solvers. The forces exerted by the fluid on the walls are the boundary conditions (loads) for the structure simulation. A displacement field is obtained from them, then transferred to the fluid calculations as the mesh displacement [25].

2.4. Aeroacoustics

The noise generated by the flow over the cavity and the oscillations of the cavity walls was computed by means of acoustic analogies. The acoustic analogy determines how the sound caused by the flow would propagate in unlimited and undisturbed domain without the flow. As the Lighthill acoustic analogy assumes that there are no walls to disrupt the flow, it was impossible to use it directly [21]. Therefore, in this work, the Ffowcs-Williams and Hawkings (FWH) acoustic analogy was used. It allows the noise generated by the flow in the vicinity of the moving walls to be calculated [22]. The FWH analogy equation is provided by [37]:

$$\begin{aligned}\frac{\partial^2(\rho' H_s)}{\partial t^2} - c_\infty^2 \frac{\partial^2(\rho' H_s)}{\partial x_i^2} &= \frac{\partial^2 (T_{ij} H_s)}{\partial x_i \partial x_j} - \frac{\partial}{\partial x_i}[(\rho v_i (v_j - V_j) + p_{ij}) n_j \delta(f) |\nabla f|] \\ &+ \frac{\partial}{\partial t}[(\rho v_j - \rho' V_j) n_j \delta(f) |\nabla f|]\end{aligned} \tag{20}$$

where ρ' repesents the (acoustic) density fluctuations, H_s is the Heaviside step function, $T_{ij} = \rho v_i v_j + (p - p_\infty) - (\rho - \rho_\infty) c_\infty^2 \delta_{ij} - \sigma_{ij}$ is the Lighthill stress tensor, c_∞ is the speed of sound, V_i is the velocity of the surface in direction i, f is the scalar function defining the surface, and δ is the Dirac delta function.

We used the solution of this equation derived by Brentner and Farassat by means of generalized functions [38], called Formulation 1A:

$$p'(\mathbf{x}, t) = p'_T(\mathbf{x}, t) + p'_L(\mathbf{x}, t), \tag{21}$$

$$4\pi p'_t(\mathbf{x}, t) = \int\limits_{f=0} \left[\frac{\rho_0 (\dot{v}_n + v_{\dot{n}})}{r|1 - M_r|^2} \right]_{\text{ret}} \mathrm{d}S + \int\limits_{f=0} \left[\frac{\rho_0 v_n (r\dot{M}_r + c_\infty M_r - c_\infty M^2)}{r^2 |1 - M_r|^3} \right]_{\text{ret}} \mathrm{d}S, \tag{22}$$

$$\begin{aligned}4\pi p'_L(\mathbf{x}, t) &= \frac{1}{c_\infty} \int\limits_{f=0} \left[\frac{\dot{l}_r}{r|1 - M_r|^2} \right]_{ret} \mathrm{d}S + \int\limits_{f=0} \left[\frac{l_r - l_M}{r|1 - M_r|^2} \right]_{\text{ret}} \mathrm{d}S \\ &+ \frac{1}{c_\infty} \int\limits_{f=0} \left[\frac{l_r (\dot{M}_r + c_\infty M_r - c_\infty M^2)}{r^2 |1 - M_r|^3} \right]_{\text{ret}} \mathrm{d}S.\end{aligned} \tag{23}$$

where ϕ_n is the dot product of variable ϕ with a unit normal vector, ϕ_r is the dot product of ϕ with a radiation vector, ϕ_M is the dot product of ϕ with a normalized surface velocity vector, ϕ_{ret} is the quantity evaluated at the retarded time $\tau = t - \frac{r}{c}$, r is the distance between the observer and source, l is the local force acting on the surface, and $\dot{\phi} = \frac{\partial \phi}{\partial \tau}$ is the source time derivative of ϕ.

The Equations (21)–(23) were implemented and solved in OpenFOAM software with the libAcoustics library [39].

3. Case Description

3.1. Computational Domain

The computational domain used in our analyses is shown in Figure 1. It is defined as a section of a rectangular ventilation duct with a cavity at its bottom. The ratio of cavity length to depth L/D was equal to 4 in all simulations; therefore, the analysed cavity can be classified as a shallow cavity ($L/D > 1$) [40]. The length of the cavity itself was chosen to represent a possible junction of the channel that has been closed. The length of the domain upstream and downstream of the cavity allows the dynamic phenomena occurring in the cavity to be captured, including vortex shedding, shear layer instabilities, and separation of the flow. The dimensions of the domain are presented in Table 1. They are based on the dimensions of a ventilation duct with a square cross-section with a depth equal to 0.125 m.

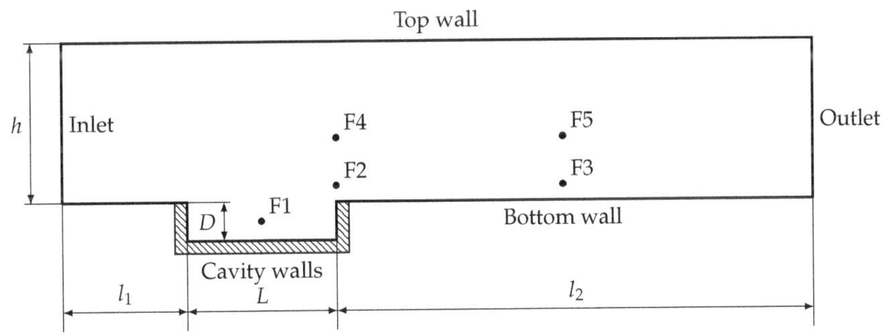

Figure 1. Analyzed model of cavity with probe locations.

Table 1. Dimensions of the model (in meters).

h	D	L	l_1	l_2
0.125	0.03	0.12	0.1	0.5

The computational mesh was generated based on the computational domain described above. The fluid mesh was generated using cfMesh and the structural mesh using GMSH open-source meshing tools. Based on the grid independence study described later in the article, fluid mesh #3, described in Table 2, and structural mesh #3, described in Table 3, were selected for all simulations.

Table 2. Mesh parameters for finite volume grid independence study.

Mesh	Element Size [mm]	Boundary Layer Thickness [mm]	Number of Elements	Aspect Ratio	Average Non-Orthogonality
#1	5	1	14,314	27.1	2.14
#2	3	1	32,754	40.8	1.52
#3	1.5	0.75	99,896	38.2	0.89
#4	1	0.5	202,230	27.1	0.78
#5	0.75	0.25	337,508	40.8	0.57

Table 3. Mesh parameters for finite element grid independence study with simulation time and mean amplitude of vibrations.

	Base Element Size [m]	Elements Along Wall	Number of Elements	Simulation Time [h]	Mean Amplitude of Vibrations [m]
Mesh #1	0.0003	2	1208	16	2.88×10^{-6}
Mesh #2	0.0001	6	10,872	34	3.01×10^{-6}
Mesh #3	0.00008	8	18,187	50	3.03×10^{-6}
Mesh #4	0.00006	10	30,200	78	3.04×10^{-6}
Mesh #5	0.00004	15	67,975	156	3.05×10^{-6}

3.2. Initial and Boundary Conditions

The boundaries of the fluid computational domain are shown in Figure 1. On each of them, the boundary conditions for each variable to be solved had to be set. The boundary conditions for the turbulent kinetic energy k and specific dissipation rate ω were estimated based on the recommendations of the author of the $k - \omega$ SST turbulence model [15]. The boundary conditions for each of the variables are shown in Table 4. Because of the implementation of the incompressible flow model in OpenFOAM, the pressure shown in the table and used for computations was scaled by density.

Moreover, the cavity walls, which acted as the interface between the fluid flow and structural simulations, were flexible and able to move. The wall movement was determined based on structural simulations.

Table 4. Fluid simulation boundary conditions.

	Pressure p/ρ [m^2/s^2]	Velocity v [m/s]	Turbulence Kinetic Energy k [m^2/s^2]	Specific Dissipation Rate ω [1/s]	Turbulent Viscosity ν_t [m^2/s]
Inlet	$\frac{\partial p}{\partial x_n} = 0$	$u_x = 15, u_{y,z} = 0$	$k = 3.375$	$\omega = 2500$	calculated
Outlet	$p = 0$	$\frac{\partial u_i}{\partial x_n} = 0$	$\frac{\partial k}{\partial x_n} = 0$	$\frac{\partial \omega}{\partial x_n} = 0$	calculated
Walls	$\frac{\partial p}{\partial x_n} = 0$	$u_i = 0$	$k = 0$	$\omega = 8 \times 10^6$	$\nu_t = 0$

The boundaries of the computational domain for structural simulations are shown in Figure 2. The same figure shows the probes where the displacement was recorded. At the boundary denoted Support, all degrees of freedom were constrained from movement in each direction. The boundary denoted Interface acted as a coupling surface between the fluid and structural simulations. In the case of structural analyses, it was the surface where the force from the fluid flow simulations was applied.

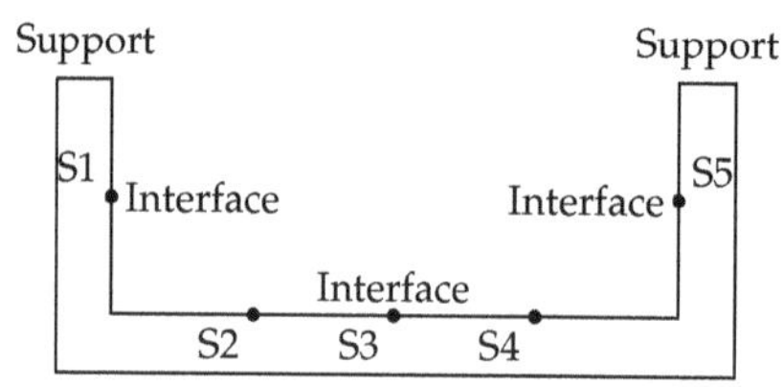

Figure 2. Boundaries of the finite element model with probe locations.

Material Parameters

Table 5 shows the parameters of the chosen materials, which are generally used for ventilation channels and ducts [41]. These materials were used for modeling the structural simulations. An air temperature of 20 °C, density $\rho = 1.23\ \mathrm{kg/m^3}$, kinematic viscosity $1.5 \times 10^{-5}\ \mathrm{m^2/s}$, and speed of sound $c_\infty = 340\ \mathrm{m/s}$ were selected for the flow simulations.

Table 5. Properties of materials used for structural modeling.

Material	Young's Modulus E [Pa]	Poisson's Ratio ν [-]	Density ρ [kg/m^3]
Polypropylene	900×10^6	0.42	1100
Polyvinyl chloride (PVC)	1.5×10^9	0.42	1400
Aluminium	70×10^9	0.33	2700
Structural steel	210×10^9	0.29	7800

3.3. *Grid Independence Study*

Two grid independence studies were performed, one for the fluid mesh and the other for the structural mesh. The main evaluation criteria for both studies were the sound pressure level evaluated at selected receivers using the FWH acoustic analogy, given by Equation (21) and the displacement of a point placed at the bottom of the cavity wall, computed from structural analyses. In addition, the simulation time and the values of the y^+ parameter (defined as a dimensionless distance from the wall) were checked and compared. In order to properly resolve the boundary layer and viscous sublayer, the values of y^+ should be less than unity. In all independence studies (both finite volume and finite element mesh), the timestep of $\Delta t = 2 \times 10^{-6}$ was used in order to keep the Courant number below 0.4. The total time of the simulated flow was 0.1 s.

For the fluid domain study, meshes with different base element sizes and boundary layer thicknesses were compared. In all cases, the mesh was refined near walls and inside the cavity, and the refinement was equal to 0.5 of the base mesh size. The mesh looked similar in all cases; a part of the mesh is shown in Figure 3. The mesh consisted of 99% hexahedral elements, while the remaining elements were of polyhedral type, mainly made up of seven-face elements. Five meshes with the parameters described in Table 2 were compared and assessed. In all cases, structural mesh #3 (described in more detail later in this section) was used. For each of them, a solution was initiated with the SIMPLE algorithm and FSI simulations were carried out. Aluminum was used as the material of the cavity walls in every grid independence study simulation.

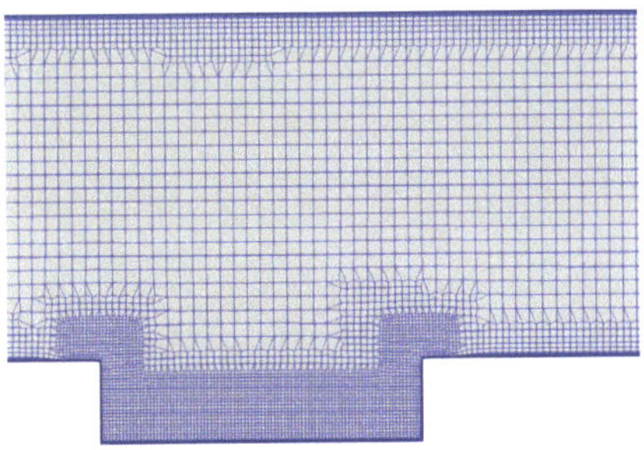

Figure 3. Computational mesh for fluid simulations (case #2).

It should be emphasized here that the results obtained during the actual simulations may differ from those obtained in the analysis of mesh independence. This is due to the fact that after 0.1 s the flow may not have fully developed, as well as to the fact that in the proper simulations the calculated flow time was much longer and a different method of initiating the flow was used, that is, by solving the potential flow.

The results of sound pressure level evaluation at a receiver located 10 m from the middle of the cavity are shown in Figure 4. The spectra of the sound pressure level were similar in all cases, with characteristic peaks for the frequencies of 2.5, 3.1, 5.3, 7.3, 8.1, and 9 kHz. A decrease in the amplitude of the noise frequency components can be seen, along with an increase in the number of elements. Additionally, for grids #1 and #2, additional peaks appear at 4.1 ÷ 4.4 kHz.

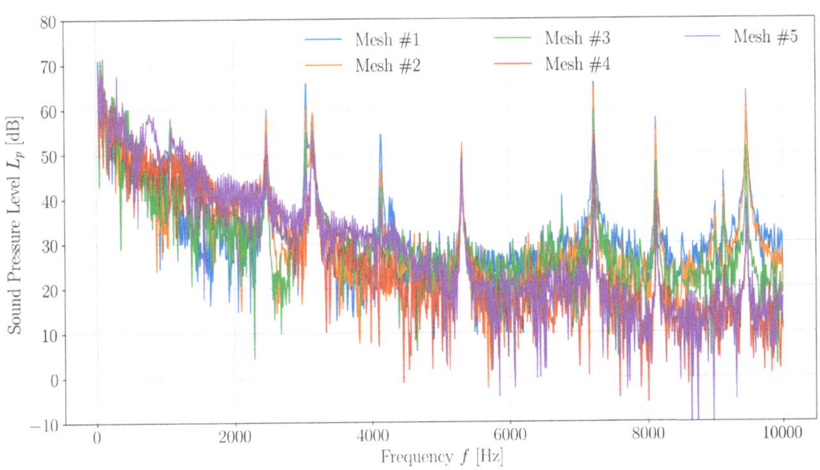

Figure 4. Sound pressure level at observer 10 m from center of the cavity for finite volume mesh independence study.

In addition, the spectra of the displacement of the cavity walls at probe S3 (Figure 2) were computed and are shown in the Figure 5. The results are again similar to each other; for the calculations on each mesh, there are peaks at the frequencies of 50 and 200 Hz. While there are additional frequency components above 1 kHz, they decrease as the number of mesh elements increases. Additional frequency components occurred below 100 Hz for meshes #1 and #2.

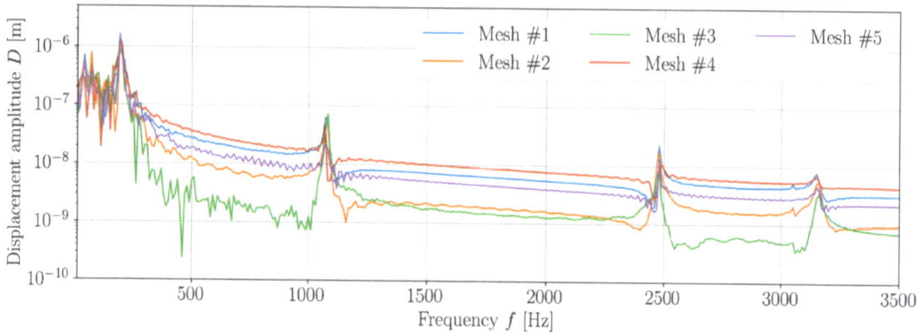

Figure 5. Spectrum of displacement along y axis at probe S3 for finite volume mesh independence study.

In Table 6, the values of the dimensionless wall distance y^+ for each wall are shown along with the time required to compute 0.1 s of the flow and the mean amplitude of vibrations at point S3. Meshes #3, #4, and #5 met the condition required by the turbulence model that this parameter should be less than one. The mean amplitude of vibrations were similar in all cases, with only the results for mesh #4 differing from the others.

Table 6. Parameter y^+, time of simulation, and amplitude of vibrations at point S3 for different meshes.

	y^+ at Top Wall	y^+ at Bottom Wall	y^+ at Cavity Walls	Simulation Time [h]	Mean Amplitude of Vibrations [m]
Mesh #1	2.7	2.1	0.86	35	3.31×10^{-6}
Mesh #2	1.45	1.2	0.6	48	2.94×10^{-6}
Mesh #3	0.63	0.49	0.27	50	3.19×10^{-6}
Mesh #4	0.43	0.42	0.16	71	1.99×10^{-6}
Mesh #5	0.29	0.26	0.14	84	2.66×10^{-6}

Visible differences between the spectra for the different meshes may be due to the large size of the elements and the thickness of the boundary layer for flow meshes #1 and #2. The thickness of the boundary layer affects how the flow is resolved in it, while the size of the elements determines whether and which vortices and eddies are modeled or resolved directly. This translates into the obtained pressure distributions and forces acting on the walls. This, in turn, affects the values of displacements and deformations of the cavity walls and the acoustic pressure values obtained by means of acoustic analogies.

Based on the convergence analysis, it was decided to choose grid #3 for the remaining simulations. This choice was due to the relatively short calculation time while maintaining the required mesh parameters (parameter y^+) and lack of significant impact on the quality of the results from further mesh refinement.

The second independence study was focused on the structural mesh. Again, five different meshes were used in the simulations and the obtained results were compared with each other. All meshes for the structural simulations consisted of hexahedral elements, with the height and width of each element the same. The meshes were generated using the Frontal-Delaunay algorithm for quads. The parameter changed during the independence study was the base size of the element, which in turn was translated into the number of elements along the wall thickness.

All simulations were performed with fluid mesh #3, and aluminium was again used as the material of the cavity walls. The parameters of the used meshes are summarized in Table 3, along with time required to simulate 0.1 s of the flow and the mean amplitude of vibrations at probe S3.

Figure 6 shows the sound pressure levels for at a receiver placed 10 m from the center of the cavity. For all meshes except #1, the results differed only slightly. The differences between the SPL spectra obtained using meshes #2–#5 increased with the frequency, although not significantly; for example, for the frequency of 7.3 kHz the maximum difference was 100 Hz, between mesh #2 and mesh #5. On the other hand, the results obtained for the first mesh differed significantly from the others, which may be due to an insufficient number of elements in the wall thickness. In addition, the sound pressure levels for mesh #2 differed from the results for the denser meshes, especially in the frequency range of 8.9–9.2 kHz.

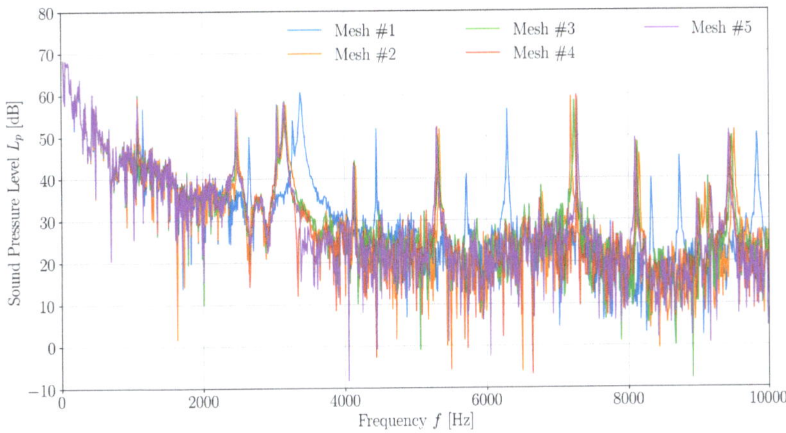

Figure 6. Sound pressure level at observer 10 m from center of the cavity for finite element mesh independence study.

Similar conclusions can be drawn from the displacement amplitude spectra presented in Figure 7. It shows the spectrum of displacement of the cavity wall at point S3 for the simulation of each mesh. Again, the results for grid 1 are significantly different from the others, and there is a shift in frequency for the peaks at 1.1 and 2.5 kHz. Moreover, as the mesh becomes denser, the amplitudes for the peak at 200 Hz decrease.

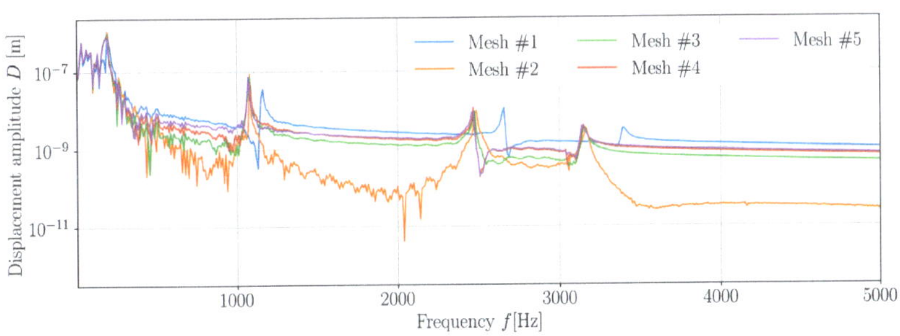

Figure 7. Spectrum of displacement along the y axis at probe S3 for finite element mesh independence study.

Differences and shifts in the presented spectra, apart from the actual difference in the results, may be due to the relatively low time resolution of the results. It was not possible to simulate a longer flow time due to the fact that they were of relatively long duration anyway.

However, even such a short time of calculations allowed us to draw conclusions about the convergence of the mesh. As can be seen from the presented data, further refinement of the mesh beyond mesh #4 does not improve the quality of the calculations. All simulations were carried out using grid #4. It was chosen for the reasons mentioned above, as well as for the fact that the computation time was only slightly longer than grid #3, and half that of grid #5. Moreover, a large number of elements in the thickness of the wall positively influenced the results.

4. Simulation Results

4.1. Flow Field

The results of the numerical simulations for the velocity and pressure fields are shown in Table 7. In all simulations, phenomena typical for flows over cavities can be seen, i.e., shear layer separation at the upstream edge, amplification of instabilities and vortices, impingement, vortex–downstream edge interactions (clipping and escape of vortices), and generation of pressure fluctuations [42]. The nature of the flows was relatively similar, and the introduction of flexible walls did not significantly affect the flow. However, in the case of the FSI model assuming walls made of polypropylene, deformation of the cavity resulted in the appearance of additional disturbances and eddies.

Table 7. Contours of velocity and pressure for reference models and FSI models with different materials at $t = 1$ s.

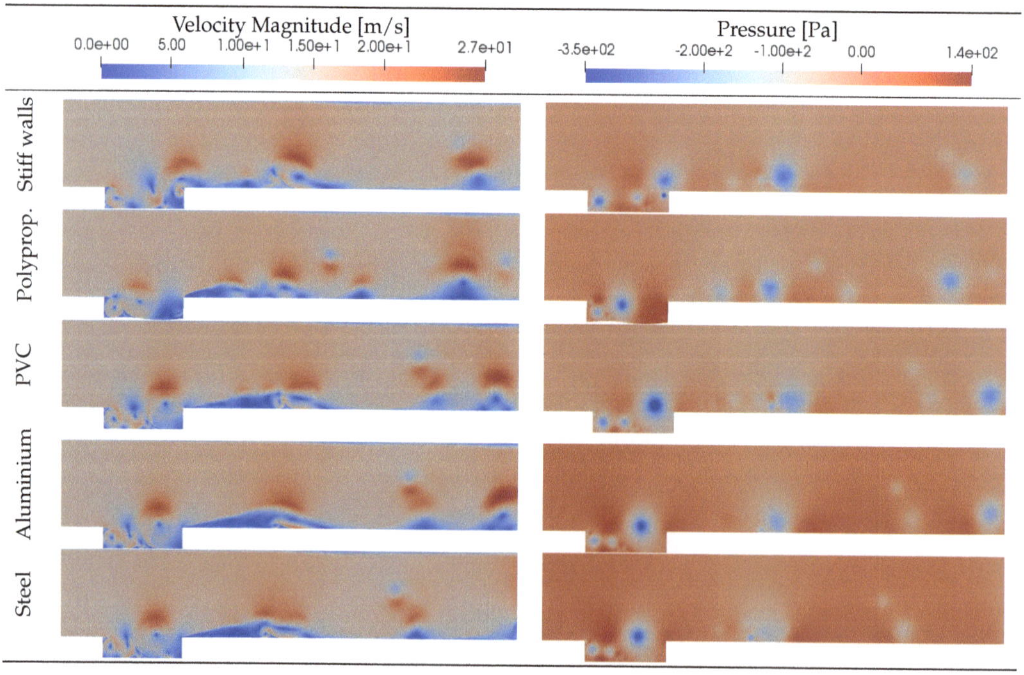

The flow pressure was sampled at points F1–F5 (Figure 1). The spectrum of pressure at point F3 is shown in Figure 8. In all cases, the highest peaks of pressure (largest vortices) appear at 34 Hz; however, their amplitudes differ. For the reference model and the model with steel as the wall material, it was 50 Pa, for the PVC and aluminium model it was 40 Pa, and for the polypropylene model it was 30 Pa. Smaller peaks appear at different frequencies depending on the material used.

As was mentioned in the introduction, Rossiter derived the equation linking the cavity dimensions, flow velocity, and frequencies of pressure fluctuations. The Rossiter model formula is provided by [6]:

$$f_m = \frac{U}{L}\frac{m-\gamma}{\frac{1}{K}+M} \tag{24}$$

where $U = 15$ m/s is velocity, $L = 0.12$ m is the length of the cavity, m is the mode number, $M = 0.44$ is the Mach number, and γ, K are constants. Rossiter assumed $\gamma = 0.25$ and $K = 0.57$, for a length to depth ratio equal to 4. In this work, the values of the constants $\gamma = 0$ and $K = 0.28$ were assumed, as the acoustic feedback can be considered as instantaneous [43,44].

The first ten computed Rossiter modes are shown in Table 8.

Table 8. Cavity Rossiter modes computed with Equation (24) for $\gamma = 0.25$ and $K = 0.57$.

Mode Number	1	2	3	4	5	6	7	8	9	10
Frequency f_R [Hz]	34.57	69.14	103.71	138.29	172.86	207.43	242.01	276.58	311.15	345.73

Rossiter stipulated that individual modes may or may not be excited. This is the case with the analyzed flows as well. The frequencies of 34 Hz and 69 Hz for which two highest peaks occured, as well as the frequency of 138 Hz, for which there is another increase in amplitude, are visible in Figure 8 and coincide with the first, second and fourth Rossiter modes.

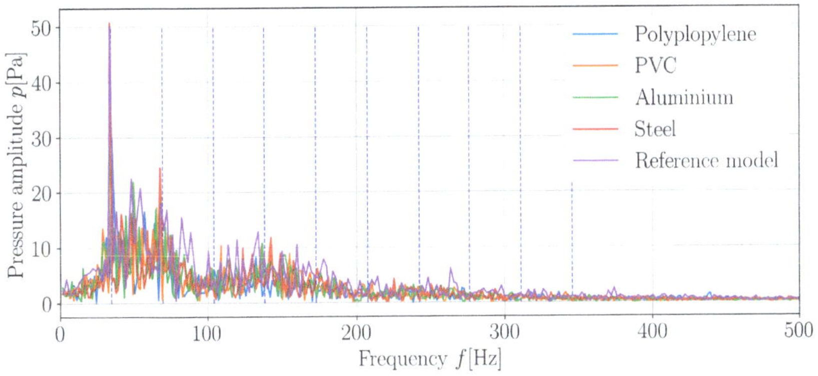

Figure 8. Spectrum of flow pressure at point F3 (Figure 1) (- - - Rossiter modes).

The values of the unsteady pressure inside and in the vicinity of the cavity consist of two main components, periodic and random [6]. The analyzed cavities have a length to depth ratio of 4. Rossiter states that this ratio is a borderline case; below this ratio, periodic components predominate in the cavity, while above this ratio random components predominate. In the analyzed case, there are strong periodic components related to the Rossiter modes, while a random signal is present as well.

Thecalculated Rossiter modes are shown in Figure 8 as blue dashed vertical lines, and are compared with the flow pressure spectrum.

4.2. Cavity Wall Vibrations

In this section, the results of the simulations of vibrations of the structure are shown. The figures presented in Table 9 show the time evolution and frequency spectrum of displacement for each of the materials used in the FSI simulations. The amplitudes of vibration differ by several orders of magnitude, from 2.5×10^{-6} m for steel to 1.25×10^{-3} m for polypropylene; this displacement reaches 5% of total cavity depth, and can be seen in the distributions shown in the second row of Table 7. These differences result from the differences in the material parameters, more specifically, the changes in the Young's modulus for each material.

Table 9. Time evolution and frequency spectra of displacement at point S3 for different materials (- - - Rossiter Modes).

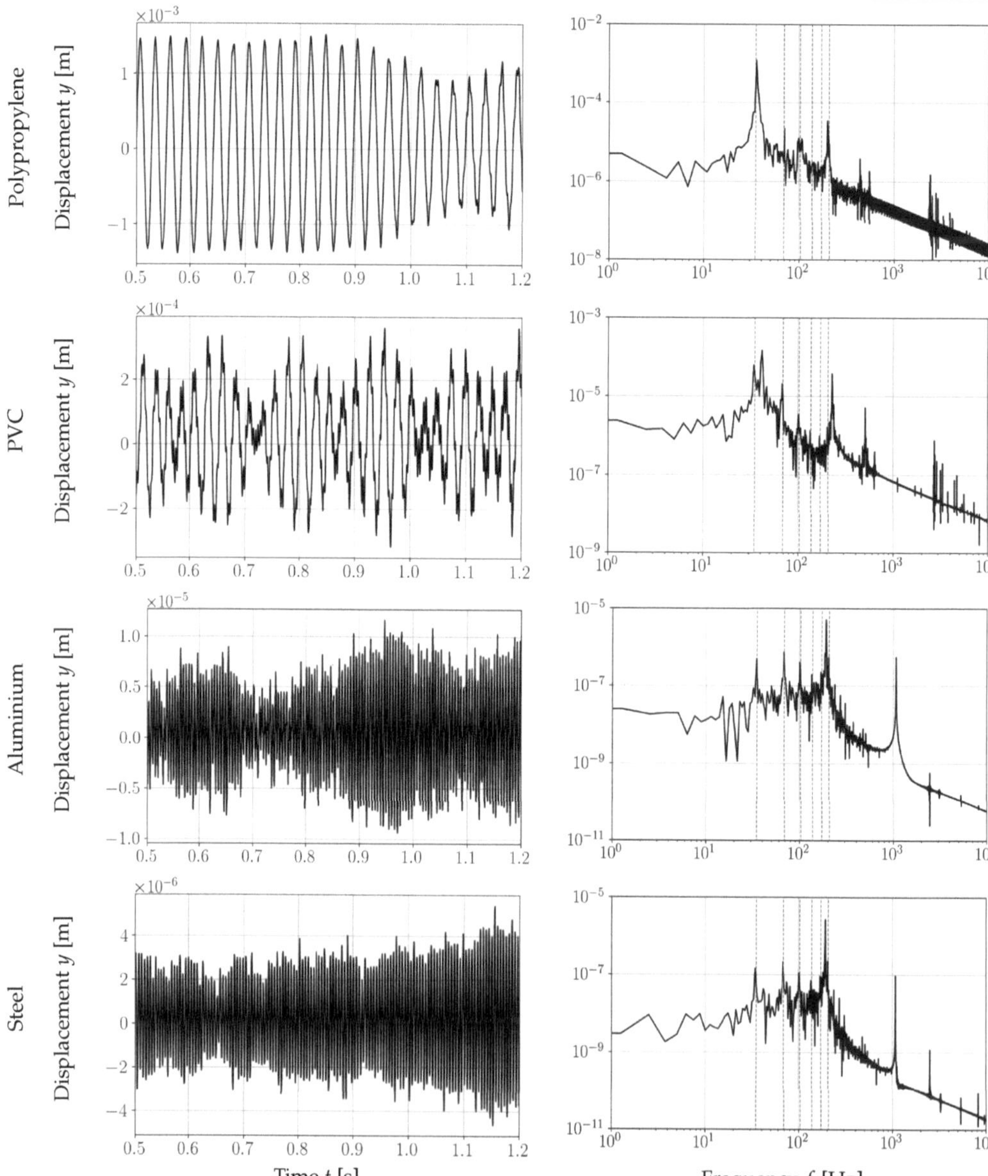

In either case, the figures show significant peaks for specific frequencies. These dominant frequencies were compared with the results of the model analyses. The model analyses were performed using the same meshes, materials, and boundary conditions used in the coupled simulations. The results are shown and compared in Table 10. The vibration frequencies that coincide with the natural frequencies are marked in italics.

Table 10. Computed natural frequencies for each material (in Hz) (*italics*—frequencies that coincide with cavity wall vibration spectrum frequency peaks (Section 4.2), **bold**—frequencies that coincide with noise spectrum frequency peaks (Section 4.3)).

Mode Number	Polypropylene	PVC	Aluminium	Steel
1	*35.4*	*40.5*	*190.8*	***192.4***
2	41.9	47.9	231.3	**234.8**
3	44.8	51.3	***241.7***	243.6
4	***103.6***	118.6	558.7	563.2
5	***109.1***	124.9	596.4	603.7
6	198.1	***226.7***	***1068.5***	***1077.0***
7	***208.7***	**238.8**	1134.5	1146.5
8	327.1	374.3	**1764.1**	**1778.2**
9	338.3	387.1	1834.0	1851.9
10	467.0	534.5	2518.3	***2538.4***
11	478.0	547.0	2587.0	2610.8
12	**560.5**	**641.4**	3030.2	3057.2
13	568.7	**650.8**	3065.6	3089.9
14	616.1	705.0	3321.4	3347.9
15	624.7	714.9	3374.7	3404.0
16	**768.6**	**879.5**	**4144.3**	**4177.6**
17	**773.3**	884.9	4179.6	4216.3
18	**989.9**	**1132.8**	5339.0	**5382.1**
19	1001.2	1145.7	5409.4	5456.2
20	**1252.4**	**1433.1**	**6754.7**	**6809.3**

In addition, the graphs shown in Table 9 compare the frequencies of vibrations with the frequencies of the first six Rossiter modes, marked with vertical blue dashed lines. In most cases, the vibration peaks coincide with the frequencies of first three (or more) frequencies of the different modes. Moreover, Table 11 shows the shape of the first four modes for each material. The shapes shown there coincide with the flow snapshots, which show the displacement of the cavity walls. This is especially evident for polypropylene, shown in second row of Table 7. In the case of walls made of other materials, the deformations were too small to be visible.

Table 11. Computed shapes of first four modes for different materials.

	Mode 1	Mode 2	Mode 3	Mode 4

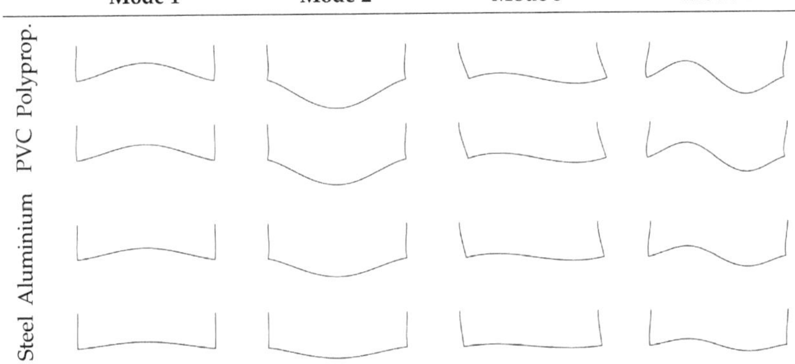

In the case of aluminium, the highest frequency of vibrations at point S3 was equal to 190 Hz, the first mode of vibrations. The same happens for PVC and steel; for both materials, the highest peak frequency occurs at the frequency of the first mode of vibrations, 40.5 Hz and 192 Hz, respectively. For polyproplylene, the highest amplitude of vibration occurred at 35 Hz, both the first Rossiter mode and the first mode of vibrations. The high amplitude of vibrations of a structure made of this material may be due to the fact that these modes overlap, meaning that the vibrations are amplified.

In Figures 9 and 10, the motion trajectories in the x-y plane of points S1–S5 are presented; the location of the points is shown in Figure 2. The time interval used to plot the trajectories was 0.8 to 0.9 s of the simulation time. Overall, it can be seen that the displacements of the bottom wall of the cavity are on average two orders of magnitude greater compared to the side walls. Moreover, the displacement of the upstream and downstream walls are similar both in order of magnitude and axisymmetrically, while the displacement amplitudes of the downstream wall are slightly larger. This is related to phenomena typical of flow over a cavity, namely, vortex shedding and its impingement on the downstream wall [45]. Depending on the material used, apart from the amplitudes of the displacements, their nature changes. The vibrations are regular for walls made of polypropylene, and the most irregular for walls made of PVC. This is especially visible on the trajectory plots of points S1 and S5. It should be noted that the trajectories for S2, S3, and S4 are not identical. This is related to various forms of natural vibrations and the fact that not every case has the highest amplitude in the first natural frequency.

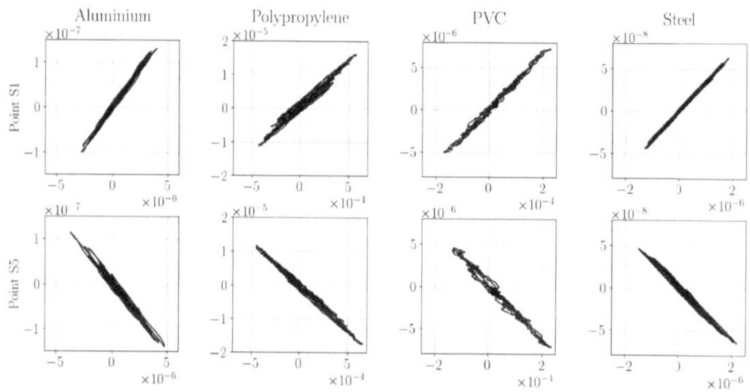

Figure 9. Displacement d_y-d_x trajectories of points S1 and S5 for each material (in meters).

4.3. Acoustic Pressure

Finally, we consider the aeroacoustic noise generated by the flow over the cavity and how flexible cavity walls affect it. In Figures 11–14, the spectrum of sound pressure levels computed with the Ffowcs-Williams and Hawkings analogy (provided by Equation (23)) are shown. In each case, the receiver was defined as being positioned directly 5 m above the cavity. The sound pressure was sampled every 50 μs, which corresponds to a sampling frequency of 20 kHz. A rectangular window was used to find the FFT of the signal.

The sound pressure levels for all cases are shown in in Figures 11–14. In each case, the results shown in black are compared with the sound pressure level computed for stiff walls, shown in red, which are repeated on each graph for clarity.

In all flow simulations involving the FSI model, significant differences between the spectra can be seen. In the low frequency range, up to 500 Hz, the spectra for all cases look similar and noise predominates. For the reference case with stiff walls, the sound pressue drops below 20 dB above 1000 Hz. A sound pressure level of 20 dB can be treated as lower than background noise for almost all rooms mentioned by the ASHRAE guide [46], and can therefore be omitted. These values may result from approximations and numerical errors. Compared to the reference case, for the frequencies below 500 Hz the sound pressure levels for the simulation with flexible walls have a similar character; however, peaks related to modal frequencies appear in the spectrum. Above 500 Hz, there is tonal noise that does not exist in the reference case. This is related to modal frequencies and natural vibrations.

As was mentioned in Section 4.2, Table 10 shows the natural frequencies of the cavity walls for all four cases. The bold text indicates the frequencies that coincide with the peaks of the spectra in Figures 11–14. Regardless of the model adopted, in each case there was an increase in the amplitude close to the frequency of Rossiter modes, while, depending on the case, different modes were excited. In each of the cases, modes 1 and 2, 5, and 6 were excited, while modes 3 and 4 were excited only for the reference case and the cavity with polypropylene walls. This is due to the high sensitivity of this phenomenon to flow parameters, as well as to its high randomness.

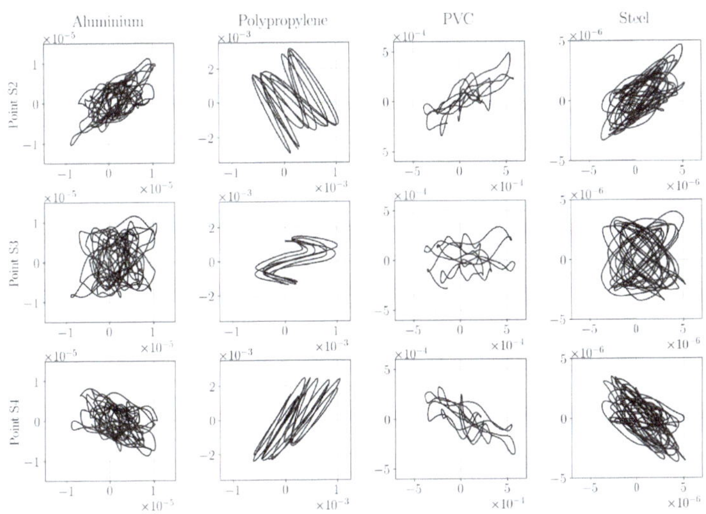

Figure 10. Displacement d_y-d_x trajectories of points S2, S3 and S4 for each material (in meters).

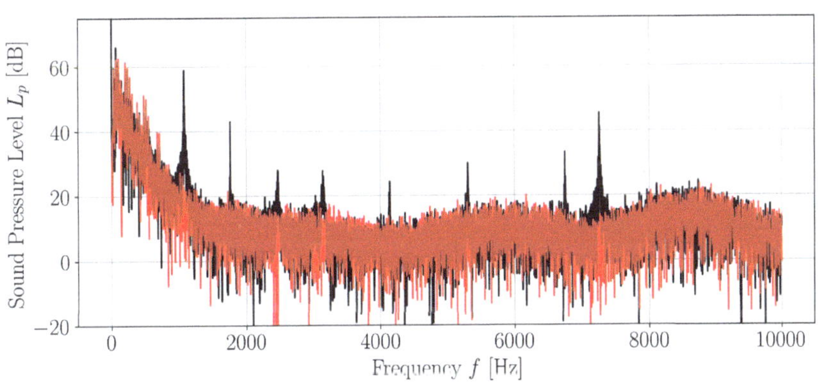

Figure 11. Sound pressure level of cavity noise with flexible aluminium walls (—) and stiff walls (—).

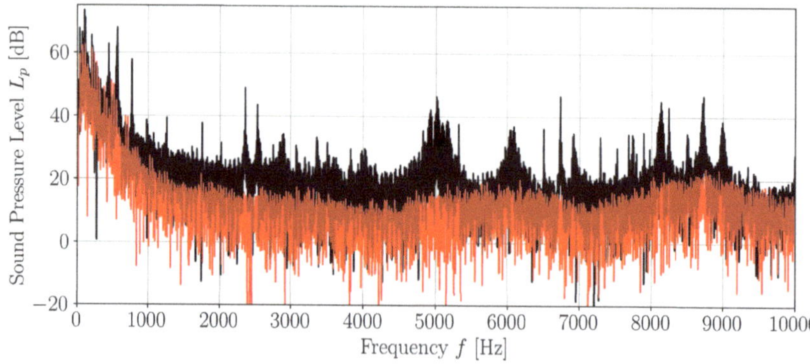

Figure 12. Sound pressure level of cavity noise with flexible polypropylene walls (——) and stiff walls (——).

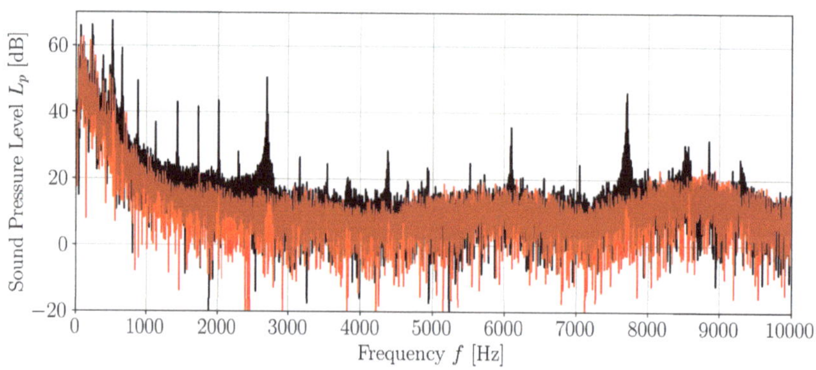

Figure 13. Sound pressure level of cavity noise with flexible PVC walls (——) and stiff walls (——).

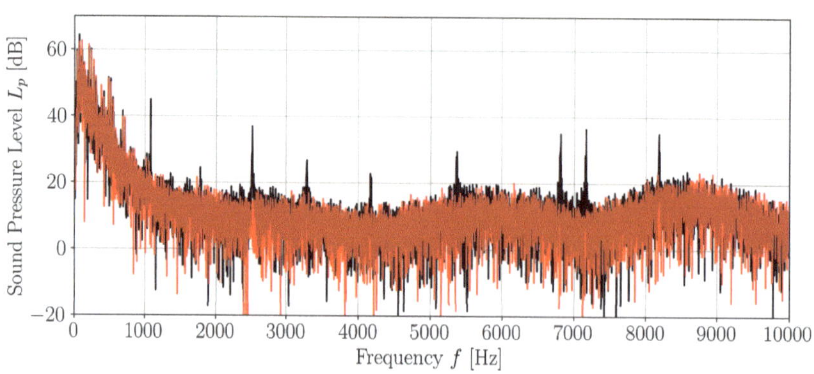

Figure 14. Sound pressure level of cavity noise with flexible steel walls (——) and stiff walls (——).

The sound pressure levels above 1000 Hz as compared to the reference increase as the Young's modulus of each material decreases. The Young's modulus was the highest for steel. In the case of walls for which steel was adopted as a material, the spectra almost coincide, except for the few peaks mentioned above. As the Young's modulus decreases, the average sound pressure level increases. The highest SPL was calculated for walls made of polypropylene; however, in this case, the result is not related only to the material used.

For this material, the first Rossiter frequency coincides with the first natural frequency, which could further amplify the noise generated by the flow over the cavity.

5. Conclusions

In this work, the effects of elastic cavity walls and fluid–structure interaction on the noise generated by flow were investigated. Simulations were carried out for one geometric model selected on the basis of mesh independence analysis, one flow velocity, and four different sets of material parameters. In addition, flow analysis was carried out for the reference model, in which only unidirectional fluid–structure interaction was assumed and the walls were treated as rigid.

The main research findings and conclusions are as follows:

- The simulations showed a significant effect of non-rigid walls and bidirectional fluid–structure interaction on flow noise.
- In each of the analysed cases, the sound pressure level calculated using the FW-H acoustic analogy was higher than for the reference case with rigid walls, and the characteristics of the sound spectrum changed as well.
- In the case of aeroacoustic analyses with flow through thin-walled channels, the fluid–structure interactions cannot be neglected.
- In the case of the flow computations themselves, fluid–structure interactions are not as important; depending on the required accuracy of the results, they may be ignored.
- Overlapping of natural and Rossiter frequencies can result in a significant increase in wall displacement amplitudes and sound pressure levels.

For the model with stiff walls, the SPL was similar throughout the band in the frequency range up to 1000 Hz, while above this range it fell below 20 dB. For the models with flexible walls, the sound pressure spectrum below 1000 Hz was similar to the reference case, while above this frequency there were additional components related to the movement of the cavity walls. Their frequencies partially coincided with the eigenfrequencies of the walls.

Overlapping between the natural and Rossiter frequencies occurred for one of the analysed materials (polypropylene). The overlapping of these frequencies at 35 Hz resulted in a significant increase in the wall displacement amplitude and sound pressure level. This should be borne in mind when designing ventilation systems. It is important to check that the Rossiter and natural frequencies do not coincide, as this can result in vibrations with high amplitudes.

The FW-H acoustic analogy was used to describe the flow-induced noise. Its limitations should be taken into account. This method does not account for the influence of the ventilation duct walls on the propagation of the acoustic wave along the duct, the propagation of the acoustic wave outside the duct, or vibroacoustic disturbances that could affect the obtained pressure level spectra.

The main purpose of this research was to verify whether the influence of the flow on the deformable walls of the cavity could affect the sound pressure levels, and this was achieved. In future research, it is necessary to investigate the influence of the remaining duct walls. In addition, it is necessary to investigate the influence of individual material parameters (as opposed to specific materials) on the flow and generated sound.

Here, it is worth mentioning several problems related to the modeling of fluid–structure interactions. Compared to calculations that do not take into account these interactions, both the computation time and the disk space needed to store the results are incomparably greater. Due to the limitations imposed by the PLGrid Infrastructure regarding the length of each task, it was necessary to perform a series of calculations in which the simulated time of each task was 0.1 s in order to compute 0.7 s of flow time. This was due to the imposed maximum duration of the simulations. The results were then combined for the purposes of this work. In the case of uncoupled CFD calculations for the same mesh and parameters, the time needed to simulate 0.1 s of the flow was 15 h. For simulations including FSI, it was over 150 h. Due to the size of the computational meshes, the simulations were performed in parallel on four and six computational nodes for the uncoupled and coupled models, respectively. Further increasing the number of nodes would not significantly affect computation time due to the time required for simulation data exchange between computational nodes. Moreover, for these simulations it was necessary to store information about both the flow field and the deformation field. Files with the results of these simulations were over 1.5 TB in size. Because of this, it would be impossible to conduct such analyses without the use of PLGrid Infrastructure and its computing resources.

Author Contributions: Conceptualization, P.Ł., I.C. and A.G.; methodology, P.Ł. and I.C.; software, P.Ł.; validation, P.Ł. and I.C.; formal analysis, P.Ł.; investigation, P.Ł.; resources, P.Ł.; data curation, P.Ł. and I.C.; writing—original draft preparation, P.Ł.; writing—review and editing, I.C. and A.G.; visualization, P.Ł. and I.C.; supervision, I.C. and A.G.; project administration, I.C.; funding acquisition, A.G. All authors have read and agreed to the published version of the manuscript.

Funding: This research was supported by national subvention no. 16.16.130.942.

Acknowledgments: This research was supported in part by PLGrid Infrastructure.

Conflicts of Interest: The authors declare no conflict of interest.

Nomenclature

$v_i = \overline{v_i} + v'_i$	ith component of velocity [m/s]
$p = \overline{p} + p'$	pressure [Pa]
ρ	density [kg/m^3]
$\overline{\tau_{ij}}$	laminar viscous stress tensor [N/m^2]
$\tau_{ij}^R = -\rho\overline{v'_i v'_j}$	Reynolds stress tensor [N/m^2]
$\overline{\psi}$	mean/filtered variable [-]
ψ'	fluctuating/sub-filter variable [-]
μ_t	eddy viscosity [m^2/s]
k	turbulent kinetic energy [m^2/s^2]
ω	specific dissipation rate [1/s]
$\beta^*, \sigma_k \sigma_\omega, \gamma, a_1$	constants of the $k-\omega$ SST turbulence model
F_1, F_2	additional functions of the $k-\omega$ SST turbulence model
$T_{ij}^S = \overline{u}_i\overline{u}_j - \overline{u_i u_j}$	subgrid-scale (SGS) stress tensor [N/m^2]
$\ddot{U} = A$	global acceleration vector [m/s^2]
U	global displacement vector [m]
M	global mass matrix [kg]
K	global stiffness matrix [N/m]
F	global force vector [N]
Σ^{KLMN}	free energy function [kg m^2/s^2]
θ	absolute temperature [K]
ρ_0	mass density [kg/m^3]
f^k	force per unit mass [N/kg]
β^{KL}	thermal stress tensor per unit temperature [N/m^{2}K]
γ^{KL}	residual stress tensor [N/m^2]
φ	shape function [-]
ξ_i, η_i, ζ_i	local coordinates of the ith point
ρ'	(acoustic) density fluctuations [kg/m^3]
H_s	Heaviside step function [-]
T_{ij}	Lighthill stress tensor [N/m^2]
c_∞	speed of sound [m/s]
V_i	velocity of surface in i direction [m/s]
f	scalar function defining the surface [-]
δ	Dirac delta function [-]

Abbreviations

The following abbreviations are used in this manuscript:

ASHRAE	American Society of Heating, Refrigerating, and Air-Conditioning Engineers
CAA	Computational Aeroacoustics
CFD	Computational Fluid Dynamics
DES	Detached Eddy Simulation
FFT	Fast Fourier Transform
FSI	Fluid–Structure Interaction
FWH	Ffowcs-Williams and Hawkings (acoustic analogy)
LES	Large Eddy Simulation
PISO	Pressure-Implicit with Splitting of Operators
PVC	Polyvinyl Chloride
RANS	Reynolds-Averaged Navier–Stokes
SGS	Sub-Grid Scale
SIMPLE	Semi-Implicit Method for Pressure-Linked Equations
SPL	Sound Pressure Level
SST	Shear Stress Transport

References

1. Crook, S.; Kelso, R.; Drobik, J. Aeroacoustics of Aircraft Cavities. In Proceedings of the 16th Australian Fluid Mechanics Conference, Gold Coast, Australia, 3–7 December 2007.
2. Wang, Z.K.; Djambazov, G.; Lai, C.H.; Pericleous, K. Numerical simulation of flow-induced cavity noise in self-sustained oscillations. *Comput. Vis. Sci.* **2007**, *10*, 123–134. [CrossRef]
3. Kim, H.; Hu, Z.; Thompson, D. Effect of cavity flow control on high-speed train pantograph and roof aerodynamic noise. *Railw. Eng. Sci.* **2020**, *28*, 54–74. [CrossRef]
4. Radavich, P.M.; Selamet, A.; Novak, J.M. A computational approach for flow–acoustic coupling in closed side branches. *J. Acoust. Soc. Am.* **2001**, *109*, 1343–1353. [CrossRef]
5. Lafon, P.; Caillaud, S.; Devos, J.; Lambert, C. Aeroacoustical coupling in a ducted shallow cavity and fluid/structure effects on a steam line. *J. Fluids Struct.* **2003**, *18*, 695–713. [CrossRef]
6. Rossiter, J.E. Wind Tunnel Experiments on the Flow over Rectangular Cavities at Subsonic and Transonic Speeds. 1964. Available online: https://reports.aerade.cranfield.ac.uk/handle/1826.2/4020 (accessed on 1 October 2022).
7. Rockwell, D.; Naudascher, E. Review—Self-Sustaining Oscillations of Flow Past Cavities. *J. Fluids Eng.* **1978**, *100*, 152–165. [CrossRef]
8. Yokoyama, H.; Hibino, K.; Nishikawara, M.; Yanada, H. Oscillations of a flexible beam in cavity flow with tonal sound. In Proceedings of the 6th International Conference of Global Network for Innovative Technology (IGNITE), Penang, Malaysia, 2–3 December 2019.
9. Thangamani, V.; Kok, F.N. Energy harvesting from cavity flow oscillations. *J. Intell. Mater. Syst. Struct.* **2022**, *33*, 400–418. [CrossRef]
10. Khanafer, K. Comparison of flow and heat transfer characteristics in a lid-driven cavity between flexible and modified geometry of a heated bottom wall. *Int. J. Heat Mass Transf.* **2014**, *78*, 1032–1041. [CrossRef]
11. Alsabery, A.; Selimefendigil, F.; Hashim, I.; Chamkha, A.; Ghalambaz, M. Fluid-structure interaction analysis of entropy generation and mixed convection inside a cavity with flexible right wall and heated rotating cylinder. *Int. J. Heat Mass Transf.* **2019**, *140*, 331–345. [CrossRef]
12. Sun, X.; Li, W.; Ye, Z. Flow-induced Vibration of Flexible Bottom Wall in a Lid-driven Cavity. *J. Vib. Test. Syst. Dyn.* **2017**, *1*. [CrossRef]
13. Sabbar, W.A.; Ismael, M.A.; Almudhaffar, M. Fluid-structure interaction of mixed convection in a cavity-channel assembly of flexible wall. *Int. J. Mech. Sci.* **2018**, *149*, 73–83. [CrossRef]
14. Strelets, M. Detached eddy simulation of massively separated flows. In Proceedings of the 39th Aerospace Sciences Meeting and Exhibit, Reno, NV, USA, 8–11 January 2001.
15. Menter, F. Zonal Two Equation k-ω Turbulence Models For Aerodynamic Flows. In Proceedings of the 23rd Fluid Dynamics, Plasmadynamics, and Lasers Conference, Orlando, FL, USA, 6–9 July 1993.
16. Wagner, C.; Huttl, T.; Sagaut, P. *Large-Eddy Simulaton for Acoustics*; Cambridge University Press: Cambridge, UK, 2007.
17. Hu, J.; Ning, X.; Zhao, W.; Li, F.; Ma, J.; Zhang, W.; Sun, S.; Zou, M.; Lin, C. Numerical simulation of the cavitating noise of contra-rotating propellers based on detached eddy simulation and the Ffowcs Williams–Hawkings acoustics equation. *Phys. Fluids* **2021**, *33*, 115117. [CrossRef]
18. Sharma, S.; Geyer, T.F.; Giesler, J. Effect of geometric parameters on the noise generated by rod-airfoil configuration. *Appl. Acoust.* **2021**, *177*, 107908. [CrossRef]
19. Liu, K.; Zhou, S.; Li, X.; Shu, X.; Guo, L.; Li, J.; Zhang, X. Flow-induced noise simulation using detached eddy simulation and the finite element acoustic analogy method. *Adv. Mech. Eng.* **2016**, *8*. [CrossRef]

20. Bungartz, H.J.; Lindner, F.; Gatzhammer, B.; Mehl, M.; Scheufele, K.; Shukaev, A.; Uekermann, B. preCICE—A fully parallel library for multi-physics surface coupling. *Comput. Fluids* **2016**, *141*, 250–258. [CrossRef]
21. Lighthill, M.J. On Sound Generated Aerodynamically. I. General Theory. *Proc. R. Soc. Lond. Ser. A Math. Phys. Sci.* **1952**, *211*, 564–587.
22. Williams, J.E.F.; Hawkings, D.L. Sound generation by turbulence and surfaces in arbitrary motion. *Philos. Trans. R. Soc. Lond. Ser. A Math. Phys. Sci.* **1969**, *264*, 321–342.
23. Rama krishna, V.; Sanaka, S.P.; Pardhasaradhi, N.; Raghava Rao, B. Hydro-elastic computational analysis of a marine propeller using two-way fluid structure interaction. *J. Ocean Eng. Sci.* **2021**. [CrossRef]
24. Ma, R.; Liu, Z.; Zhang, G.; Doolan, C.J.; Moreau, D.J. Acoustic analysis of a forced-oscillating cylinder in flow using a hybrid method. *Aerosp. Sci. Technol.* **2020**, *106*. [CrossRef]
25. Turek, S.; Hron, J. Proposal for Numerical Benchmarking of Fluid–Structure Interaction Between an Elastic Object and Laminar Incompressible Flow. *Fluid-Struct. Interact. Model. Simul. Optim.* **2007**, *53*, 371–385.
26. Chourdakis, G.; Davis, K.; Rodenberg, B.; Schulte, M.; Simonis, F.; Uekermann, B.; Abrams, G.; Bungartz, H.J.; Yau, L.; Desai, I.; et al. *preCICE v2: A Sustainable and User-Friendly Coupling Library; Open Res. Eur.* **2022**, *2*, 51.
27. Lojek, P.; Czajka, I.; Golas, A.; Suder-Debska, K. Infuence of the Elastic Cavity Walls on Cavity Flow Noise. *Vib. Phys. Syst.* **2021**, *1*.
28. III, L.N.C.; Garg, S.; Washburn, A.E. Piezoelectric actuators for fluid-flow control. In *Proceedings of the Smart Structures and Materials 1997: Industrial and Commercial Applications of Smart Structures Technologies*; Sater, J.M., Ed.; International Society for Optics and Photonics: San Diego, CA, USA; 1997; Volume 3044, pp. 147–157.
29. Chatellier, L.; Laumonier, J.; Gervais, Y. Active control of the aeroacoustics of cavity flows from the downstream edge. *C. R. Méc.* **2006**, *334*, 259–265. [CrossRef]
30. Blazek, J. *Computational Fluid Dynamics. Principles and Applications*; Elsevier Ltd.: Amsterdam, The Netherlands, 2015.
31. Weller, H.G.; Tabor, G.R.; Jasak, H.; Fureby, C. A tensorial approach to computational continuum mechanics using object-oriented techniques. *Comput. Phys.* **1998**, *12*, 620–631. [CrossRef]
32. Holzmann, T. Mathematics, Numerics, Derivations and OpenFOAM(R). 2019. Available online: https://holzmann-cfd.com/community/publications/mathematics-numerics-derivations-and-openfoam (accessed on 1 October 2022).
33. Patankar, S.; Spalding, D. A calculation procedure for heat, mass and momentum transfer in three-dimensional parabolic flows. *Int. J. Heat Mass Transf.* **1972**, *15*, 1787–1806. [CrossRef]
34. Issa, R. Solution of the implicitly discretised fluid flow equations by operator-splitting. *J. Comput. Phys.* **1986**, *62*, 40–65. [CrossRef]
35. Dhondt, G. *The Finite Element Method for Three-Dimensional Thermomechanical Applications*; John Wiley & Sons Ltd.: Hoboken, NJ, USA, 2004.
36. Lapidus, L.; Pinder, G. *Numerical Solution of Partial Differential Equations in Science and Engineerig*; John Wiley & Sons Ltd.: Hoboken, NJ, USA, 1999.
37. Glegg, S.; Devenport, W. *Aeroacoustics of Low Mach Number Flows—Fundamentals, Analysis and Measurement*; Academic Press: Cambridge, MA, USA, 2017.
38. Brentner, K.S.; Farassat, F. Modeling aerodynamically generated sound of helicopter rotors. *Prog. Aerosp. Sci.* **2003**, *39*, 83–120. [CrossRef]
39. Epikhin, A.; Evdokimov, I.; Kraposhin, M.; Kalugin, M.; Strijhak, S. Development of a Dynamic Library for Computational Aeroacoustics Applications Using the OpenFOAM Open Source Package. *Procedia Comput. Sci.* **2015**, *66*, 150–157. [CrossRef]
40. Komerath, N.; Ahuja, K.; Chambers, F., Prediction and measurement of flows over cavities—A survey. In Proceedings of the 25th AIAA Aerospace Sciences Meeting, Reno, NV, USA, 24– 26 March 1987.
41. *ASHRAE Handbook—HVAC Systems and Equipment*; ASHRAE: Atlanta, GA, USA, 2016.
42. Gloerfelt, X. Cavity Noise. 2007. Available online: https://www.researchgate.net/publication/292356579_Cavity_noise (accessed on 1 October 2022).
43. Ma, R.; Slaboch, P.; Morris, S. Fluid mechanics of the flow-excited Helmholtz resonator. *J. Fluid Mech.* **2009**, *623*. [CrossRef]
44. Verdugo, F.; Guitton, A.; Camussi, R. Experimental investigation of a cylindrical cavity in a low Mach number flow. *J. Fluids Struct.* **2012**, *28*. [CrossRef]
45. Knisely, C.; Rockwell, D. Self-sustained low-frequency components in an impinging shear layer. *J. Fluid Mech.* **1982**, *116*. [CrossRef]
46. *ASHRAE Handbook—HVAC Applications*; ASHRAE: Atlanta, GA, USA 2019.

energies

MDPI

Article

Power Consumption and Effectiveness of Azimuth Stern-Drive (ASD) Tug While Assisting at Ship's Bow

Jarosław Artyszuk * and Paweł Zalewski

Faculty of Navigation, Maritime University of Szczecin, Wały Chrobrego St. 1-2, 70-500 Szczecin, Poland
* Correspondence: j.artyszuk@pm.szczecin.pl

Abstract: The purpose of this article is to present a methodical approach to increasing the efficiency of a tug with an azimuth stern drive (ASD). The difficulties in the bow-to-bow (as a reverse tractor) operations of an azimuth stern-drive tug, while pulling at the assisted (towed) ship's bow, are investigated through a parametric study. The authors' original generic analytical model of a tug in a steady state is utilised. Various design and operation options are also compared, which cover the escort speed, hawser angle, and relative location of the towing point, among others. The latter also means a shift to stern-to-bow operation. The thruster power required and the remaining surplus for executing new alignment (steering) orders are assessed. The study is, thus, aimed at energy savings for the whole operation and, consequently, at long-term harbour-tug-fleet energy effectiveness. The basic output of the present research is the comparison of the behavioural and control patterns of various tug arrangements at a ship's bow. The results of the study show a slight advantage for a midship winch tug working through her stern for bow operation (stern-to-bow) over the reverse tractor acting bow-to-bow. Comparing various tug designs or operation options is difficult due to complicated tug dynamics and statics patterns under the hawser action. Both relative and absolute towing forces have to be considered (a high relative force may result in a low absolute force, much less than ordered by the pilot), with some geometrical and safety constraints additionally. The practical implementation of the conducted research should cover the guidance and training rendered to tug masters in the scope of the operational limitations of a particular tug and towards the design or acquisition of the tug that is best-suited for a particular harbour operation.

Keywords: azimuth stern drive; ASD tug; bow tug assistance; energy efficiency; towing effectiveness; harbour manoeuvring; full mission ship simulator; FMBS studies

1. Introduction

Large ships (often with one-screw propulsion and without lateral tunnel thrusters) cannot safely enter confined harbours and berth on their own. This is due to their high kinetic energy during an approach speed (roughly about 5 knots), and the limited number and magnitude of the control forces that can be developed by the ship to handle herself in a horizontal plane (3DOFs) and in an arbitrary way (with control over the so-called pivot point). The contemporary research (which is supported by multiple references in the literature, starting from the 1960s) of maritime towing covers the following topics:

(1) propulsive and hydrodynamic (resistance, manoeuvring, seakeeping) design of tugs, including experimental model tests, CFD computations, and force regression/analytical studies, e.g., [1], or incl. the hook location design, e.g., [2];
(2) mathematical modelling of tug manoeuvring hydrodynamics (and towing operations) for simulation and/or full-mission simulator implementation, e.g., [3,4];
(3) choice of the number and power (and type) of tugs for towing assistance, both in the open sea (ocean towing) and a harbour, in terms of operation safety and effectiveness, e.g., [5,6];
(4) tug roll stability during towline tension;

(5) ship-to-tug interaction at speed. e.g., [7];
(6) autonomous tugs and automatic control of tug manoeuvres (e.g., in a simulator environment);
(7) human-related aspects of tug handling (seamanship), e.g., [8];
(8) tug GHG emissions, e.g., [9].

There are also a number of successful conferences more or less specifically devoted to tugs, towing, and manoeuvring simulation, and the attached references list is certainly not exhaustive.

This paper is basically situated within the second topic, with some guidance for tug captains, thus relating to the application areas expressed by the last three topics as well. Partly, the present work is an application related follow-up of the authors' previous developments.

The ship and tug cooperation (and interaction) is a challenging subject, in particular while towing a ship's bow. Inadequacies of the theory concerning the advantages of a particular tug (or of a particular deployment of a tug's towing assembly or towing mode) in the ship's area and all the factors involved, especially those related to the design (the location of the towing point (*T*) vs. the propeller/thruster (*P*) installation point and the underwater hull, including the skeg, in terms of shape and size) are matter-of-fact [10,11]. It must be stressed that the present theory is similar to that of an unconventional, highly manoeuvrable cycloidal or azimuth tractor tug application for bow operation (where a towing winch is close to the end of the tug's stern) vs. a conventional tug (with aft propulsion and a central hook). A lot of discussions/views exist in this context [10,12]. The tug-to-bow operation, to a great extent, inherits the well-known hydrodynamics specific to the operation at a ship's stern, especially at a high towing (escort) speed [10–14]. In the latter, investigations often go towards the so-called direct and indirect towing (with the augmentation of a tug's performance via her underwater hull hydrodynamic force).

The azimuth stern-drive (ASD) tug is often called a reverse tractor in the common industry language, especially for bow operation that should be equivalent, at least with some generalisation (e.g., from the tug's manoeuvrability standpoint), to a tractor tug (with propellers located forward). The basic difference between a reverse tractor tug and a tractor tug comes from the human-centred difficulty while monitoring (visual viewing) and controlling the tug's movement either in an aft direction (a reverse tractor, bow-to-bow operation) or in a natural forward direction (a tractor tug, stern-to-bow operation at ship's bow). However, the main issue of the tractor and reverse tractor comparison arises from quite different factors—the mutual location of *T* (sometimes multiple locations are possible onboard a tug, depending on the design) vs. *P*. This situation necessitates the future debate and application of a new classification/redefinition of tugs or tug operational modes, since the straight/pure naming as a tractor or reverse tractor is a little confusing.

The bow tug's design and operation sensitivity to some factors (in terms of efficiency or effectiveness and safety [15]) has been rarely undertaken so far. The presented research is partly inspired by the TNI monograph of Cpt. Hensen [16] and the authors' previous study on ASD tugs used for bow operation [17] and is aimed at providing some detailed knowledge on certain fundamental aspects. The towed ship–tug hydrodynamic interaction effects were omitted in our study and postponed until future research.

In the following research, the original generic analytical model for tug pull operation statics was utilised [18]. Any type of tug's hull hydrodynamics and towing assembly arrangement can be input into this model. The towing operation at a ship's bow is considered in both the bow-to-bow and stern-to-bow configurations, especially with non-zero towing speed taken into account. This method is based on the classical equilibrium of the forces in 3DOF being solved analytically (with numerical support for lookup-table interpolation, where the hull hydrodynamic characteristics are stored), which is very fast, even for spreadsheets. The adopted dimensionless approach serves energy efficiency, and the drift angle domain ensures the uniqueness and completeness of the solution, which are novel features.

Some research centres claim they have developed special software for computing the force (power) exerted by a tug on an assisted ship (and the tug's control parameters), e.g., [10,19–22]. However, their published documentation is often insufficient, and the published results (plots, tables, analysis, discussion) are incomplete. In this paper, the capability plots and detailed steering data of a tug are, thus, provided and discussed to enhance the decision support while selecting and operating the particular mode of towing assistance. Although worldwide SI units are critical in various scientific publications and applications, in the maritime industry, especially in ship structural design and operation, and seamanship, the unit of the ton (or precisely but more rarely ton-force, t) is still commonly used for force expression. Partly, a similar situation exists with respect to the knot as a unit of speed or velocity. Sometimes, to provide direct conversion if needed, double (or parallel) units are used. Therefore, the units of ton and knot ([t], [kn]) are used throughout this paper.

The paper presents problems concerning the difficulties in the bow-to-bow operations of an ASD tug, while pulling at the assisted ship's bow. Its contents are structured as follows:

(a) problem formulation, materials and methods including a literature review, a mathematical model of an exemplary tug, and simulation cases, in Section 2;
(b) results of a parametric study of the tug's towing point vs. her propeller location, in Section 3;
(c) discussion, in Section 4;
(d) conclusions and further research plans, in Section 5;
(e) plots comparing the reference cases to other hydrodynamics and supplementary cases, in the appendices.

2. Problem Formulation, Materials, and Methods

The description (with appropriate discussion) of the harbour tug model, physics, and derived solution for tug's static equilibrium is presented in Section 2.1. Furthermore, the computational scenarios, based on the most representative, extreme (providing the deepest insight and knowledge) cases of T vs. P location are presented in Section 2.2.

2.1. Case Study Tug and Mathematical Model of Static Equilibrium

The investigated tug of generic type and environmental parameters were, for consistency with the previous research conducted by the authors [14,17,18]:

- length between perpendiculars (as a reference length), (L) 30.5 m;
- draught (as a reference draught), (T_{ref}) 5 m;
- water density (ρ), 1000 kg/m^3.

The tug described above is a typical harbour one and as such can accommodate various engine powers (bollard pull) within the ship size as presented, e.g., in [23]. The tug length and draught are used hereafter to determine the lateral underwater area (by convention being the product LT_{ref}) necessary for computing hull (H) hydrodynamic forces on a tug due to her movement through water. Knowing the basics of similitude in ship hydrodynamics (where, e.g., the applied dimensionless coefficients are strictly dependent on the reference area used in order to obtain the appropriate/measured force magnitude), one can freely and easily extrapolate or fit all the following research output to their needs.

The reference frame used to study the dynamics/statics of a tug (the most important item in our study of the two-body system—small tug and large ship) is that of the tug itself (see Figure 1). The steady-state condition can be, hence, written as follows:

$$\begin{cases} F_{xH} + F_{xP} + F_{xT} &= 0 \\ F_{yH} + F_{yP} + F_{yT} &= 0 \\ M_{zH} + M_{zP} + M_{zT} &= 0 \end{cases} \quad (1)$$

where F_x, F_y—longitudinal and lateral force, and M_z—yaw moment.

$$\begin{bmatrix} F_{xH} \\ F_{yH} \\ M_{zH} \end{bmatrix} = 0.5\rho L T_{ref} v^2 \begin{bmatrix} c_{fxh}(\beta) \\ c_{fyh}(\beta) \\ L c_{mzh}(\beta) \end{bmatrix} \quad (2)$$

where c_{fxh}, c_{fyh}, c_{mxh}—hull hydrodynamic dimensionless coefficients as functions of drift angle β in the range $(-180°, +180°\rangle$ to obtain the full general solution (the additional yaw velocity contribution disappears in a steady state), and v—towing (escort) speed.

$$\begin{bmatrix} F_{xP} \\ F_{yP} \\ M_{zP} \end{bmatrix} = F_P \begin{bmatrix} cos\delta \\ sin\delta \\ x'_P \cdot L \cdot sin\delta \end{bmatrix} \quad (3)$$

where β—propeller/thruster angle (the angle of thrust force) in the range $(-180°, +180°\rangle)$, and x'_P—propeller/thruster dimensionless location (in tug's length unit, typically ≈ -0.5 for stern-drive tug, but this study also considers the other extreme case, -0.3, which may reflect, by symmetry and similarity, the performance of forward-drive (azimuth or Voith-Schneider) tractor tug. Hereafter, because a tug usually incorporates dual/twin azimuth propulsors, they are treated as a single propulsor of double force or power, as depicted by F_P (that is positive definite, by default), thus of their parallel/combined action.

$$\begin{bmatrix} F_{xT} \\ F_{yT} \\ M_{zT} \end{bmatrix} = -F_T \begin{bmatrix} cos(\gamma-\beta) \\ sin(\gamma-\beta) \\ x'_T \cdot L \cdot sin(\gamma-\beta) \end{bmatrix} \quad (4)$$

where γ—hawser (towing line) angle in the range $(-180°, +0°)$ we assume the operation on port side of the assisted ship (6 to 12 o'clock), and x'_T—towing point (hook/winch/fairlead) location (in tug's length unit, typically $\approx +0.5$ for stern-drive tug, acting through her bow as reverse tractor, but this study also considers other intermediate cases). The starboard-side operation is to be easily derived by symmetry rules.

For consistency, the theoretical hydrodynamic data are the same as in [9]:

$$\begin{cases} c_{fxh}(\beta) & = & -0.03cos\beta \\ c_{fyh}(\beta) & = & +0.5sin\beta \\ c_{mzh}(\beta) & = & +0.1sin2\beta \end{cases} \quad (5)$$

which are used to represent the basic, reference case ('theor.'). However, to examine the validation of this theoretical case and possible actual deviations from it, our sensitivity study covers also tug response for other published hydrodynamic model data [24,25] (only for the reference T and P locations, see below), hereafter called 'Brandner' and 'Piaggio', accordingly. These are jointly presented in Figure 2. 'Brandner' case suits rather well the theoretical case, while 'Piaggio' deviates more. Both tugs differ in their underwater design.

In order to provide the final solution/algorithm in the most universal/dimensionless and compact form (e.g., denominated in propeller force) based on [18], let us define the ratios:

$$c'_{fxh}(\beta) = \frac{c_{fxh}(\beta)}{c_{fyh}(\beta)},\ c'_{mzh}(\beta) = \frac{c_{mzh}(\beta)}{c_{fyh}(\beta)} \quad (6)$$

$$F'_T = \frac{F_T}{F_P},\ F'_H = \frac{F_{yH}}{F_P} \quad (7)$$

The solution, for a given towing speed v, now constitutes the following form of four consecutive (sequential) steps:

$$\tan^{-1}\delta = \frac{-x'_P + c'_{mzh}(\beta)}{x'_T - c'_{mzh}(\beta)}[-\tan^{-1}(\gamma-\beta) + c'_{fxh}(\beta)] + c'_{fxh}(\beta) \quad (8)$$

$$F'_T = \frac{sin\delta}{sin(\gamma-\beta)} \cdot \frac{-x'_P + c'_{mzh}(\beta)}{-x'_T + c'_{mzh}(\beta)} \quad (9)$$

$$F'_H = -sin\delta + F'_T sin(\gamma-\beta) \quad (10)$$

$$F_P = \frac{0.5\rho L T_{ref} v^2 c_{fyh}(\beta)}{F'_H} \quad (11)$$

If we assume a constant hawser angle γ ordered by the pilot, of the 5 control (unknown) variables—β, δ, F'_T, F_P, and v—only the first three provide a unique solution, when this solution is being sought in β domain. Therefore, the latter domain is the basis for this study. The drift angle is the independent variable and is to be considered in full range (−180°, +180°). Although we assumed the hawser on port side of the towed ship, the tug may, at least in theory and our solution, face her port (positive β) or starboard side (negative β) with the water inflow. The absolute towing force F_P (in tons) and v are mutually related, according to (11), based on previous variables through (10). Thus, F_P can be established if one knows v and vice versa.

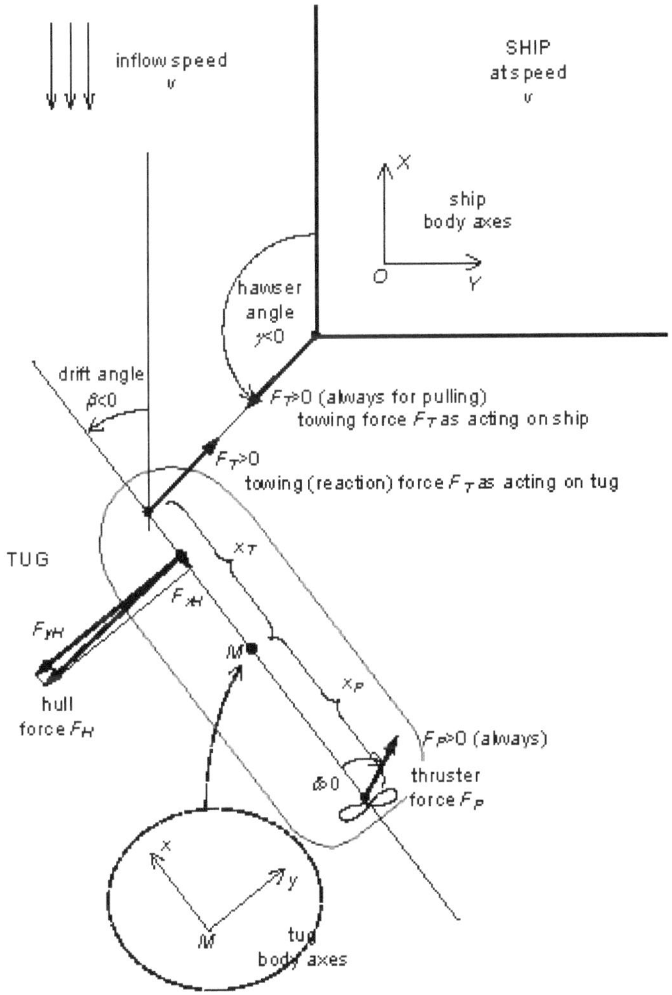

Figure 1. Tug reference frame and definition of symbols (own visualisation, based on [18]).

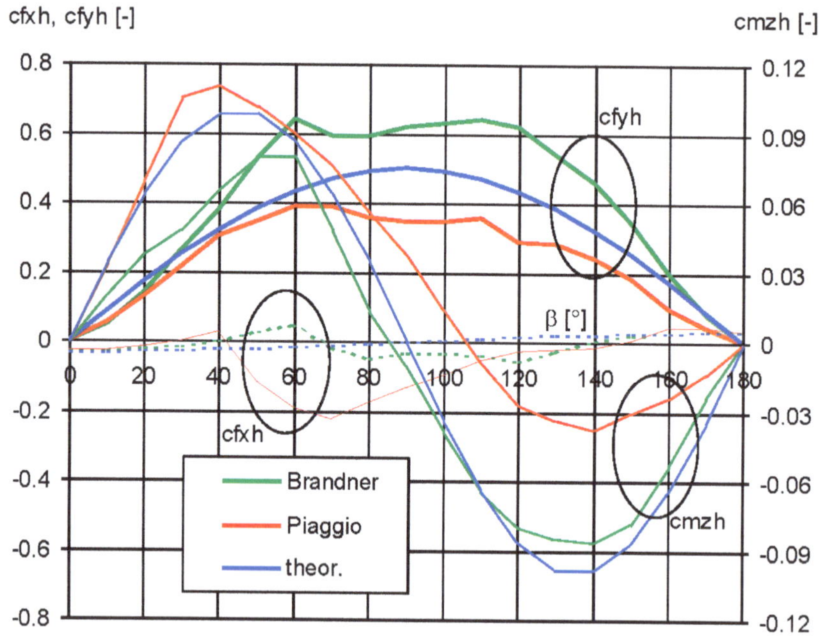

Figure 2. Hydrodynamics of tug's hull (own development, based on data from [24,25]).

There is necessity to reduce the possible mathematical solutions of the equation system (8)–(11), to acquire the valid physics. Thus, only those solutions where $F'_T > 0$ should be preserved (the hawser is being tensioned instead of compressed; in other words, it works only in one, the right, direction, thus unlike a spring acting in both directions). This can be accomplished by modifying δ from (8) and adding $\pm 180°$ (as tan^{-1} nominally returns values from $-90°$ to $-90°$), to keep δ in the original range (see explanation for (3)). Moreover, in view of hydrodynamics, we have to disregard those solutions that do not satisfy the condition of equal signs (positive or negative) for both and F'_H and β.

The full algorithm, which is flexible enough to easily accommodate other hydrodynamic input data and is ready for future improvements, was implemented in MS Excel spreadsheet. Code/numerical verification, the physical validation, and output presentation were automated.

2.2. Simulation/Computational Cases

The following cases of (x'_T, x'_P) are considered in the steady-state solution to obtain the full-range behavioural pattern for astern-drive (azimuth) tug:

- main cases: (+0.5, −0.5) ('REF'); (0, −0.5); (−0.3, −0.5);
- supplementary cases: (+0.3, −0.5); (+0.4, −0.3).

Two speeds, typical for harbour manoeuvres, of 4 and 6 knots (ca. 2 and 3 m/s) were selected to show the effective pull required and, indirectly, the margin to the tug power (for, e.g., new tug alignment/manoeuvring orders).

The main directions γ (every one hour, on port side of the towed ship, as described before) from 0° (12 o'clock) up to −180° (6 o'clock) were used. However, additional values of −45° and −135° were also included in our investigations but are not shown for plot resolution purposes; besides, they do not exhibit a specific behaviour among the surrounding main values. For $\gamma = 0°$ (the tug at 12 o'clock), the cases of $\beta \neq 0°$ do not return a valid physical solution. On the other hand, the trivial case of $\beta = 0°$, although physically justified, is numerically ill-posed, as in deriving our solution we use the function

c'_{fxh} (β)—see (6)—that assumes numerically inconvenient $\pm\infty$ for $\beta = 0°$ and terminates the computations. This is why the case $\gamma = 0°$ disappears in our next figures. The only exception is the supplementary case (+0.4, −0.3), where such a non-trivial solution really exists.

Considering the design and operational practice, e.g., [10,11], for harbour tugs, the above simulation choices seem to represent, in a simple way, the wide variety of possible solutions to towing assembly (devices) on a tug. The first ('REF') is typical for reverse tractor operation, with a hawser leaving the tug's forward centre lead (at bulwark) or close to it (as more or less directly running out of the bow winch). However, to investigate the impact of a potential higher offset in the bow winch, the extra and rather extreme case of (+0.3, −0.5) for (x'_T, x'_P) is also included in the study.

The other supplementary case (+0.4, −0.3) that is used can reflect the normal situation for tractor tugs (with forward-located propellers, mostly of Voith Schneider type, see [23], but not always, under the bow, at certain distance from the forward perpendicular). The tractor tug, with some generalisation (omitting less important details), can be essentially looked at as a reversed tractor. As for Voith Schneider tug, see, e.g., [10] (derived from the official Voith company's booklets/leaflets), the towing point is ca. 10–20% *L* from the stern, which is why it was assumed to be +0.4 above, for our sensitivity analysis.

In view of the preceding two paragraphs, the ASD tug was considered as a generic omnidirectional tug (able to provide arbitrary thrust in full 360° range), i.e., independent of the principle of operation for the propeller itself. Although such a propeller is nominally located in the aft end of δ a tug (according to the word 'stern' in the 'ASD' name), our rather general study was not bound by this fact. The name ASD was used in the most common, usual, daily, and general meaning, partly contrary to [10], where the name ASD was reserved only for a multi-tug that can pull both over her bow and stern (over stern, like a conventional tug, with aft-located fixed shaft propellers and passive rudders), dependent on a pilot or tug master's decision. However, the presented numerical simulation is close to this 'multi-purpose' ASD tug meaning, as adopted in [10]. This is why the study covered a pulling assistance on the midship (central) hook/winch—case '(0, −0.5)'—or a towing on a winch (hook/staple/fairlead) mounted more towards the tug's stern—case '(−0.3, −0.5)'. The latter is also partly equivalent to the use of staple in a conventional tug for ship-stern operation, as shifting the towing point more aft from the midship location reduces the risk of girting.

For harbour tugs (quite different from escort tugs in unsheltered waters assisting at high towing speed, although with the same physics), dependent upon many factors, the bollard pull *BP* (and, correspondingly, the engine power installed, which comes roughly almost linearly with *BP*—refer to [10,23]) is between ca. 20 and 80 t. One should also be reminded that the F_P force (computed in the course of our study), as absorbed by a tug, is the effective force provided by a tug at certain towing speed that is little different from *BP* (rated for zero speed), particularly if we assume a limiting value for this force. In our study and presentation of results, in view of chart resolution, the limit of 50 t for F_P was applied to reveal the tug's fundamental performance.

3. Results of Parametric Study

The output data in a compact form for the three main cases of the *T* (towing point) vs. *P* (propeller) locations, but with respect to the theoretic hydrodynamics of (5), are displayed in Figures 3 and 4. The rows of subfigures in Figure 3 mean different variables and the columns present different variants of *T* and *P*.

Figure 4 directly shows, based on Figure 3(a3), how to interpret the drift angle (and the possible tug's alignment to water inflow). All such detailed figures are rarely published, so this provides a full insight into the phenomena and tug performance, optimising her power/energy absorption through special operational decisions and the actions that are available at hand (which are not always directly considered during the tug design stage).

The comparison of the reference case '*T*/*P*' = '+0.5/−0.5' to other hydrodynamics (see Figure 2) is provided in the Appendix A Figure A1 (propeller angle δ, effective dimensionless/relative towing force F'_T, dimensionless hull lateral force F'_H), Figure A2 (absolute, i.e., expressed in t, towing and propeller forces, F_T and F_P, for a towing speed of 4 knots), and Figure A3 (same as before, but at 6 knots). The rows of subfigures of Figures A1–A3 in Appendix A mean different variables and the columns present different sources of hydrodynamic data.

The supplementary cases of T vs. P, for the theoretic hydrodynamics, are also illustrated in the Appendix A, although with a brief presentation: Figure A4 for case 'T/P' = '+0.3/−0.5' and Figure A5 for case 'T/P' = '+0.4/−0.3' (as aforementioned, the only case with a hawser angle of zero ($\gamma = 0°$) included).

The effective operation happens if $F'_T > 1$; in general, more is better, e.g., for a proper drift angle choice. However, such options (with a significant augmentation of the towing force over that developed by the propeller itself, e.g., $F'_T = 1.2$ means that we have only a 20% yield) are sometimes not available due to the physics involved. Moreover, a magnitude much less than 1 should be avoided as far as possible (since, paradoxically, the towing force is less than the propeller force, thus of low efficiency). The pilot always orders the best command—in terms of force and angle—for a piloted ship in the current nautical and weather situation and in view of the manoeuvring/movement goals. Then, it is on the tug master how to achieve this in a final steady-state condition (since the work of a tug usually lasts for some longer/shorter time) and in a transient phase (when changing the command, especially in the form of a new towing angle), taking into account the tug's safety, operability (effectiveness), and available power (propeller force) reserve. For example, if a tug is designed for (maximum) BP = 50 t, then she would experience a certain difficulty if forced to render assistance at a high speed (e.g., 6 knots), if the absolute towing force F_P read off our plots is close to those 50 t.

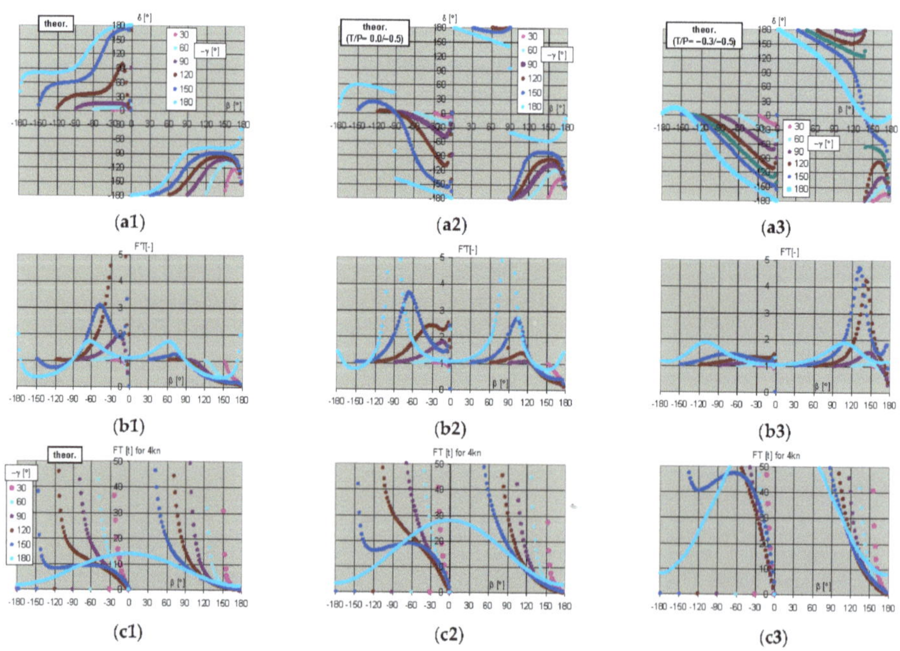

Figure 3. *Cont.*

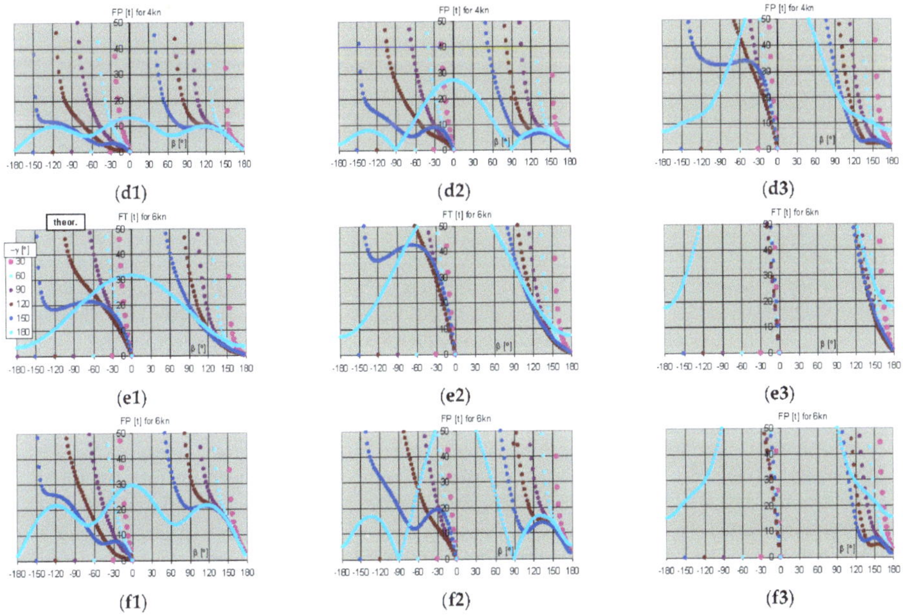

Figure 3. Generic and absolute equilibrium solution (in t) for theoretic hull in the case of variable towing point (T)'s location in the domain of drift angle β (see the legend for hawser angle γ at subfigures (a1) to (a3), '1'– reference (T at bow, '+0.5'), '2'– T at midship hook/('0.0'), and '3'– T at aft staple/deck fairlead ('−0.3')): (**a**) thruster angle $\delta[°]$; (**b**) relative towing force $F'_T[-]$; (**c**) absolute towing force $F_T[t]$ at 4 kn; (**d**) absolute propeller force $F_P[t]$ at 4 kn; (**e**) absolute towing force $F_T[t]$ at 6 kn; (**f**) absolute propeller force $F_P[t]$ at 6 kn (own study).

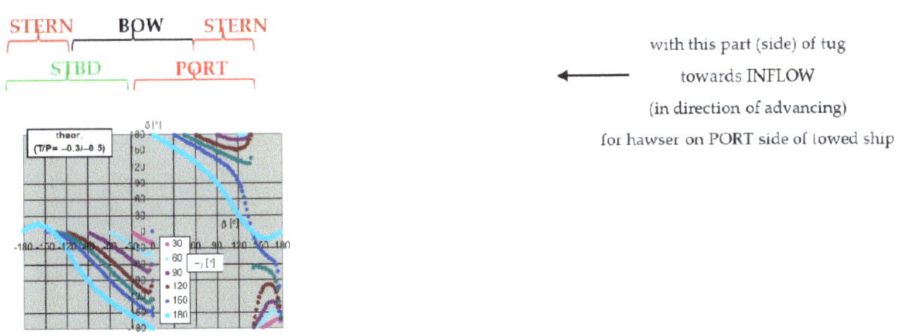

Figure 4. Graphical impression of how a tug is aligned with inflow, based on Figure 3(a3) (own study).

4. Discussion

By examining Figures A1–A3 in the Appendix A, one may conclude that the theoretic case is close to the 'Brandner' case, as should be clear from the aforementioned direct comparison of hull hydrodynamics (Figure 2). Only some minor, quantitative differences exist in the provided tug's control and effectiveness variables (steady-state solutions), thus showing a rather low sensitivity in the tug's performance. Major differences, including those that re qualitative, on the other hand, can be noticed with respect to the 'Piaggio' case. However, partly due to this research's main focus (the impact of a tug's employment mode, i.e., expressed by the towing point/winch and propeller location), this type of tug (hull hydrodynamics) is postponed to the future full mission simulator tests in dynamic scenarios using a tug master, in order to evaluate the advantages (or disadvantages) during real-time manoeuvring, before making a deeper analytical study in a steady state.

Returning to the main subject, i.e., the study and comparison of the operation effectiveness at a ship's bow via a tug's bow winch/lead ('T/P' = '+0.5/−0.5'), midship (central) winch/lead/hook ('T/P' = '0.0/−0.5'), and winch/lead/hook moved aft (or other compatible device) ('T/P' = '−0.3/−0.5')—first of all, one must notice that some points/segments of the resultant plots are not available. For example, in Figure 3, columns (1) to (3), the first tug is not technically able to sail with drift angle β around 0° (i.e., bow-first, see Figure 4, even though this is a rare case for this tug) at a hawser angle γ close to −180°. A similar situation happens for the second tug (with a central hook, most frequently advancing just bow-first), which, on the other hand, is disabled for rare drift angles ±30° around 180° (stern-first operation). The third tug (intermediate aft wind), which partly inherits the mentioned drawbacks of the second tug, would also experience a collision/contact risk (especially if deployed on a short hawser close to the hull of a towed ship) in an area of high towing-force efficiency (F'_T >> 1), i.e., for a hawser angle ca. −120° to −150° and positive drift around 120° to 150° (Figure 3(b3)). This is a shortcoming of the simple mathematical model and algorithm used, where we do not consider the geometrical restraints for the hawser, as might run across a tug's deck superstructures and fittings. This calls for some better methodical approach in the next investigations on tug performance with different winch locations. Until now, there has been no well-established standard in performing similar studies, and, thus, these are the first methodical steps in this area.

One must also observe that relative/dimensionless towing force F'_T (higher is better) is a good indicator of efficiency. If we have a look at Figure 3, subfigures (b1/F'_T), and, as an example, (e1/F_T) – (f1/F_P), corresponding to 6 knots, then we know that for practical hawser angles from −90° to −150° and a negative drift in the range from −60° to somewhere near −15°, we acquire a good towing force F_T with very little effort from the propeller (F_P) (the efficiency in terms of F'_T is much more than 1). However, if in those conditions one would set a limit to the low drift angle (abt. −30°), then the maximum towing force F_T in the steady state is only 20t. If one wants to increase this force, they have to increase the drift magnitude (within a negative range), even to set the tug aside towards the inflow (which is difficult to control) and, hence, lower the efficiency. That is why both indicators—relative and absolute towing force—are always worth being considered in parallel.

On the other hand, one should also be interested in much lower (but positive) efficiency, 20 ÷ 30% (F'_T = 1.2 ÷ 1.3), and avoid excessive periods of operating at a complete loss of efficiency (F'_T << 1), if this is the need from the pilot's point of view and their manoeuvring/steering orders). For example, for a hawser angle between −30° and −60° and the first bow winch tug (Figure 3(b1,e1,f1)), one has to take care to run stern first with a drift angle of −150° and −120° accordingly (and have $F'_T \approx 1$), which is equivalent to having a tug's centre plane in line with the hawser; otherwise, we will be at a low efficiency.

In the case of the second, central hook tug (Figure 3(e2,f2)) and 6 knots, within a negative drift angle, and independent of any hawser angle, a remarkable feature is that the tug keeps an absolute towing force (varying linearly with the drift angle—the higher the drift magnitude is, the higher the towing force is). The propeller's absolute force and, thus, the relative towing force (>>1), assumes a much more complicated pattern in this instance.

The third, intermediate aft winch tug (Figure 3(e3,f3)), at 6 knots, exhibits a very simple behavioural pattern—both the towing and propeller absolute forces vary linearly with the drift angle (in its negative range, up to −30°), independent of the hawser angle at all. In this situation, the constant medium (c.a. 25%) yield is achieved for the towing force ($F'_T \approx$ 1.25).

To summarise, comparing various tug design or operation options is difficult due to the complicated tug dynamics (statics) patterns under the hawser action, which involve multiple dimensions (parameters or factors). Both relative and absolute towing forces have to be considered (high relative force—high 'efficiency'—may produce a low absolute force, much less than ordered by the pilot—thus of slim 'effectiveness'), additionally with some geometrical and safety constraints. Generally—the better the efficiency is, the worse the effectiveness is at a significant towing speed. Both a low to medium (which are also useful in a lot of situations) as well as a high towing force yield must be examined, which partly depend on an assisted ship's speed. The provided charts show some essential potential for developing a theory for tug design and operation optimisation. Controlling drift angle β is important for keeping towing force efficiency. In general, a 30–60° hawser order is without any gain ($F'_T \approx 1$) for each tug. For the other directions, there are more differences in the gain between each tug and, thus, in the margin for executing new orders.

Roughly speaking, in view of only what is mentioned above, the midship winch tug seems to be the best solution for towed ship-bow operation (as a good compromise between all the merits and drawbacks, in terms of hydrodynamics and control). This last statement needs to be validated in a more systematic way in the future, by also including other factors that are omitted in the present research (such as, e.g., the problems of excessive list and the risk of capsizing/girting at a high towing speed, when a midship winch tug's control and safety can be lost, although this is not as serious for ASD tugs when compared to conventional tugs). As a major output of this study, both bow and midship winch tugs are almost equally efficient in terms of F'_T. By far, navigators have relied on very rough guidance and discussion on the merits of a particular tug type. Through the presented study, more detailed knowledge and insight into the phenomena have been gained. Looking at the computational results (which are not shown in the presented plots due to the vertical scale/range applied), one detail is very interesting. At 6 knots and a hawser of −90° (9 o'clock), the maximum achievable towing force for a 50 t BP differs based on the type of tug:

- 50 t ($F'_T = 1$, for either bow- or side-first movement into the water) for a bow winch tug;
- 62 t (bow-first) or 53 t (stern-first) for a midship winch tug;
- 61 t (bow-first) or 60 t (stern-first) for an intermediate aft winch tug.

The latter result partly and very qualitatively agrees with full-mission simulator tests in real time [17], conducted with a proprietary manoeuvring mathematical model of a tug (delivered by the ship-handling simulator's manufacturer). According to [26], moving the staple (winch) location from a tug's edge a little towards her midship gives rise to the towing force increase corresponding to the results presented here—the last two cases can be compared.

In general, however, there are some methodical difficulties concerning a direct comparison between the presented extensive parametric study results and the various results available in the literature, at least at this stage of research. First of all, the steering details (in terms of the thruster angle and tug's hull drift angle) are hardly given in other works. However, [25,27] are certain exceptions. Some publications also exhibit deficiencies concerning the research documentation, in terms of the input data, mathematical model, and computational/solution procedure used, so their results cannot be reliably reproduced. Secondly, many references provide a polar diagram escort capability, thus focused on the maximum achievable towing force (towing maximum performance), rather than on the actual, not maximal but effective and energy-efficient performance (such as in terms of F'_T used throughout this study or by using other possible indicators such as the towing force to power or towing force to specific fuel delivered ratio) at every moment of the towing operation. The construction of a polar capability diagram using a fully analytical method (under some simplification) was explained in [14]. Thirdly, the mentioned polar-capability diagrams (which are of some value in extreme situations and for tug selection) comprise various equilibrium solution (or equilibrium settling) techniques, including the employment of a tug's practical human control, various optimization criteria and techniques, the stability of such an equilibrium, etc. [10,16,25] (as partly reporting Brandner's Ph.D. thesis [28]). Fourthly, those data often include the advance speed and/or oblique inflow effects on the propeller (with the recent tug-oriented contributions of, e.g., [29]) and/or a tug's stability/roll issues, which were deliberately disregarded in the paper due to a lack of a reliable four-quadrant propeller flow model suited for such a parametric and dimensionless parametric study.

However, escort capability polar diagrams, for a given BP of a tug, may also provide some rough guesses for the intermediate performance of a tug, which is what we are looking for, although mainly for the comparison of bow winch to midship winch ASD tugs.

According to the presented study, both tugs seem to be rather equivalent, with only a minor advantage for the midship winch tug.

Many available references, however, favour the midship winch tug (the third solution, an intermediate aft winch tug, according to the used terminology, is rather rare) for ship bow operation much more [10,16]. However, such statements are, surprisingly, often not stated very clearly or emphasised, with only some selective data or discussion provided. Whereas this operation might be even less effective by 2–3 times or more, in terms of the maximum achievable towing force at s hawser angle in the range from $-60°$ (10 o'clock) to $-90°$ (9 o'clock) at a towing speed 6 kn—compare Hensen [10], (Figure 4.21/p. 59), [16], Figure 10/p. [16], or more recent research [30], (Figure 13/p. 8). However some other publications suggest less of a difference at 6 kn for both tugs, e.g., Brandner [25], (Figure 7/p. 346), although this difference becomes larger and converges to the previously quoted values of Hensen and others at just 8 kn [25], (Figure 8/p. 346). All of this is in significant contrast to the results of this study and necessitates further deep and systematic insight, research and development of reliable prediction tools, and industry support in decision or policy making for a tug's stern-to-bow operation at a ship's bow, in the future.

The essentially different results of other studies seem to be a matter of not only a particular tug design (and tug-design sensitivity) but also the mathematical models used for the manoeuvring hydrodynamics. In this context, it is worthwhile to refer to the fruitful discussion (written comments) made by Hutchinson, with regard to Allan [31], representing two renowned companies involved in tug design and consultancy. Both authors quote their results, with their rather comprehensive in-house mathematical models, to be different in many places by 2–3 times, in respect to the same tug.

However, surprisingly, the bow-to-bow operation with a bow winch (reverse tractor) tug is frequent in practice, e.g., while assisting LNG carriers arriving at LNG terminals. This situation happens partly because of the reasons of a particular tug design and the avail-able fleet, partly because of other advantages of this type of tug (see [10,26]), and eventually because of the varying, sometimes unclear, decision making of the respective stakeholders. The power or energy efficiency of a tug is rarely the issue in some harbour towing operations, sometimes due to a low level of scientific insight and a lack of reliable analytical tools and proper education, which the authors would like to change.

5. Conclusions and Future Research

The validity of the presented results depends on the input hydrodynamic data and assumed simplifications. The major limitation of the presented approach is, apparently, the lack of both the effect of the advance speed and (four-quadrant) oblique inflow in the propeller model. They require (particularly for the oblique inflow) a significant consolidation of the widely spread published data and, likely, extra modelling and generalisation efforts to obtain a fully useful description. However, the obtained output (propeller angle δ and propeller absolute force F_P) can be used even in this case, though they shall be treated as an 'effective' propeller thrust direction and magnitude, respectively. Moreover, the various tug designs tested, e.g., the midship winch and bow winch ones, seem to be roughly under the same influence of the two aforementioned phenomena, so their relative performance should remain as developed herein. A detailed study of this paper's results in comparison to those obtained with full models by others, as reported in a previous section, would certainly be a subject of the authors' future work. Other neglected factors, besides the aforementioned advance speed and oblique inflow, such as thruster–thruster interaction (as of dual thruster tugs), thruster–skeg interaction, independent control of both thrusters (giving the fairly good independence of the propeller total force vs. its yaw moment), or the hawser's vertical angle (especially important for short hawser) require future attention as well as (as a natural step) the formal uncertainty analysis, mostly with regard to a tug's dynamic model.

The present research is focused on a steady-state solution. One should note that a tug can provide dynamic (temporal, instant) assistance, where the steady-state solution is of less importance. Real-time (RT) simulation, with human-in-the-loop, is to be used, therefore, as more or less deliberately suggested elsewhere (also as a method for 'validating' some steady-state or analytic solutions).

After the systematic and comprehensive study of such tug physics was undertaken in the presented paper, and in view of contemporary and future energy-sensitive operations, including the GHG emission-related aspects, some guidance and training rendered to tug masters is needed. This should 'take advantage' of the operational limitations of a particular tug or give guidance towards the design or acquisition of a tug that is best-suited for a particular harbour operation.

To achieve more clear or realistic computations, a criterion (constraint) in terms of the tug-related feasible hawser angle (derived from the ship's hawser angle γ and the tug drift angle β) should be introduced. The latter is crucial for the above-deck (superstructure) and towing-assembly-related design of a tug and tug–ship safety (collision/contact free) in close quarters—this is a kind of methodical issue in investigations of tug dynamics. Moreover, as an indirect conclusion of this study (see the difficulties in comparing tugs), the overall evaluation criteria of tug design and operation need some more research, as they are not yet well-established within science and engineering.

Author Contributions: Conceptualisation, J.A. and P.Z.; methodology, J.A. and P.Z.; software, J.A.; validation, J.A.; formal analysis, J.A. and P.Z.; investigation, J.A. and P.Z.; data curation, J.A. and P.Z.; writing—original draft preparation, J.A. and P.Z.; writing—review and editing, J.A. and P.Z.; visualisation, J.A. and P.Z. All authors have read and agreed to the published version of the manuscript.

Funding: This research received no external funding.

Conflicts of Interest: The authors declare no conflict of interest.

Appendix A

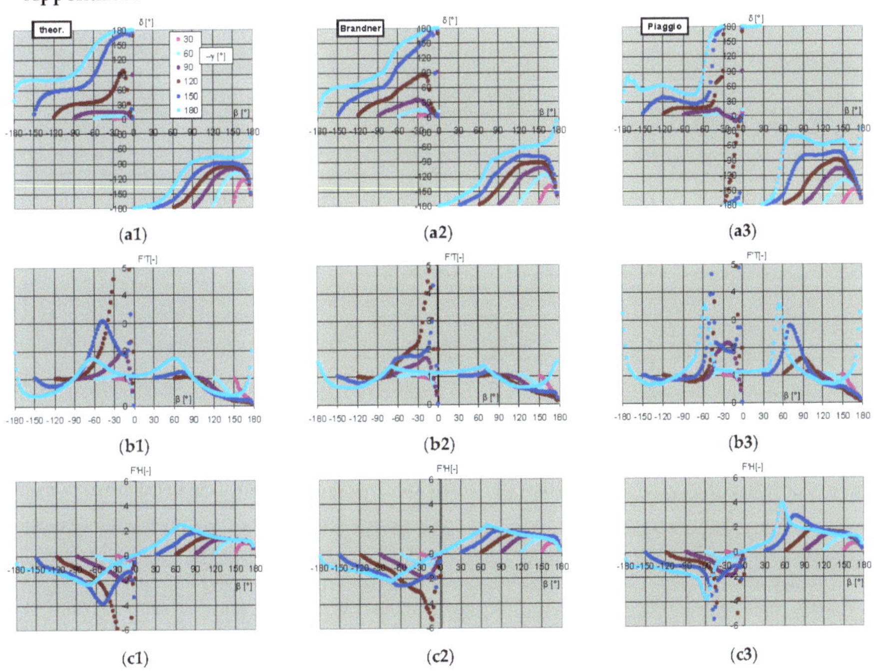

Figure A1. Equilibrium generic solution for 3 hull types in the domain of drift angle β (see the legend for hawser angle γ at subfigure (**a1**), '1'-theoretic hull, '2'-'Brandner' hull, and '3'-'Piaggio' hull): (**a**) thruster angle δ[°]; (**b**) relative towing force F'_T[−]; (**c**) relative hull force F'_H[−]. All data are for T/P = +0.5/−0.5 (see the explanation of notations in body text, own study).

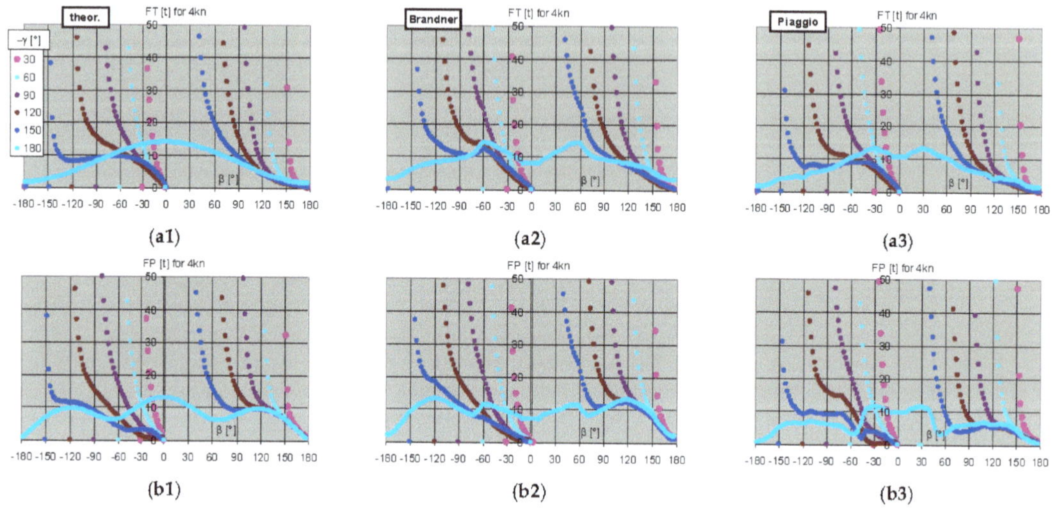

Figure A2. Absolute equilibrium solution (in t) at **4** knots for 3 hull types in the domain of drift angle β (see the legend for hawser angle γ at subfigure (a1), '1'-theoretic hull, '2'-'Brandner' hull, and '3'-'Piaggio' hull): (**a**) absolute towing force F_T[t]; (**b**) absolute propeller force F_P[t]. All data are for T/P = +0.5/−0.5 (see the explanation of notations in body text, own study).

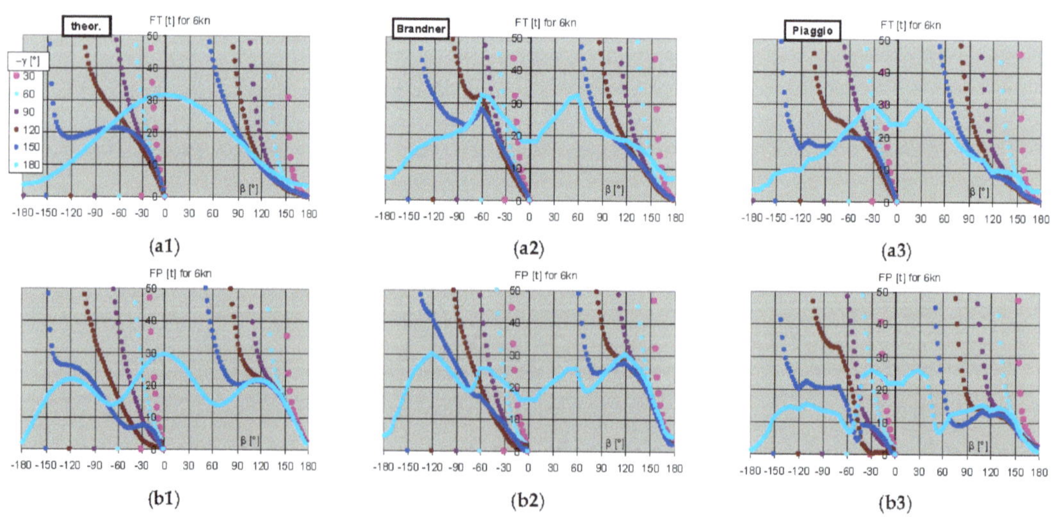

Figure A3. Absolute equilibrium solution (in t) at **6** knots for 3 hull types in the domain of drift angle β (see the legend for hawser angle γ at subfigure (a1), '1'-theoretic hull, '2'-'Brandner' hull, and '3'-'Piaggio' hull): (**a**) absolute towing force F_T[t]; (**b**) absolute propeller force F_P[t]. All data are for T/P = +0.5/−0.5 (see the explanation of notations in body text, own study).

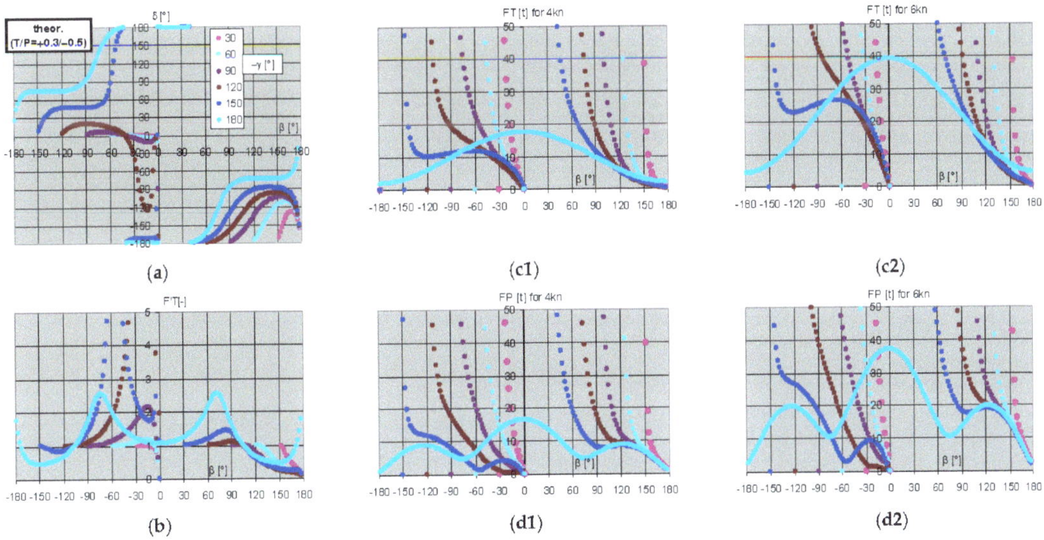

Figure A4. Generic and absolute equilibrium solution (in t) for theoretic hull in case of 'T/P' location '+0.3/−0.5' in the domain of drift angle β (see the legend for hawser angle γ at subfigure (a), '1'-4 knots and '2'-6 knots): (**a**) thruster angle $\delta[°]$; (**b**) relative towing force $F'_T[-]$; (**c1**) absolute towing force F_T[t] for 4 kn; (**c2**) absolute towing force F_T[t] for 6 kn; (**d1**) absolute propeller force F_P[t] for 4 kn; (**d2**) absolute propeller force F_P[t] for 6 kn (see the explanation of notations in main text, own study).

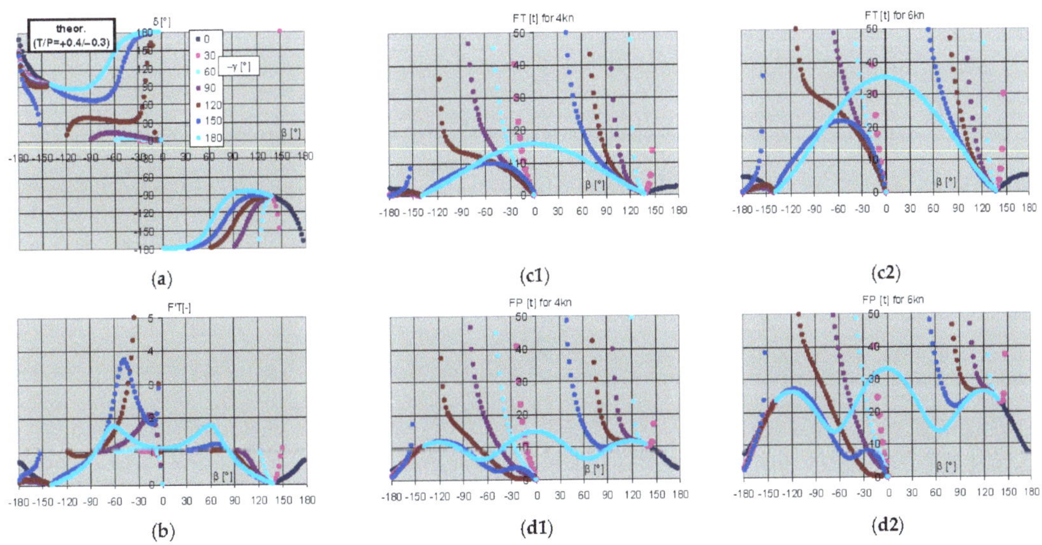

Figure A5. Generic and absolute equilibrium solution (in t) for theoretic hull in case of 'T/P' location '+0.4/−0.3' in the domain of drift angle β (see the legend for hawser angle γ at subfigure (a), '1'-4 knots and '2'-6 knots): (**a**) thruster angle $\delta[°]$; (**b**) relative towing force $F'_T[-]$; (**c1**) absolute towing force F_T[t] for 4 kn; (**c2**) absolute towing force F_T[t] for 6 kn; (**d1**) absolute propeller force F_P[t] for 4 kn; (**d2**) absolute propeller force F_P[t] for 6 kn (see the explanation of notations in main text, own study).

References

1. Molyneux, D.; Bose, N. Escort Tug at Large Yaw Angle: Comparison of CFD Predictions with Experimental Data. *Trans. RINA Part B—Int. J. Small Craft Technol.* **2008**, *150*, 13–22.
2. Baer, W. Location of the towing hook on a Voith Tractor. In Proceedings of the Second International Tug Conference, London, UK, 25–28 October 1971; Troup, K.D., Ed.; Thomas Reed Industrial Press Ltd.: Santa Ana, CA, USA, 1972.
3. Wulder, J.H.; Hoebee, W.; Kooren, A.J.; Ramsey, D.G.; Wenz, R. Applications of Interactive Tug Simulation. In Proceedings of the MARSIM 2000 Conference Proceedings: International Conference on Marine Simulation and Ship Maneuvering, Orlando, FL, USA, 8–12 May 2000.
4. Jakobsen, B.K.; Miller, E.R.; Wulder, J.H.; Hensen, H. The Simulation of Tug Operations in a Multiple Simulator Environment. In Proceedings of the MARSIM'96 Proceedings, Marine Simulation and Ship Manoeuvrability, Copenhagen, Denmark, 9–13 September 1996; Chislett, M.S., Ed.; A.A. Balkema: Rotterdam, The Netherlands, 1996.
5. Gucma, S.; Jagniszczak, I. Calculation Methods Analysis of the Number and Power of Tugs for Harbour Manoeuvres. In Proceedings of the Marine Traffic Engineering, 2nd Scientific Symposium, Szczecin, Poland, 10 March 1986. (In Polish).
6. Gray, D.L.; Reynolds, E. Engineering Methodologies Used in the Preparation of Escort Tug Requirements for the Ports of San Francisco and Los Angeles/Long Beach. *Mar. Technol.* **2001**, *38*, 51–64. [CrossRef]
7. Geerts, S.; Vantorre, M.; Eloot, K.; Huijsmans, R.; Fierens, N. Interaction forces in tug operations. In Proceedings of the 2nd International Conference on Ship Manoeuvring in Shallow and Confined Water: Ship to Ship Interaction, Conference Proceedings, Trondheim, Norway, 18–20 May 2011.
8. Slesinger, J. *ASD Tugs: Thrust and Azimuth: Learning to Drive a Z-Drive*; Cornell Maritime Press: Centreville, MD, USA; Schiffer Publishing Ltd.: Atglen, PA, USA, 2010.
9. Murphy, A.J.; Weston, S.J.; Young, R.J. Reducing fuel usage and CO2 emissions from tug boat fleets: Sea trials and theoretical modelling. *Trans. RINA Part A—Int. J. Mar. Eng.* **2012**, *154*. [CrossRef]
10. Hensen, H. *Tug Use in Port, A Practical Guide*, 2nd ed.; The Nautical Institute (TNI): London, UK, 2003.
11. Allan, R.G. Tugs and Towboats. In *Ship Design and Construction*; Lamb, T., Ed.; SNAME: Jersey City, NJ, USA, 2003; Volume 2.
12. Iglesias Baniela, S.; Platero Diaz, A. The First Escort Tractor Voith Tug with a Bulbous Bow: Analysis and Consequences. *J. Navig.* **2008**, *61*, 143–163. [CrossRef]
13. Hutchison, B.L.; Gray, D.L. New Insights into Voith Schneider Tractor Tug Capability. *Marine Technol.* **1993**, *30*, 233–242. [CrossRef]
14. Artyszuk, J. Polar plot capability of a tug in indirect and direct mode of escort towing. In *Advances in Marine Navigation and Safety of Sea Transportation*; Weintrit, A., Neumann, T., Eds.; CRC Press: Boca Raton, FL, USA; Balkema—T&F: Leiden, The Netherlands, 2019; pp. 57–64. ISBN 978-0-367-35760-3.
15. Paulauskas, V.; Simutis, M.; Plačiene, B.; Barzdžiukas, R.; Jonkus, M.; Paulauskas, D. The Influence of Port Tugs on Improving the Navigational Safety of the Port. *J. Mar. Sci. Eng.* **2021**, *9*, 342. [CrossRef]
16. Hensen, H. Bow Tug Operations with Azimuth Stern Drive Tugs. In *Risks and Effectiveness (Guidelines for Tug Captains, Ship Masters, Pilots and Tug Operators)*; The Nautical Institute (TNI): London, UK, 2006.
17. Zalewski, P. Assessment of tug assistance parameters as result of FMBS studies. *Eur. J. Navig. (EJN)* **2015**, *13*, 10–16.
18. Artyszuk, J. Steady-state manoeuvring of a generic ASD tug in escort pull and bow-rope aided push operation. *Trans. Navig. Int. J. Mar. Navig. Saf. Sea Trans.* **2014**, *8*, 449–457. [CrossRef]
19. Quadvlieg, F.; Kaul, S. Development of a calculation program for escort forces of stern drive tug boats. In Proceedings of the 19th International Tug & Salvage Convention & Exhbition, ITS 2006, Rotterdam, The Netherlands, 24–28 April 2006.
20. Renilson, M.R.; Brandner, P.; Tasker, R.L. Realistic Simulation of Tug Forces on a Manoeuvring Vessel. In Proceedings of the 2nd Conference on Manoeuvring and Control of Marine Craft, MCMC'92, Southampton, UK, 14–17 July 1992.
21. Waclawek, P.; Molyneux, D. Predicting the Performance of a Tug and Tanker During Escort Operations Using Computer Simulations and Model Tests. *SNAME Trans.* **2000**, *108*, 21–43.
22. Renilson, M.; Tuite, A. A Technique for Assessing the Difficulty of Ship Manoeuvring in Ports. In Proceedings of the International Pilotage Conference, Brisbane, UK, 11–13 March 1998.
23. Artyszuk, J. Types and Power of Harbour Tugs—The Latest Trends. *Sci. Trans. (Prace Naukowe)-Transp. Warsaw Univ. Technol.* **2013**, *98*, 9–20.
24. Piaggio, B.; Viviani, M.; Martelli, M.; Figari, M. Z-Drive Escort Tug manoeuvrability model and simulation. *Ocean Eng.* **2019**, *119*, 106461. [CrossRef]
25. Brandner, P.; Renilson, M. Hydrodynamic Aspects of Shiphandling Tugs. In Proceedings of the 3rd International Conference—Manoeuvring and Control of Marine Craft (MCMC'94), Southampton, UK, 7–9 September 1994.
26. Allan, R.G.; Molyneux, D. Escort tug design alternatives and a comparison of their hydrodynamic performance. *SNAME Trans.* **2004**, *112*, 191–205.
27. Piaggio, B.; Villa, D.; Viviani, M.; Figari, M. Numerical analysis of escort tug manoeuvrability characteristics—Part II: The skeg effect. *Appl. Ocean Res.* **2020**, *100*, 102199. [CrossRef]
28. Mauro, F. A flexible method for the initial prediction of tugs escort capability. *Ocean Eng.* **2022**, *246*, 110585. [CrossRef]
29. Yiew, L.J.; Taskar, B.; Jin, Y.; Zheng, Y.; Liu, D.; Magee, A.R. Experimental and numerical investigations on coupled thruster and hull interactions for a benchmark ASD tug. *Appl. Ocean Res.* **2022**, *120*, 103072. [CrossRef]

30. Barrera, R.D.; Schiaveto Neto, L.A.; Vieira, D.P.; Mesquita, E.S.; Tannuri, E.A. Azimuth stern drive (ASD) vector tugs positioning and towing force prediction during docking, steering and braking maneuvers. *Appl. Ocean Res.* **2021**, *110*, 102611. [CrossRef]
31. Allan, R.G. A Tale of Five Tugboats. In Proceedings of the 11th International Tug Convention and Marine Salvage Symposium (11th ITS), Halifax, NS, Canada, 24–28 September 1990; Troup, K.D., Ed.; Thomas Reed Pubs: Surrey, UK, 1990; pp. 219–246.

energies

Article

An Application of Relative Entropy in Structural Safety Analysis of Elastoplastic Beam under Fire Conditions

Marcin Kamiński * and Michał Strąkowski

Department of Structural Mechanics, Lodz University of Technology, Al. Politechniki 6, 93-590 Łódź, Poland
* Correspondence: marcin.kaminski@p.lodz.pl; Tel.: +48-42-631-35-71

Abstract: The main aim of this work is to propose a new algorithm of reliability assessment for steel civil engineering structures subjected to fire temperatures. This new algorithm is based upon the relative probabilistic entropy concept elaborated by Bhattacharyya, and this probabilistic distance is sought in-between extreme and admissible deformations of some structural beam subjected to higher temperatures. Similar to the First Order Reliability Method, this approach uses the first two probabilistic characteristics of the structural response, when structural output may be modelled with the use of Gaussian distribution. The probabilistic structural response is found here using hybrid computational technique–the Finite Element Method system ABAQUS with its fully coupled thermo-elastic analysis with 3D solid elements and probabilistic modules implemented in the computer algebra system MAPLE. The probabilistic response is determined via a triple stochastic analysis, which is based on the classical Monte-Carlo simulation, iterative generalized stochastic perturbation technique, and also using semi-analytical probabilistic calculus. Final determination of the relative entropy in the Serviceability Limit State of the given structure and its comparison with the results obtained using the FORM analysis enables to calibrate this new technique to numerical values proposed in the engineering designing codes. Hence, a more accurate probabilistic method may use some experimental-based admissible values included in the existing design of legal provisions.

Keywords: stochastic perturbation technique; stochastic finite element method; fire simulation; coupled thermal-stress analysis; relative entropy; reliability analysis

1. Introduction

Structural safety of civil engineering structures [1] concerning possible fire accidents is a very important practical problem, especially in the area of steel structures, where the demands and expectations are unusually high [2]. Hence, the reliability of engineering structures under fire conditions [3,4] remains always a very challenging and practically important area of knowledge [5,6]. This is due to a partial lack of experimental data, the complexity of the numerical simulation, even while using the Finite Element Method (FEM) or the Finite Volume Method (FVM) as well as the difficulty in stochastic simulations. The full-scale experiments necessary to build and calibrate efficient numerical models [7,8] need to include both realistic fire scenarios, temperature-dependent paths of deformations, and stresses as well as detailed information concerning mechanical and physical characteristics of the structural materials. A fundamental minority in the experimental methods is that individual large-scale structures may be subjected to the fire only once; some structural elements (such as the cold-formed, for instance [9]) or their connections (cf. [10]) may be tested with a few times, but their qualitative and quantitative results have limited application for the framed-structures [11,12], buildings, bridges [13], or steel large-surface halls [14]. It has been widely documented that Computational Fluid Dynamics (CFD) is one of the most powerful numerical simulation tools in fire propagation prediction [15], which is commonly used with some FEM systems to predict the failure [16], and where fire–thermomechanical interface may play a very important role [17].

There is no doubt that deterministic numerical simulations of fire accidents may be very useful in engineering practice, taking into account the dramatic cost of such full-scale experiments, and may have repeatable character; nevertheless, these models are very sensitive to mechanical and thermal boundary conditions, temperature-dependent material parameters, and also the details in their algorithms [18,19]. It can be realized by some hybrid computer systems with an option of the gas–solid interaction or just by the coupled solid mechanics numerical analyses (e.g., thermo-mechanical) available in many well-known FEM commercial systems such as ABAQUS [3,14,19] or ANSYS. Such a simulation could serve as an efficient prediction of the structural failure time to obtain more specific information concerning fire resistance and evacuation time (decisive especially for high-rise buildings safety or large bridges). Let us underline that all these methods and case studies have purely traditional deterministic character and do not allow directly for any structural reliability assessment.

Stochastic models in fire safety analyses are traditionally related to the models of fire spread in woodlands [20], numerical simulation related to this issue [21], as well as to fire outbreaks following earthquake disasters [22]. The Monte-Carlo simulation approach was traditionally the first numerical technique to analyze structural response under a fire [23], and to deliver risk analysis for steel beams [24]. It is well-known that the widely accessible and relatively easily programmable Monte-Carlo simulation needs enormously huge computer time and power consumption [25]; a semi-analytical technique availability depends upon the initial choice of the input uncertainty type [3,26,27], whereas various expansion techniques (Karhunen-Loeve and polynomial chaos [28] or Taylor [26]) may exhibit limited applicability in terms of the larger initial uncertainty level or multivariable (and state-dependent) character of the material and physical characteristics. Some Stochastic Finite Element Method (SFEM) studies are available in the literature [29–31], but their connection with fire simulation is rather scarce [3], so no well-documented experiments and related conclusions can be found. The most popular engineering tool in this area was the Second Order Second Moment (SOSM) [32,33], which has been further generalized to the higher order approach engaging the Least Squares Method (LSM) [34,35] recovery of polynomial bases [36] relating the desired structural response with a given input of random parameter(s); it is in fact similar to the response surface methodology [37,38]. The main goal of such an approach would be a final calculation of the reliability index, which in civil engineering designing codes is still based upon the First Order Reliability Method (FORM) [1,39]. Despite the stochastic computer method chosen, the main difficulty would be the collection of basic statistical parameters and corresponding probability distributions for material/physical characteristics of structural materials subjected to high temperatures. It is known that some alternative stochastic approach could be based on a probabilistic distance or probabilistic divergence, but their application is still not quite straightforward in reliability assessment. One of the alternatives in this area could be the so-called Bhattacharyya divergence [14,27,40], but many other models can be useful including Shannon, Renyi, or Tsallis entropies [41,42] and probabilistic distances [43] including Hellinger theory [44], Jeffreys model [45], Kullback–Leibler theory [46], and also Jensen–Shannon entropy [47]. Let us note that certain engineering uncertainty analyses have been delivered in the context of various entropies in the literature [48–51], but they are a little bit distant from engineering reliability index determination.

The main aim of this paper is to present some stochastic numerical analysis schemes of the fire scenario and to apply them to analyze the reliability of some popular steel structures of the hot-rolled I beam being a part of the structural roof. A very important aspect of this model is its fully coupled character, capturing of temperature variations of all material parameters, and also the usage of 3D finite elements, which enables decisively higher numerical accuracy than the Euler–Bernoulli, Timoshenko, or shell elements applicable in engineering practice; some sequential coupling with ABAQUS has been demonstrated by the authors in [3] before. The stochastic scheme is based upon a triple stochastic methodology–with (i) Monte-Carlo simulation, (ii) semi-analytical approach as well as

(iii) iterative generalized higher (the 10th) order stochastic perturbation technique. The entire computational implementation has been carried out with the use of the FEM system ABAQUS® as well as the computer algebra system MAPLE 2019.2® and has a general character, independent of the engineering structure type. Determination of the first four probabilistic moments and coefficients available in this algorithm allows for a calculation of the reliability index based on the First Order Reliability Method (FORM), and this index is presented as a function of the fire duration time (and its mean temperature). Additionally, a concept of the probabilistic divergence (relative entropy) usage to approximate structural reliability has been presented and discussed here by a contrast of this entropy to the FORM index [3]. This concept has been successfully employed before to study the reliability of statically uploaded linear elastic steel truss [27], and also in the stochastic dynamic response of some steel halls [14]. The most creative work is an extension of the entropy-based approach from elastic problems towards fully coupled thermo-elasticity Stochastic Finite Element Method structural analysis. Contrary to the previous studies, input uncertainty in fire gas temperatures induces multiple random variability in mechanical and thermal characteristics of the given structure. A methodology proposed here may serve for further fire (and not only) structural safety analysis of the steel structures and the very important aspect of this study is application of the 3D finite elements for detection of structural behavior of steel thin-walled elements. The authors have invented the new reliability index, which reflects probabilistic divergence in-between admissible and extreme deformations depending both on higher temperatures. It has been shown that probabilistic entropy may be efficiently used in engineering analyses not only in the context of the maximum entropy principle but also as a direct function of the input uncertainty and may contain key information concerning structural reliability. The essential innovative aspect of this work is to apply the relative entropy-based reliability index in fully coupled thermo-elastic FEM analysis for simulation of fire accidents in some popular civil engineering structures.

2. Physical Model and Its Implementation

Let us consider a transient thermo-mechanical boundary initial value problem defined on isotropic and homogeneous domain Ω shown schematically below in Figure 1.

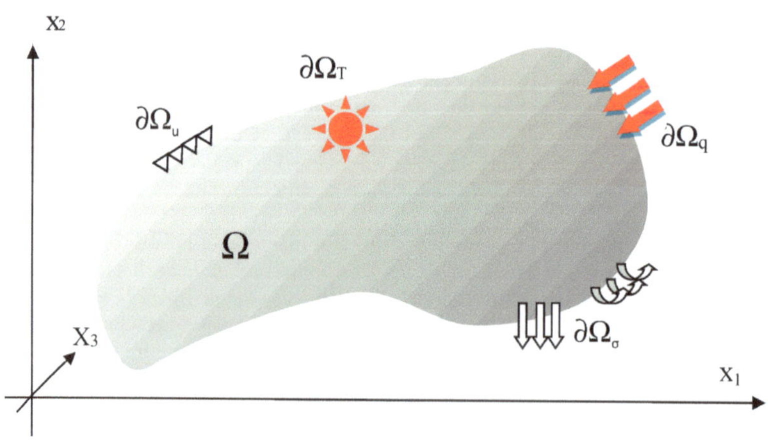

Figure 1. Boundary-initial thermo-mechanical problem of the homogeneous domain Ω.

Its mechanical part is driven by the following incremental static equilibrium equations [52]:

$$\Delta\sigma_{kl,l} + \rho\Delta f_k = 0;\ \mathbf{x} \in \Omega \tag{1}$$

$$\Delta\sigma_{kl} = C_{klmn}(T)\Delta\varepsilon_{mn} + \delta_{kl}\ \alpha(T)\Delta T\ E(T);\ \mathbf{x} \in \Omega \tag{2}$$

$$\Delta\varepsilon_{mn} = \frac{1}{2}[\Delta u_{k,l} + \Delta u_{l,k} + u_{i,k}\Delta u_{i,l} + \Delta u_{i,k}u_{i,l} + \Delta u_{i,k}\Delta u_{i,l}];\ \mathbf{x} \in \Omega \tag{3}$$

with the following essential and natural boundary conditions:

$$\Delta\sigma_{kl}n_l = \Delta\hat{t}_k;\ \mathbf{x} \in \partial\,\Omega_\sigma,\ k = 1,2,3 \tag{4}$$

$$\Delta u_k = \Delta\hat{u}_k;\ \mathbf{x} \in \partial\,\Omega_u,\ k = 1,2,3 \tag{5}$$

This problem is solved for the displacement vector $u_k(\mathbf{x})$, the strain tensor $\varepsilon_{kl}(\mathbf{x})$ and the stress tensor $\sigma_{kl}(\mathbf{x})$, where the stress tensor increments $\Delta\sigma_{kl}(\mathbf{x})$, $\Delta\widetilde{\sigma}_{kl}(\mathbf{x})$ and denote the first and the second Piola–Kirchhoff tensors

$$\Delta\sigma_{kl} = \Delta F_{km}\Delta\widetilde{\sigma}_{ml} + F_{km}\Delta\widetilde{\sigma}_{ml} + \Delta F_{km}\widetilde{\sigma}_{ml};\ \mathbf{x} \in \Omega \tag{6}$$

with

$$\Delta F_{km} = \Delta u_{k,m};\ \mathbf{x} \in \Omega. \tag{7}$$

All static state variables, i.e., displacements, strains and stresses are temperature-dependent, but this dependence is omitted for a brevity of presentation in all equilibrium equations. Simultaneously, a transient heat flow problem for the temperature field $T = T(\mathbf{x},\tau)$ is solved from the following differential equation [53]:

$$\rho(T)\,c(T)\dot{T} - \left(\lambda_{ij}(T)T_{,j}\right)_{,i} - g = 0;\ x_i \in \Omega; \tau \in [0,\infty), \tag{8}$$

where $c(T)$ is the temperature-dependent heat capacity of the region Ω, $\rho(T)$ is the temperature-dependent material density of Ω, $\lambda_{ij}(T)$ is temperature-dependent second-order tensor thermal conductivity, and g is the rate of heat generated per unit volume. Traditionally, T and τ denote temperature field values and time, respectively.

This equation should fulfil the boundary conditions of the $\partial\Omega$, which are given as follows:

(1) temperature (essential) boundary conditions

$$T = \hat{T};\ x \in \partial\Omega_T, \tag{9}$$

and for $\partial\Omega_q$ part of the total $\partial\Omega$:

(2) heat flux (natural) boundary conditions

$$\frac{\partial T}{\partial n} = \hat{q};\ x \in \partial\Omega_q, \tag{10}$$

where $\partial\Omega_T \cup \partial\Omega_q = \partial\Omega$ and $\partial\Omega_T \cap \partial\Omega_q = \{\varnothing\}$. The initial conditions have been introduced as

$$T^0 = T(x_i;0);\ x_i \in \Omega,\ \tau = 0. \tag{11}$$

The following functional defined on Δu_k is introduced to obtain a numerical solution to the deformation problem:

$$J(\Delta u_k) = \int_\Omega \left(\frac{1}{2}C_{klmn}(T)\Delta\varepsilon_{kl}\Delta\varepsilon_{mn} + \frac{1}{2}\widetilde{\sigma}_{kl}\Delta u_{i,k}\Delta u_{i,l} - \rho\Delta f_k\Delta u_k\right) d\Omega - \int_{\partial\Omega} \Delta\hat{t}_k\Delta u_k d(\partial\,\Omega) \tag{12}$$

whose solution is determined from the minimization of the incremental version of the potential energy stationarity principle

$$\delta J(\Delta u_k) = \frac{\partial J}{\partial\Delta u_k}\delta(\Delta u_k) \tag{13}$$

Analogously, one considers some continuous temperature variations $\delta T(x_i)$ defined in the interior of the region Ω and vanishing on $\partial\Omega_T$. A variational formulation may be proposed here as

$$\int_{\Omega} \left(\rho(T)c(T)\dot{T}\delta T + \lambda_{ij}(T)T_{,j}\delta T_{,i} - g\delta T \right) d\Omega - \int_{\partial\Omega_q} \hat{q}\,\delta T\, d(\partial\Omega) = 0; \quad x_i \in \Omega; \tau \in [0,\infty). \tag{14}$$

Let us recall the classical Finite Element Method formulation, where the displacements increments $\Delta u_k(\mathbf{x})$ being a continuous and differentiable function over the region Ω consisting of the geometrically continuous subsets (finite elements) Ω_e, where $e = 1, \ldots, E$ gives a complete representation of the set Ω. Let us consider the following approximation of the displacement increments [52,54]:

$$\Delta u_k(\mathbf{x}) = \sum_{\zeta=1}^{N_e} \phi_{\zeta k}(\mathbf{x})\, \Delta q_{\zeta}^{(N)}, \tag{15}$$

where $\phi_{\zeta k}(\mathbf{x})$ are the shape functions in the node k, $\Delta q_{\zeta}^{(N)}$ represent the nodal degrees of freedom vector, while N_e is the total number of those degrees of freedom in the considered node. Starting from the proposed approximation it is possible to express the gradients of the displacement vector as well as the strain tensor components as

$$\Delta u_{k,l}(\mathbf{x}) = \phi_{k,l}^{\zeta}(\mathbf{x})\, \Delta q_{\zeta}^{(N)}, \tag{16}$$

$$\Delta \bar{\varepsilon}_{kl}(\mathbf{x}) = [\overline{B}_{kl}^{(1)\zeta} + \overline{B}_{kl}^{(2)\zeta}]\, \Delta q_{\zeta}^{(N)} = \overline{B}_{kl}^{\zeta} \Delta q_{\zeta}^{(N)}, \tag{17}$$

$$\Delta \bar{\bar{\varepsilon}}_{kl}(\mathbf{x}) = \overline{\overline{B}}_{kl}^{\zeta\xi} \Delta q_{\zeta}^{(N)} \Delta q_{\xi}^{(N)} \tag{18}$$

and finally

$$\Delta \varepsilon_{kl}(\mathbf{x}) = \Delta \bar{\varepsilon}_{kl}(\mathbf{x}) + \Delta \bar{\bar{\varepsilon}}_{kl}(\mathbf{x}). \tag{19}$$

The following notation has been applied in the above equations:

$$\overline{B}_{kl}^{(1)\zeta}(\mathbf{x}) = \phi_{k,l}^{\xi}(\mathbf{x}),\ \overline{B}_{kl}^{(2)\zeta}(\mathbf{x}) = \phi_{i,k}^{\zeta}(\mathbf{x})\phi_{i,l}^{\xi}(\mathbf{x}) q_{\xi}^{(N)},\quad \overline{\overline{B}}_{kl}^{\zeta\xi}(\mathbf{x}) = \frac{1}{2}\phi_{i,k}^{\zeta}(\mathbf{x})\phi_{i,l}^{\xi}(\mathbf{x}). \tag{20}$$

Now, the following elemental stiffness matrices are introduced

$$k_{\zeta\xi}^{(\sigma)e} = \int_{\Omega_e} \tilde{\sigma}_{kl} \phi_{i,k}^{\zeta}(\mathbf{x})\, \phi_{i,l}^{\xi}(\mathbf{x})\, d\Omega, \tag{21}$$

$$k_{\zeta\xi}^{(con)e} = \int_{\Omega_e} \frac{1}{2} C_{klmn}(T) \overline{B}_{kl}^{(1)\zeta} \overline{B}_{mn}^{(1)\xi} d\Omega, \tag{22}$$

$$k_{\zeta\xi}^{(u)e} = \int_{\Omega_e} \frac{1}{2} C_{klmn}(T) \left(\overline{B}_{kl}^{(1)\zeta} \overline{B}_{mn}^{(2)\xi} + \overline{B}_{kl}^{(2)\zeta} \overline{B}_{mn}^{(1)\xi} + \overline{B}_{kl}^{(2)\zeta} \overline{B}_{mn}^{(2)\xi} \right) d\Omega, \tag{23}$$

while the second and the third order stiffnesses are equal, respectively

$$k_{\zeta\xi}^{(2)e} = \int_{\Omega_e} \frac{3}{2} C_{klmn}(T) \left(\overline{B}_{kl}^{\zeta} \Delta u_{\zeta}^{(N)} \overline{\overline{B}}_{mn}^{\mu\nu} \Delta q_{\mu}^{(N)} \Delta q_{\nu}^{(N)} + \overline{\overline{B}}_{kl}^{\zeta\xi} \Delta q_{\zeta}^{(N)} \Delta q_{\xi}^{(N)} \overline{B}_{mn}^{\mu} \Delta q_{\mu}^{(N)} \right) d\Omega, \tag{24}$$

$$k_{\zeta\xi}^{(3)e} = \int_{\Omega_e} 2C_{klmn}(T) \left(\overline{\overline{B}}_{kl}^{\zeta\xi} \Delta q_{\zeta}^{(N)} \Delta q_{\xi}^{(N)} \overline{\overline{B}}_{mn}^{\mu\nu} \Delta q_{\mu}^{(N)} \Delta q_{\nu}^{(N)} \right) d\Omega. \tag{25}$$

Introducing $k^{(i)}_{\zeta\xi}$ for i = 1,2,3 into the functional $J(\Delta u_k)$ in Equation (8), applying a transformation from the local to global coordinates system, one may obtain from the stationarity condition that

$$K^{(1)}_{\alpha\beta}(T)\Delta q_\beta + K^{(2)}_{\alpha\beta\gamma}(T)\Delta q_\beta \Delta q_\gamma + K^{(3)}_{\alpha\beta\gamma\delta}(T)\Delta q_\beta \Delta q_\gamma \Delta q_\delta = \Delta Q_\alpha \tag{26}$$

Fulfilled for any configuration of the region Ω, where $\hat{K}_{\alpha\beta}$, q_β, and Q_α are stiffness matrix, displacement vector, and nodal loads vector, respectively. The same discretization serves for the discretization of the temperature field by the nodal temperatures vector θ_α as [52,54]

$$T(x_i) = H_\alpha(x_i)\,\theta_\alpha;\ i = 1,2;\ \alpha = 1,2,\ldots,\mathrm{N}, \tag{27}$$

The temperature gradients can be rewritten in the form

$$T_{,i} = H_{\alpha,i}\,\theta_\alpha,\ \mathrm{i} = 1,2. \tag{28}$$

We introduce the capacity matrix $C_{\alpha\beta}(T)$, the heat conductivity matrix $\hat{K}_{\alpha\beta}(T)$ and the vector P_α as

$$C_{\alpha\beta}(T) = \int_\Omega \rho(T)c(T)\,H_\alpha\,H_\beta\,d\Omega,\ \ \hat{K}_{\alpha\beta}(T) = \int_\Omega \lambda_{ij}(T)\,H_{\alpha,i}\,H_{\beta,j}\,d\Omega \tag{29}$$

and also the R.H.S. vector in the following way:

$$P_\alpha = \int_\Omega g\,H_\alpha d\Omega + \int_{\partial\Omega} \hat{q}\,H_\alpha d\Omega. \tag{30}$$

Next, let us introduce these matrixes into the variational formulation (14) to obtain the following algebraic equations system:

$$C_{\alpha\beta}(T)\dot{\theta}_\beta + \hat{K}_{\alpha\beta}(T)\theta_\beta = P_\alpha \tag{31}$$

Equations (26) and (31) are finally solved simultaneously by the system ABAQUS to obtain time fluctuations of the deformations and stresses into the given boundary thermo-elasticity problem. This coupled problem is solved in a homogeneous continuous domain with no initial stresses and strains. A steel material occupying this domain shows temperature-dependent material and physical characteristics presented in Figures 2 and 3 below. It is reported here, after experimental works included in Eurocode 3 [2], that Young modulus as well as heat conductivity decrease while increasing temperature induced into the steel volume, quite opposite to thermal elongation, which increases. Heat capacity shows very complex behavior in this situation, especially in terms of the singularity reached close to 700 °C. This singularity results in remarkable difficulty during a solution of Equation (31), which is practically omitted by listing heat capacity at each 50 °C and numerical interpolation in-between these discrete data.

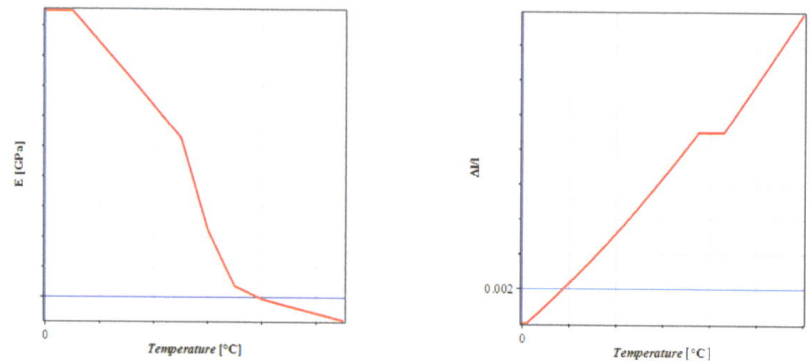

Figure 2. Young modulus (left graph) and relative elongation (right graph) temperature variations.

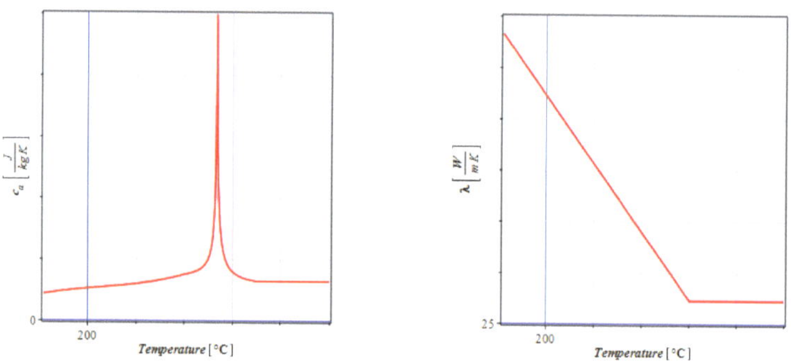

Figure 3. Specific heat (left graph) and heat conductivity (right graph) of temperature variations.

3. Probabilistic Aspects and Relative Entropy

Let us introduce the random variable b and its probability density function as $p_b(x)$. Then, the first two probabilistic moments of this variable are defined as [55]

$$E[b] \equiv b^0 = \int_{-\infty}^{+\infty} b p_b(x) dx \tag{32}$$

where b^0 means the average value of b itself and

$$Var[b] = \int_{-\infty}^{+\infty} [b - E(b)]^2 p_b(x) dx \tag{33}$$

Higher probabilistic moments and related coefficients may be defined according to the classical definitions from the probability theory. The basic idea of the stochastic perturbation approach employed here is to expand all input random variables and all the resulting state functions of the given boundary initial problem via the Taylor series about their spatial expectations using the perturbation parameter ε. The random function $f(b)$ with respect to its parameter b about its mean value is given as follows

$$f = f^0 + \sum_{n=10}^{1} \frac{\varepsilon^n}{n!} \frac{\partial^n f}{\partial b^n} \Delta b^n, \text{ where } \varepsilon \Delta b = \varepsilon\left(b - b^0\right) \tag{34}$$

is the first variation of a variable b about its expected value and the symbol $(.)^0$ represents the value of a function calculated for its mean. Let us analyze further the expected values of displacement state function $u(b)$ defined by its expansion via the Taylor series as follows

$$E[u(b)] = \int_{-\infty}^{+\infty} u(b)p_b(x)dx = \int_{-\infty}^{+\infty} \left(u^0\left(b^0\right) + \sum_{n=10}^{1} \frac{\varepsilon^n}{n!} \frac{\partial^n u(b)}{\partial b^n} \Delta b^n \right) p_b(x)dx \quad (35)$$

This expansion is valid only if the state function is analytic in ε and the series converges and, therefore, any criteria of convergence should include the magnitude of the perturbation parameter. A perturbation parameter here as usually in practical engineering computations is equal 1. It yields for the input random variable with symmetric probability density function in the tenth order approach

$$E[u(b)] = u^0\left(b^0\right) + \frac{1}{2!}\frac{\partial^2 u(b)}{\partial b^2}\mu_2(b) + \cdots + \frac{1}{10!}\frac{\partial^{10} u(b)}{\partial b^{10}}\mu_{10}(b) \quad (36)$$

$\mu_n(b)$ denotes the nth order central probabilistic moment of variable b. All the terms with odd orders are equal to 0 for the symmetric random variable and the orders higher than the 10th have been simply neglected. Similar considerations lead to the expressions, such as the variance, for instance

$$Var[u(b)] = \mu_2(u(b)) = \int_{-\infty}^{+\infty} (u(b) - E[u(b)])^2 p(b)db \quad (37)$$

Quite similarly, it is possible to derive higher-order central probabilistic moments from their definitions

$$\mu_3(u(b)) = \int_{-\infty}^{+\infty} (u(b) - E[u(b)])^3 p(b)db, \quad \mu_4(u(b)) = \int_{-\infty}^{+\infty} (u(b) - E[u(b)])^4 p(b)db \quad (38)$$

Let us mention that it is necessary to multiply each of these equations by the relevant order probabilistic moments of the input random variable to obtain the algebraic form convenient for any symbolic computations. Based on the classical definition of the variance we can calculate the coefficient of variation, skewness, and kurtosis as follows

$$\alpha(u(b)) = \frac{\sqrt{\mu_2(u(b))}}{E[u(b)]}, \quad S(u(b)) = \frac{\mu_3(u(b))}{\left[\sqrt{\mu_2(u(b))}\right]^3}, \quad \kappa(u(b)) = \frac{\mu_4(u(b))}{\left[\sqrt{\mu_2(u(b))}\right]^4} - 3 \quad (39)$$

It should be mentioned that at this stage the proposed procedure is still independent of a choice of the initial probability distribution function, however, a satisfactory probabilistic convergence of the final results may demand various lengths of the expansions for random variables with different distributions. A common implementation of the Monte-Carlo simulation, semi-analytical approach, as well as higher order stochastic perturbation technique in the system MAPLE was possible thanks to the Least Squares Method polynomial recovery of the structural displacements with respect to the input uncertain parameter b.

Finally, the reliability index is to be determined to measure the structural safety of the given case study of the beam under fire conditions. The engineering codes (such as Eurocode 0) advise applying the FORM theory, where one calculates

$$\beta_{FORM} = \frac{E[R] - E[E]}{\sqrt{Var(R) + Var(E)}}; \quad (40)$$

E[R] denotes here the maximum acceptable displacement of a beam according to Eurocode 3, and E[E] is the expected value of extreme displacement calculated based upon the iterative generalized stochastic perturbation technique. It is well known that the FORM approach has some limitations, so that higher order theories have been developed such as the Second and even Third Order Reliability Methods (SORM and TORM), especially because of the linear character of the limit function in the FORM technique. This seems to be unjustified in many practical applications and undoubtedly fire simulation is one of them. Further, the main mathematical methodological difficulty is that both distributions, of R and of E, are assumed as Gaussian here, which may be far from the experimental evidence and even misleading in the view of many theoretical studies and computer simulations.

This was a reason to seek for another concept to assess structural reliability in the presence of fire and to define the reliability index using probabilistic distance in-between two distributions of structural resistance and of structural extreme effort. One of the available mathematical models is the Bhattacharyya theory [40], which proposes for two different PDFs, namely $p_R(x)$ and $p_E(x)$, the following probabilistic distance measure:

$$\beta' = \int_{-\infty}^{+\infty} \sqrt{p_R(x)\, p_E(x)}\, dx, \tag{41}$$

where associated with structural resistance and probability function related to structural effort, respectively. Such formula enables the application of two different probability distributions of practically any nature and the given set of their parameters; further application of the analytical integration using some computer algebra system may lead to the desired numerical value. However, this formula can be simplified in the case of two Gaussian distributions with the given expectations and standard deviations ($\mu(R)$, $\mu(E)$, and also $\sigma(R)$, $\sigma(E)$, correspondingly). There holds

$$\beta' = \frac{1}{4}\frac{(E[R]-E[E])}{\sigma^2(R)+\sigma^2(E)} + \frac{1}{2}\ln\left(\frac{\sigma^2(R)+\sigma^2(E)}{2\sigma(R)\sigma(E)}\right) \tag{42}$$

Another problem is the upscaling of the variability interval of reliability indices obtained as the result of Equation (41) (or (42)) to contrast them with the values resulting from the FORM approach. The main idea would be to apply this mathematical apparatus with its theoretical importance, but to retain the existing engineering codes recommended minimum values of the reliability indices. Without such an upscaling direct determination of the reliability index, it would never show any extremes separating safe and unsafe design and/or exploitation of the given structure. Some preliminary computational experiments enable to propose the following rescaling of such a reliability index to numerical values comparable with the existing codes:

$$\beta = \frac{1}{2}\sqrt{\frac{1}{4}\frac{(E[R]-E[E])}{\sigma^2(R)+\sigma^2(E)} + \frac{1}{2}\ln\left(\frac{\sigma^2(R)+\sigma^2(E)}{2\sigma(R)\sigma(E)}\right)} \tag{43}$$

Both square root function as well as the multiplier preceding this function has been taken to make the first component, free from logarithmic function as close as possible to Equation (40); further numerical analysis would check an efficiency of such a modified reliability index. Computational implementation of the aforementioned approach has hybrid character and uses both the FEM system ABAQUS Standard, and also computer algebra package MAPLE. ABAQUS Standard fully coupled thermoelastic analysis is run multiple times to have the series of temperatures, stresses, and displacements obtained for the few different input fire temperatures. Then, these discrete results are in vector formats transferred to the MAPLE system, where the Least Squares Method procedures enable to recover polynomial approximations of these state functions with respect to fire temperature. Finally, this fire temperature is randomized according to the Gaussian distribution with the given mean value and given variability interval of the coefficient of variation. Polynomial approximations of temperatures, stresses, and displacements together with a triple implementation of probabilistic analysis return the basis probabilistic characteristics (i.e., expectations and standard deviations) of the state functions. Finally, two different reliability indices presented in Equations (40) and (43) are determined as the functions of fire exposure time.

4. Computational Illustration

The numerical example consists of the 6m both ends fixed steel I-profile beam with linear load applied to top flange equal 2.0 kN/m (Figure 4), whereas its cross section and the FEM discretization have been shown in Figure 5. The mesh consisting of 34,800 brick finite elements has been used; they are denoted in the system ABAQUS as C3D8T: the 8-node thermally coupled brick, tri-linear displacement, and temperature. The edge length of each brick element is less than 10 mm and it guarantees quite smooth approximation of stress distribution throughout the web. The thermal load has been adopted from the Standard ISO fire curve (Figure 6). According to Eurocode fire fumes after time $t_f = 15$ min are about 740 °C. This thermal load has been applied to each side of the cross-section except the top flange where the concrete slab lays according to the given fire scenario (Figure 6). Surface radiation and surface film conditions as thermal loads have been applied in this model (Figure 7). All material parameters are fully temperature-dependent and the fully coupled transient temperature-displacement incremental analysis has been used with the full Newton solution technique. The total time period of numerical simulation has been set as 900 s. The initial increment size is 0.01 step time, in this case, the minimum increment size is 0.0001 of the step time, and the maximum increment has been fixed as a single second.

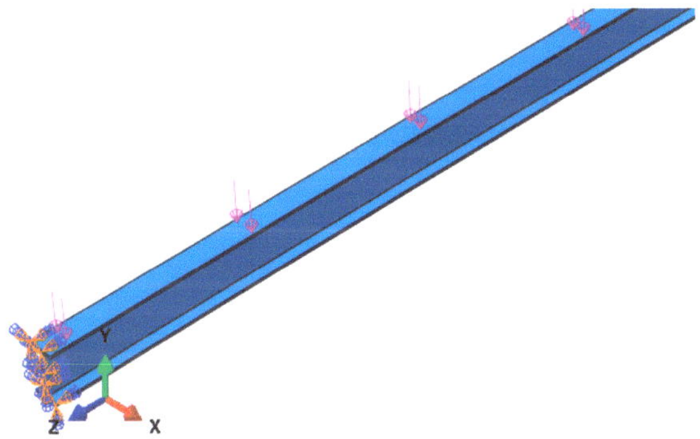

Figure 4. Geometry, loads and mechanical boundary conditions applied to the given beam.

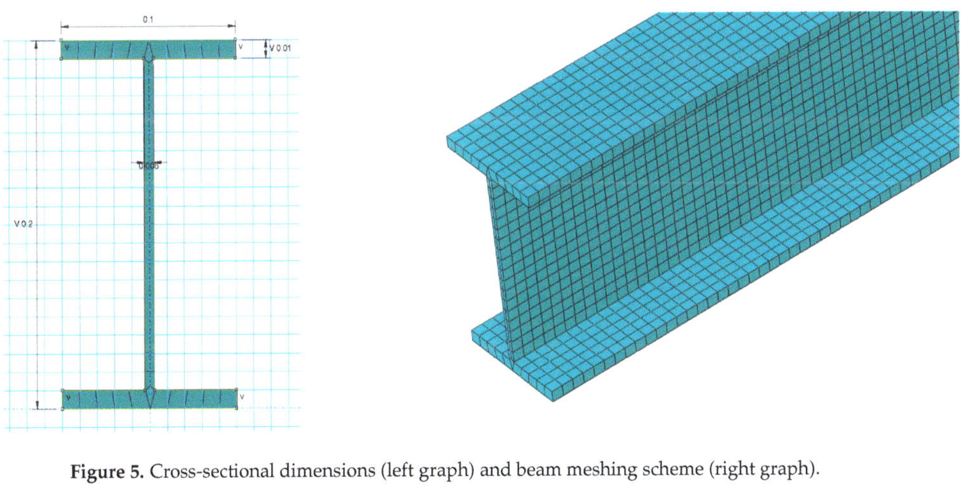

Figure 5. Cross-sectional dimensions (left graph) and beam meshing scheme (right graph).

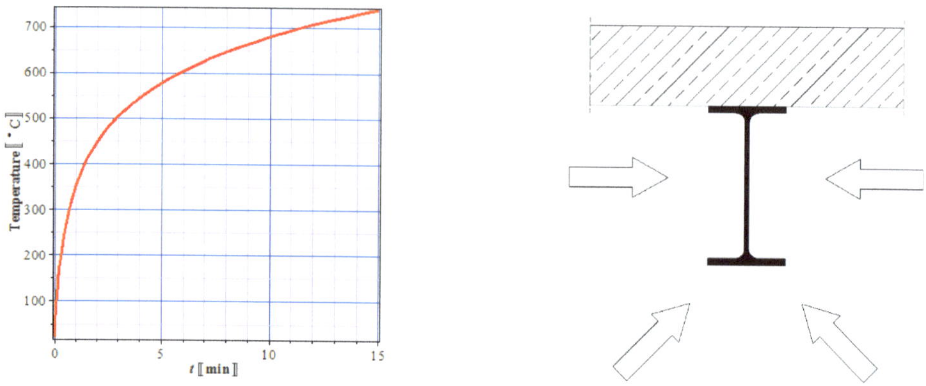

Figure 6. ISO-Standard fire curve (left graph) and fire scenario (right graph).

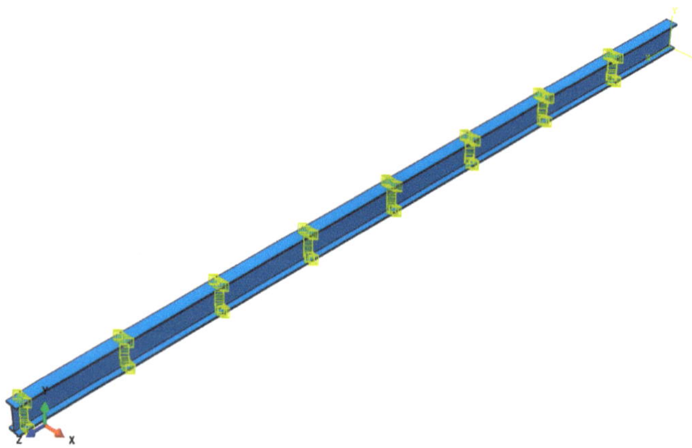

Figure 7. Thermal boundary conditions.

Figure 8 shows a comparison of this beam deflection under the given fire conditions (left diagram) and without a fire (right diagram). It is noticeable that fire heating increases the extreme deflection at half of the beam span more than four times (13.0 cm) than its admissible value (3.0 cm); maximum deflection in this case of no fire is less than 2.0 mm. One concludes that extreme deformation in the beam under fire is two orders larger than for the beam with no temperature attached (analytical approximation following basic equations of strength of materials appears to be quite efficient). The second major difference in-between these two models is that a beam subjected to fire shows remarkable deformations close to the supported area, whereas the beam without fire loadings has these deformations negligibly small. Next, Figure 9 reports normal stress marked as S33 (left diagram) and the reduced von Mises stress (right diagram). Von Mises stresses are almost 4 times larger than the longitudinal stresses in this specific case and both stress fields exhibit quite different distributions along the beam and throughout its cross-section. Figure 10 shows for a completeness temperature distribution at the steady-state (at the end of the beam heating by a fire) and one compares with an engineering intuition that the thinner web accumulates definitely more heat than the relatively thicker rest of the beam.

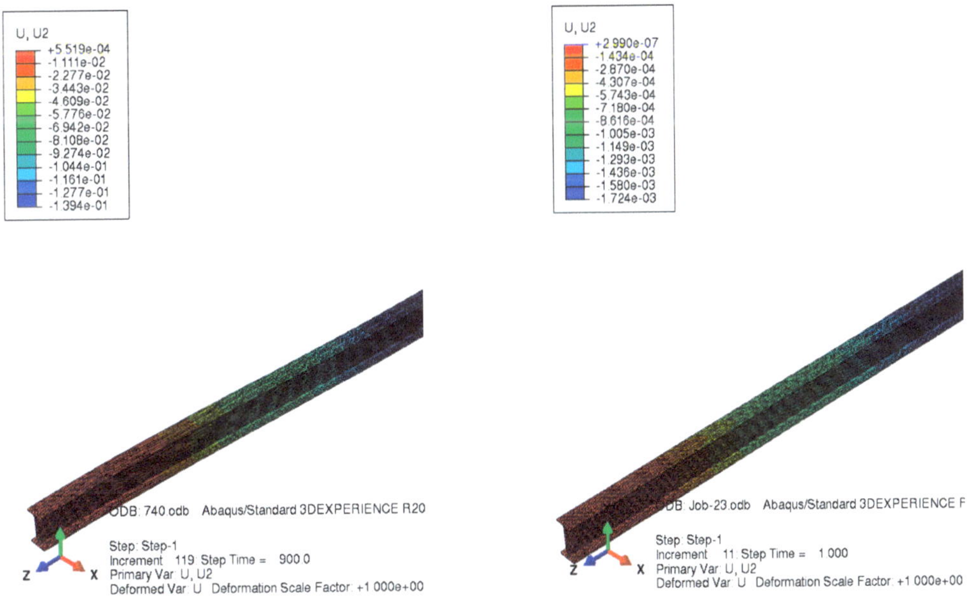

Figure 8. Deflections of the beam at t = 900 s (left–under, right–without a fire).

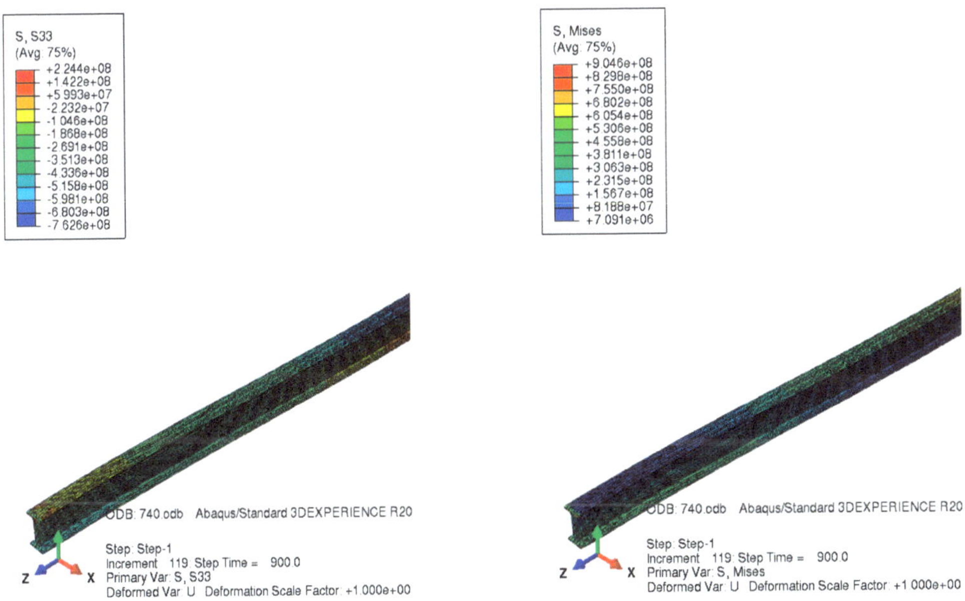

Figure 9. Normal stress S33 (left) and the reduced stress (right) distributions at t = 900 s.

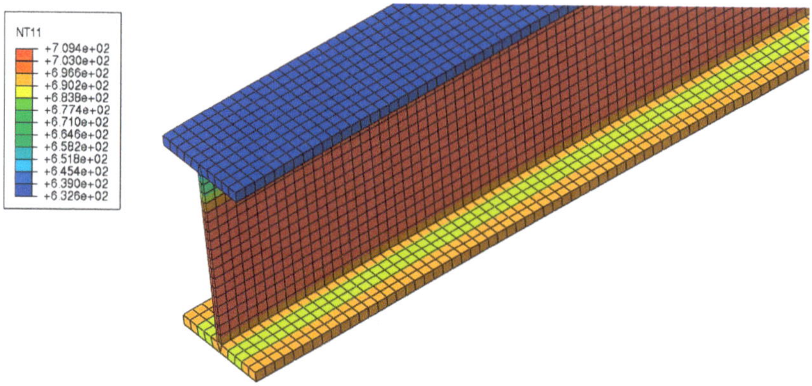

Figure 10. Temperatures distribution at t = 900 s.

Further results concern probabilistic characteristics of the structural response, which include expectations, coefficients of variation, skewness, and kurtosis of displacements shown in Figures 11–14, correspondingly. They have been computed using three different probabilistic methods, namely the perturbation method (abbreviated as a PM here), the semi-analytical method (SAM), and also using the Monte-Carlo simulation (MCS); they have been shown all as the functions of the input coefficient of variation ranging from 0.0 until 0.25. They concern extreme vertical displacements obtained at the half of the beam structure for the needs of further reliability assessment according to the Serviceability Limit State (SLS). First of all, it is seen that the first two moments are stable until t = 500 s, and then they start to diverge; initially, both moments equal almost 0. Generally, three different probabilistic numerical methods coincide in the case of the first two probabilistic characteristics with each other until $\alpha = 0.20$. Higher order statistical characteristics are more dispersed—they exhibit both numerical values very close to 0, which start to diverge at about t = 500 s. This can be interpreted that until this time moment of the fire exposure, the resulting displacements can be approximated as Gaussian, which means that both the FORM expression for the reliability index as well as its relative entropy counterpart proposed by Equation (43) are justified very well. Finally, one can notice that all three probabilistic methods coincide very well until $\alpha(t_f) = 0.10$.

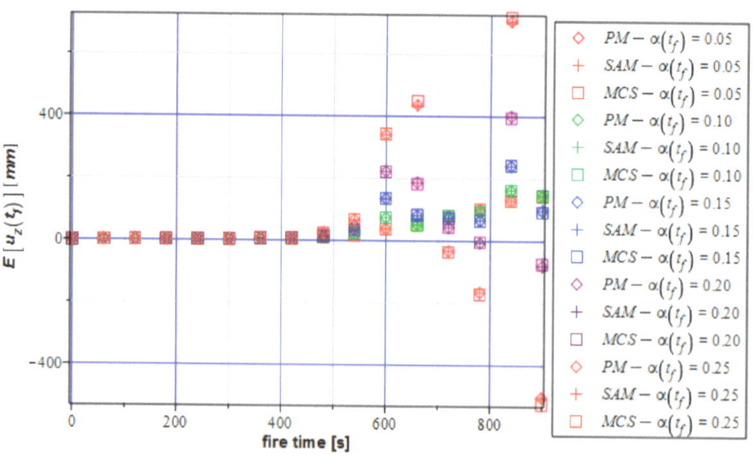

Figure 11. Expectations of the vertical deflections.

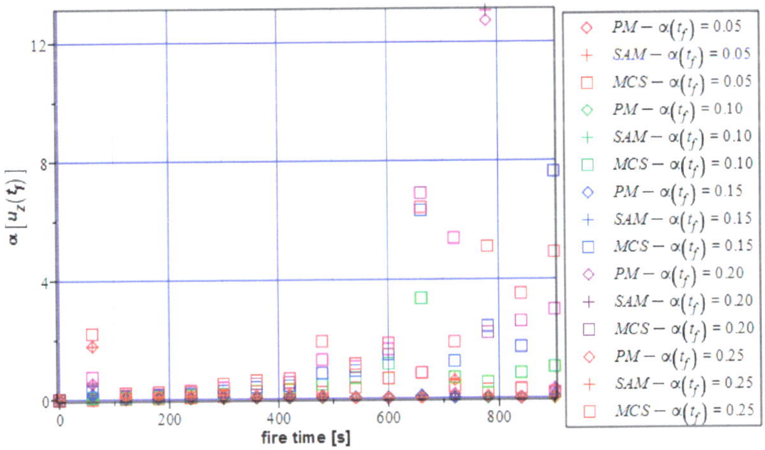

Figure 12. Coefficient of the variation of the vertical deflections.

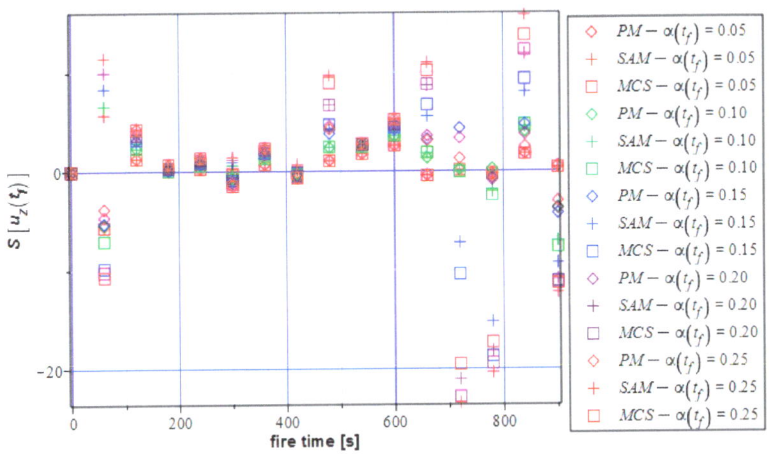

Figure 13. The skewness of the vertical deflections.

The next set of probabilistic characteristics computed for the reduced von Mises stresses is given similarly in Figures 15–18. They have been collected here as decisive in the calculation of the reliability index in the Ultimate Limit State (ULS). The expected values increase monotonously from 0 until the same moment t = 500 s; then, they exhibit large fluctuations, although all three probabilistic methods return the same numerical values. The reduced von Mises stresses start to highly depend upon the input coefficient of variation for the fire exposure time t > 500 s, which is an observation quite unusual for elastic problems with any uncertainty. A little bit different conclusions can be drawn from the results contained in Figure 16—the output CoVs keeps very close to 0 during the entire fire heating process showing some numerical discrepancies almost at the beginning of this process (t = 50 s) as well as its end (t = 800 s). Higher order statistics are rather distant from 0 (see Figures 17 and 18), so they cannot be efficiently modeled as Gaussian and need larger numerical effort. A coincidence of Monte-Carlo simulation, the semi-analytical method as well as the stochastic perturbation technique is worse and can be assumed $\alpha(t_f) = 0.05$. This is a quite expected result because probabilistic characteristics of the stresses are calculated based on probabilistic moments of displacements (since displacement-version of the FEM is used) and of probabilistic characteristics of the constitutive tensor (additionally depending on the nodal temperatures).

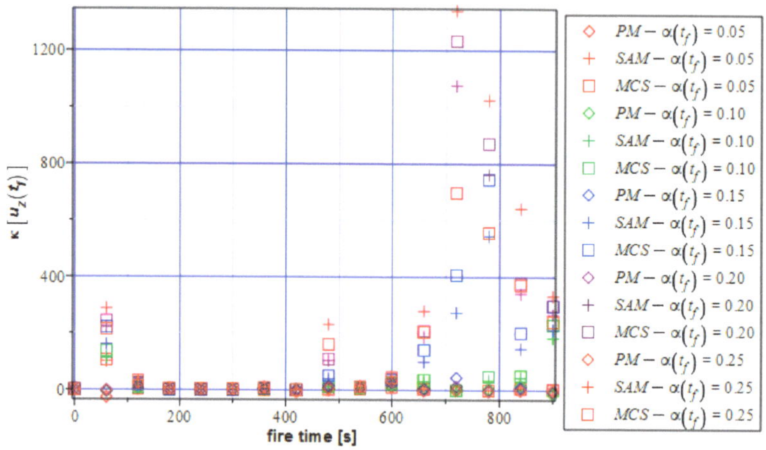

Figure 14. Kurtosis of the vertical deflections.

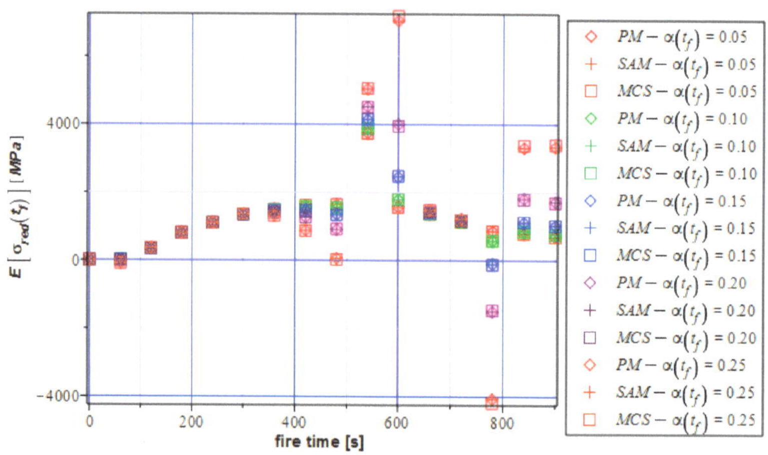

Figure 15. Expectations of reduced stress.

Higher order statistics of the reduced stresses are definitely distant from 0 during fire exposure time, so that these stresses cannot be approximated with the Gaussian distribution. Therefore, the FORM index given in Equation (40) includes a remarkable modelling error, whereas relative entropy should be calculated here using the general analytical formula (42). Further numerical results concern reliability analysis of the given structure, so that Figure 19 contains the key results in this study and presents a comparison of the reliability indices determined using the FORM approach (left graph) with these calculated thanks to the Bhattacharyya relative entropy. They are both plotted as the functions of the fire exposure time and are related to the Serviceability Limit State. This is because previous results document well approximation of the distribution of the displacement by the Gaussian PDF; this is not the case with the reduced stresses, so their reliability has been discussed in Figure 20 using the FORM index only. As one could expect, these time fluctuations of reliability indices have typical exponential decay in the first stage of fire heating, but then, after reaching critical and 0 values, exhibit further oscillatory character. This part is interesting from the numerical point of view and has no practical importance and the structure lost its reliability. Such a wavy character is totally absent for civil engineering structures subjected to, i.e., static loads and some stochastic ageing process, when reliability decreases monotonously.

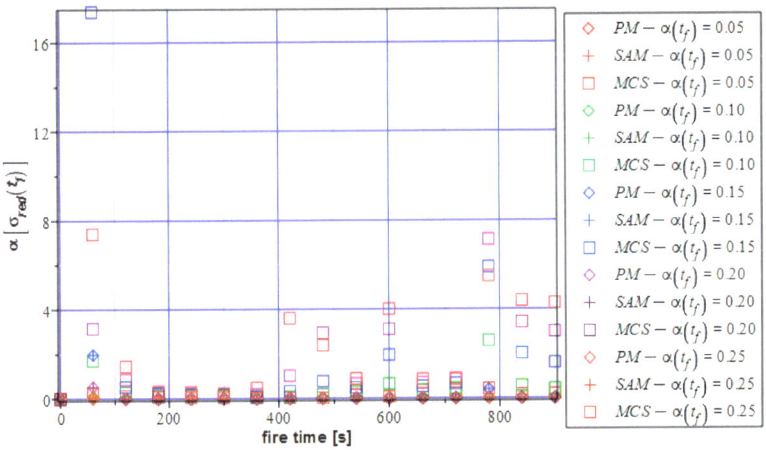

Figure 16. Coefficient of variation of the reduced stress.

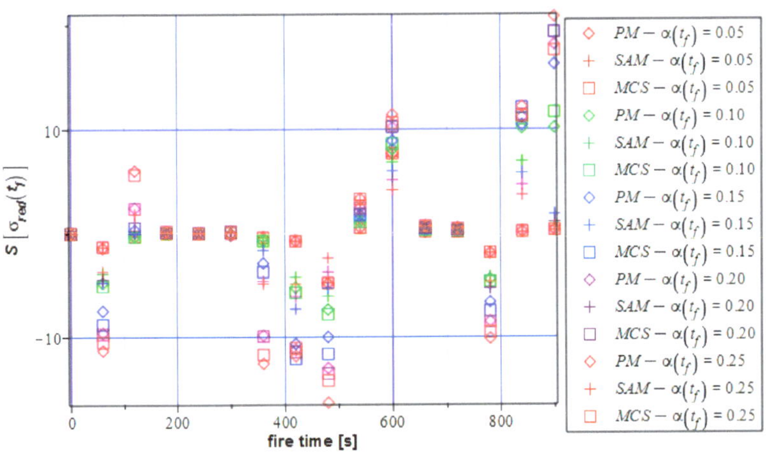

Figure 17. The skewness of the reduced stress.

The very important conclusion following the comparison of the two graphs in Figure 19 is that the FORM index and relative entropy variations show the same character and extreme values location. The most important conclusion is that they cross the admissible value at almost the same time as a fire accident, which means that structural safety analysis results in both cases in the same evacuation time. Further, it is seen that the FORM analysis results in a few times higher reliability index at the very beginning of the fire accident. This can be meant as some overestimation of the realistic reliability, but this difference is meaningless for structural design. A very interesting result is that the FORM index is more sensitive to the input uncertainty level, whereas the relative entropy shows almost no such sensitivity.

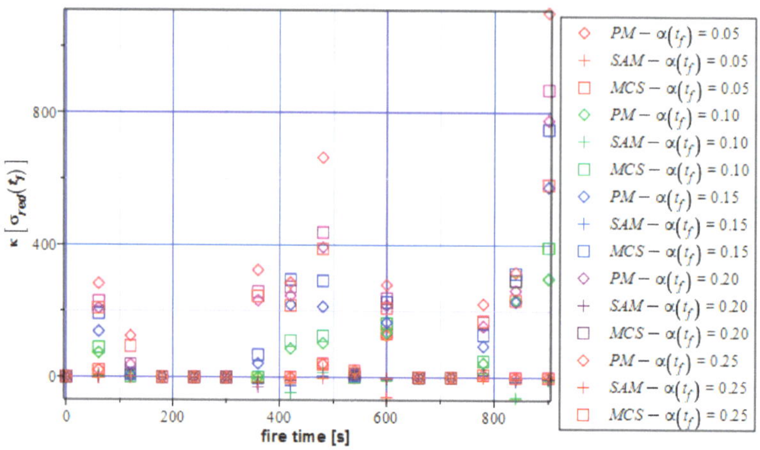

Figure 18. Kurtosis of the reduced stress.

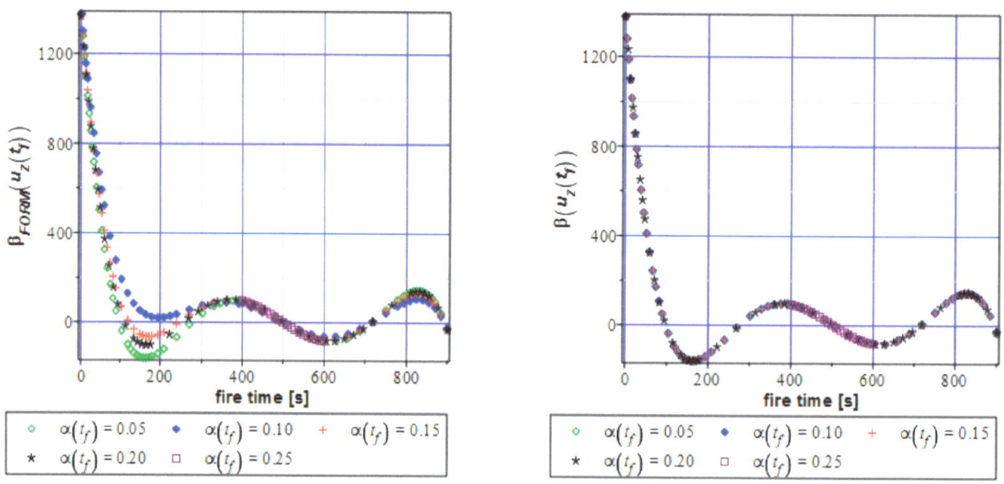

Figure 19. Reliability index for the vertical displacements as a function of fire duration according to the FORM theory (left) and the relative entropy approach (right).

The reliability index in the ULS analysis shows less waviness above the reliability limit. It does not affect the conclusion and structural safety—it concerns computational aspects rather. A contrast of Figures 19 and 20 results in a conclusion that the ULS is decisive for the overall safety of this element—it seems that the given beam faster falls into the plastic regime than approaches the admissible deformations at half of its span. This conclusion is a little bit out of the probabilistic analysis, nevertheless, the stochastic reliability study confirms an engineering observation. This notice is supported mainly by the reduced von Mises stress time fluctuations. Evaluation of the beam safety from the normal longitudinal stresses may lead to an improper conclusion that the ULS and the SLS exhibit almost the same failure time.

A more detailed comparison of the classical FORM approach with the proposed new one based upon the relative entropy has been provided in Table 1 below ($\beta_{FORM}(u_z(t_f))$) and relative entropy approach ($\beta(u_z(t_f))$). Both reliability indices have been compared with each other throughout the entire fire accident simulation time for the few different values of input coefficient of variation. There is no doubt that both methods return almost the same values (positive and negative also) for any input statistical scattering. Furthermore, a failure time defined as the moment of fire duration when the reliability index falls down below the admissible values suggested in the designing codes (t = 80 s) in both methods is also almost the same. It seems that the rescaled relative entropy calculated from Equation (43) enables to predict fire safety with the use of the existing engineering codes.

Table 1. Reliability index for the vertical displacements as a function of fire duration values comparison for the FORM theory and the relative entropy approach.

t_f [sec.]	$\beta_{FORM}(u_z(t_f))$					$\beta(u_z(t_f))$				
	$\alpha(t_f)$					$\alpha(t_f)$				
	0.05	0.10	0.15	0.20	0.25	0.05	0.10	0.15	0.20	0.25
0	1378.65	1379.21	1367.60	1369.04	1371.28	1378.93	1378.87	1378.81	1378.75	1378.65
60	312.18	508.92	397.98	359.21	340.79	312.26	312.29	312.34	312.32	312.18
120	−97.00	130.42	10.54	−34.75	−57.39	−96.87	−96.84	−96.81	−96.85	−97.00
180	−153.26	23.50	−63.06	−98.32	−116.80	−152.95	−152.98	−153.04	−153.12	−153.26
240	−65.15	36.38	−6.63	−26.80	−38.25	−64.67	−64.79	−64.91	−65.03	−65.15
300	38.66	74.53	66.69	59.70	54.77	39.26	39.07	38.90	38.76	38.66
360	93.80	89.41	100.40	101.46	100.87	94.41	94.19	94.00	93.87	93.80
420	84.44	67.40	80.84	84.37	85.56	84.97	84.74	84.58	84.48	84.44
480	27.98	18.82	23.52	25.38	26.31	28.35	28.16	28.05	28.00	27.98
540	−40.04	−32.90	−40.25	−41.77	−42.08	−39.87	−40.00	−40.05	−40.04	−40.04
600	−80.72	−61.70	−76.53	−80.50	−81.90	−80.75	−80.81	−80.80	−80.75	−80.72
660	−66.04	−49.14	−61.33	−64.89	−66.32	−66.24	−66.24	−66.19	−66.10	−66.04
720	7.35	5.74	7.06	7.25	7.16	7.06	7.09	7.15	7.26	7.35
780	104.99	77.01	96.39	101.92	103.94	104.74	104.71	104.78	104.88	104.99
840	142.07	102.81	129.23	137.05	140.10	142.02	141.82	141.89	141.98	142.07
900	−28.24	−23.07	−27.43	−28.62	−29.03	−27.94	−28.46	−28.37	−28.29	−28.24

The data presented in both Table 1, and also in Figures 19 and 20 demonstrate that steel beams without any fire protection are extremely sensitive to fire temperatures. A sensitivity of its deformation is of course a few times higher than of the ULS limit function, which is confirmed by the negative reliability indices detected in case of the SLS. It is important that both reliability approaches return the same qualitative results.

Finally, it should be noticed that quite a satisfactory accuracy of both probabilistic simulations and reliability analysis shows that the methodology presented could be applied to other types of steel structures and further applications towards aluminum alloys may be taken into account. Undoubtedly, this approach would be closer to industrial applications when the phase change (from solid to liquid) could be accounted for; this needs brand new implementations in the system ABAQUS. It should be mentioned that the civil engineering designing codes still do not include any statements enabling efficient engineering reliability concerning fire safety (neither for steel nor for concrete or traditional structures). On the other hand, closer interoperability of the FEM system ABAQUS, with the computer algebra packages such as MAPLE should be also achieved.

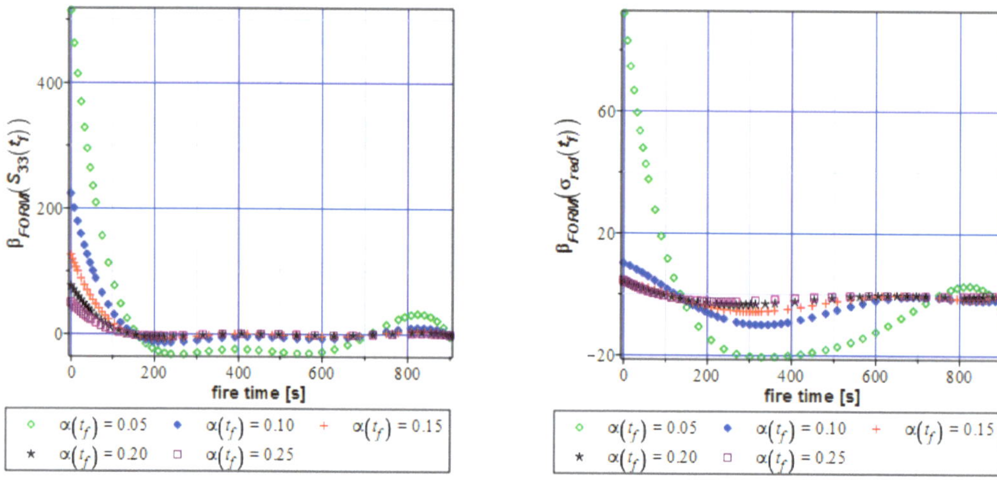

Figure 20. Reliability index for the normal stress S_{33} (left graph) versus von Mises reduced stresses (right graph) according to the FORM.

5. Conclusions

(1) The major original contribution of this study is a development of new reliability assessment procedure for steel structures subjected to fire conditions. It has been completed with the use of the relative probabilistic entropy model developed by Bhattacharyya and contrasted after some numerical rescaling with the FORM index inherent in the current designing civil engineering codes. The key research finding in this work is that the Bhattacharyya relative entropy sensitivity with respect to the given fire temperature curve (and the additional duration time) is the same as in the case of the FORM reliability index. The innovative scaling procedure enables to use the new mathematical theory to verify the structural safety according to the existing procedures and recommended values of the reliability indices. This is exactly the same scaling as in the case of linear elastic models of steel trusses and spatial structures of the halls [14,20], which means that this procedure may have general character. It should be mentioned that such a new verification procedure may be relatively easily implemented in any hybrid computational FEM and computer algebra system numerical environment.

(2) A triple probabilistic approach based on the Monte-Carlo simulation, generalized iterative stochastic perturbation technique, and also the semi-analytical approach programmed in conjunction with the polynomial bases of the fire temperature returns very consistent and stable numerical results of expected values, coefficients of variations, as well as skewness and kurtosis—all as the functions of the fire exposure time. Application of such a technique enabled to discover that extreme deformations of the given system have a probability distribution very similar to the Gaussian one, and hence, determination of their probabilistic divergence could proceed with the use of the analytical formula. Probabilistic divergence of the admissible and extreme stresses needs to be proceeded by semi-analytical integration of the general formula. Despite the rather simple character of the structural beam investigated in this work, such a coincidence of three different probabilistic techniques is expected for other structures and their fire scenarios, which deserves further numerical simulations.

Author Contributions: Conceptualization, M.K. and M.S.; methodology, M.K.; software, M.S.; validation, M.K. and M.S.; formal analysis, M.K.; investigation, M.S.; resources, M.K.; data curation, M.S.; writing—original draft preparation, M.K. and M.S.; writing—review and editing, M.K.; visualization, M.S.; supervision, M.K.; project administration, M.K.; funding acquisition, M.K. All authors have read and agreed to the published version of the manuscript.

Funding: This paper has been written in the framework of the research grant OPUS no. 2021/41/B/ST8/02432 "Probabilistic entropy in engineering computations" sponsored by the National Science Center in Cracow, Poland.

Informed Consent Statement: The informed consent was obtained from all subjects involved in the study.

Data Availability Statement: Not applicable.

Conflicts of Interest: The authors declare no conflict of interest.

References

1. Melchers, R.E. *Structural Reliability. Analysis and Prediction*, 2nd ed.; Wiley: Chichester, UK, 1999.
2. *EN 1993–1-9*; Eurocode 3: Design of Steel Structures. CEN: Brussels, Belgium, 1993.
3. Kamiński, M.; Strąkowski, M. On sequentially coupled thermo-elastic stochastic finite element analysis of the steel skeletal towers exposed to fire. *Eur. J. Mech. Part A Solids* **2017**, *62*, 80–93. [CrossRef]
4. Maślak, M. *Fire Durability of Steel Bar Structures*; Technical University of Cracow Press: Cracow, Poland, 2008. (In Polish)
5. Aziz, E.M.; Kodur, V.K.; Glassman, J.D.; Moreyra Garlock, M.E. Behavior of steel bridge girders under fire conditions. *J. Constr. Steel Res.* **2015**, *106*, 11–22. [CrossRef]
6. Burgess, I.W.; Najjar, S.R. A simple approach to the behaviour of steel columns in fire. *J. Constr. Steel Res.* **1994**, *31*, 77–89. [CrossRef]
7. Dwaikat, M.M.S.; Kodor, V.K.R.; Quiel, S.E.; Garlock, M.E.M. Experimental behaviour of steel beam-columns subjected to the fire-induced thermal gradients. *J. Constr. Steel Res.* **2011**, *67*, 30–38. [CrossRef]
8. Wald, F.; Simões da Silva, L.; Moore, D.B.; Lennon, T.; Chladná, M.; Santiago, A.; Beneš, M.; Borges, L. Experimental behaviour of a steel structure under natural fire. *Fire Safety J.* **2006**, *41*, 509–522. [CrossRef]
9. Laím, L.; Rodrigues, J.P.C. Fire design methodologies for cold-formed steel beams made with open and closed cross-sections. *Engrg. Struct.* **2018**, *171*, 759–778. [CrossRef]
10. Hadianfard, M.; Razani, R.A. Effects of semi-rigid behaviour of connections in the reliability of steel frames. *Struct. Saf.* **2011**, *25*, 123–138. [CrossRef]
11. Landesmann, A.; Batista, E.M.; Alves, J.L.D. Implementation of advances analysis method for steel-framed structures under fire conditions. *Fire Safety J.* **2005**, *40*, 339–366. [CrossRef]
12. Liew, Y.R.; Tang, L.K.; Holmaas, T.; Chao, Y.S. Advances analysis for the assessment of steel frames in fire. *J. Constr. Steel Res.* **1998**, *47*, 19–45. [CrossRef]
13. Kodur, V.K.; Aziz, E.M.; Naser, M.Z. Strategies for enhancing fire performance of steel bridges. *Engrg. Struct.* **2017**, *131*, 446–458. [CrossRef]
14. Bredow, R.; Kamiński, M. Structural safety of the steel hall under dynamic excitation using the relative probabilistic entropy concept. *Materials* **2022**, *15*, 3587. [CrossRef] [PubMed]
15. McGrattan, K.; McDermott, R.; Floyd, J.; Hostikka, S.; Forney, G.; Baum, H. Computational fluid dynamics modelling of fire. *Int. J. Comput. Fluid Dyn.* **2012**, *26*, 349–361. [CrossRef]
16. Alos-Moya, J.; Paya-Zartofeza, I.; Garlock, M.E.M.; Loma-Ossorio, E.; Schiffner, D.; Hospitaler, A. Analysis of a bridge failure due to fire using computational fluid dynamics and finite element models. *Engrg. Struct.* **2014**, *68*, 96–100. [CrossRef]
17. Silva, J.C.G.; Landesmann, A.; Ribeiro, F.L.B. Fire-thermomechanical interface model for performance-based analysis of structures exposed to fire. *Fire Safety J.* **2016**, *83*, 66–78. [CrossRef]
18. Najjar, S.R.; Burgess, I.W. A nonlinear analysis for three-dimensional steel frames in fire conditions. *Engrg. Struct.* **1996**, *18*, 77–89. [CrossRef]
19. Sokołowski, D.; Kamiński, M. FEM Study of a Steel Corrugated Web Plate Girder Subjected to Fire. *Int. J. Appl. Mech. Engrg.* **2021**, *26*, 201–218. [CrossRef]
20. Morales, J.M.; Mermoz, M.; Gowda, J.H.; Kitzberger, T. A stochastic fire spread model for north Patagonia based on fire occurrence maps. *Ecol. Model.* **2015**, *300*, 73–80. [CrossRef]
21. Braun, W.J.; Woolford, D.G. Assessing a stochastic fire spread simulator. *J. Env. Inf.* **2013**, *22*, 1–12. [CrossRef]
22. Zhao, S.J.; Xiong, L.Y.; Ren, A.Z. A spatial-temporal stochastic simulation of fire outbreaks following earthquake based on GIS. *J. Fire Sci.* **2006**, *24*, 313–339. [CrossRef]
23. Shi, K.; Guo, Q.; Jeffers, A. Stochastic analysis of structures in fire by Monte-Carlo simulation. *J. Struct. Engrg.* **2013**, *4*, 37–46. [CrossRef]

24. Huang, P.T.; Delichatsios, M.A. Quantitative risk analysis for the response of steel beams in fire. *J. Struct. Engrg.* **2010**, *1*, 231–241. [CrossRef]
25. Keese, A.; Matthies, H.G. Hierarchical parallelization for the solution of stochastic finite element equations. *Comput. Struct.* **2005**, *83*, 1033–1047. [CrossRef]
26. Kamiński, M. *The Stochastic Perturbation Method for Computational Mechanics*; Wiley: Chichester, UK, 2013.
27. Kamiński, M.; Błoński, R. Analytical and numerical reliability analysis of certain Pratt steel truss. *Appl. Sci.* **2022**, *12*, 2901. [CrossRef]
28. Ghanem, R.G.; Spanos, P.D. *Stochastic Finite Elements: A Spectral Approach*; Springer: New York, NY, USA, 1991.
29. Kamiński, M.; Świta, P. Generalized stochastic finite element method in elastic stability problems. *Comput. Struct.* **2011**, *89*, 1241–1252. [CrossRef]
30. Stefanou, G.; Papadrakakis, M. Stochastic finite element analysis of shells with combined random material and geometrical properties. *Comput. Method. Appl. Mech. Eng.* **2004**, *93*, 139–160. [CrossRef]
31. Waarts, P.H.; Vrouwenvelder, A.C.W.M. Stochastic finite element analysis of steel structures. *J. Constr. Steel Res.* **1999**, *52*, 21–32. [CrossRef]
32. Hien, T.D.; Kleiber, M. Stochastic finite element modelling in linear transient heat transfer. *Comput. Method. Appl. Mech. Eng.* **1997**, *144*, 111–124. [CrossRef]
33. Kleiber, M.; Hien, T.D. *The Stochastic Finite Element Method*; Wiley: Chichester, UK, 1992.
34. Bjőrck, A. *Numerical Methods for Least Squares Problems*; SIAM: Philadelphia, PA, USA, 1996.
35. Wolberg, J. *Data Analysis Using the Method of Least Squares*; Springer: Berlin/Heidelberg, Germany, 2005.
36. Chen, X.; Tung, Y.K. Investigation of polynomial normal transform. *Struct. Saf.* **2005**, *25*, 423–445. [CrossRef]
37. Allaix, D.L.; Carbone, V.I. An improvement of the response surface method. *Struct. Saf.* **2011**, *33*, 165–172. [CrossRef]
38. Gavin, H.P.; Yau, S.C. High-order limit state functions in the response surface method for structural reliability analysis. *Struct. Saf.* **2008**, *30*, 162–179. [CrossRef]
39. *EN 1990:2002+A1*; Eurocode 0: Basis of Structural Design. CEN: Brussels, Belgium, 1990.
40. Bhattacharyya, A. On a measure of divergence between two statistical populations defined by their probability distributions. *Bull. Calcutta Math. Soc.* **1943**, *35*, 99–109.
41. Cover, T.M.; Thomas, J.A. *Elements of Information Theory*, 2nd ed.; Wiley & Sons: New York, NY, USA, 2006.
42. Shannon, C.E. A mathematical theory of communication, Part I & II. *Bell Sys. Tech. J.* **1948**, *27*, 379–423, 623–656.
43. Donald, M.J. On the relative entropy. *Comm. Math. Phys.* **1986**, *105*, 13–34. [CrossRef]
44. Hellinger, E. Neue begrűndung der theorie quadratischer Formen von unendlichvielen veränderlichen. *J. Reine Angew. Math.* **1909**, *136*, 210–271. [CrossRef]
45. Jeffreys, H. An invariant form for the prior probability in estimation problems. *Proc. R. Soc. Lond. Ser. A Math. Phys. Sci.* **1946**, *186*, 453–461.
46. Kullback, S.; Leibler, R.A. On information and sufficiency. *Annals Math. Stat.* **1951**, *22*, 79–86. [CrossRef]
47. Nielsen, F. On a variational definition for the Jensen-Shannon symmetrization of distances based on the information radius. *Entropy* **2021**, *23*, 464. [CrossRef]
48. Batou, A.; Ritto, T.G.; Sampaio, R. Entropy propagation analysis in stochastic structural dynamics: Application to a beam with uncertain cross sectional area. *Comput. Mech.* **2014**, *54*, 591–601. [CrossRef]
49. Soize, C. Maximum entropy approach for modeling random uncertainties in transient elasto-dynamics. *J. Acoust. Soc. Amer.* **2001**, *109*, 1979–1996. [CrossRef]
50. Tao, K.; Zhen, W.; Jiang, D. Entropy method for structural health monitoring based on statistical cause and effect analysis of acoustic emission and vibration signals. *IEEE Access* **2019**, *7*, 172515–172525. [CrossRef]
51. Xu, Y. A maximum entropy method for the dynamical structural optimization design. *Engrg. Mech.* **1996**, *2*, 598–601.
52. Oden, J.T. *Computational Methods in Nonlinear Mechanics*; North-Holland: Amsterdam, The Netherlands, 1980.
53. Nowacki, W. *Thermoelasticity*; Ossolineum: Wrocław, Poland, 1972. (In Polish)
54. Zienkiewicz, O.C.; Taylor, R.L. *The Finite Element Method*; Elsevier: Amsterdam, The Netherlands, 2005.
55. Vanmarcke, E. *Random Fields. Analysis and Synthesis*; MIT Press: Cambridge, MA, USA, 1983.

energies

Article

Determination of 12 Combustion Products, Flame Temperature and Laminar Burning Velocity of Saudi LPG Using Numerical Methods Coded in a MATLAB Application

Roberto Franco Cisneros * and Freddy Jesús Rojas *

Faculty of Mechanical Engineering, Pontificia Universidad Católica del Perú, Lima 15088, Peru
* Correspondence: a20150535@pucp.pe (R.F.C.); fjrojas@pucp.edu.pe (F.J.R.)

Abstract: The characterization of a specific fuel has always been an important point for developing and designing new components or systems with the maximum efficient possible. Studying the laminar burning velocity can lay a necessary prerequisite for the accurate poststudy of the turbulent range and to understand how the combustion process takes place. The study of the combustion products from a specific reaction is a requisite for any system in order to understand the elements that are taken in the process and if it is possible to improve it. In this study, a new open code methodology was developed for the determination of combustion products, flame temperature and laminar burning velocity using numerical methods (Newton–Raphson, Taylor series and Gaussian elimination) in an application codified in MATLAB. The MATLAB application was applied for the study of Saudi LPG setting parameters such as initial temperature, pressure and equivalence ratio that are meaningful because they have a great effect on the results. In addition, simulation in Ansys Chemkin using San Diego and RedSD mechanisms was carried out. The results from the MATLAB application were compared with other experimental research and Ansys Chemkin simulation. These are presented in different plots and it is shown that: (1) For the laminar burning velocity results, the numerical method agrees with the experimental results for ratios (0.6–1.2) by other authors and the simulation in Ansys Chemkin. (2) For the highest studied equivalence ratios (1.3–1.7) the laminar burning velocity results between all the resources have more difference. (3) The combustion products calculated by the MATLAB application agree with those simulated in Ansys Chemkin except N and NO. (4) The MATLAB application gives a maximum value of 40.35 cm/s, that is greater than 35 $\pm$ 0.91, the one determined by Bader A. Alfarraj. (5) The flame temperature calculated by the MATLAB application overestimates that simulated in Ansys Chemkin but has the same behavior for all the calculated ratios. (6) The MATLAB application has also been developed for the study and analysis of other fuels.

Keywords: flame temperature; burning velocity; equivalence ratio; combustion products; product enthalpy; reactant enthalpy; Saudi LPG; Newton–Raphson; Taylor series; Gaussian elimination

1. Introduction

The analysis and study of the most used fuels in the world allow us to continue extracting and improving all the information collected to date. Liquefied petroleum gas is one of the fuels with the most applications in all fields, for example: combustion engines, ovens, burners, industries, etc. To better understand this fuel, it is necessary to determine and study the properties that characterize it, which a posteriori allow optimizing the devices that use it in order to obtain a balance between performance and efficiency. Among the main characteristics and properties are the flame temperature, the laminar burning velocity and the products of combustion. The laminar burning velocity is one of the most fundamental characteristics of the combustion of a certain mixture, likewise, it contains fundamental information on reactivity, diffusivity and exotherm, therefore, it is used to characterize the flames and precise knowledge is important for the equipment design, turbulent combustion model and validation of chemical kinetic mechanisms [1]. The burning velocity for any type of fuel mixture is affected by different factors, the main ones are pressure, temperature, equivalence ratio (Φ) and the properties of the fuel components [2].

To obtain the flame speed theoretically and experimentally, studies and methods were developed many years ago. On one hand, the oldest ones date back to 1949, when Powling and Edgerton developed a flat flame burner which allowed a close approximation to the ideal of a one-dimensional flat flame, but it was limited to low flame velocities (0.15–0.2 m/s) [1]. On the other hand, more recent studies such as that of Samahat Samim in 2016, which investigated the determination of the laminar flame speeds of gas to liquid (GTL) fuel with conventional diesel (50–50 by volume) in a cylindrical pump, were able to measure the flame speed at initial temperatures and atmospheric pressure equivalence ratios, by analyzing the pressures just after combustion, which were detected by a pressure transducer installed in the pump and, as a conclusion, the blended fuel was determined to have the lowest flame velocity at the highest temperature (89.7 cm/s at 250 °C) [1,3].

Regarding the investigations on the laminar flame speed of LPG, in 2004 Kihyung Lee and Jeaduk Ryu conducted a study of the flame propagation and combustion characteristics of this fuel using an optical technique for measurement of flame velocity in a constant volume combustion chamber (CVCC) and a heavy duty liquefied petroleum liquid injection (LPLi) port [4]. Later, in 2007, A.S. Huzayyin, H.A. Moneib, M.S. Shehatta and A.M.A. Attia determined the variations of the laminar flame velocity and the explosion rates of LPG–air mixtures in a wide range of equivalence ratios (Φ = 0.7–2.2), initial temperatures (Ti = 295–400 K) and pressures (Pi = 50–400 kPa) where they used a cylindrical combustion pump [5]. Three years later, Ajay Tripathi, H. Chandra and M. Agrawal studied the flame speed of LPG–air and LPG–air–diluent (CO_2) mixtures at different values of equivalence ratio and diluent concentration using two experimental methods: Bunsen burner and cylindrical tube [6]. In the same year, Domnina Razus, Venera Brinzea, Maria Mitu and Dumitru Oancea determined the laminar flame speeds of LPG–air and LPG–air–exhaust mixtures from pressure vs. time graphs obtained from a spherical container with central ignition, making use of a correlation based on the cubic law of the pressure increase during the initial phase of the explosion at different fuel/oxygen ratios and in different environmental conditions [7]. Later, in 2016, Ahmed Yasiry and Haroun Shahad carried out an experimental study of the laminar flame speed of Iraqi LPG using a constant volume chamber with central ignition at different initial pressures (0.1–0.3 MPa) and an initial temperature of 308 K. Likewise, a range of equivalence ratios from 0.8 to 1.3 were used [8]. Recently, in the year 2022, Bader Alfarraj, Ahmed Al-Harbi, Saud A. Binjuwair and Abdullah Alkhedhair carried out the characterization of the Saudi LPG using a Bunsen burner and a modified Bunsen burner, which allowed them to work in a range of equivalence ratios of 0.68 to 1.3 [9].

In this study, mathematical methodology coded in a MATLAB application was developed with the main idea of giving a contribution to the scientific community of an open code program in an easy tool that can be used by any researcher or student involved in this kind of study. It was found that most of the studies carried out by researchers on numerical methods are quite complex and do not include an interactive way to see, compare and analyze the results, which is one of the most valuable characteristics that the MATLAB application has. In addition, the application has the option to be improved by anyone by adding other functions or assumptions that can grow its range of application and accuracy. The MATLAB application was used for the determination of the main combustion characteristics of Saudi LPG (flame temperature, combustion products and laminar burning velocity) studied in the investigation by Bader Alfarraj, Ahmed Al-Harbi, Saud A. Binjuwair and Abdullah Alkhedhair [9]. The limitations of the application are that it can be only applied to hydrocarbon fuels, including oxygen and/or nitrogen or not; the flame temperature cannot exceed 3725 °C and cannot be lower than 327 °C; and the equivalence ratio cannot be too high so it does not allow the formation of free carbon. Moreover, as part of the research, simulation was carried out in Ansys Chemkin using the San Diego [10] and

RedSD [11] mechanisms to obtain these properties, with the aim of extending the research carried out by these authors by using a simple numerical method.

2. Materials and Methods

2.1. Materials and Procedure

A computer with an AMD Ryzen 7 5700 G with Radeon Graphics 3.80 GHz processor with 32 GB RAM and Windows 11 Enterprise operating system was used for coding in MATLAB version R2021b and simulation in Ansys Chemkin R2 2022. For the Ansys Chemkin simulation the San Diego mechanism and RedSD mechanism were used. The general procedure can be seen in Figure 1.

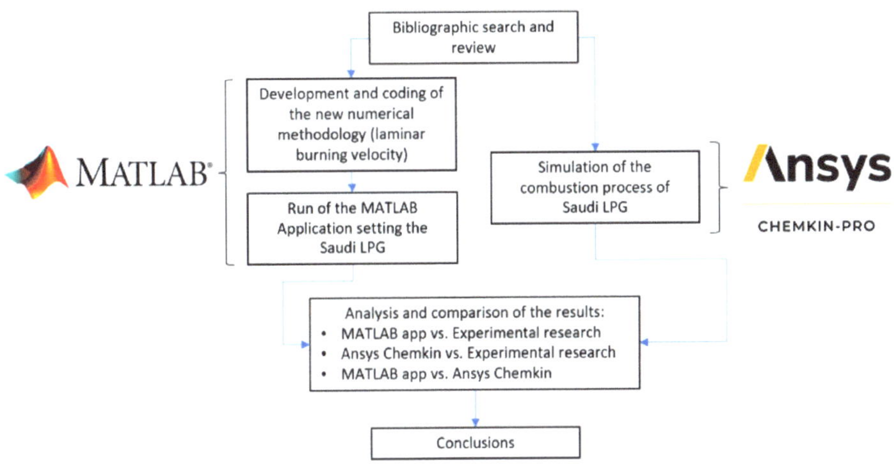

Figure 1. General procedure of the present work.

2.2. Numerical Methodology for the Determination of Combustion Products and Flame Temperature

For the determination of the combustion products and their molar fractions, the numerical methodology developed in FORTRAN by Olikara and Borman in 1975 [12] was coded in MATLAB, using numerical methods such as Taylor series, Newton–Raphson and Gaussian elimination. To start with the methodology, Equation (1) is proposed as the equation corresponding to the combustion reaction and starting point for the formulation of the following equations, with the fuel being $C_n + H_m + O_l + N_k$, the equivalence ratio Φ, $X_1 \rightarrow X_{12}$ as the mole fractions of the products and x_{13} as the number of moles from fuel that give 1 mol of products.

$$X_{13}[C_n + H_m + O_l + N_k + ((n + 0.25 * m - 0.5 * l)/\Phi)\,(O_2 + 3.7274N_2 + 0.0444Ar)] \rightarrow \\ X_1H + X_2O + X_3N + X_4H_2 + X_5OH + X_6CO + X_7NO + X_8O_2 + X_9H_2O + X_{10}CO_2 + X_{11}N_2 + X_{12}Ar \quad (1)$$

The reactant part can also be written as

$$X_{13}[nC + mH + rO_2 + r'N_2 + r''Ar], \quad (2)$$

with the equivalences for r, r′, r″ as

$$r = l/2 + r_o, \quad (3)$$

$$r' = k/2 + 3.7274\ r_o, \quad (4)$$

$$r'' = 0.0444\ r_o, \tag{5}$$

$$r_o = (n + 0.25 * m - 0.5 * l)/\Phi. \tag{6}$$

Setting the balance of the elements of the reaction and considering that all the products must add up to 1, five equations for each element are obtained: carbon, hydrogen, oxygen, nitrogen and argon, respectively, and an additional one for the sum of fractions of the products equivalent to unity.

$$X_6 + X_{10} = n * X_{13}, \tag{7}$$

$$X_1 + 2X_4 + X_5 + 2X_9 = m * X_{13}, \tag{8}$$

$$X_2 + X_5 + X_6 + X_7 + 2X_8 + X_9 + 2X_{10} = 2r * X_{13}, \tag{9}$$

$$X_3 + X_7 + 2X_{11} = 2r' * X_{13}, \tag{10}$$

$$X_{12} = r'' * X_{13}, \tag{11}$$

$$X_1 + X_2 + X_3 + X_4 + X_5 + X_6 + X_7 + X_8 + X_9 + X_{10} + X_{11} + X_{12} = 1. \tag{12}$$

In order to solve the equations system, 7 additional equations are required so 7 chemical reactions are considered in this situation, and they were selected for the benefit of the products that were set in Equation (1).

$$1/2H_2 \leftrightarrow H, \tag{13}$$

$$1/2\,O_2 \leftrightarrow O, \tag{14}$$

$$1/2\,N_2 \leftrightarrow N, \tag{15}$$

$$1/2\,H_2 + 1/2\,O_2 \leftrightarrow OH, \tag{16}$$

$$1/2\,O_2 + 1/2\,N_2 \leftrightarrow NO, \tag{17}$$

$$H_2 + 1/2\,O_2 \leftrightarrow H_2O, \tag{18}$$

$$CO + 1/2\,O_2 \leftrightarrow CO_2, \tag{19}$$

and each chemical reaction has a partial pressure equilibrium constant, which are presented in the same order as their chemical reaction:

$$K_1 = X_1 * p^{1/2}/{X_4}^{1/2}, \tag{20}$$

$$K_2 = X_2 * p^{1/2}/{X_8}^{1/2}, \tag{21}$$

$$K_3 = X_3 * p^{1/2}/{X_{11}}^{1/2}, \tag{22}$$

$$K_5 = X_5/(X_4{}^{1/2} * X_8{}^{1/2}), \quad (23)$$

$$K_7 = X_7/(X_8{}^{1/2} * X_{11}{}^{1/2}), \quad (24)$$

$$K_9 = X_9/(X_4 * X_8{}^{1/2} * p^{1/2}), \quad (25)$$

$$K_{10} = X_{10}/(X_6 * X_8{}^{1/2} * p^{1/2}). \quad (26)$$

It is important to mention that each of the equilibrium constants at partial pressure (Equations (20)–(26)) was adjusted in the range of 327 °C to 3725 °C using the tables and properties from JANAF Thermodynamical Tables 1985 [13], so the MATLAB application has this limitation. The general equation that adjusts all the constants and their respective coefficients can be found in the paper of Olikara and Borman [12]. Using the equations of the constants for each reaction and of Equations (20)–(26), the respective constants (C_n) can be set for each combustion product, which will remain for every equation with X_4, X_6, X_8 and X_{11} pressure as the only variables. In this way,

$$X_1 = C_1 * X_4{}^{1/2}, \text{ where } C_1 = K_1/p^{1/2}, \quad (27)$$

$$X_2 = C_2 * X_8{}^{1/2}, \text{ where } C_2 = K_2/p^{1/2}, \quad (28)$$

$$X_3 = C_2 * X_8{}^{1/2}, \text{ where } C_3 = K_3/p^{1/2}, \quad (29)$$

$$X_5 = C_5 * X_4 * X_8{}^{1/2}, \text{ where } C_5 = K_5, \quad (30)$$

$$X_7 = C_7 * X_8{}^{1/2} * X_{11}{}^{1/2}, \text{ where } C_7 = K_7, \quad (31)$$

$$X_9 = C_9 * X_4 * X_8{}^{1/2}, \text{ where } C_9 = K_9 * p^{1/2}, \quad (32)$$

$$X_{10} = C_{10} * X_6 * X_8{}^{1/2}, \text{ where } C_{10} = K_{10} * p^{1/2}, \quad (33)$$

are the equations to set the equation system. In order to set the system, it is necessary to define

$$X_{12} = r''X_{13} = r''(X_6 + X_{10})/n, \quad (34)$$

using (7) and (11). With (34) it is possible to eliminate X_{12} and X_{13} from (7)–(12) and form

$$X_1 + 2X_4 + X_5 + 2X_9 - m * (X_6 + X_{10})/n = 0, \quad (35)$$

$$X_2 + X_5 + X_6 + X_7 + X_9 + 2X_{10} - 2r * (X_6 + X_{10})/n = 0, \quad (36)$$

$$X_3 + X_7 + 2X_{11} - 2r'/n\,(X_6 + X_{10}) = 0, \quad (37)$$

$$X_1 + X_2 + X_3 + X_4 + X_5 + X_6 + X_7 + X_8 + X_9 + X_{10} + X_{11} + r''(X_6 + X_{10})/n - 1 = 0. \quad (38)$$

Using (27)–(33) in (35)–(38) it is possible to obtain the system equation with just 4 variables (X_4, X_6, X_8 and X_{11}). The equation system is a non-linear system with 4 variables whose general representation is

$$f_j\ (X_4, X_6, X_8, X_{11}) = 0, j = 1, 2, 3, 4. \tag{39}$$

In order to solve the equation system, a Taylor series is applied to linearize the system. For applying the Taylor series, it is a requisite to know a vector,

$$[X_4^{(1)}, X_6^{(1)}, X_8^{(1)}, X_{11}^{(1)}], \tag{40}$$

that is near to the solution vector

$$[X_4^*, X_6^*, X_8^*, X_{11}^*], \tag{41}$$

For each of the equations on the left side of the system (Equation (39)), they can be expanded around the solution vector to obtain

$$\Delta X_i = X_i^* - X_i^{(1)}, i = 4, 6, 8, 11. \tag{42}$$

Deriving each expression and ignoring partial derivatives of which the order is higher than 1, it is possible to obtain the linear equation

$$f_j + (\partial f_j/\partial X_4)\ \Delta X_4 + (\partial f_j/\partial X_6)\ \Delta X_6 + (\partial f_j/\partial X_8)\ \Delta X_8 + (\partial f_j/\partial X_{11})\ \Delta X_{11} \cong 0, j = 1, 2, 3, 4, \tag{43}$$

in which the derivatives are evaluated in the known vector. Using the Gauss elimination method, it is possible to solve the new linear system and obtain the values of ΔX_4, ΔX_6, ΔX_8, ΔX_{11} that will be useful to bring the known vector closer to the solution vector by applying

$$X_i^{(2)} = X_i^{(1)} + \Delta X_i, i = 4, 6, 8, 11. \tag{44}$$

in which $X_i^{(2)}$ is the improved vector which is entered into (43) in order to obtain another improvement. This process is repeated the necessary times until the values of ΔX_4, ΔX_6, ΔX_8, ΔX_{11} are less than or equal to 0.0001, which is the maximum error margin considered for each molar fraction.

Moreover, for the calculation of the adiabatic flame temperature, the equivalence between the reactants and the products is applied as

$$h(T, p_o, F_o) - h_r(T_o, p_o, F_o) = 0. \tag{45}$$

It is necessary to obtain an improved temperature, so the Newton–Raphson is defined as

$$T_{n+1} = T_n - (h(T_n, p_o, F_o) - h_r)/(\partial h/\partial T)_n. \tag{46}$$

in which T_n is the first assumed temperature (preferably greater than the final temperature), h_r is the reactant enthalpy at initial conditions of temperature (T_o), pressure (p_o) and equivalence ratio (F_o), $h(T_n, p_o, F_o)$ is obtained once the combustion products are calculated at the first assumed temperature (T_n) as

$$h(T_n, p_o, F_o) = \sum X_i h_i/M, \tag{47}$$

this last equation can be derived with respect to temperature and it will give

$$\partial h/\partial T = 1/M\ [\sum(X_i\ dh_i/dT + h_i\ dX_i/dT) - \partial M/\partial T\ h(T_n, p_o, F_o)], \tag{48}$$

in which dh_i/dT is the specific heat at a constant pressure of each element, so

$$dh_i/dT = Cp_i, \tag{49}$$

where dX_i/dT is calculated using (27)–(33) and $\partial M/\partial T$ is the molar mass of the mixture with respect to the temperature, that is,

$$\partial M/\partial T = \sum dX_i/dT\ (M_i). \tag{50}$$

Finally, the Newton–Raphson iteration concludes when

$$(h(T_n, p_o, F_o) - h_r)/(\partial h/\partial T)_n \leq 1 \tag{51}$$

In order to start this procedure, it is necessary to know the vector (40) that is near to the solution vector (41), and the steps to obtain it are mentioned in detail in the paper of Olikara and Borman [11]. Furthermore, all the partial derivatives with respect to temperature, pressure and equivalence ratio of each element can be found in the same source.

2.3. *Determination of the Laminar Burning Velocity*

For the determination of the laminar burning velocity, the adiabatic flame temperature previously calculated is directly replaced in

$$S_L = (e^{-Ea/RuT})^{1/2} \tag{52}$$

that corresponds to the Mallard and Le Chateller theory. Ea is the activation energy, R_u is the universal gas constant (1.987 cal/mol·K) and T is the flame temperature calculated using the method. The value for Ea for every calculation is taken from the average of two values. The first one is set according to Kenneth Kuo, who said that for most hydrocarbon reactions the energy of activation is around 120 kJ/mol [2]. The second one is from Markatou, who studied methane–air oxidation and determined an activation energy of approximately 130 kJ/mol [14], so the average is 125 kJ/mol or 29,675.7 cal/mol.

2.4. *Development of the MATLAB Application*

2.4.1. First Part: New_Code.m

All of the procedure and steps described in Sections 2.2 and 2.3 were codified in the MATLAB application to be directly used just by setting the fuel and, in some cases, the diluents available (CO_2). The first part of the code corresponds to the archive called New_code.m, which can be found in Code S1, having defined as inputs the equivalence ratio, percentage of diluent if applicable, name of diluent and name of fuel. The indices of each element of the selected fuel (C, H, O, N) are stored in an Excel archive called Reactants_Enthalpy.xlsx and those of the diluents in Reactants_Diluents.xlsx, and they can be found in Archives S1 and S2. In order to improve the application range of the MATLAB application, it is possible to add more fuels and diluents to Archives S1 and S2.

Moreover, the determination of the reactant enthalpy (h_r) is carried out, for which the initial temperature and pressure of work will always be 298 K and 1 atm, respectively. After the first steps are carried out, New_code.m enters into a loop in order to obtain the combustion products by calling the function Fractions_Derivatives.m and, after it, the first assumed temperature is adjusted with the value calculated in (51) or "DELTA" which is the name used in the code.

In case the DELTA is less than or equal to 1, the laminar burning velocity is calculated and the iteration process finished. On the other hand, in case is not is not less than or equal to 1, the loop continues adjusting the combustion products and the adiabatic flame temperature.

2.4.2. Second Part: Fractions_Derivatives.m

This second part corresponds to the archive called Fractions_Derivatives.m, whose code can be found in Code S2, which determines the twelve combustion products by setting a matrix equation system and solving it by using a for loop in order to apply the Gaussian elimination method using row interchange. In other words, it uses Equations (3)–(6) to carry out the simplification of the combustion reaction elements, Equations (7)–(12) for the balance of the elements and Equations (20)–(33) for setting all the molar fractions with respect to X_4, X_6, X_8 and X_{11} using the 7 chemical reactions as a starting point (Equations (13)–(19)). Additional to this, Equations (35)–(38) are set in order to obtain the equation system (39) in which the Taylor series is applied to linearize the system and the Gaussian elimination method is applied to solve the linearized system.

By the same method, a matrix equation system is set and solve for the derivatives of each molar fraction with respect to temperature, equivalence ratio and pressure using a maximum pivot strategy.

When the molar fractions are adjusted, the product enthalpy, the derivative of the enthalpy with respect to temperature and the combustion products' molar fractions are given to New_code.m that calculates by using the Newton–Raphson method in the enthalpy equation balance, that is, Equation (45), in order to obtain

$$(\partial h/\partial T)_n, \tag{53}$$

The process runs until it fulfills the tolerance for the final flame temperature and laminar burning velocity is calculated. For solving both matrix equation systems, constants and some variables are set in order to form the system, such as the partial pressure equilibrium constant for each reaction. Archive S4 (Products.xlsx) includes the thermodynamical properties of the combustion products in order to obtain the correct DELTA.

2.4.3. Third Part: Developing the MATLAB Application in the MATLAB App Designer Interface

In order to create the app archive, the App Designer interface is used. The MATLAB code of the application can be found in Code S3. It is possible to see in Figure 2 some tools and buttons that will help the user to obtain and analyze the information. Table 1 details the information about each tool available in the application.

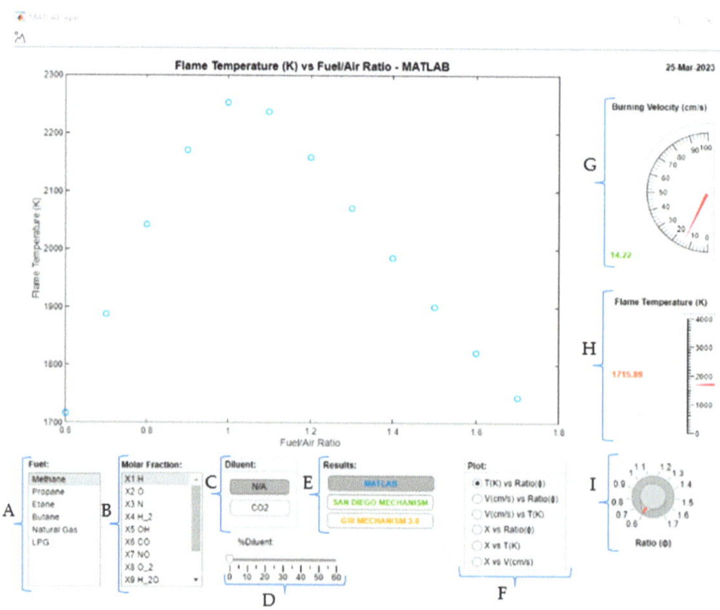

Figure 2. Identification of tools available in the application: (A) Fuel list; (B) Molar fraction list; (C) Diluent button list; (D) Percentage bar of diluent; (E) Results button list; (F) Plot button list; (G) Equivalence ratio knob; (H) Laminar burning velocity gauge; (I) Flame temperature gauge.

Table 1. Tools available in the MATLAB application.

Position in Figure 2	Name of Tool	Description	Type of Selection
A	Fuel list	List of fuels available for calculus in the application	Unique
B	Molar fraction list	List of molar fractions available for plotting results	Unique
C	Diluent button list	List of diluents available to use in the calculus	Unique
D	Percentage bar of diluent	Percentage of diluent by volume to be considered in fuel	Unique
E	Results button list	Results from resources available to be shown in the plot	Multi
F	Plot button list	Type of plot to be shown on the screen	Unique
G	Equivalence ratio knob	Knob that shows the value of laminar burning velocity and flame temperature in their respective gauges	Unique
H	Laminar burning velocity gauge	Laminar burning velocity value at the knob equivalence ratio value selected	NA
I	Flame temperature gauge	Flame temperature value at the knob equivalence ratio value selected	NA

Figure 3 gives an example of the fuel and results selection; it shows the plot results for the molar fraction of the hydrogen (H_2) vs. equivalence ratio for the LPG with 10% CO_2 obtained just by two resources (MATLAB code and San Diego mechanism). The equivalence ratio knob is also set at the stoichiometric value so the flame temperature and laminar burning velocity values are shown for that ratio value in their respective gauges (H and I).

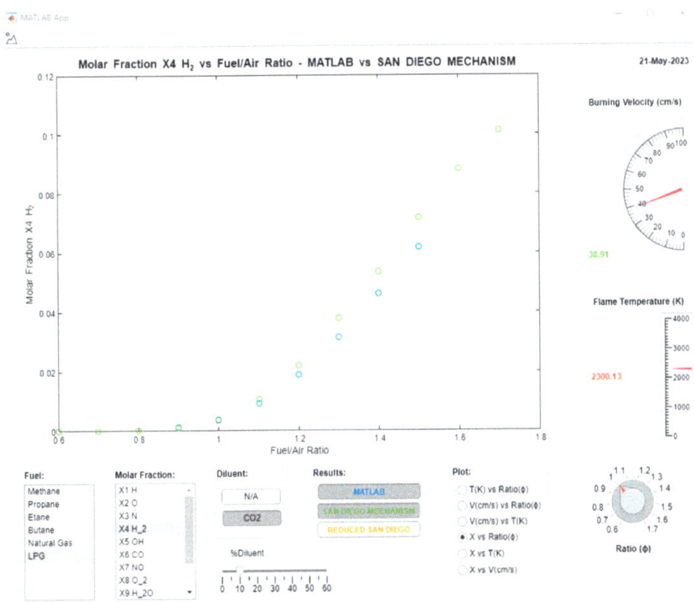

Figure 3. Results for the molar fraction of hydrogen (X_4-H_2) vs. equivalence ratio for the LPG with 10% CO_2 determined by MATLAB code (blue circles) and San Diego mechanism (green circles).

2.5. Composition of Fuel and Mixtures to Be Tested

The fuel to be analyzed and set in the application as a newly available option is the Saudi LPG, whose composition is 50% propane (C_3H_8) and 50% butane (C_4H_{10}) and is the same used by Bader Alfarraj et al. [9]. The tested mixtures are from 0.6 to 1.7, increasing by 0.1. As was mentioned previously, the initial conditions are 298 K for temperature and 1 atm of pressure.

2.6. Simulation in Ansys Chemkin

The simulation is carried out in Ansys Chemkin 2022 R2 using the San Diego mechanism (57 species) and RedSD mechanism (47 species). The reason for choosing these two mechanisms is the good agreement for simulations using propane and butane. Both follow the idea of having a small number of species and reactions for the scope of combustion, and this gives accurate results. In addition to this, the limitation of 70 species in the Ansys Chemkin 2022 R2 has an influence. Reduced San Diego (RedSD) is a reduced mechanism developed in the Department of Mechanical Engineering, Indian Institute of Technology Madras, Chennai, India from the San Diego mechanism [10] by S.M. Kumaran, D. Shanmugasundaram, K. Narayanaswamy and V. Raghavan in 2021 [11]. This compact mechanism was comprehensively validated against experimental data on flames using 1D and 2D computations. The 1D computations of premixed and non-premixed flames are carried out using Flame Master and Chemkin-Pro, while 2D axisymmetric calculations are performed within Ansys Fluent, taking into account multicomponent diffusion, thermal diffusion and radiation sub-models [11]. In general, the results obtained show that the 45 species mechanism is able to predict the desired combustion characteristics in different types of flames of propane, n-butane and their mixtures satisfactorily [11]. The principal parameters of initial temperature, pressure of work and equivalence ratios are configured with the same values of the calculations in the MATLAB application. Figure 4 presents the flow diagram in the Ansys Chemkin software for the simulation. All the results are in Ansys Chemkin Results (0.6–1.7).xlsx, which can be found in Archive S3.

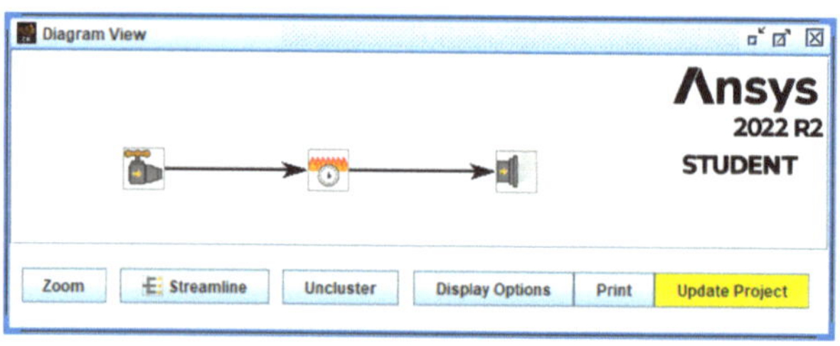

Figure 4. Diagram in Ansys Chemkin for the simulation.

3. Results and Discussion

The results for the numerical method run by the MATLAB application are presented for the twelve combustion products, flame temperature and laminar burning velocity and compared with the Ansys Chemkin simulation.

3.1. Combustion Products

The numerical method run by the MATLAB application for the mixtures of Saudi LPG gives the molar fractions of the combustion products values presented in Table 2 (X_1–X_6) and Table 3 (X_7–X_{12}).

It is possible to see from Tables 2 and 3 that the most predominant fractions in lean mixtures are X_8O_2, X_9H_2O, $X_{10}CO_2$ and $X_{11}N_2$. This is an expected result and it is the first proof that the MATLAB application method agrees with the general combustion theory statements (lean mixture, excess of oxygen, greater percentage of oxygen in the products). As the equivalence ratio increases, X_8O_2 decreases and $X_{10}CO_2$ decreases while X_6CO

increases, which agrees with rich mixtures, in other words, there is not enough oxygen for the combustion process. Considering Equation (19), it is not possible to transform CO to CO_2. Moreover, the results obtained by carrying out the simulation in Ansys Chemkin using the San Diego mechanism are presented in Table 4 (X_1–X_6) and Table 5 (X_7–X_{12}). It is important to mention that there were more combustion products in the results given by Ansys Chemkin such as propane and butane that did not manage to react and this can be seen principally at higher equivalence ratios (1.2–1.7). The results using RedSD mechanism are presented in Table 6 (X_1–X6) and Table 7 (X_7–X12).

Table 2. Combustion Products of Saudi LPG (X_1–X_6) calculated by MATLAB application.

Equivalence Ratio (Φ)	X_1H	X_2O	X_3N	X_4H_2	X_5OH	X_6CO
0.6	9.2644×10^{-7}	2.3773×10^{-5}	1.6175×10^{-11}	1.2640×10^{-5}	0.00041207	3.5819×10^{-5}
0.7	1.0249×10^{-5}	0.00010408	3.4726×10^{-10}	8.5230×10^{-5}	0.0011626	0.00028795
0.8	6.5024×10^{-5}	0.00028288	3.4003×10^{-9}	0.00039475	0.00240522	0.0015017
0.9	0.00025914	0.00048031	1.6695×10^{-8}	0.00138214	0.00369457	0.00556398
1	0.00066624	0.00046349	4.0295×10^{-8}	0.00393763	0.00401577	0.0154476
1.1	0.00108218	0.00021577	4.3271×10^{-8}	0.00962653	0.00281794	0.0331991
1.2	0.00116468	5.8248×10^{-5}	2.3838×10^{-8}	0.0194133	0.00141014	0.0549746
1.3	0.00099712	1.3753×10^{-5}	9.9015×10^{-9}	0.0323431	0.00063771	0.075101
1.4	0.00075851	3.2136×10^{-6}	3.6705×10^{-9}	0.047353	0.00028167	0.0922573
1.5	0.00053636	7.5281×10^{-7}	1.2757×10^{-9}	0.0636388	0.00012302	0.106798
1.6	0.00035653	1.7192×10^{-7}	4.1393×10^{-10}	0.0806081	5.2459×10^{-5}	0.119274
1.7	0.00022589	3.8207×10^{-8}	1.2688×10^{-10}	0.097815	2.1858×10^{-5}	0.13018

Table 3. Combustion Products of Saudi LPG (X_7–X_{12}) calculated by MATLAB application.

Equivalence Ratio (Φ)	X_7NO	X_8O_2	X_9H_2O	$X_{10}CO_2$	$X_{11}N_2$	$X_{12}AR$
0.6	0.00232901	0.0786287	0.0935573	0.0729014	0.743231	0.00886708
0.7	0.00353978	0.0575011	0.107843	0.084112	0.73656	0.00879482
0.8	0.00436595	0.0373029	0.121284	0.094098	0.729583	0.00871664
0.9	0.00425856	0.0196666	0.133438	0.100834	0.721799	0.00862329
1	0.00300759	0.0071197	0.143566	0.101098	0.712176	0.0085012
1.1	0.0013496	0.00141652	0.150092	0.0925431	0.699319	0.00833818
1.2	0.00044049	0.00019234	0.151587	0.0790268	0.683587	0.00814537
1.3	0.00013844	2.6966×10^{-5}	0.148982	0.0665658	0.667245	0.0079489
1.4	4.4765×10^{-5}	4.1781×10^{-6}	0.143582	0.056652	0.651305	0.00775848
1.5	1.4838×10^{-5}	6.9691×10^{-7}	0.136333	0.0489914	0.635987	0.00757583
1.6	4.9101×10^{-6}	1.1887×10^{-7}	0.127913	0.0430682	0.621321	0.00740107
1.7	1.6125×10^{-6}	2.0403×10^{-8}	0.118823	0.0384122	0.607287	0.00723388

Table 4. Combustion Products of Saudi LPG (X_1–X_6) using San Diego mechanism.

Equivalence Ratio (Φ)	X_1H	X_2O	X_3N	X_4H_2	X_5OH	X_6CO
0.6	6.6093×10^{-7}	1.8192×10^{-5}	3.5006×10^{-12}	1.0017×10^{-5}	0.00039345	2.7661×10^{-5}
0.7	8.3166×10^{-6}	8.693×10^{-5}	1.3135×10^{-10}	7.4804×10^{-5}	0.00117059	0.00024538
0.8	4.9654×10^{-5}	0.00022725	1.4973×10^{-9}	0.00033309	0.00236175	0.00123292
0.9	0.00019441	0.00038015	7.3197×10^{-9}	0.00114976	0.00356367	0.00454945
1	0.00055523	0.00032513	1.6402×10^{-8}	0.00388401	0.00365408	0.0149326
1.1	0.00085203	0.00010131	8.8505×10^{-9}	0.0107692	0.00205185	0.0351553
1.2	0.00083450	2.1024×10^{-5}	2.6195×10^{-9}	0.0221748	0.00088483	0.058047
1.3	0.00064603	3.7060×10^{-6}	5.9878×10^{-10}	0.0380662	0.00034043	0.0792958
1.4	0.00047560	8.6550×10^{-7}	1.3136×10^{-10}	0.0536471	0.00014997	0.0948882
1.5	0.00031061	1.7192×10^{-7}	1.7254×10^{-11}	0.0717481	6.0068×10^{-5}	0.109207
1.6	0.00020575	4.1360×10^{-8}	1.9497×10^{-12}	0.0882793	2.6349×10^{-5}	0.120062
1.7	0.00013802	1.1794×10^{-8}	2.8318×10^{-13}	0.101292	1.2937×10^{-5}	0.12784

Table 5. Combustion Products of Saudi LPG (X_7–X_{12}) using San Diego mechanism.

Equivalence Ratio (Φ)	X_7NO	X_8O_2	X_9H_2O	$X_{10}CO_2$	$X_{11}N_2$	$X_{12}AR$
0.6	1.2659×10^{-6}	0.0778787	0.0957267	0.0734329	0.0743441	0.00906312
0.7	1.0758×10^{-5}	0.0564053	0.111239	0.0849199	0.73685	0.00898358
0.8	7.4972×10^{-5}	0.0364389	0.124853	0.0951511	0.730366	0.00890481
0.9	0.00028322	0.0190562	0.136546	0.102239	0.723214	0.00881881
1	0.00036887	0.00517981	0.147765	0.102416	0.712228	0.00868567
1.1	8.0115×10^{-5}	0.00060559	0.15336	0.0911168	0.697399	0.00850336
1.2	1.8270×10^{-5}	6.1474×10^{-5}	0.153406	0.076219	0.680035	0.0082915
1.3	3.9921×10^{-6}	6.2188×10^{-6}	0.149864	0.0628266	0.660878	0.00805869
1.4	7.1151×10^{-7}	9.9641×10^{-7}	0.143506	0.0540122	0.645431	0.00786988
1.5	8.0017×10^{-8}	1.4057×10^{-7}	0.135334	0.0466537	0.628933	0.00766906
1.6	2.5797×10^{-8}	2.5968×10^{-8}	0.127036	0.0415919	0.614249	0.00749027
1.7	1.1576×10^{-8}	6.0938×10^{-9}	0.119289	0.0384194	0.602665	0.00734881

Table 6. Combustion Products of Saudi LPG (X_1–X_6) using RedSD mechanism.

Equivalence Ratio (Φ)	X_1H	X_2O	X_3N	X_4H_2	X_5OH	X_6CO
0.6	6.9492×10^{-7}	1.8833×10^{-5}	3.5006×10^{-12}	1.0326×10^{-5}	0.00039894	2.8759×10^{-5}
0.7	8.4643×10^{-6}	8.8159×10^{-5}	1.3135×10^{-10}	7.5337×10^{-5}	0.00117376	0.00024771
0.8	5.5236×10^{-5}	0.00024429	1.4973×10^{-9}	0.00035698	0.00244458	0.00132638
0.9	0.00022744	0.00041572	7.3197×10^{-9}	0.0012917	0.00374339	0.00510691
1	0.00059085	0.00034823	1.6402×10^{-8}	0.00399172	0.00377842	0.015321
1.1	0.00088504	0.00010743	8.8505×10^{-9}	0.0108747	0.00211538	0.0354616
1.2	0.00084480	2.0644×10^{-5}	2.6195×10^{-9}	0.0226464	0.00087630	0.0588178
1.3	0.00066097	3.9565×10^{-6}	5.9878×10^{-10}	0.0377389	0.00035226	0.0789434
1.4	0.00047605	8.4477×10^{-7}	1.3136×10^{-10}	0.0542954	0.00014795	0.0952242
1.5	0.00032336	1.9393×10^{-7}	1.7254×10^{-11}	0.0706793	6.4263×10^{-5}	0.108435
1.6	0.00021240	4.3565×10^{-8}	1.9497×10^{-12}	0.0887154	2.7351×10^{-5}	0.120091
1.7	0.00014340	1.2649×10^{-8}	2.8318×10^{-13}	0.101579	1.3416×10^{-5}	0.127924

Table 7. Combustion Products of Saudi LPG (X_7–X_{12}) using RedSD mechanism.

Equivalence Ratio (Φ)	X_7NO	X_8O_2	X_9H_2O	$X_{10}CO_2$	$X_{11}N_2$	$X_{12}AR$
0.6	1.2659×10^{-6}	0.0778787	0.0957267	0.0734329	0.0743441	0.00906312
0.7	1.0758×10^{-5}	0.0564053	0.111239	0.0849199	0.73685	0.00898358
0.8	7.4972×10^{-5}	0.0364389	0.124853	0.0951511	0.730366	0.00890481
0.9	0.00028322	0.0190562	0.136546	0.102239	0.723214	0.00881881
1	0.00036887	0.00517981	0.147765	0.102416	0.712228	0.00868567
1.1	8.0115×10^{-5}	0.00060559	0.15336	0.0911168	0.697399	0.00850336
1.2	1.8270×10^{-5}	6.1474×10^{-5}	0.153406	0.076219	0.680035	0.0082915
1.3	3.9921×10^{-6}	6.2188×10^{-6}	0.149864	0.0628266	0.660878	0.00805869
1.4	7.1151×10^{-7}	9.9641×10^{-7}	0.143506	0.0540122	0.645431	0.00786988
1.5	8.0017×10^{-8}	1.4057×10^{-7}	0.135334	0.0466537	0.628933	0.00766906
1.6	2.5797×10^{-8}	2.5968×10^{-8}	0.127036	0.0415919	0.614249	0.00749027
1.7	1.1576×10^{-8}	6.0938×10^{-9}	0.119289	0.0384194	0.602665	0.00734881

Figure 5a–f show the behavior differences between the results from the MATLAB application and the simulation in Ansys Chemkin from X_1 to X_6, while Figure 6a–f show the results from X_7 to X_{12}. From Figure 5d,f it can be affirmed for all cases that only hydrogen (X_4H_2) and carbon monoxide (X_6CO) had an increase as the equivalence ratio increased. This is mainly a consequence of the increase in the amount of fuel in the mixture and the lack of oxygen in it to finish transforming these variables into hydrogen hydroxide (X_9H_2O) and carbon dioxide ($X_{10}CO_2$), respectively, which is shown by the fall in the fractions corresponding to those products at a higher equivalence ratio. On the other hand, nitrogen ($X_{11}N_2$) and argon ($X_{12}AR$) presented a decrease for all cases as the equivalence ratio increased because of the opposite effect that this change has on the elements that compound the oxidization, in this case, air.

Likewise, another product that presents the same trend, but in a more predominant way, is oxygen (X_8O_2), whose value, in cases of a higher ratio, tends to practically zero. Similarly, in the case of hydroxide (X_5OH), there is a tendency to zero for its values at the highest equivalence ratio and this is due to the lack of molecular oxygen to complete the whole reaction (16). Regarding the similarity and difference in the results for all cases, in general, there exists a correlation between those determined by the MATLAB application and simulation in Ansys Chemkin using the San Diego mechanism and RedSD mechanism, specifically for X_4H_2, X_5OH, X_6CO, X_8O_2, X_9H_2O, $X_{10}CO_2$, $X_{11}N_2$, $X_{12}AR$, whose values presented the same evolutionary curve between all the sources as the value of the equivalence ratio varied.

Furthermore, taking as a reference the results of the simulation in Ansys Chemkin using the San Diego mechanism, because it is the most complete mechanism (greater number of species and reactions) of the two tested, the smallest differences for the cases X_9, X_{10}, X_{11}, X_{12} were found, being 3.05%, 5.95%, 1.15%, 2.22%, respectively.

3.2. Flame Temperature

Saudi LPG/air mixtures tested using the MATLAB application and simulation in Ansys Chemkin using San Diego and RedSD mechanisms gave the results presented in Table 8 for flame temperature.

Figure 7 shows the results for the flame temperature and the differences between the three methods. It was found that the flame temperatures of the mixtures tested using the MATLAB application were higher than those simulated in Ansys Chemkin using San Diego and RedSD mechanisms. In addition to this, it can be observed that the flame temperature was the maximum in the stoichiometric mixture for the Ansys Chemkin test cases, however, in the case of the MATLAB application, the maximum value was found for an equivalence ratio of 1.1.

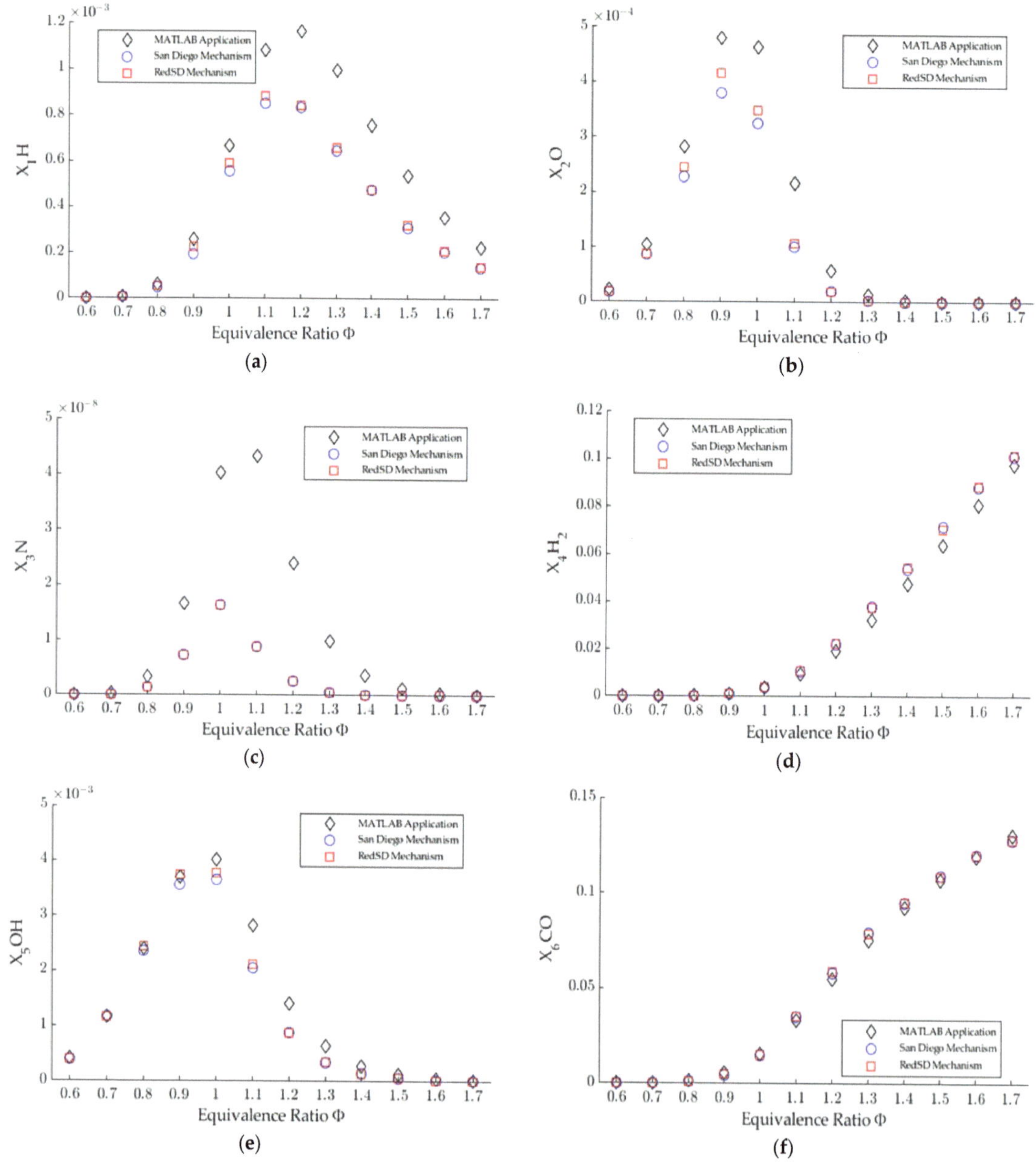

Figure 5. Results for molar fractions from X_1–X_6: (**a**) Molar fraction versus equivalence ratio of X_1H; (**b**) molar fraction versus equivalence ratio of X_2O; (**c**) molar fraction versus equivalence ratio of X_3N; (**d**) molar fraction versus equivalence ratio of X_4H_2; (**e**) molar fraction versus equivalence ratio of X_5OH; (**f**) molar fraction versus equivalence ratio of X_6CO.

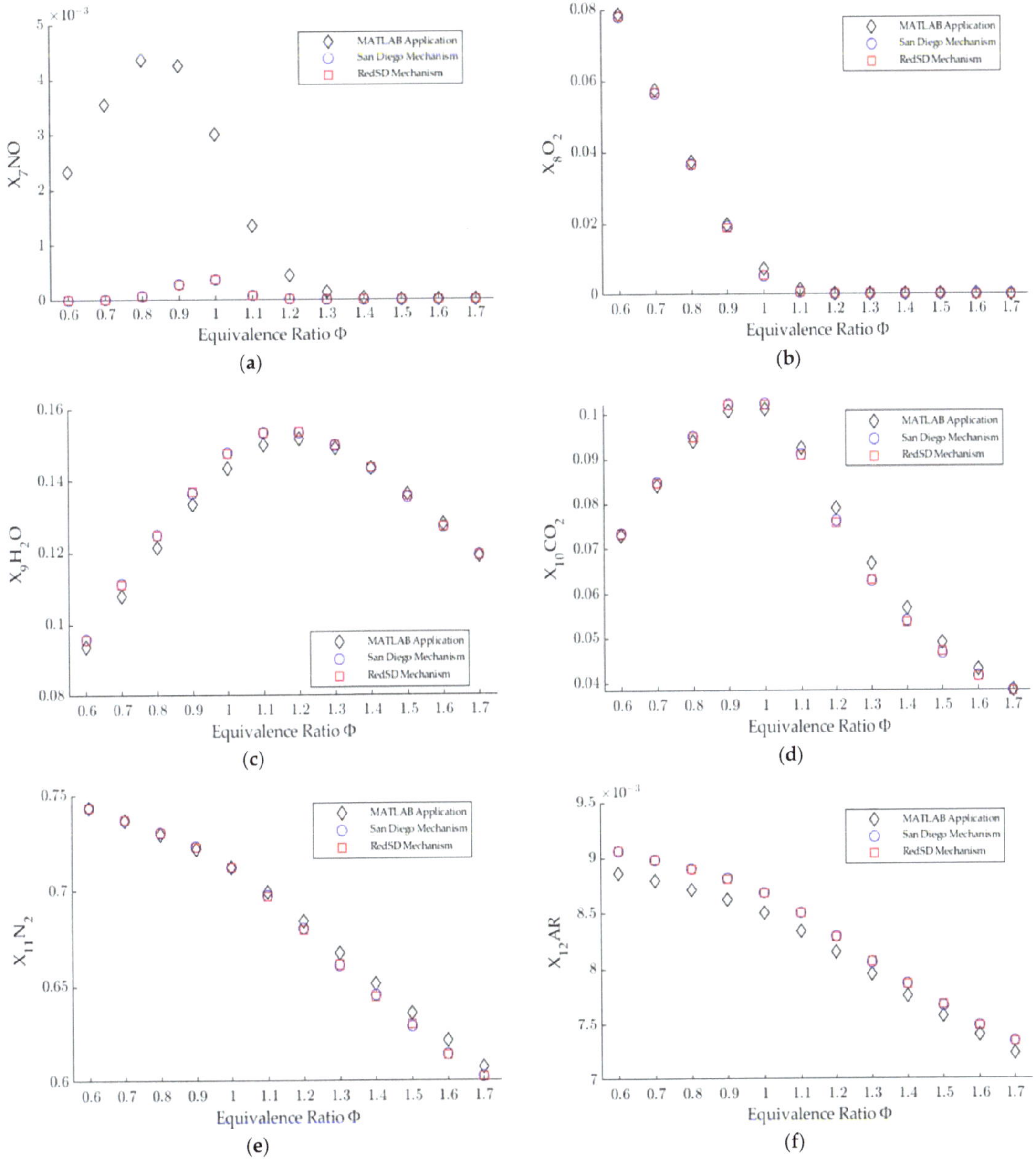

Figure 6. Results for molar fractions from X_7–X_{12}: (**a**) Molar fraction versus equivalence ratio of X_7NO; (**b**) molar fraction versus equivalence ratio of X_8O_2; (**c**) molar fraction versus equivalence ratio of X_9H_2O; (**d**) molar fraction versus equivalence ratio of $X_{10}CO_2$; (**e**) molar fraction versus equivalence ratio of $X_{11}N_2$; (**f**) molar fraction versus equivalence ratio of $X_{12}AR$.

The gap between the results from the MATLAB application and simulation in Ansys Chemkin is correct according to the MATLAB application calculation of the adiabatic temperature, considering a balance between enthalpies of reactants and products, while Ansys Chemkin simulation included the loss of heat from the mixture to the to the burner walls by convection, the subsequent loss by convection of the walls to the outside and the heat loss from the flame to the outside, therefore, the flame temperature determined by the two mechanisms must be lower. Moreover, it is possible to validate the formation of N and NO that was presented in Figures 5c and 6a, respectively, which occurs at higher temperatures, and therefore a higher fraction was obtained for the results using the MATLAB application in comparison with the simulation in Ansys Chemkin. In addition, the absolute difference and its percentage between the results for the three cases are presented in Table 9. It can be affirmed that, on average, there was a difference of 57.89 K and 2.86% for the results between the MATLAB application and those obtained through simulation in Ansys Chemkin using the San Diego mechanism. In addition to this, an average difference of 57.7 K and 2.85% was obtained between the MATLAB application results and simulation in Ansys Chemkin using the RedSD mechanism.

Table 8. Results for flame temperature.

Equivalence Ratio (Φ)	MATLAB Application T (K)	San Diego Mechanism T (K)	RedSD Mechanism T (K)
0.6	1763.1	1723.7	1721.8
0.7	1945.7	1914.3	1911.5
0.8	2108.1	2076.8	2075.6
0.9	2238.6	2202.8	2206.0
1	2318.5	2283.9	2283.3
1.1	2326.2	2267.3	2268.9
1.2	2272.7	2198.0	2196.9
1.3	2197.8	2109.6	2111.1
1.4	2118.8	2033.4	2031.4
1.5	2040.1	1956.1	1961.1
1.6	1963.1	1889.2	1889.0
1.7	1888.0	1831.0	1831.7

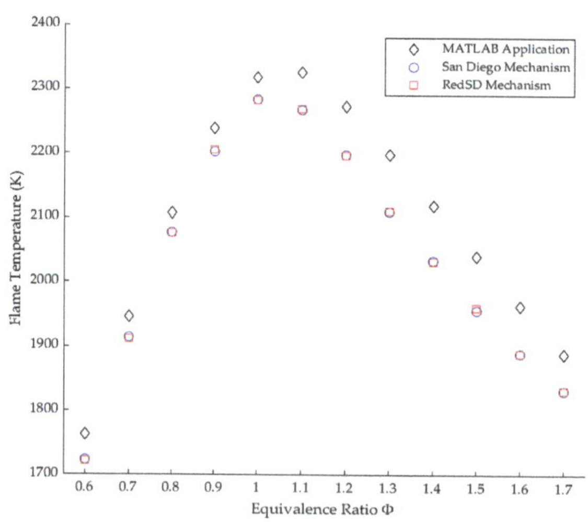

Figure 7. Results for flame temperature (K) versus equivalence ratio.

For the San Diego mechanism, the minimum difference was found at an equivalence ration of 0.8 with a difference value of 31.21 K and a percentage of 1.50%, while the maximum difference was found at an equivalence ratio of 1.3 with a difference value of 88.19 K and a percentage of 4.18%. For the RedSD mechanism, the minimum difference was found at an equivalence ratio of 0.8 with a difference value of 32.41 K and a percentage of 1.56%, while the maximum difference was found at an equivalence ratio of 1.4 with a difference value of 87.42 K and a percentage of 4.30%. With these two statements, one for each mechanism, it can be affirmed for flame temperature calculations that the MATLAB application has a percentage of confidence greater than or equal to 95%.

Table 9. Flame temperature differences between the MATLAB application and simulations in Ansys Chemkin.

Equivalence Ratio (Φ)	MATLAB Application—San Diego Mechanism		MATLAB Application—RedSD Mechanism	
	ΔT (K)	%	ΔT (K)	%
0.6	39.4	2.2	41.2	2.4
0.7	31.3	1.6	34.1	1.7
0.8	31.2	1.5	32.4	1.5
0.9	35.8	1.6	32.6	1.4
1	34.5	1.5	35.1	1.5
1.1	58.5	2.6	57.3	2.5
1.2	74.7	3.4	75.8	3.4
1.3	88.1	4.1	86.7	4.1
1.4	85.4	4.2	87.4	4.3
1.5	84.0	4.3	79.0	4.0
1.6	73.8	3.9	74.0	3.9
1.7	57.0	3.1	56.3	3.0

3.3. Laminar Burning Velocity

Saudi LPG/air mixtures tested using the MATLAB application and simulation in Ansys Chemkin using San Diego and RedSD mechanisms gave the results presented in Table 10 for laminar burning velocity.

Table 10. Results for laminar burning velocity.

Equivalence Ratio (Φ)	MATLAB Application S_L (cm/s)	San Diego Mechanism S_L (cm/s)	RedSD Mechanism S_L (cm/s)
0.6	14.4	16.7	17.0
0.7	21.5	26.3	26.4
0.8	28.9	33.8	33.9
0.9	35.5	38.6	39.1
1	39.9	41.7	41.6
1.1	40.3	41.3	40.9
1.2	37.4	37.5	36.4
1.3	33.4	29.0	27.3
1.4	29.4	19.5	17.1
1.5	25.7	12.9	12.2
1.6	22.2	9.9	9.5
1.7	19.1	8.0	7.7

From Figure 8, it can be seen that with the increase in equivalence ratio until the stoichiometric condition, the adiabatic temperature and laminar burning velocity increase and, afterward, the adiabatic temperature and laminar burning velocity start decreasing according to the results obtained by using the MATLAB application. On the other hand, with the increase in equivalence ratio until 1.1, the adiabatic temperature and laminar burning velocity increase and, afterward, the adiabatic temperature and laminar burning velocity start decreasing according to the results obtained by simulation in Ansys Chemkin using San Diego and RedSD mechanisms. Furthermore, according to the results for the laminar burning velocity and the differences between the three methods, it was found that the laminar burning velocities of the mixtures tested using the MATLAB application were lower than those simulated in Ansys Chemkin for equivalence ratios less than or equal to 1.2, but they were higher than those simulated in Ansys Chemkin for higher equivalence ratios. In addition to this, the laminar flame speed is the maximum at an equivalence ratio of 1 for both simulations and at 1.1 for the MATLAB application.

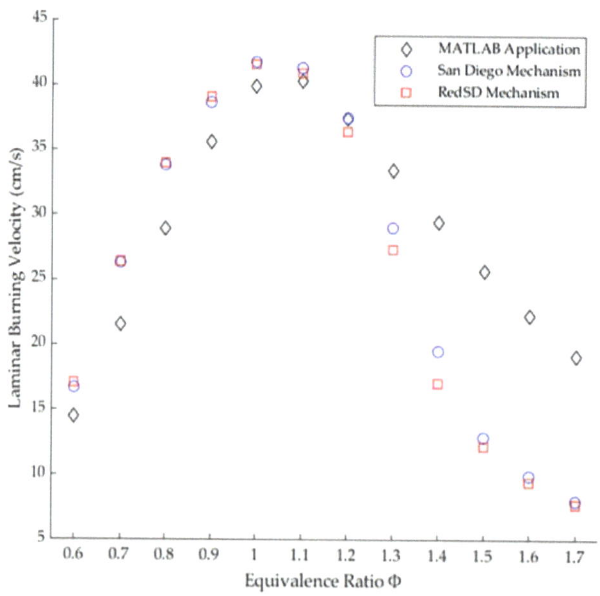

Figure 8. Results for laminar burning velocity (cm/s) versus equivalence ratio.

It is worth noting that the laminar flame speed is not only influenced by flame temperature, it also has a stretch relationship with parameters such as ambient temperature and pressure, kinetic mechanisms and reactions that occur during combustion. None of these additional factors is considered in the Mallard and Le Chateller method and are considered in the simulation in Ansys Chemkin, so the differences between the results are mainly caused by these factors. The activation energy in the equation of the Mallard and Le Chateller method gives an additional percentage of error for the results obtained by the MATLAB application, which happens because, for each reaction, there is a specific energy of activation, and this is defined in the kinetic data of the mechanisms used in Ansys Chemkin and not in the MATLAB application. In addition to this, for higher equivalence ratios (higher than 1.2) there will exist other combustion products (unburned fuel—C_3H_8 or C_4H_{10}, NO_X, among others, that appeared in the Ansys Chemkin simulations) that cause the difference to increase, as can be seen in Figure 8. This does not happen for lower ratios. A better approximation for the calculation and another improvement for this application is to define an activation energy as a function of the equivalence ratio and temperature, taking as a reference the reactions considered in the MATLAB application and adding the ones involved directly with the tested fuel.

Moreover, the differences between the results and their percentages for the three cases are presented in Table 11. It is possible to affirm that the results of the MATLAB application are similar to what is simulated in Ansys Chemkin between ratios of 0.6 to 1.2, according to which percentage differences are not higher than 20. The average found for the absolute differences between the values of the MATLAB application and simulation in Ansys Chemkin using the San Diego mechanism is 5.70 cm/s with a percentage average difference of 40.59%, while the simulation in Ansys Chemkin using the RedSD mechanism gave an average difference of 6.28 cm/s and a percentage of average difference of 45.96%.

Table 11. This is a table. Tables should be placed in the main text near to the first time they are cited.

Equivalence Ratio (Φ)	MATLAB Application—San Diego Mechanism		MATLAB Application—RedSD Mechanism	
	ΔS_L (cm/s)	%	ΔS_L (cm/s)	%
0.6	−2.2	−13.3	−2.5	−15.1
0.7	−4.7	−18.1	−4.9	−18.6
0.8	−4.8	−14.4	−5.0	−14.8
0.9	−3.0	−7.9	−3.5	−9.0
1	−1.8	−4.4	−1.7	−4.0
1.1	−0.9	−2.3	−0.5	−1.4
1.2	−0.1	−0.2	0.9	2.6
1.3	4.3	15.0	6.0	22.1
1.4	9.9	50.7	12.3	72.3
1.5	12.8	99.0	13.5	110.6
1.6	12.3	123.5	12.7	134.6
1.7	11.1	137.7	11.3	145.9

Furthermore, the minimum and maximum absolute difference values found were 0.1 cm/s (0.26%) at an equivalence ratio of 1.2 and 12.8 cm/s (99.06%) at an equivalence ratio of 1.5 for the comparative case between the MATLAB application and simulation in Ansys using the San Diego mechanism, while, for the comparative case between the MATLAB application and the simulation in Ansys Chemkin using the RedSD mechanism, the minimum value was 0.59 cm/s (1.45%) at an equivalence ratio of 1.1 and the maximum value was 13.52 cm/s (110.69%) at an equivalence ratio of 1.5.

Figure 9 shows the comparison between the results for laminar burning velocity results obtained in this work and the ones obtained in the investigations of B. A. Alfarraj et al. [9], B. Yang [15], Miao et al. [16], Huzayyin et al. [5], Ahmed Sh. Yasiry et al. [8] and Chakraborty et al. [17]. The composition and details of the LPG used in each work are presented in Table 12.

From Figure 9, it can be seen that results obtained using the MATLAB application and simulation in Ansys Chemkin present the same plot shape as the results obtained by Miao et al. [16], B. Yang [15] and B.A. Alfarraj et al. (2022) [9] for ratios under the stoichiometric and fuel-rich conditions. It has to be considered that Miao et al. [16] used a different LPG composition (30% propane and 70% butane) while B. Yang [15] and B.A. Alfarraj et al. (2022) [9] used the same composition in the mixture. On the other hand, for stoichiometric and fuel-rich conditions, the results of the MATLAB application present the same plot shape as the results of B. Yang (2006) [15].

It is important to consider the type of study and instruments the authors of these experimental studies used. Of those who have been mentioned in the previous paragraph, Miao et al. [16] and B. Yang [15] used the constant volume chamber method, which is the most controlled condition for carrying out the experiment (regarding temperature and pressure values) from all of those mentioned in Table 12. Meanwhile, B.A. Alfarraj et al. (2022) [9] used the modified Bunsen burner method in order to have more controlled conditions during the experiments.

Using the results from the previous paragraph, it is possible to affirm that the MATLAB application has similar results to those found using the constant volume chamber method for the whole range of equivalence ratios.

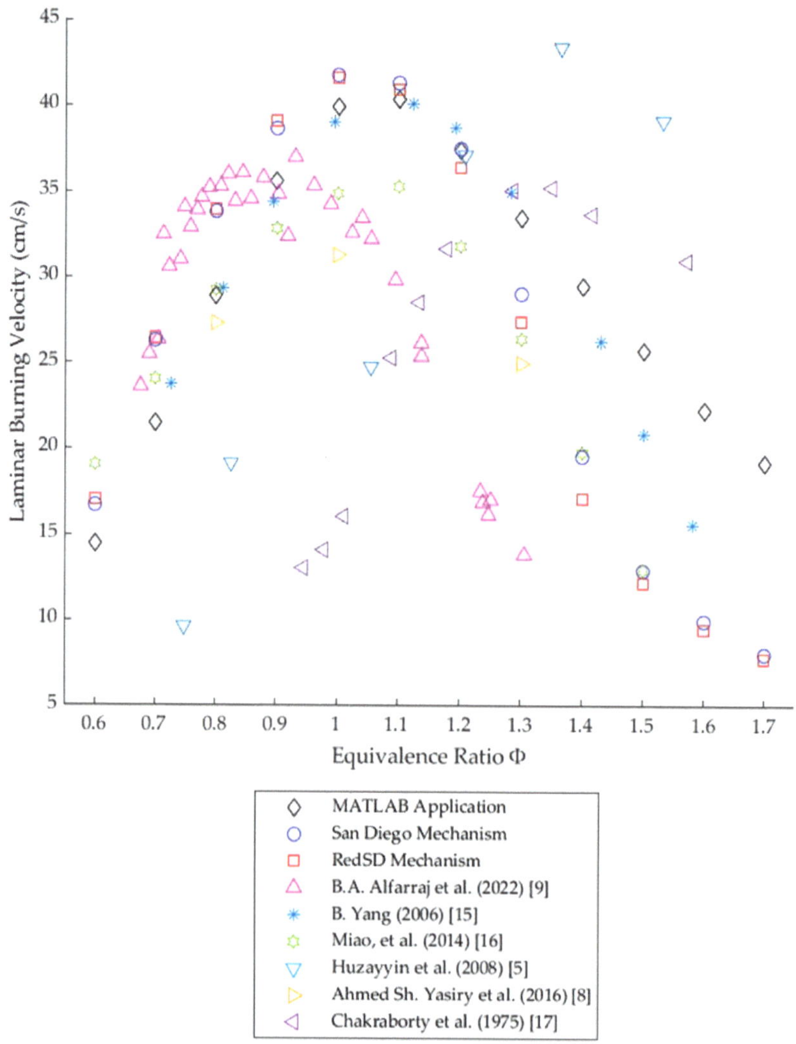

Figure 9. Results for laminar burning velocity (cm/s) versus equivalence ratio [5,8,9,15–17].

Table 12. LPG laminar burning velocity study specifications.

Work	Type of Study	Pressure (atm)	Initial Temperature (K)	LPG Composition			
				Propane C_3H_8	Butane C_4H_{10}	Ethane C_2H_6	Pentane C_5H_{12}
MATLAB Application—Current Work	Numerical methodology in MATLAB	1	298.15	50%	50%	0%	0%
San Diego Mechanism—Current Work	Numerical Simulation in Ansys Chemkin	1	298.15	50%	50%	0%	0%
RedSD Mechanism—Current Work	Numerical Simulation in Ansys Chemkin	1	298.15	50%	50%	0%	0%
B.A. Alfarraj et al. [9]	Experimental—Modified Bunsen Burner Method	1	298.15	50%	50%	0%	0%
B. Yang [15]	Experimental—Constant Volume Bomb Method	1	298.15	50%	50%	0%	0%
Huzayyin et al. [5]	Experimental—Constant Volume Chamber Method	1	294 ± 3	26.41%	73.54%	0.04%	0%
Ahmed Sh. Yasiry et al. [8]	Experimental—Constant Volume Chamber Method	1	308	36.3%	62.3%	0.9%	0.5%
Miao et al. [16]	Experimental—Constant Volume Chamber Method	1	298.15	30%	70%	0%	0%
Chakraborty et al. [17]	Experimental—Flat Flame Burner Method	1	298	30.1%	67.7%	1.4%	0%

4. Conclusions

In this work, a numerical method was coded in a MATLAB application for the study of Saudi LPG combustion characteristics (products of combustion, the adiabatic flame temperature and the laminar flame velocity), and the results were compared with the results of a simulation in Ansys Chemkin using the San Diego and RedSD mechanisms and other investigations. The following conclusions were reached:

1. For the laminar burning velocity results, the numerical method agrees with the experimental results for ratios (0.6–1.2) used by other authors and the simulation carried out in Ansys Chemkin, while, for the highest studied equivalence ratios (1.3–1.7) the laminar burning velocity results show a greater difference between all the resources.
2. The numerical method used in the MATLAB application agrees with the simulation in Ansys Chemkin for the Saudi LPG combustion products, except for N and NO, for the whole range of equivalence ratios.
3. The Saudi LPG maximum laminar burning velocity determined by the modified Bunsen burner method [9] was 35 ± 0.91 while that determined by the MATLAB application was 40.3 cm/s, having a difference of 5.35 ± 0.91 and an overestimate of 15.2% in favor of the MATLAB application.
4. The Saudi LPG maximum laminar burning velocity determined by the MATLAB application was 40.3 cm/s, corresponding to a ratio of 1.1, with an underestimate of 2.3% with respect to the simulated values in Ansys Chemkin using the San Diego mechanism and an underestimate of 1.4% using the RedSD mechanism.
5. The maximum adiabatic flame temperature of the Saudi LPG determined using the MATLAB application was 2326.2 K, corresponding to a ratio of 1.1, with an overestimate of 2.6% and 2.5% with respect to the simulated values in Ansys Chemkin using the San Diego and RedSD mechanisms, respectively.
6. The MATLAB application, compared with previous experimental studies, presents the same behavioral results as those obtained by Miao et al. [16], B. Yang [15] and B.A. Alfarraj et al. (2022) [9] for lean mixture conditions. Meanwhile, for stoichiometric and fuel-rich conditions, it presents the same plot shape as that of B. Yang (2006) [15].

7. The new code in the MATLAB application determined the experimental results more accurately, for equivalence ratios from 0.7 to 1.4, compared to Ansys Chemkin, taking B. Yang (2006) [15] as the experimental result reference.
8. The MATLAB application has been developed for additional fuels such as methane, propane and natural gas and has the possibility of adding extra fuels, diluents and tools to improve the analysis of the results. It could be also applied to further studies using different kinds of mixtures.

Supplementary Materials: The following supporting information can be downloaded at: https://www.mdpi.com/article/10.3390/en16124688/s1, Code S1: New_code.m, Archive S1: REACTANTS ENTHALPY.xlsx; Archive S2: DILUENTS ENTHALPY.xlsx; Code S2: Fractions_Derivatives.m; Code S3: MATLAB_Application.mlapp. Archive S3: Ansys Chemkin Results.xlsx, Archive S4: Products.xlsx.

Author Contributions: Conceptualization, F.J.R.; methodology, R.F.C.; software, R.F.C.; validation, F.J.R.; investigation, R.F.C. and F.J.R.; resources, R.F.C.; writing–original draft, R.F.C.; writing–review & editing, F.J.R.; supervision, F.J.R.; project administration, F.J.R. All authors have read and agreed to the published version of the manuscript.

Funding: This research (PI0735) was funded by DFI PUCP.

Data Availability Statement: Not applicable.

Acknowledgments: Thanks are due to the DFI of the Pontificia Universidad Católica del Perú that, through the annual competition for applied research projects CAP PUCP 2022, financed research PI0735.

Conflicts of Interest: The authors declare no conflict of interest.

Nomenclature

Φ	is the equivalence ratio;
n	is the carbon coefficient value of the fuel;
m	is the hydrogen coefficient value of the fuel;
l	is the oxygen coefficient value of the fuel;
k	is the nitrogen coefficient value of the fuel;
X_1	is the molar fraction of hydrogen (H) in the products;
X_2	is the molar fraction of oxygen (O) in the products;
X_3	is the molar fraction of nitrogen (N) in the products;
X_4	is the molar fraction of hydrogen (H_2) in the products;
X_5	is the molar fraction of hydroxide (OH) in the products;
X_6	is the molar fraction of carbon monoxide (CO) in the products;
X_7	is the molar fraction of nitric oxide (NO) in the products;
X_8	is the molar fraction of oxygen (O_2) in the products;
X_9	is the molar fraction of dihydrogen oxide (H_2O) in the products;
X_{10}	is the molar fraction of carbon dioxide (CO_2) in the products;
X_{11}	is the molar fraction of nitrogen (N_2) in the products;
X_{12}	is the molar fraction of argon (Ar) in the products;
X_{13}	is the number of moles from fuel that give 1 mol of product;
K_i	is the partial pressure equilibrium constant of a chemical reaction;
p	is the pressure;
f_j	is the equation system with 4 variables (X_4, X_6, X_8 and X_{11})
X_i^*	is the molar fraction real solution value used in Taylor series;
$X_i^{(1)}$	is the molar fraction approximate value to the real one used in Taylor series;
ΔX_i	is the difference between the molar fraction real value and the approximate one;
$\partial f_j/\partial X_i$	is the equation system derivative with respect to molar fraction (X_4, X_6, X_8 and X_{11});
$X_i^{(2)}$	is the improved molar fraction value after the first iteration;
h_r	is the reactant enthalpy;
h	is the product enthalpy;
p_o	is the initial pressure;

F_o	is the initial equivalence ratio;
T_o	is the initial temperature;
T	is the adiabatic flame temperature;
T_n	is the first assumed flame temperature (n = 1) or a current temperature iteration (n > 1);
T_{n+1}	is the improved temperature after an iteration using the Newton–Raphson method;
$(\partial h/\partial T)_n$	is the enthalpy derivative with respect to temperature at n iterations;
M	is the molar mass of the mixture;
dh_i/dT	is the specific heat at constant pressure of an element (i);
Cp_i	is the specific heat at constant pressure of an element (i);
dX_i/dT	is the partial derivative of a molar fraction with respect to temperature;
$\partial M/\partial T$	is the molar mass of the mixture with respect to temperature;
S_L	is the laminar burning velocity;
Ea	is the activation energy;
Ru	is the universal gas constant.

References

1. Khudhair, O.; Shahad, H.A.K. A Review of Laminar Burning Velocity and Flame Speed of Gases and Liquid Fuels. *Int. J. Curr. Eng. Technol.* **2017**, *7*, 183–197.
2. Kuo, K.K. *Principles of Combustion*, 2nd ed.; John Wiley & Sons: Hoboken, NJ, USA, 2005; pp. 437–457.
3. Samim, S.; Sadeq, A.M.; Ahmed, S.F. Measurements of Laminar Flame Speeds of Gas-to-Liquid-Diesel Fuel Blends. *J. Energy Resour. Technol.* **2016**, *138*, 052213. [CrossRef]
4. Lee, K.; Ryu, J. An experimental study of the flame propagation and combustion characteristics of LPG fuel. *Fuel* **2005**, *84*, 1116–1127. [CrossRef]
5. Huzayyin, A.S.; Moneib, H.A.; Shehatta, M.S.; Attia, A.M.A. Laminar burning velocity and explosion index of LPG-air and pro-pane-air mixtures. *Fuel* **2008**, *87*, 39–57. [CrossRef]
6. Tripathi, A.; Chandra, H.; Agrawal, M. Effect of mixture constituents on the laminar burning velocity of LPG-Co2-air mixtures. *J. Eng. Appl. Sci. (Asian Res. Publ. Netw.)* **2010**, *5*, 16–21.
7. Razus, D.; Brinzea, V.; Mitu, M.; Oancea, D. Burning Velocity of Liquefied Petroleum Gas (LPG)−Air Mixtures in the Presence of Exhaust Gas. *Energy Fuels* **2010**, *24*, 1487–1494. [CrossRef]
8. Yasiry, A.S.; Shahad, H.A. An Experimental Study for Investigating the Laminar Flame Speed and Burning Velocity for LPG. *Int. J. Therm. Technol.* **2016**, *6*, 7–12.
9. Alfarraj, B.A.; Al-Harbi, A.A.; Binjuwair, S.A.; Alkhedhair, A. The Characterization of Liquefied Petroleum Gas (LPG) Using a Modified Bunsen Burner. *J. Combust.* **2022**, *2022*, 1–9. [CrossRef]
10. Chemical-Kinetic Mechanisms for Combustion Applications. San Diego Mechanism Web Page. Available online: https://web.eng.ucsd.edu/mae/groups/combustion/mechanism.html (accessed on 30 November 2022).
11. Kumaran, S.M.; Shanmugasundaram, D.; Narayanaswamy, K.; Raghavan, V. Reduced mechanism for flames of propane, n-butane, and their mixtures for application to burners: Development and validation. *Int. J. Chem. Kinet.* **2021**, *53*, 731–750. [CrossRef]
12. Olikara, C.; Borman, G. A Computer Program for Calculating Properties of Equilibrium Combustion Products with Some Applications to I.C. Engines. *SAE Tech. Pap.* **1975**, 23. [CrossRef]
13. Chase, M.W. *JANAF Thermochemical Tables*, 3rd ed.; American Chemical Society and American Institute of Physics: New York, NY, USA, 1986; pp. 1–1856.
14. Markatou, P.; Pfefferle, L.D.; Smooke, M.D. A computational study of methane-air combustion over heated catalytic and non-catalytic surfaces. *Combust. Flames* **1993**, *93*, 185–201. [CrossRef]
15. Yang, B. Laminar Burning Velocity of Liquefied Petroleum Gas Mixtures. Doctoral Thesis, Loughborough University, Loughborough, UK, 2005.
16. Miao, J.; Leung, C.W.; Huang, Z.; Cheung, C.S.; Yu, H.; Xie, Y. Laminar burning velocities, Markstein lengths, and flame thickness of liquefied petroleum gas with hydrogen enrichment. *Int. J. Hydrogen Energy* **2014**, *39*, 13020–13030. [CrossRef]
17. Chakraborty, S.K.; Mukhopadhyay, B.N.; Chanda, B.C. Effect of inhibitors on flammability range of flames produced from LPG/air mixtures. *Fuel* **1975**, *54*, 10–16. [CrossRef]

energies

MDPI

Article

Electrodynamic Forces in Main Three-Phase Busbar System of Low-Voltage Switchgear—FEA Simulation

Sebastian Łapczyński [1,2], Michał Szulborski [1,3,4], Łukasz Kolimas [1,*], Przemysław Sul [5], Maciej Owsiński [6], Przemysław Berowski [6], Tomasz Żelaziński [7] and Andrzej Lange [8]

1 Institute of Electrical Power Engineering, Warsaw University of Technology, ul. Koszykowa 75, 00-662 Warsaw, Poland; seb.lapczynski@gmail.com (S.Ł.); mm.szulborski@gmail.com or szulborskim@etipolam.com.pl (M.S.)
2 ELKO-BIS Systemy Odgromowe, ul. Swojczycka 38e, 51-501 Wroclaw, Poland
3 Symkom Sp. z o.o., Ansys Channel Partner, ul. Głogowa 24, 02-639 Warsaw, Poland
4 ETI Polam Sp. z o.o., Jana Pawła II 18, 06-100 Pułtusk, Poland
5 Institute of Theory of Electrical Engineering, Measurement and Information Systems, ul. Koszykowa 75, 00-662 Warsaw, Poland; przemyslaw.sul@pw.edu.pl
6 Institute of Power Engineering, Research Institute, ul. Mory 8, 01-330 Warsaw, Poland; maciej.owsinski@ien.com.pl (M.O.); przemyslaw.berowski@ien.com.pl (P.B.)
7 Department of Production Engineering, Institute of Mechanical Engineering, Warsaw University of Life Sciences (SGGW), Nowoursynowska 164, 02-787 Warsaw, Poland; tomasz_zelazinski@sggw.edu.pl
8 Department of Electrical Engineering, Power Electronics and Automation, University of Warmia and Mazury in Olsztyn, ul. Michała Oczapowskiego 2, 10-719 Olsztyn, Poland; andrzej.lange@uwm.edu.pl
* Correspondence: lukasz.kolimas@pw.edu.pl; Tel.: +48-537-884-042

Abstract: This paper concerns the effects of electrodynamic forces that act on current paths that are part of high-grade industrial distribution switchgear. This work is composed of experimental and simulation sections. In the experimental section, the short-circuit tests are presented and the occurrence of electrodynamic forces are shown in a visible way. The formation of electrodynamic forces in the current circuits of electrical energy distribution systems is related to the flow of high currents, but mostly it is related to short-circuit currents. In order to highlight these phenomena, the detailed specification of the parameters during tests is displayed. In the simulation section, the physical phenomenon of electrodynamic forces is being captured by employing a detailed real-scale model of switchgear and current paths. Therefore, the authors proposed employment of the FEM (finite element method) in order to obtain values of electrodynamic forces acting on the current paths by executing the detailed 3D coupled simulation. The analysis of the results and aftermath effects of their interactions provided interesting conclusions that concerned the operation of such power distribution layouts in critical short-circuit conditions.

Keywords: electrodynamic forces; FEM; simulation; current paths; experiment; short-circuit tests

1. Introduction

The occurrence of electrodynamic forces in high-current circuits during the flow of a short-circuit current is a common phenomenon. Nevertheless, determining the exact electrodynamic load capacity for single- and multi-strand copper current paths in switchgear is a real challenge even for an experienced constructor. In general, the obtained analytical results are burdened with a certain amount of approximation through the use of calculation simplifications. For each calculation for individual current circuits, the calculation is based on knowledge of the permissible values of bending stress for the material from which the considered section of the current circuit is made. In order to determine the moments and electrodynamic forces affecting the current paths, the equations of Biot Savart, Lorentz, or Maxwell are used. Maxwell's equations are used to calculate the electrodynamic forces in current path systems for which analytical expressions of inductance are known. The

Lorentz and Biot Savart equations are generally used to calculate the electrodynamic moments and forces that act on the rectilinear parts of the current paths. When designing current circuits of switchgears, attention should be paid to, and necessary calculations should be related to the following (Figure 1):

- Stresses arising in the current circuits at the time of the flow of short-circuit currents;
- Reaction forces acting on the fastening elements and other brackets with which the current paths are fixed; this allows the selection of an appropriate insulator of the required strength and number of these elements;
- Moments acting on moving parts, i.e., moving contacts of switches, in the closed state and at the moment of switching on short-circuit currents (shown in Figure 1 below);
- Forces acting on the electric arc.

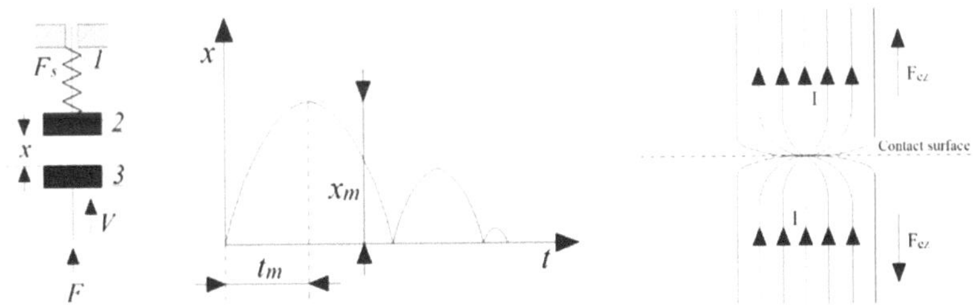

Figure 1. Spring bounces of contacts during flow of current and currentless conditions: (**a**) arrangement of contacts in currentless conditions: 1—spring contact, 2—fixed contact, 3—movable contact, V—contact velocity, F—movable contact force, Fs—spring force; (**b**) diagram of the relative motion of the contacts, tm—contact movement time, Xm—travelled distance by movable contact; (c) electrodynamic interactions during flow of current through electrical contacts layout, Fez—electrodynamic repulsion force, I—current flow.

Performing this type of calculation poses significant difficulty, and it is easy to insert a calculation error that will make the result useless. During analytical calculations, the constructor must determine such parameters as the following:

- Electrodynamic stress distribution along the current path;
- The resultant force acting on the considered current path;
- Moment of resultant force;
- Point of application of the resultant force.

Analytical methods do not allow precise calculations concerning complex switchgear systems, such as the main low- or medium-voltage switchgears. Currently, calculation results are obtained through the use of FEA (finite element analysis) methods with the use of a structural model with 3D technology, which allows very accurate mapping of the analyzed structure and the performance of accurate calculations of the electrodynamic forces in the virtually considered element of the switchgear structure, both of low and medium voltages. In this work, authors focused on confirming the thesis that the use of FEA numerical analysis employing the ANSYS software 2023 provides accurate calculation results regarding the determination of electrodynamic forces in low-voltage switchgears.

2. State of the Art

In the literature, it is possible to find research dealing with the interactions of electrodynamic forces in overhead lines. The authors of [1] conducted tests of short-circuit stresses in 145 kV busbar systems with flexible conductors. The influence of various parameters, such as span length, sag, cable type, cable spacing, and the spring effect in poles was tested. RMS currents of 20–50 kA and short-circuit times of 0.1–0.5 s were tested analytically and numerically. Dynamic short-circuit stresses in HV (high-voltage) substations with flexible conductors are also presented in [2]. Tensile forces, stresses, and cable displacements were tested for short-circuit currents of 20–40 kA (RMS) in test systems representing stressed cables of typical 123 kV and 420 kV substations over a span of 40 m. A simple calculation method was developed based on the approach from IEC 865 and DIN publications VDE 0103, but taking into account the short-circuit duration, leading to reduced forces and displacements in the event of short-term current flow.

In [3], the author considers the calculation of the axial electromagnetic forces on the windings of a step-up transformer in unbalanced MMFs (modal properties in multi-mode fibers) of the same phase using finite element code. The unbalance between FRPs (fluctuation—relaxation paths) of the same phase, which can be on the order of the nominal FRP, is due to the high saturation of the iron core that occurs during the transient due to the improper paralleling of the transformer with the very low-impedance network. These FRPs are assessed using a non-linear circuit model of a three-phase transformer. At MMF peaks, the magnetic field in the transformer window is calculated using a two-dimensional non-linear model and then the Lorentz forces acting on the windings are evaluated. The result of the calculation is expressed as the ratio of the axial forces with and without the unexpressed MMF as a function of the impedance of the out of parallel network. The paper presents a comparison of the axial force acting on the concentric windings of a step-up transformer with and without MMF, during false parallelism with a phase error of 180 degrees. The analysis showed that the axial force acting on the outer winding with an unbalanced FRP can be much greater—2 to 10 times—for the two power transformers under consideration than assessed without MMF unbalance of the same phase, as is usually completed to verify the mechanical strength of transformers at the design stage. The authors point out, therefore, that designers should be aware of the importance of magnetic-circuit saturation during certain short-circuit transients that cause MMF asymmetry between windings of the same phase.

In item [4], the author extended the model from item [3] with a field interpretation. The use of a magnetostatic three-dimensional non-linear finite element model for the calculation of axial forces acting on the windings of a high-power autotransformer caused by short-circuit currents is described. The analyzed autotransformer with a five-arm core has an internal delta-connected tertiary winding. A comparison was made of the axial forces caused by the peak currents of both a time-varying fault and two conventional faults—a two-phase earth fault and a three-phase earth fault. The obtained results show that the time-varying fault causes greater axial forces than conventional faults, especially on the internal delta-connected tertiary winding.

One of the most pressing problems is the phenomenon of the flow of large currents during internal short circuits caused by failures of semiconductor elements [5,6]. This situation may result in damage to the device housing and other damage caused by the destructive effects of electrodynamic forces. The proposal to eliminate fast-acting fuses, commonly used as short-circuit protection, requires the determination of the strength characteristics of the device (peak value of the current that destroys the housing as a function of time). In such a case, the load level of the converter should be designed so that the current of the device does not exceed the permissible value before the AC power switch is tripped. The article analyzes overcurrent protection methods. Some experiences and research results in this field are given. Mechanisms of destruction of semiconductor power devices under the influence of very high short-circuit currents are discussed. The summary provides guidelines for the design of power electronic devices without fast overcurrent protection with a semiconductor fuse.

In publication [7], during the conditioning pulse, the toroidal field in RFX machines is connected to a capacitor bank, conditioning the oscillating current, and protecting the winding from an event by short-circuiting the winding with a pry bar, created from a set of end igniters. The efficiency of this pry bar is highly dependent on its application, which is associated with a device that can be held level and placed when applied to the winding. On the other hand, the additional current in the crowbar must be low enough to withstand one more ignition to avoid an accident in the event of it starting. A compromise on the value of the crowbar was obtained based on the analysis of electrodynamic forces and stresses in the toroidal field coils during the operation of the crowbar. The operation of the model also includes functions that are available in the stress values in the calculations, which are triggered by the stress from the crowbar.

In works [8–10], a three-phase busbar system with straight, rigid conductors carrying short-circuit currents was developed. Calculations were completed with an alternating current mains voltage with a peak value equal to the peak value of the short-circuit current. This assumption is used in the IEC 865/92 standards. In this article, the work force and electric current are calculated by numerically solving the electromagnetic field diffusion equations using finite elements. Results were compared to calculation results with the IEC 865/86 standards and the revision manual of the IEC 865/92 standards. The comparison concerns parallel cross-sections in indoor basic and control AC installations. The calculated value in the case of multiple subconductors was larger than the value calculated using the standards. Using the older standards, the calculations reach fifty percent, while using the amendment there is a problem. They are likely to be removed from proximity. The authors analyzed a three-phase definite current and short-circuit current busbar, controlled using FEA, which was previously commissioned for rotating machines. The bus consists of aluminum conductors, three switches for each phase, in aluminum channels. This following technique is required: it includes distributions of individual phase currents between wires and conductors, which are not marked in advance. In the application of the calculation results, these are determined in terms of the frequency and unsteady state (in time).

The authors of [11] investigated the arrangement of three-phase copper busbars in a low-voltage network. Each main phase conductor consisted of one to four bars of rectangular cross sections. It has been pointed out that a short circuit can introduce large, destructive forces. In order to ensure safe operation of the busbars and the power system, the dynamic stability of the short circuit must be verified. Three methods of calculating short-circuit electrodynamic forces of busbars are presented and analyzed. Valuable analytic expressions are proposed. For example, three methods with calculations were compared.

In publication [12], the electromagnetic forces affected by the short-circuit current in a three-phase busbar in a vertical and horizontal arrangement were calculated. The short-circuit current densities are calculated mathematically. Calculations are made assuming that the steady-state AC peak value is equal to the short-circuit current peak value. The electromagnetic forces caused by the short-circuit current are calculated according to the equation introduced by the IEC 865/1993 standards. The electromagnetic force generated in the vertical arrangement is compared with the level of the busbar. The result showed that the busbar layout in the vertical arrangement has about twice the electromagnetic force compared to the horizontal arrangement. The arrangement of the busbars obviously affects the electromagnetic force caused by the short-circuit current. In addition, the electromagnetic force obtained from the vertical finite element simulation was consistent with the calculations obtained using the IEC 865/1993 standards.

Manuscript [13] deals with EDFs (electrodynamic forces) research and development for the ERDF in the field of energy efficiency, focusing on amorphous metal distribution transformers (AMDT). Many prototypes built on Metglas 2605SA1, rated from 250 to 630 kVA from various manufacturers, were tested under short-circuit conditions according to IEC and ERDF standards. Some prototypes exceeded the acceptable value of 4% change in impedance after the short-circuit test. Rectangular windings provide poor resistance to short-circuit stresses. At the bottom of the tank, small particles of metal were found, detached from the amorphous core. In this case, the dielectric strength is not sufficient. The results show that the active part should be strong enough to withstand the high electrodynamic forces generated during the short-circuit test.

Knowledge of electromagnetic forces in various winding topologies provides designers and researchers [2,14] with a good basis to meet some of the requirements for new-generation transformers. The location of the control coil in the transformer determines the influence of electrodynamic forces on the behavior of the transformer. The article presents the principles of designing the control coil in transformers and presents the electrodynamic forces by comparing the analytical results with the results of the finite element analysis (FEM). The article analyzes the electrodynamics and the magnetic force dissipation field of 25 MVA power transformers in the conditions of short-circuiting high-voltage windings. The electrodynamic forces generated during short circuits must be well characterized [15] in power grid applications because those can threaten the integrity of grid systems, in particular power transformers, but also inductive superconducting fault current limiters (SFCLs). These stresses can damage the windings, causing equipment failures, and affecting the operation of the power grid. The article presents an analysis of electrodynamic forces arising in an inductive SFCL of the transformer type in such extreme conditions based on the finite element method (FEM). The secondary of the planned devices is made of conductors coated with high-temperature superconductors (HTS), and various configurations (radial and axially arranged windings) are analyzed to determine the most suitable one for minimizing electromagnetic forces. FEM results for the developed radial and axial forces are evaluated and compared, as well as the current distribution and normal magnetic induction in each winding [15].

The article [16] presents numerical models obtained in the QuickField program for the analysis of three-phase rectangular busbar systems of low or medium voltage, in the steady state and in the short-circuit state. Using a magnetic harmonic model coupled with a fixed thermal model, the distribution of the thermal field for the rated current is determined. By combining the steady-state thermal model for the rated current and the magnetic model for the short-circuit current with the thermal model in the transient state, the time evolution of the temperature distribution in the short-circuit regime (not developed in this paper) can be determined. The transient magnetic model is used to determine the electrodynamic forces. Various types of busbar systems are being studied, with one or more conductors per phase in a balanced arrangement [16].

Paper [17] presents a coupled electrical, magnetic, thermal, and mechanical analysis of busbar systems under the influence of short-circuit currents. The analysis is executed using the finite element method, which allows accurate modeling of bidirectional interactions between separate continuum physicists. Unlike previous work, which only considered the peak value of the short-circuit current, this method evaluates the magnetic force, temperature rise, mechanical displacement, and their interactions over the simulation time of interest. Mechanical displacements are obtained using three-dimensional analysis. It was found that the type of busbar support can significantly affect the displacement of the conductor during the flow of short-circuit current. It was found that the increase in temperature caused by the flow of short-circuit current has little effect on the displacement of the busbar conductors.

In the publication [18], it was noted that in the current carrying parts of electrical apparatus, under certain conditions, large currents are transferred, which also define large electrodynamic forces tending to distort the current contours. The most severe is the short-circuit regime in which the electrodynamic forces are proportional to the square of the short-circuit-breaking current. As a result of their operation, the temperature of the elements increases, the mechanical strength decreases and a very large regime of mechanical loads arises. This also places increased demands on materials for conductive busbars. From a practical point of view, the question of how a given coating affects the magnitude of the electrodynamic force is interesting. The use of protective coatings on heterogeneous materials can result in both a modification of the surface layer material that improves the performance properties of the workpiece, and the creation of some new composite materials with both high strength and sufficient ductility, as well as increased resistance to wear and corrosion. This leads to the improvement of some operating parameters in the contact area of current carrying busbars. It is also important to assess the influence of coatings on the magnitude of the electrodynamic force, which is also the subject of this study.

In [19], the electrodynamic forces generated in transformers during a short circuit in power systems can be dangerous for the windings; therefore, transformers must be designed with certainty that the windings will withstand these forces. The article presents a computational model of a transformer made for calculations of electrodynamic forces in windings in the Comsol Multiphysics modeling environment.

Arrangement of the busbars obviously affects the electromagnetic force caused by the short-circuit current. In addition, the electromagnetic force obtained from the vertical finite element simulation was consistent with the calculations obtained using the IEC 865/1993 standards.

In [19], the electrodynamic forces generated in transformers during a short circuit can be dangerous for the windings; therefore, transformers must be designed with the certainty that the windings will withstand these forces. The article presents a computational model of a transformer made for calculations of electrodynamic forces in windings in the Comsol Multiphysics modeling environment.

The two most dangerous operating modes of transformers in terms of electrodynamic force are discussed. The first mode of operation is a short circuit, and the second is the process of switching the transformer to the non-load state at the moment of the voltage passing through zero. The distribution of the magnetic field near the magnetic core, currents, voltages, and electrodynamic forces for the high- and low-voltage windings are calculated. The results for these two modes are compared. This article can be used as a basis for the creation of similar models by engineers. These models can be used to check the electrodynamic strength of new transformers in order to reduce the total costs of field tests [20].

In publication [21], the authors consider the issue of transformer resistance to wind from a point of view that has not been mentioned before. The authors compare the forces in the wind between short-circuit and non-load switching modes. The risk of winding damage caused by the inrush current occurring during the no-load state switching process was assessed. The inrush current and short-circuit current values were calculated for the ORTS417000/750/24/24 transformer. Conclusions were drawn as to the possibility of reaching a value close to the short-circuit current by inrush currents in specific situations. Calculations of magnetic fields and electrodynamic forces in the transformer for these operating modes were performed. Conclusions were drawn regarding the influence of the transformer operating mode on the value and place of application of electrodynamic forces. It was found that when checking the electrodynamic wind brace of transformers, the process of turning on the idle state should be taken into account.

The article [4,9,22–25] focused on finding the short-circuit duration for the calculation of dynamic effects in HV distribution substations with flexible conductors using the probabilistic method. To determine the short-circuit duration for selected substation points, a risk criterion was adopted, which is the expected annual frequency of exceeding the calculated values of dynamic forces. A dedicated simulation model based on the Monte Carlo method was used to analyze the dynamic effects of short-circuit current flow in the substation during short circuits within and in its vicinity. The fault duration model is a key part of the simulation model. As a result of discussing the influence of selected factors on the estimated short-circuit duration, guidelines were formulated to help determine its value and examples for 220 kV switchgear were provided.

A lot of interesting and valuable research is presented above. It should be noted that the presented models refer to standards that treat a three-phase short-circuit current as two-phase from an analytical point of view. This applies to the electrodynamic standards. The result of this approach is the finding that time-consuming and labor-intensive tests are difficult to procure in short-circuit laboratories. The results of these tests are usually burdened by large discrepancies between the proposed prototype and the final product. There is also the aspect of test system to consider [26]. There are not many systems dedicated to testing electrodynamic forces in design or in accredited laboratories. Therefore, the authors of this publication focused on the analysis of physical phenomena related to the flow of a short-circuit current. Only the phenomenon of electrodynamic forces was selected. Real waveforms obtained from the short-circuit laboratory were used for the simulation. The whole waveforms were made for the actual runs, for the prototype of the low-voltage switchgear, which was proposed by the authors along with the arrangement of current paths within.

3. Electrodynamic Forces in FEM Simulations

The purpose of the publication was not to test the mechanical and electromechanical strength of electrotechnical equipment. Model validation was executed for critical states without any short-circuit-protection device. Manufacturer and equipment quality were not assessed. The conducted tests were not the type of tests used for the associated equipment. It is also worth mentioning that the research was critical to determining the value of the flowing current. The quality of the casing and insulators was not analyzed, and this has not been researched. The purpose of the research is clearly indicated at the beginning of the publication. The authors focused on interactions and phenomena in current paths.

Low-voltage switchgears are constructed of materials that are able to withstand mechanical, thermal, and dynamic exposure from short-circuit currents. Requirements as to mechanical strength are imposed not only on enclosures, but also on all partitions, brackets, hinges, and locks, which should have adequate mechanical endurance to withstand the stresses arising during short circuits in the low-voltage switchgear. The real object prepared for experiments and its 3D model are shown in Figure 2 below.

(a) (b)

Figure 2. Low-voltage switchgear used for experimental works and for modelling in real scale in order to procure FEA: (**a**) real object armed with arranged current paths; (**b**) 3D detailed model representing the real object used for FEM simulations.

During the flow of the short-circuit current through the current paths and switchgear wires, electrodynamic forces are generated, which generate stresses on the support insulators of the current paths and cable fastenings, mainly between the following:

- Current paths;
- Current wires;
- Ferromagnetic materials, conductors, and busbars;
- Boundary surfaces of materials with different magnetic permeability.

Based on the real object, low-voltage switchgear, a precise 3D model of the analyzed switchgear was created in the ANSYS SpaceClaim environment. The prepared real test object was transported to the Distribution Equipment Laboratory at the Institute of Power Engineering in Warsaw. There, it was carefully checked and prepared for short-circuit tests. The short-circuit system in the Distribution Equipment Laboratory the Institute of Power

Engineering in Warsaw is powered from the medium-voltage network, protected by a main power switch. Short-circuit current values are set using appropriate MV shunt reactors connected directly to short-circuit transformers, from which the outputs of secondary windings are connected to the tested object. By using current transformers and appropriate recorders, it was possible to record the exact values of short-circuit current waveforms. The short-circuit system is shown in Figure 3 below.

(a) (b)

Figure 3. Short-circuit system in the Distribution Equipment Laboratory of the Institute of Power Engineering where short-circuit tests of the LV switchgear were performed: (**a**) short-circuit reactors; (**b**) short-circuit transformers.

After preparing the test object and the short-circuit test station, it was possible to perform short-circuit tests. In the performed test formula, it was assumed that the trials would be performed for two and for one jumper of a horizontal rail bridges. This made it possible to analyze the distribution of the short-circuit current in the current path in both cases. Tests for both one and two armatures were performed for different values of three-phase short-circuit currents with a time of 150 ms, steadily increasing the value: 5 kA, 15 kA, 20 kA, and 30 kA RMS (root mean square). After each test, insulators and brackets were analyzed and checked for damage. If necessary, insulators and support elements were replaced when they showed any signs of damage. The first considered case was the performance of a short-circuit test for two jumpers installed on both horizontal current paths (Figure 4 below). All performed short-circuit tests for the values of 5 kA, 15 kA, and 20 kA did not cause any damage to the insulators or plastic deformations of the supporting structures (metal brackets and holders). The last test for the value of 30 kA RMS caused damage to the first two lower insulators of the current path. The peak current value in the L1 phase busbar was over 67 kA, in the L2 phase busbar over 57 kA, and in the L3 phase busbar, the peak value was over 48 kA. As it turned out, these values were destructive for the lower insulators of the switchgear, because the short-circuit current was closed by the lower horizontal current path, which did not cause destructive stresses in the upper insulators of the vertical current path. An example of a short-circuit current waveform from the tests is shown in Figure 4 below. The damage caused is shown in Figure 5 below.

Photos of the dismantled insulators are shown in Figure 5 below; as it is possible to see, after the short-circuit test, the insulators were unusable, and they were completely destroyed. The moment of arcing was shown in Figure 6 below. For the test object with two jumpers, no more tests were performed in the laboratory. All damaged insulators and one deformed insulator bracket was replaced. The tested object was restored to its original state in order to perform subsequent short-circuit tests.

The recorded waveforms of the surge current were imported to the 3D model of the tested LV switchgear during short-circuit calculations using the Ansys Maxwell 3D software 2023.

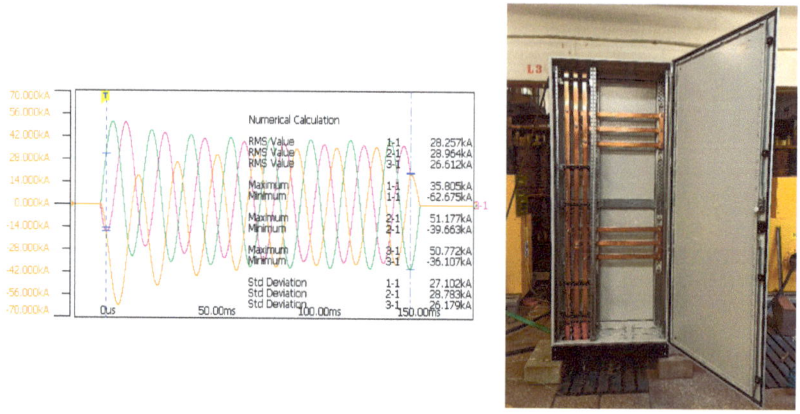

(**a**) (**b**)

Figure 4. Short-circuit test preparation: (**a**) example of short-circuit current waveform; (**b**) test object mounted and prepared in short-circuit station—two jumpers.

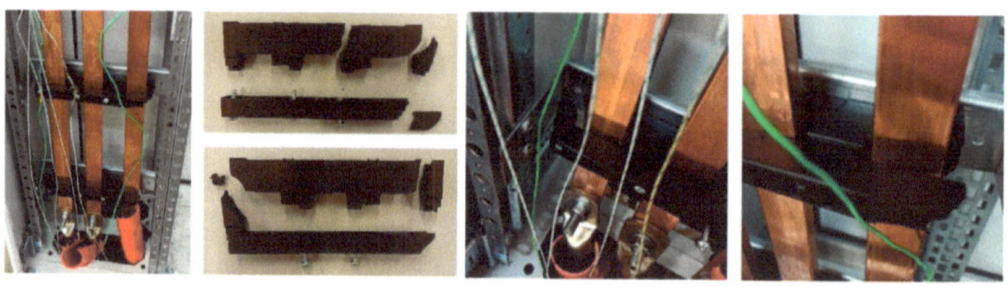

Figure 5. Short-circuit test results with two jumpers mounted—damaged support insulators, before dismounting and after dismounting.

Figure 6. Short circuit tests results with two jumpers mounted—30 kA RMS: moment of recorded arcing and displacement of the busbar.

The next stage of the short-circuit tests was to execute tests for the system of current paths with one copper jumper at the end of the horizontal upper current path. It was assumed that in this way the path of the short-circuit current flowing to the fault point in the current path will become longer. For this purpose, the lower armature was dismantled and the test object was prepared for short-circuit tests. The tests were performed for the same values of three-phase short-circuit currents as in the previous tests: 5 kA, 15 kA, 20 kA, and 30 kA RMS. After each test, the insulators and brackets were analyzed and checked for damage. If necessary, insulators and support elements were replaced when they showed any signs of damage. As was the case in previous tests, for the values of three-phase RMS short-circuit currents, 5 kA, 15 kA, and 20 kA, they did not cause any damage to the insulators or plastic deformation of the supporting structure (metal brackets and holders). The only phenomenon that could be seen was the "tightening" of the busbars and the cable as a result of electrodynamic interactions caused by the flow of the short-circuit current. It can be said that the results were similar to those obtained in previous short-circuit tests. The extent of damage in the low-voltage switchgear was different for the three-phase short-circuit current of RMS 30 kA. Due to electrodynamic impacts, the entire vertical current path with a rail spacing of 60 mm was damaged, in which practically all supporting insulators were broken. The horizontal bridges were not affected due to the use of stronger insulators with rail spacing of 100 mm. Examples of damage to support insulators are shown in Figure 7 below. In order to illustrate the moment of damage to the support insulators, the short-circuit test was recorded with a high-speed camera, and the recorded frame is shown in Figures 7 and 8 below. It is noticeable that the copper conductor rail was torn out of the insulators and caused a metallic short circuit with the metal supporting elements of the switchgear. Two insulators of the vertical rail bridge were damaged. For the test object with two jumpers, no more tests were performed in the laboratory.

Figure 7. Short-circuit tests results with one jumper mounted—during the test for the peak values of the short-circuit current: L1 = 67 kA, L2 = 57 kA, L3 = 48 kA: damaged current path.

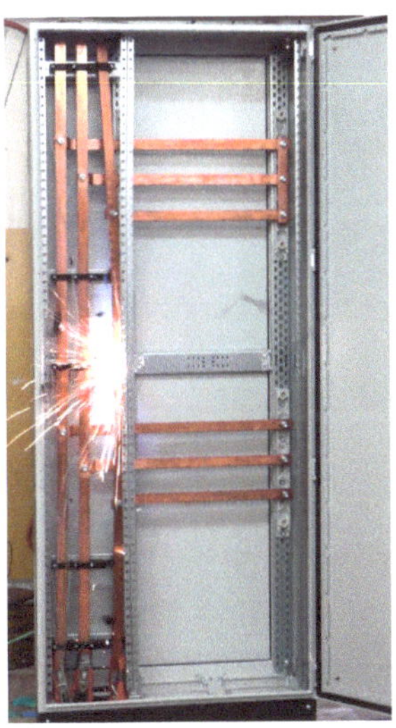

Figure 8. The recorded moment of damage to the support insulators of the vertical busbar bridge during the peak values of the short-circuit current: L1 = 67 kA, L2 = 57 kA, L3 = 48 kA in the tested LV switchgear with one jumper mounted.

4. FEM Simulations Assumptions

In order to perform simulation work for this manuscript, a coupled finite element method analysis was executed. Coupled FEM analysis is a technique that allows us to study the interactions of various physical phenomena, including structural, mechanical, and electromagnetic phenomena. It is carried out by corelating various nodes of a numerical program in order to fulfil specific conditions and recreate physical phenomena in a digital environment that can be witnessed before concluding the experiment. Coupled FEM analysis is based on the transformation of conjugate partial differential equations in a set of unrelated ordinary differential equations using the modal decomposition technique. It follows that the system is represented by a linear combination of its modes, which are the characteristic patterns of vibration or oscillation of each domain. Modal decomposition is a solution to solve complex numerical problems in the most direct way. The modal coefficients can then be used to reproduce the application in each domain. Coupled analysis offers several advantages over other methods of solving coupled problems using the FEM, such as the ability to deal with non-linearities and varying material properties, taking into account mechanical interactions, energy dissipation, and inter-domain energy transfer, gaining insight into the physical mechanisms and dominant modes of the system, and reducing computational costs and memory requirements. Below are the most important equations that were inserted into a module solver during calculations.

Faradays law:

$$\nabla \times \vec{E} = -\frac{\partial \vec{B}}{\partial t} \quad (1)$$

Ampere law:

$$\nabla \times \vec{H} = \vec{J} + \frac{\partial \vec{D}}{\partial t} \quad (2)$$

Gauss law for electricity:

$$\nabla \cdot \vec{D} = \rho \quad (3)$$

Gauss law for magnetism:

$$\nabla \cdot \vec{B} = 0 \quad (4)$$

where $\vec{D}$—electric induction [C/m^2]; $\vec{B}$—magnetic induction [T]; $\vec{E}$—electric field intensity [V/m]; $\vec{H}$—magnetic field intensity [A/m]; $\vec{J}$—current density [A/m^2]; ρ—electric charge density [C/m^3]; ∇—divergence operator [1/m]; ∇x—rotation operator [1/m].

In Figure 9 below, functions of the module are presented alongside the governing equations behind them.

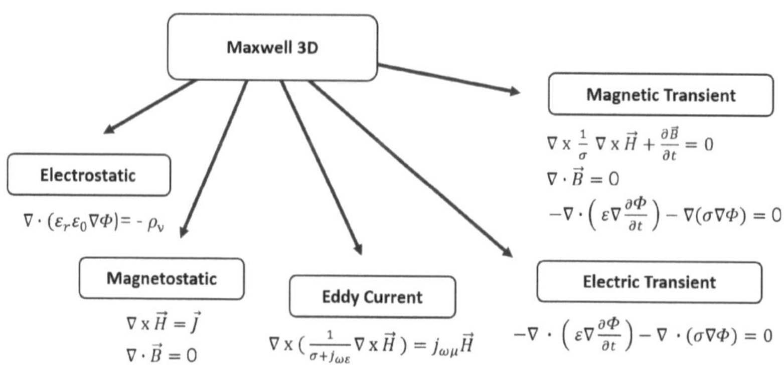

Figure 9. Functions for analyzing physical phenomena of Maxwell's module and mathematical formulas behind them.

Deformations of busbar during simulations were described by Lorentz as a force that can be expressed using the equation below:

$$F = \int J \cdot B \, dvol \tag{5}$$

Thermal interactions were not taken into consideration during the numerical analysis as the given simulation time was too short—150 ms. Moreover, taking into account the thermal interactions would made simulations much more complex and lead to results without physical meaning.

Boundary conditions used at least one of the following sources for their magnetic field:

- Stranded or solid windings with voltage or current supply;
- A permanent magnet;

Outer boundary conditions were defined as follows:

- The default boundary conditions;
- An odd symmetry boundary;
- An even symmetry boundary.

5. Simulations Results

Following the short-circuit tests that were executed in the Distribution Equipment Laboratory at the Institute of Power Engineering in Warsaw, simulation analyses were performed to compare to the obtained experimental results. In order to perform numerical analyses, it was necessary to prepare a detailed 3D model for the tested switchgear structure. The model was developed in the ANSYS SpaceClaim software 2023, where all elements of the switchgear structure were accurately reproduced with some simplifications, such as housing, support brackets, insulators, or a single-strip three-phase busbar bridge with a vertical/horizontal arrangement. The model was prepared analogically to the short-circuit tests of the switchgear for one and two jumpers at the ends of horizontal busbar bridges. The geometries prepared in this way were imported into the ANSYS Maxwell 3D environment, as shown in Figure 10 below.

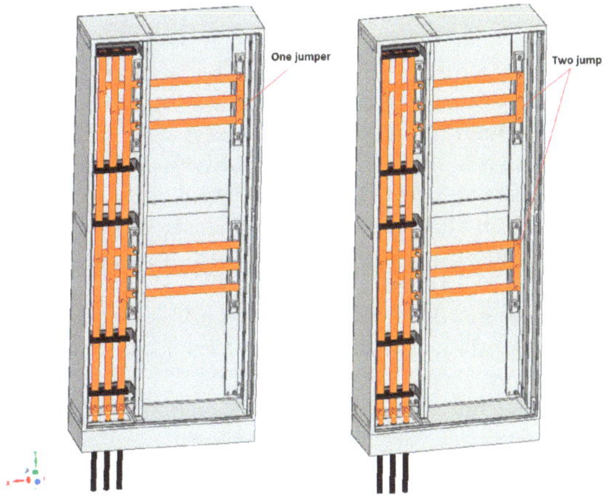

Figure 10. Prepared detailed 3D models of the switchgear for two tested cases: with two and one jumpers in horizontal current paths.

In both cases, after importing the 3D models into the ANSYS Maxwell 3D software 2023, the real material properties were provided. Boundary conditions corresponding to those prevailing during the short-circuit tests in the Distribution Equipment Laboratory were provided, such as temperature, and the ambient environment in this case was air. In the next stage, the most important operation was to load into the model the short-circuit current waveforms previously recorded during short-circuit tests for individual phases. For this purpose, a file in the .csv format was used, from which data in the form of tables were downloaded to the Maxwell 3D module. The short-circuit current values prepared in this way for each phase were assigned to individual busbars in the current path of the switchgear, as was the case during the actual short-circuit tests. The imported short-circuit currents for individual phases are shown in the waveform generated from the Maxwell 3D software 2023 in Figure 11 below. At a later stage of simulation preparation, it was necessary to create an appropriate mesh, which required additional densification in places most exposed to electrodynamic forces: current path busbars, insulators, or support brackets.

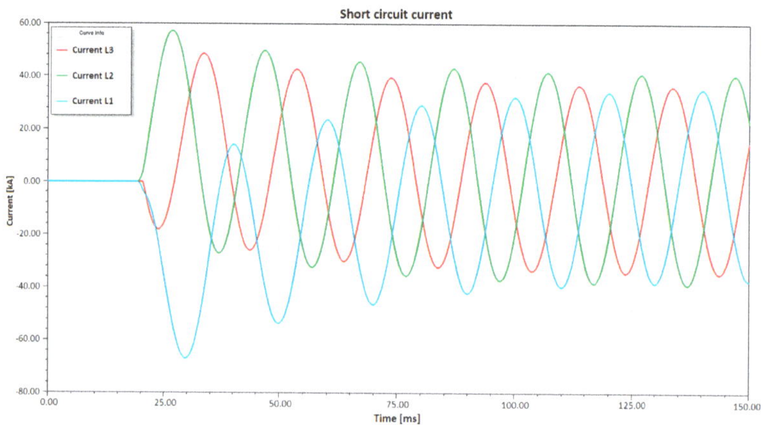

Figure 11. Short-circuit current waveform for L1, L2 and L3 imported to ANSYS Maxwell 3D 2023–recorded during short circuit tests in the Distribution Equipment Laboratory at the Institute of Power Engineering in Warsaw.

For the case of the model with one armature, the mesh size consisted of 687,264 tetrahedral elements. However, for the model with two armatures, the mesh had more tetrahedral elements, with 698,629 elements. In the next step, the simulation time was defined, and, due to the fact that the imported short-circuit current was for 150 ms, the simulation time was assumed to be the same. The same settings were adopted for both analyzed cases. The main purpose of the simulation analyses performed was to determine the values of the resulting electrodynamic forces on individual elements of the switchgear structure. In addition, deformations and stresses that arose at the time of occurrence of electrodynamic forces were analyzed. In the analysis, special attention was paid to comparing the actual deformations with the deformations obtained during the performed numerical analyses.

After performing the simulation calculations in the Maxwell 3D module, it was possible to generate waveforms of the obtained values of electrodynamic forces for selected points. The maximum values of electrodynamic forces were obtained on the vertical part of the current track, due to the smaller rail spacing (60 mm) compared to the horizontal parts of the current paths, where the spacing was 100 mm between each busbar. The obtained maximum values of electrodynamic forces for the analyzed cases are presented in the graphs below (Figure 12). The maximum values of electrodynamic forces in the current path differ for both analyzed cases. For switchgear with two jumpers, the maximum obtained values of electrodynamic forces are lower by half compared to the case for one jumper in the upper part of the horizontal current path, though the current path closes at a different point for both cases, which is clearly shown in Figure 13 below.

Other fault locations, as can be seen from simulations and experimental tests, are of great importance to the strength of insulators and the supporting structure of the switchgear under the influence of electrodynamic interactions. For the model, for one armature, it is possible to observe very high values of electrodynamic forces in the vertical current path. During the experimental tests, as is known, all support insulators in the vertical part of the current circuit with a rail spacing of 60 mm were damaged. For the system with two jumpers, much less damage to the switchgear can be observed. Electrodynamic forces affect only a part of the vertical current path, which is precisely presented in the analysis performed.

The distribution of stresses on the switchgear supporting the structures for both the considered cases of the same switchgear is completely different. Different stress values can be observed, which is of great importance if we consider the strength of insulators and supporting elements in impacting electrodynamic forces. In order to illustrate how both cases look, Figure 14a shows the distribution of "Body Force Density" imported from the Maxwell 3D module to Ansys Transient Structural Mechanical. As shown in Figure 14b below, in the case with two jumpers at the ends of the horizontal parts of the current path, the highest force values expressed in N/m^2 are at the bottom of the vertical current path and the middle busbar of the lower horizontal current path. This is because the flowing fault current follows the shortest possible path to the fault point where there is a lower resistance value. In the case of one armature, the short-circuit current flows only to one short-circuit point, where the electrodynamic forces act along the entire length of the vertical part of the current path and on its upper horizontal part. This causes very high values for electrodynamic forces at the level of 50 kA in the current peak of the analyzed case. Performing additional coupled analysis in Ansys Transient Structural Mechanical allowed us to determine the values of stresses and strains occurring on the insulators and supporting elements of the switchgear. In order to perform this type of analysis in the ANSYS Workbench environment, a module for performing mechanical analyses over time was used. Geometry and calculation results from the ANSYS Maxwell 3D electromagnetic module were imported into this module. In the module for mechanical calculations over time, a new "mesh" had to be additionally generated for this type of calculation. In the case of the analyzed model, for one jumper, the grid amounted to 2,409,405 elements and 4,737,480 computational nodes. For the second case with two jumpers, the grid was more extensive and amounted to 2,519,877 elements and 4,854,431 computational nodes. An example of the density of the mesh computational grid to illustrate the scale is shown in Figure 15 below.

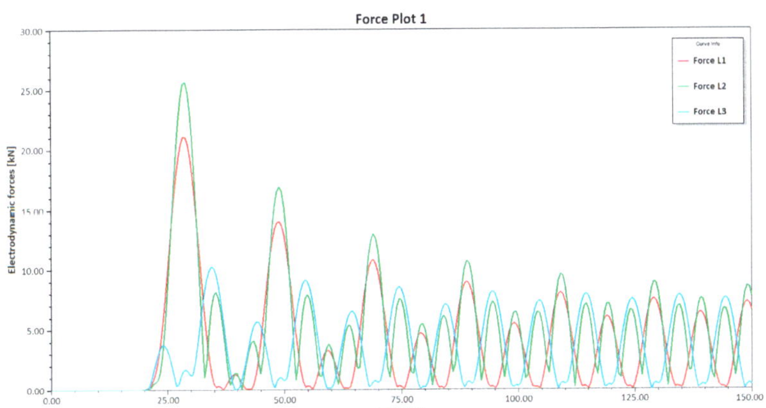

(**a**)

Figure 12. *Cont.*

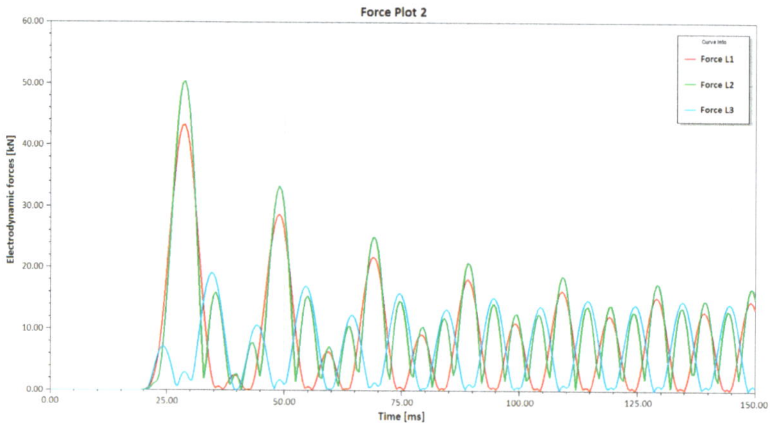

(**b**)

Figure 12. Waveforms obtained for the maximum values of electrodynamic forces for the switchgear with (**a**) two jumpers at the ends of the horizontal parts of the current circuit; (**b**) one jumper at the end of the horizontal parts of the current circuit.

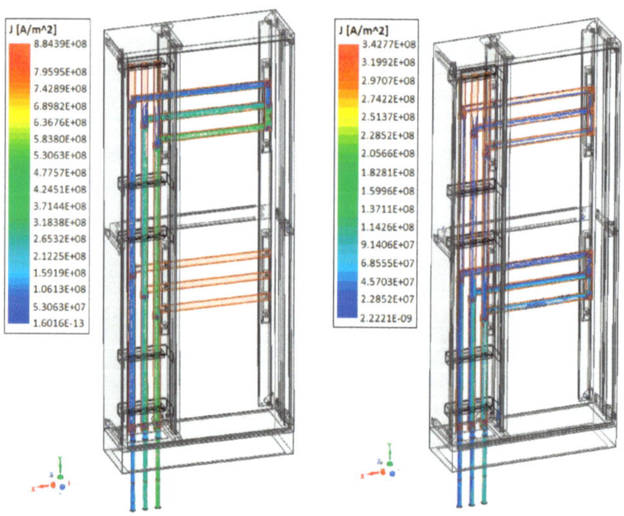

Figure 13. Maximum values of electrodynamic forces obtained for switchgear variant with two jumpers at the ends of the horizontal parts of the current circuit.

After setting the simulation time to 150 ms, it was possible to perform computational analyses for both analyzed cases. As a result of the obtained results, it was possible to analyze stresses and strains in the tested switchgear models. For this purpose, the "Equivalent Elastic Strain" option was used to generate the places most exposed to stress in the support insulators; an example for a selected time point for a system with one armature is shown in Figure 15 below. For the system with two jumpers, it looks analogous, and the highest stress values arise in the places where the busbars are fastened in the insulators to which they are screwed; an example for this system is shown in Figure 16 below.

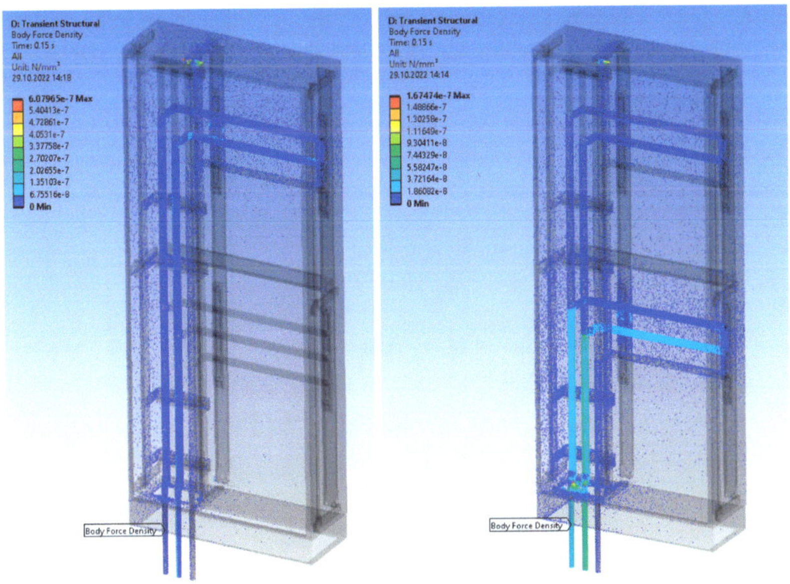

(**a**) (**b**)

Figure 14. Generated "Body force density" values for: (**a**) one jumper of horizontal current paths; (**b**) two jumpers of horizontal current paths.

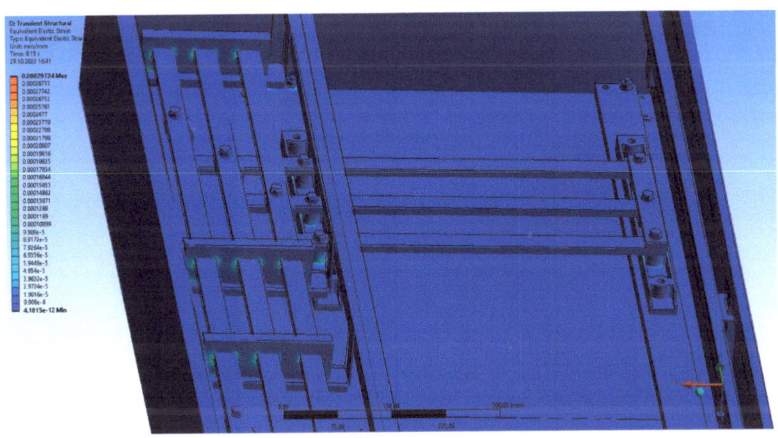

Figure 15. Stress distribution on support insulators for a selected time point of 28 ms during the short-circuit current flow for a system with one jumper in the upper horizontal current path.

Another dependence confirming that the obtained simulation and experimental results have a high level of convergence is the deformation of the current rail of the lower horizontal busbar bridge. This relationship is shown in Figure 17 below.

The experimental and simulation studies conducted in the Distribution Equipment Laboratory at the Institute of Power Engineering in Warsaw, presented the possibility of using coupled analyses to determine the value of electrodynamic forces in low-voltage switchgears. Using a real object and a virtual 3D model for research, a very high level of convergence and accuracy of the obtained results was proven. The results of deformations during the analysis showed high levels of convergence and correctness of the proposed methodology for determining the exact values of electrodynamic forces in low-voltage switchgears. The results obtained from non-standard and expensive research at an accredited research institute were used to verify the correctness of the assumptions made in this work. The presented methodology allows for the analysis of the values of generated electrodynamic forces in each millisecond of the duration of the flowing short-circuit current, in each element, and at each point of the structure of the analyzed low-voltage switchgear.

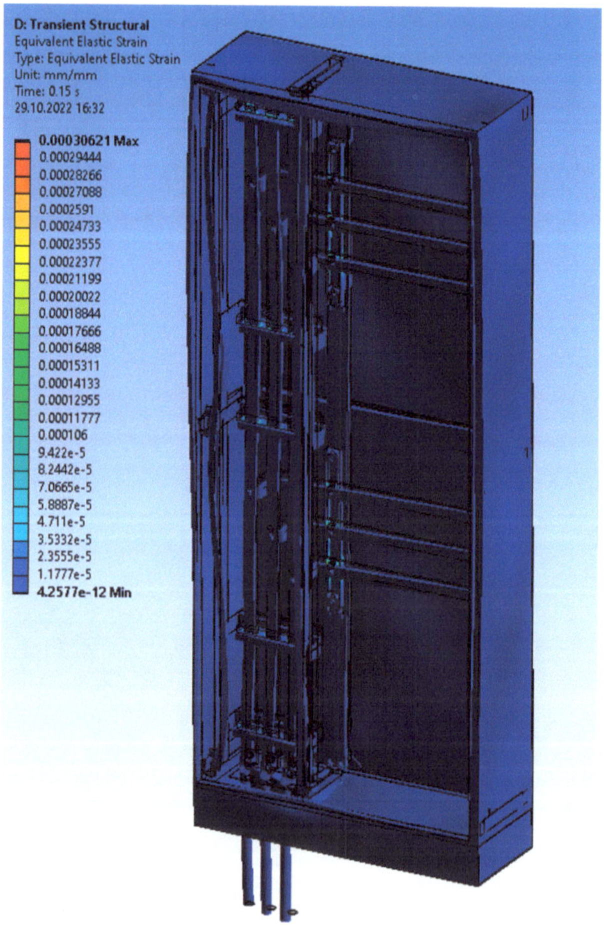

Figure 16. Stress distribution on support insulators for a selected time point of 28 ms during the short-circuit current flow for a system with two jumpers in horizontal current paths.

This type of methodology with the use of accurate, structural 3D models of the designed low-voltage switchgears can be successfully used in R&D departments. Each modification and change of the structure can be quickly validated and allows you to quickly obtain information on the generated electrodynamic exposures in the designed current circuits of low-voltage switchgears. At the same time, each change in the structure in the 3D construction model of the switchgear can be quickly validated through simulation, which, in the process of creating a prototype of current circuits or selection of a supporting structure, is extremely important in terms of cost reduction and time reduction or the number of necessary prototypes before performing type tests in accredited laboratories during the certification of the manufactured product.

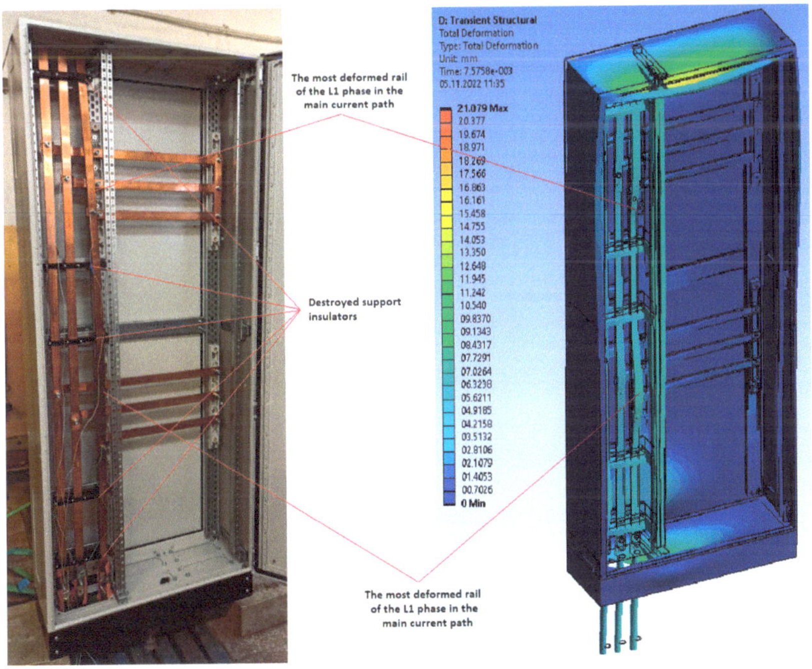

Figure 17. Distribution of deformations on the support insulators for the time of 28 ms during the short-circuit current flow for the system with one jumper in the upper horizontal current path of the switchgear.

The switchgear and insulators under normal conditions and their rated parameters would not be damaged. Additionally, damaging them was not the purpose or aim of this publication. The electrodynamics tests were carried out during abnormal conditions without switchgear or any main circuit breaker that interrupts short-circuit currents.

6. Conclusions

We performed simulation analyses in the Ansys Transient Structural Mechanical module which also allowed for a thorough analysis of deformations and displacements of both the copper current path and the elements of the supporting structure of the entire switchgear. In this case, special attention was paid to the compliance of the results obtained with the analyzed deformations in the simulations with deformations of the current path obtained after short-circuit tests executed in the Distribution Equipment Laboratory at the Institute of Power Engineering for the tested switchgear. For the case of the tested LV switchgear with one armature, the obtained simulation results with high precision reflect the deformations caused during the actual short-circuit test for a short-circuit current of 30 kA RMS (peak current values: L1 = 67 kA, L2 = 57 kA, L3 = 48 kA). In order to compare the obtained results, two drawings showing the largest deformations in the L1 phase are presented in Figure 17 above.

Analyzing and comparing the results obtained from the experimental and simulation tests, the first basic dependence is the occurrence of the largest deformation on the vertical busbar of the L1 phase. In the figure above (on the left), it can be seen that during the numerical analysis the busbar of the L1 phase is the most deformed, but the insulators limit its further plastic deformation. The drawing from the experimental tests (on the right) shows that the rail has been torn out of all the support insulators in the vertical part of the current path.

Author Contributions: Conceptualization, S.Ł., M.S., Ł.K., P.S., M.O., P.B., T.Ż. and A.L.; data curation, S.Ł., M.S., Ł.K., P.S., M.O., P.B., T.Ż. and A.L.; formal analysis, S.Ł., M.S., Ł.K., P.S., M.O., P.B., T.Ż. and A.L.; investigation, S.Ł., M.S., Ł.K., P.S., M.O., P.B., T.Ż. and A.L.; methodology, S.Ł., M.S., Ł.K., P.S., M.O., P.B., T.Ż. and A.L.; project administration, S.Ł., M.S., Ł.K., P.S., M.O., P.B., T.Ż. and A.L.; resources, S.Ł., M.S., Ł.K., P.S., M.O., P.B., T.Ż. and A.L.; supervision, S.Ł., M.S., Ł.K., P.S., M.O., P.B., T.Ż. and A.L.; validation, S.Ł., M.S., Ł.K., P.S., M.O., P.B., T.Ż. and A.L.; visualization, S.Ł., M.S., Ł.K., P.S., M.O., P.B., T.Ż. and A.L.; writing—original draft preparation, S.Ł., M.S., Ł.K., P.S., M.O., P.B., T.Ż. and A.L.; writing—review and editing, S.Ł., M.S., Ł.K., P.S., M.O., P.B., T.Ż. and A.L. All authors have read and agreed to the published version of the manuscript.

Funding: This research received no external funding.

Institutional Review Board Statement: Not applicable.

Informed Consent Statement: Not applicable.

Data Availability Statement: The original contributions presented in the study are included in the article, further inquiries can be directed to the corresponding author.

Conflicts of Interest: Author Sebastian Łapczyński was employed by the company ELKO-BIS Systemy Odgromowe. Author Michał Szulborski was employed by the company Symkom Sp. z o.o. Author Michał Szulborski was employed by the company ETI Polam Sp. z o.o. The remaining authors declare that the research was conducted in the absence of any commercial or financial relationships that could be construed as a potential conflict of interest.

References

1. Bun, R.; Krzyszkowski, A. Mathematical models of electrodynamic influences on an example of the flat inductor. In Proceedings of the Modern Problems of Radio Engineering, Telecommunications and Computer Science (IEEE Cat. No.02EX542), Lviv-Slavsko, Ukraine, 18–23 February 2002; pp. 24–25.
2. Triantafyllidis, D.G.; Dokopoulos, P.S.; Labridis, D.P. Parametric short-circuit force analysis of three-phase busbars-a fully automated finite element approach. *IEEE Trans. Power Deliv.* **2003**, *18*, 531–537. [CrossRef]
3. Shen, B.; Zhang, X.; Zhou, D. Analysis and calculation of short-circuit electro-dynamic forces on rectangular bus bars. In Proceedings of the 2009 IEEE 6th International Power Electronics and Motion Control Conference, Wuhan, China, 17–20 May 2009; pp. 2618–2621.
4. Yusop, F.M.; Jamil, M.K.M.; Ishak, D.; Masri, S. Study on the electromagnetic force affected by short-circuit current in vertical and horizontal arrangement of busbar system. In Proceedings of the International Conference on Electrical, Control and Computer Engineering 2011 (InECCE), Kuantan, Malaysia, 21–22 June 2011; pp. 196–200.
5. Mouhamad, M.; Elleau, C.; Mazaleyrat, F.; Guillaume, C.; Jarry, B. Short-Circuit Withstand Tests of Metglas 2605SA1-Based Amorphous Distribution Transformers. *IEEE Trans. Magn.* **2011**, *47*, 4489–4492. [CrossRef]
6. Komurgoz, G.; Gundogdu, T. Effect of the tap winding configurations on the electromagnetic forces acting on the concentric transformer coils. In Proceedings of the 2013 3rd International Conference on Electric Power and Energy Conversion Systems, Istanbul, Turkey, 2–4 October 2013; pp. 1–6.
7. Vilhena, N.; Taillacq, A.; Pronto, A.; Murta-Pina, J.; Álvarez, A. Analysis of Electromagnetic Forces in Superconducting Fault-Current Limiters Under Short-Circuit Condition. *IEEE Trans. Appl. Supercond.* **2016**, *26*, 1–4. [CrossRef]
8. Popa, I.A.; Dolan, D.I. Numerical modeling of three-phase busbar systems: Calculation of the thermal field and electrodynamic forces. In Proceedings of the 2016 International Conference on Applied and Theoretical Electricity (ICATE), Craiova, Romania, 6–8 October 2016; pp. 1–9.
9. Kadkhodaei, G.; Sheshyekani, K.; Hamzeh, M. Coupled electric–magnetic–thermal–mechanical modelling of busbars under shortcircuit conditions. Generat. *Transmiss. Distrib.* **2016**, *10*, 955–963. [CrossRef]
10. Aprahamian, B.; Marinova, M.; Dimova, T. Assessment of the Influence of Magnetron Sputtered Coatings on the Electrodynamic Forces of Current-Carrying Busbars. In Proceedings of the 2018 20th International Symposium on Electrical Apparatus and Technologies (SIELA), Bourgas, Bulgaria, 3–6 June 2018; pp. 1–4.
11. Popov, I.V.; Evdokunin, G.A.; Syutkin, A.V. Computation Modelling of Electrodynamic Forces in Transformers and Reactors. In Proceedings of the 2020 IEEE Conference of Russian Young Researchers in Electrical and Electronic Engineering (EIConRus), Petersburg, Russia, 27–30 January 2020.
12. Popov, I.V.; Evdokunin, G.A.; Popov, V.V. Influence of Transformer's Operating Mode for Value of Electrodynamic Force. In Proceedings of the 2020 Ural Smart Energy Conference (USEC), Ekaterinburg, Russia, 13–15 November 2020; pp. 88–92.
13. Piechocki, P.; Frackowiak, R.; Dombek, G.; Chmielewska, I. Estimation of Short Circuit Duration for the Assessment of Dynamic Effects in Distribution Substations with Flexible Conductors. *Energies* **2020**, *13*, 970. [CrossRef]
14. Yang, J.; Liu, Y.; Hu, D.; Wu, B.; Che, B.; Li, J. Transient electromagnetic force analysis of GIS bus based on FEM. In Proceedings of the International Conference on Condition Monitoring and Diagnosis (CMD), Xi'an, China, 25–28 September 2016; pp. 554–557.

15. Maksymiuk, J.; Pochanke, Z. *Calculations and Diagnostic Tests of Switchgear*; WNT: Warsaw, Poland, 2001; ISBN 83-204-2523-9.
16. Kulas, S. *Current Paths and Contact System*; OWPW: Warsaw, Poland, 2008; ISBN 978-83-7207-774-5.
17. Jiaxin, Y.; Ruichao, W.; Huimin, L.; Longqing, B.; Hongjian, W. Research on the calculation methods of enclosed isolated phase busbar in short-circuit condition. In Proceedings of the IEEE 62nd Holm Conference on Electrical Contacts (Holm), Clearwater Beach, FL, USA, 9–12 October 2016; pp. 111–114.
18. Park, S.W.; Cho, H. A practical study on electrical contact resistance and temperature rise at at the connections of the copper busbars in switchgears. In Proceedings of the IEEE 60th Holm Conference on Electrical Contacts (Holm), New Orleans, LA, USA, 12–15 October 2014; pp. 1–7.
19. Jamali-Abnavi, A.; Hashemi-Dezaki, H. Evaluating the electromagnetic forces on the electric arc furnaces power cables under various operation conditions. *Electr. Power Syst. Res.* **2021**, *201*, 107553. [CrossRef]
20. Gatherer, J.; Jackson, R.L. A multi-variable parametric study on the performance of bolted busbar contacts. In Proceedings of the IEEE 61st Holm Conference on Electrical Contacts (Holm), San Diego, CA, USA, 11–14 October 2015; pp. 124–131.
21. Borkowski, P.; Bartosik, M.; Rodak, M.; Sienicki, A.; Wójcik, F.; Panev, B.; Rodrig-Mateos, F.; Siemko, A. Ultrafast, Redundant, and Unpolarized DC Energy Extraction Systems for the Protection of Superconducting Magnet Circuits. *IEEE Trans. Ind. Electron.* **2021**, *69*, 9895–9904. [CrossRef]
22. Szulborski, M.; Łapczyński, S.; Kolimas, Ł.; Tyryk, M. Electrodynamic Forces in a High Voltage Circuit Breakers with Tulip Contact System—FEM Simulations. *IEEE Access* **2022**, *10*, 99299–99320. [CrossRef]
23. Zhong, J.Y.; Wu, S.J.; Wang, Z.; Guo, Y.J.; Qin, X.Y. A 3-D steady-state analysis of thermal behavior in EHV GIS busbar. *J. Electr. Eng. Technol.* **2016**, *11*, 781–789.
24. Foroughi Nematollahi, A.; Vahidi, B. 3D FEM analysis of induced current in the shield of the buried cable by non-vertical channel. *Electr. Power Syst. Res.* **2022**, *212*, 108256. [CrossRef]
25. Szulborski, M.; Łapczyński, S.; Kolimas, Ł. Increasing Magnetic Blow-Out Force by Using Ferromagnetic Side Plates inside MCB. *Energies* **2022**, *15*, 2776. [CrossRef]
26. Scott, A.; Xu, W.; Tefferi, M.; Laso, A.; Uzelac, N. Conference documents: Design of robust, safe and reliable switchgear with electro-thermal numerical simulations. In Proceedings of the 22nd International Symposium on High Voltage Engineering (ISH 2021), Hybrid Conference, Xi'an, China, 21–26 November 2021; pp. 1351–1356. [CrossRef]

computers

MDPI

Article

Observer-Based Suboptimal Controller Design for Permanent Magnet Synchronous Motors: State-Dependent Riccati Equation Controller and Impulsive Observer Approaches

Nasrin Kalamian [1], Masoud Soltani [2], Fariba Bouzari Liavoli [3] and Mona Faraji Niri [4,*]

[1] Department of Electrical Engineering, Tafresh University, Tafresh 3951879611, Iran; nkalamian@tafreshu.ac.ir
[2] Department of Electrical Engineering, Pooyesh Institute of Higher Education, Qom 8776637116, Iran; m.soltani@pooyesh.ac.ir
[3] Department of Electrical Engineering, Danesh Alborz University, Qazvin 3441152278, Iran; fbouzari@alborzu.ac.ir
[4] Warwick Manufacturing Group, University of Warwick, Coventry CV4 7AL, UK
* Correspondence: mona.faraji-niri@warwick.ac.uk

Abstract: Permanent Magnet Synchronous Motors (PMSMs) with high energy efficiency, reliable performance, and a relatively simple structure are widely utilised in various applications. In this paper, a suboptimal controller is proposed for PMSMs without sensors based on the state-dependent Riccati equation (SDRE) technique combined with customised impulsive observers (IOs). Here, the SDRE technique facilitates a pseudo-linearised display of the motor with state-dependent coefficients (SDCs) while preserving all its nonlinear features. Considering the risk of non-available/non-measurable states in the motor due to sensor and instrumentation costs, the SDRE is combined with IOs to estimate the PMSM speed and position states. Customised IOs are proven to be capable of obtaining quality, continuous estimates of the motor states despite the discrete format of the output signals. The simulation results in this work illustrate an accurate state estimation and control mechanism for the speed of the PMSM in the presence of load torque disturbances and reference speed changes. It is clearly shown that the SDRE-IO design is superior compared to the most popular existing regulators in the literature for sensorless speed control.

Keywords: permanent magnet synchronous motor; state-dependent Riccati equation; pseudo-linearisation; impulsive observer; speed control

1. Introduction

Permanent Magnet Synchronous Motors (PMSMs) have advantages such as constant speed, an adjustable power factor, high efficiency, field winding elimination, DC power supply rejection, Joule loss deletion in excitation circuits, omission in brushes and slip rings, a short radial length, and easy maintenance [1–5]. They have simple configurations and easy-to-understand operation mechanisms which make them ideal to be used as actuators in many applications [6,7].

The control of PMSMs has progressed very well in the last few decades. Examples include works such as [8], where Nonlinear Model Predictive Speed Controllers (NMPSCs) are proposed to predict and control the positions and speed at each sampling period through the variable forecast horizon. In [9], to estimate and control the rotor speed and position, an Extended Kalman Filter (EKF) is designed for a sensor-free vector control drive. In [10,11], linear model prediction controllers (MPCs) with the addition of least mean square (LMS) identification strategies and online parameter detection are presented for permanent magnet synchronous generators (PMSGs). In [12], a Robust Adaptive Backstepping Control (RABC) is suggested for High-Speed Permanent Magnet Synchronous Motors (HSPMSM). In [13], a Sliding Mode Observer (SMO) and Sliding Mode Control (SMC) are, respectively, employed to estimate and control the motor speed, torque, and current with third-degree harmonic current. In [14], a model reference adaptive system (MRAS) for estimating and controlling the velocity in internal PMSMs with bandwidth modulation is offered. Furthermore, in [15], an MRAS controller is utilised for sensorless speed control. The authors used the pseudo-linearisation technique and produced state-dependent matrices. Then, the state-dependent matrices were applied to an MRAS mechanism. It should be noted that only the pseudo-linearisation concept was used, and the Riccati equation technique was not investigated in [15]. Moreover, an adaptive, interconnected observer was suggested in order to estimate the speed. In [16], a robust adaptive observer-based approach for the sensorless speed control of PMSMs is studied to estimate the rotor shaft speed and position. In [17], an Adaptive Nonlinear Continuous Observer (ANCO) is used to estimate the self-induction coefficient, stator resistance, rotor velocity/position, and load torque by using the current measurements for the PMSM. In [18], adaptive fuzzy RST digital control (AFRSTD), digital control, and fuzzy digital RST control are generated for a PMSM system. In [19], a Sliding Mode focused on an Extended State Observer (SMESO) and Fast Terminal Sliding Mode Control (FTSMC) is recommended to develop disturbance elimination capacity and increase the dynamic performance of PMSM drive systems.

Reviewing the literature as summarised above reveals that most of the controllers and observers designed for PMSMs have been based on approaches that depend on linear or linearised models. These approaches, although simple, are not ideal for dealing with the inherent nonlinearities of a PMSM. Conventional linearisation methods such as Jacobian approximations do not preserve critical information about the nonlinear mechanisms and are usually time-invariant to reduce the controller design efforts. However, these aspects all reduce the control performance. In the same context, Linear Quadratic Regulators (LQRs) have been shown to perform very well for many linear systems. Therefore, numerous efforts have been made to expand them to nonlinear dynamical systems. The introduction of the state-dependent Riccati equation (SDRE) [20–23] technique was an effort to fulfil this desire. The SDRE technique is a state-dependent control method based on a pseudo-linearisation display which produces a optimal plan for the cost function of a suboptimal solution [20–23]. In the SDRE technique, the nonlinear system is transformed into a pseudo-linearised matrix depending on states without any approximation. This pseudo-linearisation representation might not be unique, and there might be countless versions that would make the degrees of freedom of the controller infinite. This is one of the aspects that requires research for its application in different systems.

The SDRE technique has been applied in many applications, such as robotics, navigation systems, Unmanned Aerial Vehicles (UAVs), submarines, Heating Ventilation and Air Conditioning Systems (HVACs), healthcare, and power systems [21–23]. In [24], a suboptimal SMC was obtained from a mix of the SDRE and SMC for a class of nonlinear closed-loop systems for the Scout robot. In [25], an SDRE controller was designed for UAVs using the Artificial Potential Field (APF) approach. In [26], the nonlinear SDRE was applied for the fault-tolerant control of a DC micro-grid (MG) in the power system.

On the other hand, the SDREs face challenges regarding access to the system's states for feedback design. Here, state observers can provide a solution. In [27], the SDRE method is utilised for both control and observation purposes for an Air Handling Unit (AHU) system. The Riccati equation is solved at every sample time, and a suboptimal control law and an estimated state vector are produced based on its solution. It should be noted that the SDRE observer needs continuous data from output sensors at each sample time. In this regard, impulsive systems and observers are well suited for simulating real-world systems in which internal states change instantaneously. Compared to classical continuous and discrete systems, impulsive systems can contain sudden, abrupt, and intermittent changes at variable time intervals [28,29]. In such a mechanism, an impulse could model the sudden change that happens in the system state. The time intervals between these impulses can be fixed or variable with time, known or unknown, and deterministic or stochastic. In [30], an impulsive observer (IO) is introduced based on state-dependent matrices for nonlinear time-delay systems. Furthermore, an adaptive IO (AIO) is suggested for a class of nonlinear time-delay systems with unknown parameters in [31]. Moreover, in [32], an AIO is investigated based on the feasibility of a centralisation perspective for a class of uncertain nonlinear systems.

Based on the literature review summarised and the challenges listed above, in this paper, a state-dependent, centred pseudo-linearisation representation is combined with an impulsive observer to not only estimate motor speed/position in a so-called "sensorless" scheme but also to control these parameters. The proposed observer can provide continuous estimates of the output states of the system, even in the presence of discrete output measurements. Through this approach, by utilising a pseudo-linearisation display instead of linearisation, not only is the design of a linear control scheme facilitated, but also all the nonlinear specifications of the system are preserved along with an accurate estimate of the states.

It should be mentioned that in [27], the authors used an SDRE controller and SDRE observer for an AHU system. The SDRE observer used continuous output signals to estimate the states, and its structure is of the Luenberger type but with state-dependent matrices. Both controller and observer gains are obtained by solving the SDRE at every sample time. In the present paper, the SDRE controller is used with an observer state-dependent impulsive observer for the PMSM when no speed sensors are available. This observer only utilises the output signal at impulse times and does not require continuously estimated states (between the impulses). In [15], the authors used an MRAS to control an IPMSM. The purpose of the mentioned paper was to use the pseudo-linearisation presentation in the structure of the MRAS controller. To estimate the speed, stator resistance, winding inductance, and torque load, an augmented adaptive interconnected observer (AAIO) was used with a similar structure in [33].

In [30], the basic structure of the state-dependent impulsive observer is introduced for nonlinear time-delay systems. The Congo Ebola disease is addressed and presented as a case study without any input control signal criteria. The combination of the SDRE and an impulsive observer was introduced for the first time by the authors of [34] for a manipulator robot case study. The major difference between the present paper and this previous one is that each joint of the manipulator arm was assumed to have a control actuator, and there were the same number of output and control input signals in this system. Thus, the input matrix was an orthogonal one. Unlike the previous case study, the PMSM system in the present paper has only states that are measurable. Therefore, for the speed and position signals, no direct control is available. Also, here, it is assumed that the angles are measured and available only in impulse time stamps, and only angular velocities are estimated by the observer, i.e., there is no access to physical measurements. Addressing this state access problem while there is a nonlinear dependency between the outputs and speed signals is the key challenge addressed in this work. This challenge is addressed through the following:

1. Developing a pseudo-linearised representation of the PMSM system.
2. Designing a controller for optimal tracking of the PMSM's reference speed with high accuracy and a quick speed without a speed sensor.
3. Estimating motor speed in a sensorless framework.
4. Addressing the challenge of disturbances during the course of the control.
5. Maintaining the function of estimation and speed control during all times of sampling and not just at the impulse sample times.
6. Quantifying the effects of impulse intervals (the sample rate) and load torque.

The simulated examples in this work demonstrate the performance of the controlled system despite the changes in the reference speed and load torque and the variable sample rates. To show the superiority of the proposed estimation and control approach, the results are compared with the linear control approach of LQRs. It is worth noting that the PMSM here is only considered as a case study, and the proposed work has the potential to be applied to systems with similar challenges and requirements without a loss in generalisability.

Given the introduction and motivation for this work given above, the rest of the paper is organised as follows: Section 2 is dedicated to the state-dependent Riccati equation principles, and Section 3 summarises the fundamentals of the impulsive state estimator. Section 4 includes the main results and the proposed work; the stability theorems and the proofs are given in this section. Section 5 is dedicated to the PMSM control case study, with simulations and visualisations given in Section 6; this section includes a comprehensive discussion of the results, the sensitivity analysis, and practical considerations as well. Comparisons are also included in this section. The last section includes the conclusion and potential future work.

2. State-Dependent Riccati Equation

In the SDRE method, the nonlinear system is first transformed into a pseudo-linearisation form without any approximation, and then the benefits of the linear representation are used to design the controller, observer, or any nonlinear filters. Its ability to be implemented in a wide range of nonlinear systems, the existence of a methodical process for the transformation, and the possibility of obtaining optimal solutions for control purposes are among the advantages of the SDRE technique [21–23].

The dynamics of a nonlinear system are considered as follows [21]:

$$\begin{aligned} \dot{x}(t) &= f(x(t), u(t)) \\ y(t) &= h(x(t)), \end{aligned} \tag{1}$$

where $x \in \mathbf{R}^n$, $u \in \mathbf{R}^p$, and $y \in \mathbf{R}^q$ are, respectively, the states, control, and output vectors of the system. $f : \mathbf{R}^n \times \mathbf{R}^p \to \mathbf{R}^n$ and $h : \mathbf{R}^n \to \mathbf{R}^q$ are the nonlinear functions here. The pseudo-linearised display of the system (1) is as follows:

$$\begin{aligned} \dot{x}(t) &= A(x(t))x(t) + B(x(t))u(t) \\ y(t) &= C(x(t))x(t), \end{aligned} \tag{2}$$

$A(x(t)) \in \mathbf{R}^{n\times n}$, $B \in (x(t)) \in \mathbf{R}^{n\times p}$, and $C(x(t)) \in \mathbf{R}^{q\times n}$ are the system's state-dependent coefficient matrices, inputs, and outputs, respectively. To track the reference path of the system, the cost function with the SDRE's infinite time horizon is defined as follows:

$$J = \int_0^\infty \Big\{ (Cx(t) - r(t))^T Q(x)(Cx(t) - r(t)) + u^T(t)R(x)u(t) \Big\} dt, \tag{3}$$

where the weight coefficients $Q(x)$ and $R(x)$ are state-dependent [24] and can determine state or input preferences. The control input to minimise the cost function (4) is calculated as follows [20]:

$$u(t) = -K(x)x(t) + R^{-1}(x)B^T(x)v(x), \tag{4}$$

such that $K(x) = R^{-1}(x)B^T(x)P(x)$ and $P(x)$ are the solutions to the Riccati equation in the form of (5):

$$A^T(x)P(x) + P(x)A(x) - P(x)B(x)R^{-1}(x)B^T(x)P(x) + C^T Q(x)C = 0, \tag{5}$$

where $v(x)$ is the solution to the following algebraic equation:

$$\begin{aligned} &(A(x) - B(x)K(x))^T v(x) + C^T Q(x)r(t) = 0 \\ &\quad \to v(x) = -\Big((A(x) - B(x)K(x))^T \Big)^{-1} \big(C^T Q(x)r(t) \big). \end{aligned} \tag{6}$$

In the process presented above, optimal tracking can be achieved without worrying about the rank reduction in the controllability and observability matrices.

3. Impulsive Observer

In this section, it is described how the state observer is designed to estimate the states of the system. Through this framework, the controller can perform based on the estimated states. Impulsive systems have two continuous and discrete dynamic behaviours. The continuous dynamic behaviours are described by differential equations that are continuous in time, describing the intervals between impulses. However, the discontinuous dynamic behaviour is described by differential equations that denote the impulse moments when system states abruptly change. According to these features, impulsive systems are very suitable for describing real-world variable processes in which system states change instantaneously at certain points in time. Impulsive dynamic systems can be seen as a subset or class of hybrid systems with precipitate jumps. There is a main continuous or discrete dynamic in impulsive systems, in which moments also occur between impulses and change the states of the system. The time interval between these impulses can be fixed or variable with time, known or unknown, and deterministic or stochastic.

In general, for a nonlinear system, the system is denoted as in Equation (7) [28]:

$$\begin{cases} \dot{x}(t) = f(t, x(t)); & t \neq t_k \\ \Delta x(t) = f_I(x(t)); & t = t_k \end{cases} \tag{7}$$

with $x \in \mathbf{R}^n$ as the continuous state vector and t as the time variable. With, $f : \mathbf{R}^+ \times \mathbf{R}^n \to \mathbf{R}^n$ and $f_I : \mathbf{R}^n \to \mathbf{R}^n$ as nonlinear functions with appropriate dimensions, the state variable jumps can be expressed as $\Delta x(t_k) = x(t_k^+) - x(t_k)$, in discrete intervals of $k = 1, 2, \ldots t_k$, where $t_k > t_{k-1} > 0$. $\Delta x(t_k)$ is also known as the Impulsive Differential Equation (IDE). There are two main approaches to studying stability issues in impulsive systems. The classical approach is based on the positive definite Lyapunov function, with a negative derivative for non-impulsive systems. It should be noted that most of the existing articles have used the classical perspective to examine the stability of a system or its observer. Unlike non-impulsive systems, impulsive systems can have a positive derivative at some point in time but have a descending trend overall [28].

A system with impulsive equations is defined as follows:

$$\begin{cases} \dot{w}(t) = g(t, w(t)); & t \neq t_k \\ w(t_k^+) = \psi_k(w(t_k)), \end{cases} \tag{8}$$

where $V \in V_0$ and $g : \mathbf{R}^+ \times \mathbf{R}^+ \to \mathbf{R}$ are continuous, and $\psi_k : \mathbf{R}^+ \to \mathbf{R}^+$ is non-descending. Also, the following equations are established:

$$\begin{cases} D^+ V(t, x) \leq g(t, V(t, x)); & t \neq t_k \\ V(t, x + \Delta x) \leq \psi_k(V(t, x)); & t = t_k \end{cases} \tag{9}$$

Hypothesis 1. *It is assumed that $f(t, 0) = 0$, $f_I(0) = 0$, and $g(t, 0) = 0$ for all k with $t > 0$.*

Therefore, the obvious response of the main system and the comparison system is in $(t_{k-1}, t_k]$. The theorem of the comparison system is presented as Theorem 1.

Theorem 1. *The conditions are assumed to be as follows:*

- $V : \mathbf{R}^+ \times S_\rho \to \mathbf{R}^+$, $V \in V_0$, $\rho > 0$ *in* $(t_{k-1}, t_k]$, *and* $D^+ V(t, x) \leq g(t, V(t, x))$.
- *There is a* $\rho_0 > 0$, *so* $x \in S_{\rho_0}$ *and* $x + \Delta x \in S_{\rho_0}$ *for all k, and in* $t = t_k$, $V(t, x + \Delta x) \leq \psi_k(V(t, x))$.
- $b(\|x\|) \leq V(t, x) \leq a(\|x\|)$ *on* $\mathbf{R}^+ \times S_\rho$ *when* $a, b \in \kappa$.

Moreover, the stability characteristics of the obvious solution to the comparison system will result in the stability of the main system [28].

Outcome 1. *Suppose* $g(t, V(t,x)) = \dot{\xi}(t)V(t,x)$ *when* $\xi \in C^1[\mathbf{R}^+, \mathbf{R}^+]$ *and* $\psi_k(V(t,x)) = d_k V(t,x)$ *The origin of the main system will be asymptotically stable if the following conditions are met:*

$$\begin{cases} \dot{\xi}(t) \geq 0 \\ d_k \geq 0; \qquad k = 1,2,\ldots \\ \xi(t_k) - \xi(t_{k-1}) + \ln(\gamma d_k) \leq 0; \qquad \gamma > 1 \end{cases} \tag{10}$$

State-Dependent Impulsive Observer

A state-dependent impulsive observer (SDIO) is suggested as follows:

$$\begin{cases} \dot{\hat{x}} = A(\hat{x})\hat{x} + B(\hat{x})u; \quad t \neq t_k \\ \hat{y} = C\hat{x} \\ \Delta\hat{x} = F(\hat{x})(y - \hat{y}); \qquad t = t_k \end{cases} \tag{11}$$

where $\hat{x}$ and $\hat{y}$ are the estimated state and output vectors, respectively. $\Delta\hat{x}$ is the jump vector of the estimated states. The observer gain is F, which is also determined according to the closed-loop stability theorem [29].

4. Main Design: SDRE Controller Based on State-Dependent Impulsive Observer

The impulsive observer is assumed to be state-dependent (11). The control input when using the state-dependent Riccati controller is as follows:

$$\begin{aligned} u(t) &= -R^{-1}(x)B^T(x)P(x)x(t) + R^{-1}(x)B^T(x)v(x) \\ v(x) &= -\left((A(x) - B(x)K(x))^T\right)^{-1}\left(C^T Q(x)r(t)\right). \end{aligned} \tag{12}$$

Hypothesis 2. *The following Lipshitz conditions are assumed to be valid for the system:*

$$\begin{aligned} \|A_c(x)x - A_c(\hat{x})\hat{x}\| &\leq K_A\|e\| \\ \|B_c(x)r - B_c(\hat{x})r\| &\leq K_B\|e\| \end{aligned} \tag{13}$$

where K_A *and* K_B *are constant and positive.*

Theorem 2. *The proposed state-dependent impulsive observer's estimation error for the SDRE controller is* $e = x - \hat{x}$, *which will converge to zero asymptotically when the following assumptions and conditions are met:*

$$\begin{bmatrix} \Sigma_i & \left(\varepsilon_1^{-1} + \varepsilon_2^{-1}\right)M_i \\ \left(\varepsilon_1^{-1} + \varepsilon_2^{-1}\right)M_i & -\left(\varepsilon_1^{-1} + \varepsilon_2^{-1}\right)I \end{bmatrix} \leq 0; \qquad i = 1,2 \tag{14}$$

$$\sum_i = A_c^T(\hat{x})M_i + M_i A_c(\hat{x}) + (M_1 - M_2)/\Delta_k + \left(2\varepsilon_1 K_A^2 + \varepsilon_2 K_B^2\right)I + 2\varepsilon_1 A_c^T(\hat{x})A_c(\hat{x}) - \alpha M_i \tag{15}$$

$$\begin{bmatrix} -\sigma M_1 & M_2 - \left(\overline{F}(\hat{x})C^T\right) \\ M_2 - \overline{F}(\hat{x})C & -M_2 \end{bmatrix} \leq 0 \tag{16}$$

$$\alpha\Delta_k + \ln(\gamma\sigma) \leq 0 \tag{17}$$

where $\alpha \geq 0$, $\sigma \geq 0$, $\gamma \geq 1$, *and* $\gamma\sigma \leq 1$ *are the design parameters, and* $F(\hat{x}) = \overline{F}(\hat{x})M_2^{-1}$.

Proof of Theorem 2. The SDRE closed-loop system is written as follows:

$$\begin{aligned} \dot{x} &= A(x)x + B(x)(-k(x)x + v(x)r) \\ &= (A(x) - B(x)k(x))x + B(x)v(x)r = A_c(x)x + B_c(x)r, \end{aligned} \tag{18}$$

where $A_c(x) = A(x) + B(x)k(x)$ and $B_c(x) = B(x)\nu(x)$. It can also be written for the observer in $t \neq t_k$:

$$\dot{\hat{x}} = (A(\hat{x}) - B(\hat{x})k(\hat{x}))\hat{x} + B(\hat{x})\nu(\hat{x})r = A_c(\hat{x})\hat{x} + B_c(\hat{x})r \tag{19}$$

Therefore, the dynamics of the estimation error in $t \neq t_k$ are rewritten as follows:

$$\begin{aligned} \dot{e} &= A_c(x)x + B_c(x)r - A_c(\hat{x})\hat{x} + B_c(\hat{x})r \\ &= A_c(\hat{x})(x - \hat{x}) + (A_c(x) - A_c(\hat{x}))x + (B_c(x) - B_c(\hat{x}))r \\ &= A_c(\hat{x})e + \widetilde{A}x + \widetilde{B}r \end{aligned} \tag{20}$$

where $\widetilde{A} = A_c(x) - A_c(\hat{x})$ and $\widetilde{B} = B_c(x) - B_c(\hat{x})$. In $t = t_k$,

$$\Delta e = F(\hat{x})Ce(t_k) \rightarrow e\left(t_k^+\right) = (I - F(\hat{x})C)e(t_k). \tag{21}$$

The positive definite Lyapunov function is introduced as follows:

$$\begin{aligned} &V(x,t) = e^T Me; \qquad M = (1-\rho)M_1 + \rho M_2, \\ &\rho = \tfrac{t_k - t}{\Delta_k}; \qquad \Delta_k = t_k - t_{k-1}; \qquad t \in (t_{k-1}, t_k], \end{aligned} \tag{22}$$

where M is an alternating matrix in $t \in (t_{k-1}, t_k]$, and M_1 and M_2 are defined as symmetric, positive definite matrices.

$$M(t_k) = M_1; \qquad M\left(t_k^+\right) = M_2, \tag{23}$$

So, the derivative of the Lyapunov function in the time interval $t \in (t_{k-1}, t_k]$ is equal to

$$\dot{V}(x,t) = e^T\left(A_c^T(\hat{x})M + MA_c(\hat{x}) + \frac{M_1 + M_2}{\Delta_k}\right)e + x^T\widetilde{A}^T Me + e^T M\widetilde{A}x + r^T\widetilde{B}^T Me + e^T M\widetilde{B}r. \tag{24}$$

Using the Yang inequality [30], the result is as follows:

$$\begin{aligned} &x^T\widetilde{A}^T Me + e^T M\widetilde{A}x \leq \varepsilon_1 x^T\widetilde{A}^T\widetilde{A}x + \varepsilon_1^{-1}e^T M^2 e \\ &r^T\widetilde{B}^T Me + e^T M\widetilde{B}r \leq \varepsilon_2 r^T\widetilde{B}^T\widetilde{B}r + \varepsilon_2^{-1}e^T M^2 e, \end{aligned} \tag{25}$$

And, according to Hypothesis 2, the following result is obtained:

$$\begin{aligned} x^T\widetilde{A}^T\widetilde{A}x = \left\|\widetilde{A}x\right\|^2 &\leq 2\|A_c(x)x - A_c(\hat{x})\hat{x}\|^2 + 2\|A_c(\hat{x})(x - \hat{x})\|^2 \\ &\leq 2K_A^2 e^T e + 2e^T A_c^T(\hat{x})A_c(\hat{x})e \end{aligned} \tag{26}$$

$$r^T\widetilde{B}^T\widetilde{B}r = \left\|\widetilde{B}r\right\|^2 \leq K_B^2 e^T e. \tag{27}$$

Hence, the derivative of Lyapunov's function will be taken as follows:

$$\begin{aligned} \dot{V} \leq e^T(&A_c^T(\hat{x})M + MA_c(\hat{x}) + (M_1 + M_2)/\Delta_k + (2\varepsilon_1 K_A^2 + \varepsilon_2 K_B^2)I \\ &+ \left(\varepsilon_1^{-1} + \varepsilon_2^{-1}\right)M^2 + 2\varepsilon_1 A_c^T(\hat{x})A_c(\hat{x}))e. \end{aligned} \tag{28}$$

By applying $\pm\alpha V(x,t)$ to (28), the following result is obtained:

$$\dot{V}(x,t) \leq e^T\sum e + \alpha V(x,t), \tag{29}$$

which can be written as follows:

$$
\begin{aligned}
\overline{\Sigma} &= \overline{\Sigma}_1 + \rho\overline{\Sigma}_2 + \rho^2\overline{\Sigma}_3, \\
\overline{\Sigma}_1 &= A_c^T(\hat{x})M_1 + M_1A_c(\hat{x}) + (M_1 - M_2)/\Delta_k + (2\varepsilon_1 K_A^2 + \varepsilon_2 K_B^2)I + \left(\varepsilon_1^{-1} + \varepsilon_2^{-1}\right)M_1^2 + 2\varepsilon_1 A_c^T(\hat{x})A_c(\hat{x}) - \alpha M_1, \\
\overline{\Sigma}_2 &= A_c^T(\hat{x})(M_2 - M_1) + (M_2 - M_1)A_c(\hat{x}) + 2\left(\varepsilon_1^{-1} + \varepsilon_2^{-1}\right)M_1(M_2 - M_1) - \alpha(M_2 - M_1), \\
\overline{\Sigma}_3 &= \left(\varepsilon_1^{-1} + \varepsilon_2^{-1}\right)(M_2 - M_1)^2.
\end{aligned} \tag{30}
$$

If $0 \leq \rho(t) \leq 1, \overline{\Sigma}_3 \geq 0$, and the following conditions are satisfied by the LMIs (14), as well as the definition of $\dot{\lambda}(t) = \alpha \geq 0$, the first condition of the comparison system theorem is established as follows:

$$
\begin{aligned}
&\overline{\Sigma}_1 \leq 0 \\
&\overline{\Sigma}_1 + \overline{\Sigma}_2 + \overline{\Sigma}_3 \leq 0.
\end{aligned} \tag{31}
$$

In $t = t_k^+$, the Lyapunov function is presented as follows:

$$
V(x, t_k^+) = e^T(t_k^+)M(t_k^+)e(t_k^+) = e^T(t_k)(I - F(\hat{x})C)^T M_2(I - F(\hat{x})C)e(t_k). \tag{32}
$$

This is the case if the following condition is met:

$$
(I - F(\hat{x})C)^T M_2(I - F(\hat{x})C) \leq \sigma M_1 \tag{33}
$$

which is the same condition for the LMI (16) as $V(x, t_k^+) \leq \sigma V(x, t_k)$ and $d_k = \sigma \geq 0$.
Therefore, the second condition of the comparison system theorem is established. □

According to the third condition, the upper limit of the impact distance is obtained as follows:

$$
\Delta_{k\max} = -\ln(\gamma\sigma)/\alpha \tag{34}
$$

5. Case Study: Permanent Magnet Synchronous Motor

The dynamic model of the permanent magnet synchronous motor used in this paper on the $d - q$ axis is as follows [35]:

$$
\begin{cases}
\frac{di_d}{dt} = -\frac{R_s}{L_d}i_d + \omega_r i_q + \frac{V_d}{L_d} \\
\frac{di_q}{dt} = -\frac{R_s}{L_q}i_q - \omega_r i_d - \lambda_f\omega_r + \frac{V_q}{L_q} \\
\frac{d\omega_r}{dt} = \frac{3}{2}\frac{N_p}{J}\lambda_f i_q - \frac{B}{J}\omega_r - \frac{T_L}{J} \\
\frac{d\theta}{dt} = \omega_r
\end{cases} \tag{35}
$$

where i_d and i_q are the currents and V_d and V_q are the voltages of the d and q axes, respectively. R_s is the stator resistance. L_q and L_d are, respectively, the inductances of the winding on the q and d axes. λ_f and ω_r are the magnetic flux of the permanent magnet and the mechanical speed of the motor, respectively. θ and J are the angular position of the rotor and the moment of inertia. B is the coefficient of friction. N_p is the number of poles, and T_L is the load torque. Table 1 presents the values of the engine parameters according to [35]. The pseudo-linearised representation of the PMSM is obtained as follows:

$$
A(x) = \begin{bmatrix} -\frac{R_s}{L_d} & x_3 & 0 \\ -x_3 & -\frac{R_s}{L_q} & -\lambda_f \\ 0 & \frac{3}{2}\frac{N_p}{J}\lambda_f & -\frac{B}{J} \end{bmatrix}, B(x) = \begin{bmatrix} \frac{1}{L_d} & 0 \\ 0 & \frac{1}{L_q} \end{bmatrix}, C(x) = \begin{bmatrix} 1 & 0 & 0 \\ 0 & 1 & 0 \end{bmatrix} \tag{36}
$$

where $x = \begin{bmatrix} i_d & i_q & \omega_r \end{bmatrix}^T$ is the state vector, $u = \begin{bmatrix} V_d & V_q \end{bmatrix}$ is the input vector, and $y = \begin{bmatrix} i_d & i_q \end{bmatrix}^T$ is the output vector. The advantage of this presentation is that it is both controllable and observable.

Table 1. The PMSM's parameters [35].

Parameter	Value
Number of poles (N_p)	4
Stator resistance (R_s)	0.0875 Ω
Permanent magnetic flux $\left(\lambda_f\right)$	1 Wb
Inductance $(L_d = L_q)$	0.2155 H
Moment of inertia (J)	0.0151 kgm^2
Friction coefficient (B)	0.0378 $\text{kgm}^2\text{s}^{-1}$

An overview of the control, speed estimation, and identification of the synchronous motor's parameters is shown in Figure 1. In this block diagram, the voltage in the *d–q* plane is applied to the motor, and the currents are captured as the response. These current values are not necessarily available at each sample point, but to reduce the cost of measurements, they are assumed to be available only for a few time stamps. The impulse observer receives the current and voltage values and estimates the motor speed for all of the time stamps. This speed is then used as an input to the SDRE controller for tracking purposes. Here signals with * reference to the setpoint of the system.

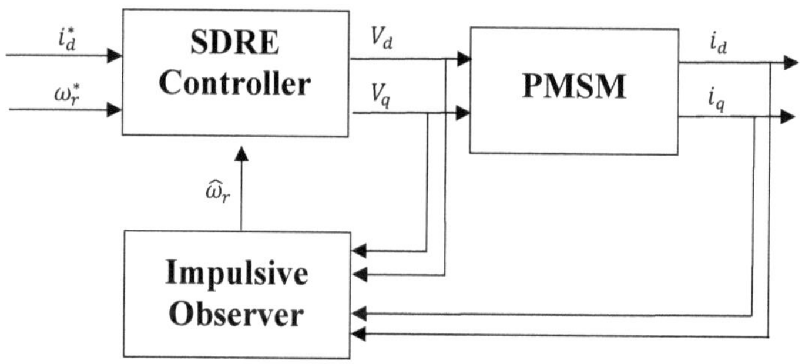

Figure 1. Block control diagram of the proposed method.

6. Simulation Results and Discussions

6.1. The Main Simulation Results

The design parameters are considered as follows:

$$R(x) = diag(1,1),\ Q(x) = diag\left(10^3 + x_1^2,\ 10^3 + 10x_2^2,\ 10^2 + x_3^2\right),\ \alpha = 10,\ \gamma = 1.001,\ \sigma = 0.95.$$

The initial conditions of the states and their estimations are as follows:

$$x(0) = [0,0,1]^T,\ \ \hat{x}(0) = [10,10,50]^T.$$

Also, the desired values of the speed and current on the d-axis are assumed to be zero. According to the values of the design parameters, the maximum distance between impulses is calculated as $\Delta_{k\max} = -\ln(1.001 \times 0.95)/10 = 0.005$ s. The sampling time $T_s = 10^{-4}$ s, and the interval between impulses is also assumed to be $\Delta = 5 \times 10^{-3}$ s, which means that an output is available every five sampling times. The tracking error is defined as $x^{ref} - \hat{x}$ and the estimation error is called as $x - \hat{x}$.

The results were simulated by MATLAB 2023b software with an ASUS laptop (Ausu Laptop Model: K556CPU COREi7/7500u, RAM DDR4, Graphic GFORCE 940MX). These results are displayed in Figure 2 for the speed tracking error, the speed estimation error, and the estimated speed jumps (speed impulses). The figures on the top row are the tracking performance and its zoomed-in version. It is very clear that the reference speed was successfully tracked after only the first 0.02 s. The figures in the second row demonstrate the tracking errors, which are very small after convergence happens. According to the results, it is clear that the proposed observer can estimate the motor speed very well. The proposed controller also tracks reference paths with appropriate accuracy. At first, it is seen that, despite the error between the actual and estimated values, these jumps have a greater amplitude, which gradually decreases.

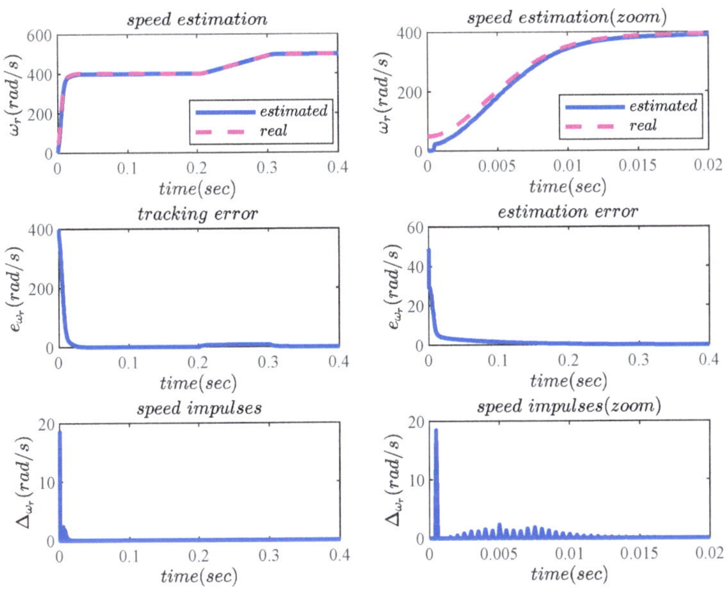

Figure 2. Speed profiles of the PMSM, the state estimation, the tracking error, and the speed impulses, together with the zoomed-in areas.

Further results are plotted in Figure 3 to show the d-current estimation, the d-current tracking error, the d-current estimation error, and the estimated d-current jumps (d-current impulses).

Figure 4 illustrates the q-current estimation, the q-current estimation error, and the estimated q-current jumps (q-current impulses). Eventually, as the signal converges to the true value, the jump amplitude also becomes zero.

The control inputs are shown in Figure 5. As can be seen, all inputs and their changes are smooth and slow. The allowed range for the input control signals was between −250 and +250, so it is clear that the signals are within the prescribed range. Therefore, these inputs can be implemented in practice. Generally, in the proposed method, the control signals can be forced to stay within a range by tuning the controller's parameters. Also, by making a trade-off between the performance and the control input variations, a design that can be implemented in real applications can be achieved.

The observer gains are also plotted in Figure 6, which are obtained at any time by solving the LMIs of Theorem 2. Obviously, the gains remain unchanged after the conversion.

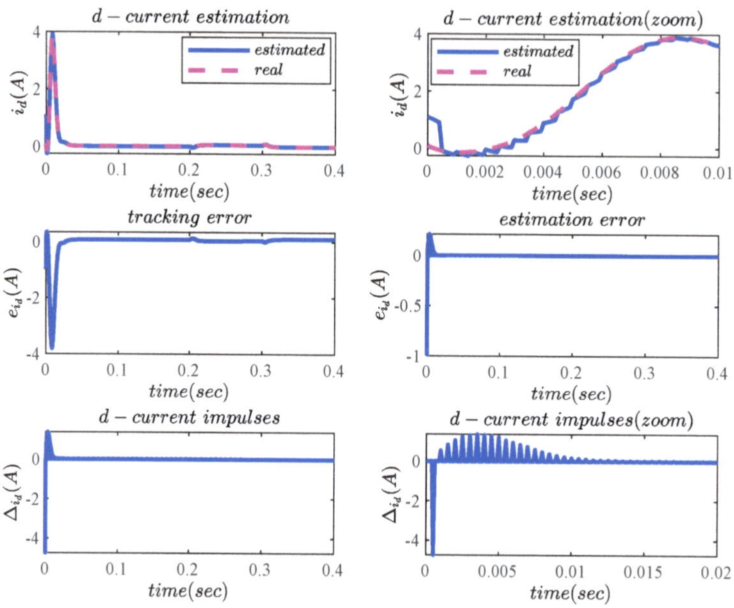

Figure 3. Current on the d-axis of the PMSM, the estimated states, the errors, and the impulses.

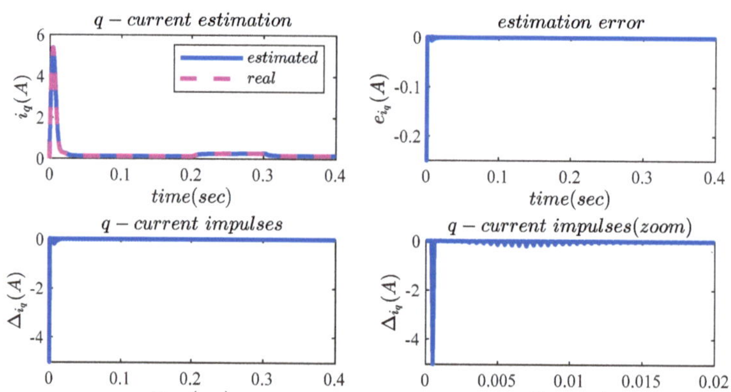

Figure 4. Current on the q-axis of the PMSM, the estimated states, the errors, and the impulses.

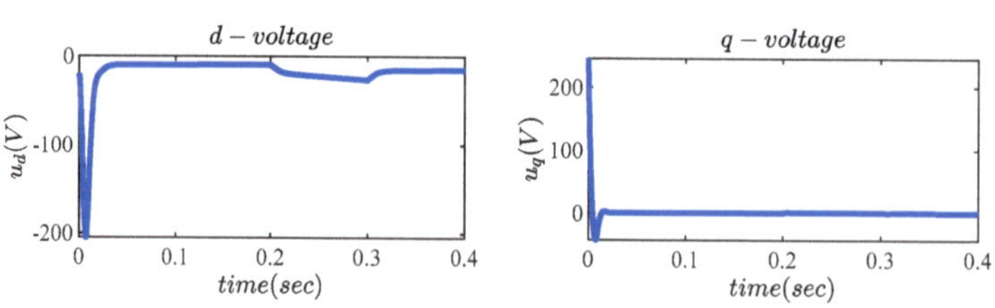

Figure 5. Voltage along the *d*-axis and *q*-axis of the PMSM, along with the *d*–*q* axis.

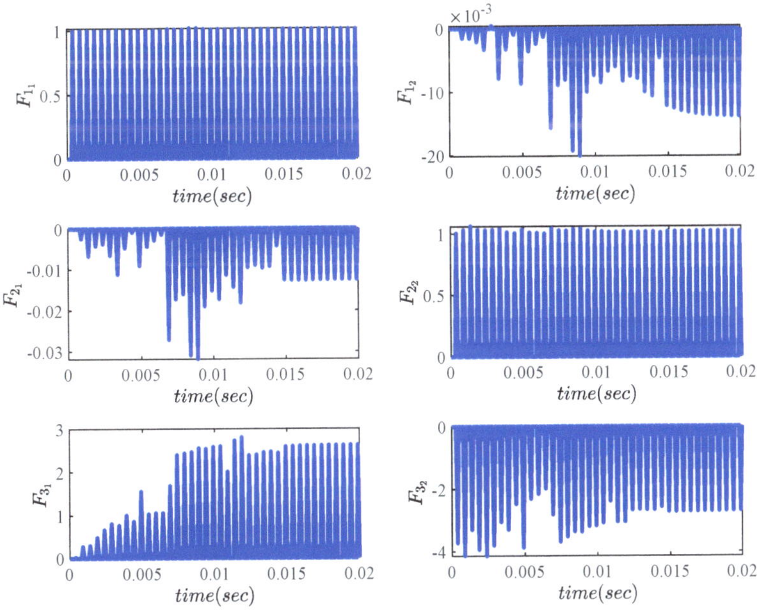

Figure 6. Impulsive observer gains over the course of estimation and control.

6.2. The Effect of Impulse Intervals

To show the effects of impulse intervals on the estimation error of the states by the proposed observer, the two criteria of the normalised mean square error (*NMSE*) and the correlation coefficient (*CC*) were used. These two criteria can be calculated as follows:

$$NMSE = \frac{1}{N}\sum_{j=1}^{N}\sum_{i=1}^{3}\left(\frac{e(i,j)}{\max_{j}|e(i,j)|}\right)^2 \tag{37}$$

$$cc_i = \frac{\sum_{j=1}^{N} x(i,j)\hat{x}(i,j)}{\sqrt{\sum_{j=1}^{N} x^2(i,j)}\sqrt{\sum_{j=1}^{N} \hat{x}^2(i,j)}} \tag{38}$$

where $e(i,j)$ is the error of the ith state. The *NMSE* metric shows the strength of the estimation and the correlation coefficient of the similarity of the real and estimated states. If the two signals are completely similar, the correlation coefficient is 1, and if they are completely different, the correlation coefficient is -1.

Table 2 shows that the accuracy of the estimation decreases with increasing impulse intervals, although the values are still acceptable up to the maximum impulse intervals determined by the proposed Theorem 2. In the last column, where the impulse interval is selected above the maximum value, a high error is witnessed, and the similarity between the actual and estimated signals is reduced.

Table 2. Summary of the impulse interval effect.

Metrices	**Impulse Intervals (Δ(s))**				
	10^{-4}	5×10^{-4}	10^{-3}	5×10^{-3}	6×10^{-3}
$NMSE$	45.6892	128.6873	177.1134	301.8423	2521.4897
CC_1	0.9992	0.9968	0.9925	0.9725	0.6850
CC_2	1	0.9999	0.9998	0.9994	0.7982
CC_3	1	1	0.9999	0.9812	0.8241

6.3. The Effect of Load Torque

In this section, to investigate the effect of the load torque as a disturbance to the performance of the closed-loop system, a load torque of 5 Nm at an equilibrium point from 0.25 to 0.35 per second is applied to the motor. The load torque impulsive observer gains, speed tracking and estimation with load torque, speed tracking error with load torque, and speed estimation error with load torque are plotted in Figure 7.

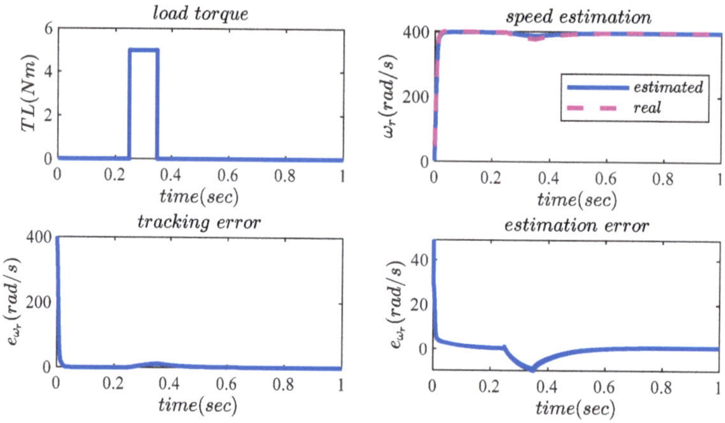

Figure 7. Load torque impulsive observer gains, speed tracking and estimation with load torque, speed tracking error with load torque, and speed estimation error with load torque.

According to the results, it is clear that despite the load torque acting on the system as a disturbance, the performance of the proposed controller and observer in tracking reference paths and estimating the states is satisfactory. The tracking error is directly related to increasing or decreasing load torque amplitude. It is worth noting that the load torque was considered such that a change in the speed or current response is visible. Robust approaches should be used if there is more disturbance or higher accuracy is required.

6.4. Comparisons

In this section, a comparison between the SDRE and LQR optimal controllers is provided. In this comparison, the SDRE controller examines the system nonlinearly, and the LQR, which is a linearised version, examines the system linearly. This comparison shows the better performance of the SDRE method. In the SDRE approach, a pseudo-linearisation matrix and SDC are generated, which will preserve the nonlinear properties of the system, whereas in the LQR version, the system is first linearised around the equilibrium point. So, this linearisation cannot include nonlinear features of the system and eliminates them. Hence, the system does not have a nonlinear nature and will have the same system matrices at all times. Meanwhile, the system matrices in the SDRE controller update their state-dependent system matrices at each sampling time. This solves the state-dependent Riccati equation for each time sample. But in LQRs, the Riccati equation is checked only once. The linearisation of the motor equations around the equilibrium point is performed using the Jacobian method as follows:

$$A(x) = \begin{pmatrix} \frac{\partial(\dot{x}_1)}{\partial u_1} & \cdots & \frac{\partial(\dot{x}_1)}{\partial u_3} \\ \vdots & \ddots & \vdots \\ \frac{\partial(\dot{x}_3)}{\partial u_1} & \cdots & \frac{\partial(\dot{x}_3)}{\partial u_3} \end{pmatrix} = \begin{bmatrix} -\frac{R_s}{L_d} & x_{e3} & x_{e2} \\ -x_{e3} & -\frac{R_s}{L_q} & -\lambda_f - x_{e1} \\ 0 & \frac{3}{2}\frac{N_p}{J}\lambda_f & -\frac{B}{J} \end{bmatrix},$$
$$B = \begin{pmatrix} \frac{\partial(\dot{x}_1)}{\partial u_1} & \cdots & \frac{\partial(\dot{x}_1)}{\partial u_3} \\ \vdots & \ddots & \vdots \\ \frac{\partial(\dot{x}_3)}{\partial u_1} & \cdots & \frac{\partial(\dot{x}_3)}{\partial u_3} \end{pmatrix}_{x=x_e} = \begin{bmatrix} \frac{1}{L_d} & 0 \\ 0 & \frac{1}{L_q} \\ 0 & 0 \end{bmatrix}. \tag{39}$$

The weighting coefficients R and Q are considered to be the same in both the LQR and SDRE methods, and the calculation of these coefficients is arbitrary. The response of the algebraic Riccati equation for the matrices (39) and the LQR state feedback gain is obtained as follows:

$$P = \begin{bmatrix} 2.1259 & 0.0012 & -0.5347 \\ 0.0012 & 1.5945 & 0.0693 \\ -0.5347 & 0.0693 & 0.5487 \end{bmatrix}, \; K = \begin{bmatrix} 9.8650 & 0.0055 & -2.4810 \\ 0.0055 & 7.3993 & 0.3217 \end{bmatrix}.$$

The following figures show the outputs and states of the SDRE and LQR controllers. Both controllers assume that all states are available, and they track both paths optimally. Figure 8 displays the speed estimation, currents (d-axis and q-axis), and voltages (d-axis and q-axis). It is clear that the SDRE controller performs much better than the LQR, especially when changing the equilibrium point. Because the LQR controller is linearised around the equilibrium point, by changing the equilibrium point, the linearisation accuracy is significantly reduced. Furthermore, Table 3 shows a numerical comparison between the proposed method and the LQR. The numerical results verified the efficiency of the proposed framework compared to the LQR technique in varying the reference speed.

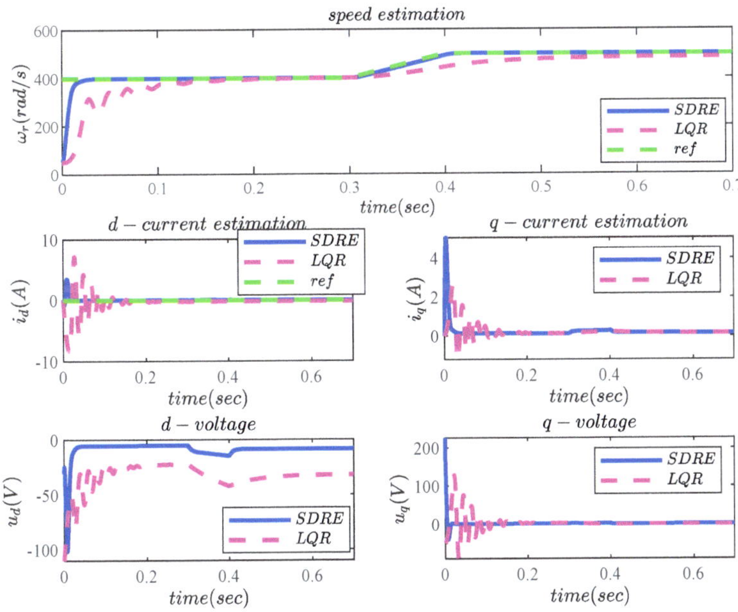

Figure 8. Comparison of PMSM parameters between the SDRE and the LQR.

Table 3. Summary of the comparison between the proposed method and the LQR.

$\Delta(s) = 5 \times 10^{-3}$	The Proposed Method	LQR
$NMSE$	301.8423	3128.2159
CC_1	0.9725	0.6014
CC_2	0.9994	0.7102
CC_3	0.9812	0.7371

As Table 3 shows, for both cases, the distance between the impulses (where the output signals become available) is 0.005 s. In both cases, a similar impulse observer was designed. However, the LQR performs more weakly compared to the SRDE controller. All the performance measures of the NMSE, $c\,c_1$, $c\,c_2$, and $c\,c_3$, which are the normalised mean squared errors and correlation coefficients across both the d and q axes, are lower for the LQR compared to the SDRE. Based on the performance metrics, this superiority is due to the SDRE being able to take care of the system's intrinsic nonlinearities, especially when the working point changes during tracking.

7. Conclusions

In this article, an observer-based suboptimal controller for the PMSM according to the SDRE controller and impulsive observer approaches was applied, and a pseudo-linearised representation and SDRE strategy procedure were presented. Moreover, a mathematical formulation of the impulsive state observer was presented. The proposed SDRE technique guarantees the tracking of reference paths with acceptable accuracy and speed. Despite the change in the reference speed, the performance of the closed-loop system was maintained, and the reference paths were well tracked. The results illustrate that despite the load torque, the performance of the system decreases to some extent, but it is still acceptable. A comparison was applied between the SDRE method and an LQR, which was designed based on the Jacobian linearisation approach. According to the results obtained, it is clear that when changing the reference speed, the LQR controller cannot guarantee the performance of the system in tracking the reference path. In the future, suggestions can be made for robust and adaptive SDRE controllers to handle disturbances and uncertainty in system models.

Author Contributions: Conceptualization, N.K. and M.F.N.; methodology, N.K.; software, N.K. and M.S.; validation, N.K., F.B.L. and M.S.; formal analysis, N.K.; investigation, M.F.N.; resources, M.F.N.; data curation, F.B.L. and M.S.; writing—original draft preparation, N.K.; writing—review and editing, M.F.N. and F.B.L.; visualization, M.S.; supervision, M.F.N.; project administration, N.K.; funding acquisition, M.F.N. All authors have read and agreed to the published version of the manuscript.

Funding: This research received no external funding.

Data Availability Statement: Data is contained within the article.

Conflicts of Interest: Authors declare not conflicts of interests.

References

1. Li, S.; Liu, H.; Ding, S. A speed control for a PMSM using finite-time feedback control and disturbance compensation. *Trans. Inst. Meas. Control* **2010**, *32*, 170–187. [CrossRef]
2. Li, S.; Zong, K.; Liu, H. A composite speed controller based on a second-order model of permanent magnet synchronous motor system. *Trans. Inst. Meas. Control* **2011**, *33*, 522–541. [CrossRef]
3. Yan, Y.; Zhu, J.G. A survey of sensorless initial rotor position estimation schemes for permanent magnet synchronous motors. In Proceedings of the Australasian Universities Power Engineering Conference, Brisbane, Australia, 26–29 September 2004; pp. 26–29.
4. Benjak, O.; Gerling, D. Review of position estimation methods for IPMSM drives without a position sensor, Part I: Nonadaptive Methods. In Proceedings of the International Conference on Electrical Machines, Rome, Italy, 6–8 September 2010; pp. 1–6.
5. Benjak, O.; Gerling, D. Review of position estimation methods for IPMSM drives without a position sensor, Part II: Adaptive Methods. In Proceedings of the International Conference on Electrical Machines and Systems, Incheon, Republic of Korea, 10–13 October 2010; pp. 1–6.

6. Benjak, O.; Gerling, D. Review of position estimation methods for IPMSM drives without a position sensor, Part III: Methods based on saliency and signal injection. In Proceedings of the International Conference on Electrical Machines, Rome, Italy, 6–8 September 2010; pp. 873–878.
7. Zhao, Y. Position/Speed Sensorless Control for Permanent-Magnet Synchronous Machines. Ph.D. Thesis, University of Nebraska, Electrical Engineering Department, Lincoln, NE, USA, 2014.
8. Li, S.; Li, J.; Tang, Y.; Shi, Y.; Cao, W. Model-based model predictive control for a direct-driven permanent magnet synchronous generator with internal and external disturbances. *Trans. Inst. Meas. Control* **2020**, *42*, 386–397.
9. Wei, Y.; Wei, Y.; Yuan, S.; Qi, H.; Guo, X.; Li, M. Nonlinear Model Predictive Speed Control with Variable Predictive Horizon for PMSM Rotor Position. *J. Control Eng. Appl. Inform.* **2021**, *23*, 86–94.
10. Walambe, R.A.; Joshi, V.A. Closed Loop Stability of a PMSM-EKF Controller-Observer Structure. *IFAC-Pap.* **2018**, *51*, 249–254. [CrossRef]
11. Sawma, J.; Khatounian, F.; Monmasson, E.; Idkhajine, L.; Ghosn, R. Analysis of the impact of online identification on model predictive current control applied to permanent magnet synchronous motors. *IET Electr. Power Appl.* **2017**, *11*, 864–873. [CrossRef]
12. El-Sousy, F.F.; El-Naggar, M.F.; Amin, M.; Abu-Siada, A.; Abuhasel, K.A. Robust Adaptive Neural-Network Backstepping Control Design for High-Speed Permanent-Magnet Synchronous Motor Drives: Theory and Experiments. *IEEE Access* **2019**, *7*, 99327–99348. [CrossRef]
13. Hezzi, A.; Bensalem, Y.; Elghali, S.B.; Abdelkrim, M.N. Sliding Mode Observer based sensorless control of five phase PMSM in electric vehicle. In Proceedings of the International Conference on Sciences and Techniques of Automatic Control and Computer Engineering (STA), Sousse, Tunisia, 24–26 March 2019; pp. 530–535.
14. Khlaief, A.; Boussak, M.; Châari, A. A MRAS-based stator resistance and speed estimation for sensorless vector controlled IPMSM drive. *Electr. Power Syst. Res.* **2014**, *108*, 1–15. [CrossRef]
15. Izadinasab, A.; Ghanbari, M. Control of sensorless PMSM using state dependent model reference adaptive system adaptive augmented observer. *J. Model. Eng.* **2020**, *18*, 85–95.
16. Usama, M.; Kim, J. Robust adaptive observer-based finite control set model predictive current control for sensorless speed control of surface permanent magnet synchronous motor. *Trans. Inst. Meas. Control* **2021**, *43*, 1416–1429. [CrossRef]
17. Hamida, M.A.; De Leon, J.; Glumineau, A.; Boisliveau, R. An Adaptive Interconnected Observer for Sensorless Control of PM Synchronous Motors With Online Parameter Identification. *IEEE Trans. Ind. Electron.* **2013**, *60*, 739–748. [CrossRef]
18. Toumi, D.; Mihoub, Y.; Hassaine, S.; Moreau, S. Design and implementation of adaptive fuzzy-RST digital speed control of PMSM drive. *Asian J. Control* **2021**, *24*, 2534–2547. [CrossRef]
19. Xu, W.; Junejo, A.K.; Liu, Y.; Hussien, M.G.; Zhu, J. An Efficient Antidisturbance Sliding-Mode Speed Control Method for PMSM Drive Systems. *IEEE Trans. Power Electron.* **2021**, *36*, 6879–6891. [CrossRef]
20. Lewis, F.L. *Optimal Control*, 3rd ed.; John Wiley: Hoboken, NJ, USA, 2012.
21. Çimen, T. State-Dependent Riccati Equation (SDRE) Control: A Survey. *IFAC Proc. Vol.* **2008**, *41*, 3761–3775. [CrossRef]
22. Çimen, T. Systematic and effective design of nonlinear feedback controllers via the state-dependent Riccati equation (SDRE) method. *Annu. Rev. Control* **2010**, *34*, 32–51. [CrossRef]
23. Çimen, T.; Manikandan, R.; Saha, N.; Korayem, A.H.; Korayem, M.H.; Nekoo, S.R.; Lin, L.-G.; Xin, M.; Lee, J.; Lee, Y.; et al. Survey of State-Dependent Riccati Equation in Nonlinear Optimal Feedback Control Synthesis. *J. Guid. Control Dyn.* **2012**, *35*, 1025–1047. [CrossRef]
24. Korayem, A.H.; Nekoo, S.R.; Korayem, M.H. Optimal sliding mode control design based on the state-dependent Riccati equation for cooperative manipulators to increase dynamic load carrying capacity. *Robotica* **2018**, *37*, 321–337. [CrossRef]
25. Nekoo, S.R.; Acosta, J.A.; Ollero, A. Collision Avoidance of SDRE Controller using Artificial Potential Field Method: Application to Aerial Robotics. In Proceedings of the 2020 International Conference on Unmanned Aircraft Systems (ICUAS), Athens, Greece, 1–4 September 2020.
26. Batmani, Y.; Takhtabnus, M.; Mirzaei, R. DC microgrid fault-tolerant control using state-dependent Riccati equation techniques. *Optim. Control Appl. Methods* **2021**, *43*, 1–15. [CrossRef]
27. Liavoli, F.B.; Fakharian, A. Sub-optimal observer-based controller design using the state dependent riccati equation approach for air-handling unit. In Proceedings of the 2019 27th Iranian Conference on Electrical Engineering (ICEE), Yazd, Iran, 30 April–2 May 2019; pp. 991–996.
28. Lakshmikantham, V.; Bainov, D.D.; Simeonov, P.S. *Theory of Impulsive Differential Equations*; World Scientific: London, UK, 1989.
29. Li, Z.; Soh, Y.; Wen, C. *Switched and Impulsive Systems: Analysis, Design and Applications*, 1st ed.; Springer: Berlin, Germany, 2005.
30. Kalamian, N.; Khaloozadeh, H.; Ayati, M. Design of state-dependent impulsive observer for nonlinear time-delay systems. *IET Control Theory Appl.* **2019**, *13*, 3155–3163. [CrossRef]
31. Kalamian, N.; Khaloozadeh, H.; Ayati, M. Adaptive state-dependent impulsive observer design for nonlinear deterministic and stochastic dynamics with time-delays. *ISA Trans.* **2020**, *98*, 87–100. [CrossRef]
32. Kalamain, N.; Khaloozadeh, H.; Ayati, M. On design of adaptive impulsive observer based on comparison system: Modifications in stability theory and feasibility centralization. *Int. J. Dyn. Control* **2022**, *11*, 149–161. [CrossRef]
33. Kivanc, O.C.; Ozturk, S.B. Sensorless PMSM drive based on stator feedforward voltage estimation improved with MRAS multiparameter estimation. *IEEE/ASME Trans. Mechatron.* **2018**, *23*, 1326–1337. [CrossRef]

34. Kalamian, N.; Niri, M.F.; Mehrabizadeh, H. Design of a Suboptimal Controller based on Riccati Equation and State-dependent Impulsive Observer for a Robotic Manipulator. In Proceedings of the 2019 6th International Conference on Control, Instrumentation and Automation (ICCIA), Sanandaj, Iran, 30–31 October 2019; pp. 1–6.
35. Liu, Z.H.; Nie, J.; Wei, H.L.; Chen, L.; Li, X.H.; Zhang, H.Q. A newly designed VSC-based current regulator for sensorless control of PMSM considering VSI nonlinearity. *IEEE J. Emerg. Sel. Top. Power Electron.* **2020**, *9*, 4420–4431. [CrossRef]

energies

MDPI

Article

Study of a Numerical Integral Interpolation Method for Electromagnetic Transient Simulations

Kaiyuan Sun [1], Kun Chen [1], Haifeng Cen [1], Fucheng Tan [2] and Xiaohui Ye [2,*]

[1] Guangzhou Power Supply Bureau, Guangdong Power Grid Co., Ltd., Guangzhou 510510, China
[2] School of Electrical Engineering, Yanshan University, Qinhuangdao 066004, China
* Correspondence: yexiaohui@ysu.edu.cn; Tel.: +86-18611692240

Abstract: In the fixed time-step electromagnetic transient (EMT)-type program, an interpolation process is applied to deal with switching events. The interpolation method frequently reduces the algorithm's accuracy when dealing with power electronics. In this study, we use the Butcher tableau to analyze the defects of linear interpolation. Then, based on the theories of Runge–Kutta integration, we propose two three-stage diagonally implicit Runge–Kutta (3S-DIRK) algorithms combined with the trapezoidal rule (TR) and backward Euler (BE), respectively, with TR-3S-DIRK and BE2-3S-DIRK for the interpolation and synchronization processes. The proposed numerical integral interpolation scheme has second-order accuracy and does not produce spurious oscillations due to the size change in the time step. The proposed method is compared with the critical damping adjustment method (CDA) and the trapezoidal method, showing that it does not produce spurious numerical oscillations or first-order errors.

Keywords: algorithm accuracy; Butcher tableau; electromagnetic transient simulations; interpolation method; Runge–Kutta integration

1. Introduction

In the development and construction of ultra-high-voltage DC lines, the proportion of power electronic components in the power system has been increasing year by year. Therefore, electromagnetic transient (EMT) simulations have become an important tool for analyzing the stability of AC/DC hybrid power systems [1–4]. EMT simulations can essentially be solved as differential–algebraic equations (DAEs), which are more complex, with stricter algorithmic requirements than ordinary differential equations (ODEs) [5]. The error in the values of algebraic equations causes numerical oscillation, and this should be suppressed by applying an L-stable algorithm to solve the DAE problem [6], especially when dealing with a circuit's sudden changes [7].

The most commonly used EMT-type program [8,9] employs the nodal analysis method, which uses a fixed step-size integration method to discretize the differential equation of the circuit elements into a current source with a shunt resistor. The admittance matrix of the network is unchanged during the computation process, which can improve the speed of the computation [10]. The EMT-type program is specially designed for power system simulation, provides a variety of power system models, and has been used in engineering computations for many years [11].

Trapezoidal rule (TR) integration is an implicit method with the feature of symmetrical A-stability, which can show the unstable modes in the system. Further, it is most widely used in power system simulation. This algorithm enables the original instability patterns in the system to be manifested and eliminates the problem of superstability, but causes numerical oscillations [12]. The numerical oscillation of EMT can be suppressed by adding a buffer circuit [13] or using the damping trapezoid method, critical damping adjustment method (CDA) [12], two-stage diagonally implicit Runge–Kutta method (2S-DIRK) [14–16], root-matching method [17], trapezoidal method with the second-order backward difference formula (TR-BDF2) [17], or Rosenbrock numerical integration method [18]. The CDA method uses a two-step backward Euler method [19], which does not need to modify the admittance matrix and has good adaptability. As a result, it is widely used in EMT-type software [20].

On the other hand, switching operation events do not occur exactly at the constant-step intervals, and delays in the switching operation cause the emergence of unreal "spikes" [21] in the waveform and, thus, harmonics, which requires an accurate simulation of the switching operation through the variable time-step algorithm or interpolation method [22,23]. This requires multiple interpolations [22,24] or flexible integration for readjustment in the simulation of transients [25] to achieve resynchronization during real-time simulations. The interpolation method is a good choice because it does not change the admittance matrix of the network. However, linear interpolation reduces the simulation accuracy depending on when exactly the discontinuity occurs [26]. And interpolation requires a smaller time step [24]. In addition, exponential matrix methods can solve this problem [27,28], but their implementation requires significant modification of the CDA architecture, rendering it a less practical choice. Therefore, to develop an enhanced algorithm, the following conditions should be met:

(1) L-stable, to effectively eliminate oscillations;
(2) Variable step size, to handle switching events at any time;
(3) High computational accuracy, at least secondary-order, during interpolation;
(4) Consistent equivalent impedance, and the conductivity matrix does not need to re-triangulated;
(5) It is better to construct it on the CDA scheme to minimize modifications of EMT-type tools.

Recent developments in DIRK methods have advanced the field of computational mathematics. Kennedy et al. [29,30] evaluated DIRK methods for stability and accuracy in solving differential equations, despite increasing computational demands. Da Silva et al. [31] introduced iso-spectral Runge–Kutta methods using the reduced Hamiltonian principle. Kamiński et al. [32] applied a novel Runge–Kutta–Fehlberg algorithm to the stochastic Duffing equation, enhancing structural response calculations. Despite these advancements, Runge–Kutta methods should be redesigned to meet all above conditions for EMT computations.

In this study, we use the Butcher tableau [33] to analyze the computation error caused by interpolation in the CDA method and discuss the defects of linear interpolation algorithms in Section 2. It is found that the linear interpolation algorithm cannot maintain the high-order simulation accuracy. Based on the Runge–Kutta theories, two three-stage diagonally implicit Runge–Kutta (3S-DIRK) methods combined with TR and BE are proposed: TR-3S-DIRK and BE2-3S-DIRK. Then, a new integration-based interpolation scheme is proposed, and the integral formulas for inductors, capacitors, and mutual inductors are derived. The proposed algorithms and interpolation scheme have similar advantages to CDA, such as an invariant admittance matrix and numerical oscillation suppression. Moreover, the proposed algorithms have second-order accuracy throughout the whole simulation.

To verify the feasibility of the proposed method, a special case was constructed combined with the line-commutated converter-based high-voltage direct-current (LCC-HVDC) system and the IEEE9 system, in which the calculation algorithm needs to be switched to obtain accurate LCC-HVDC results, while the errors in algorithm switching are reflected in the curves of IEEE9. Then, the proposed method is compared with the CDA method in a well-constructed case.

2. Interpolation in CDA and Its Defects

2.1. Switching Interpolation Architecture

The CDA method is the most commonly used method in EMT simulation tools, which adopts the TR method in the normal computation process. As is partly shown in Figure 1, when a switching action is detected, stage 1 involves interpolating to the moment of the switching action, point K, using linear interpolation; stage 2 and stage 3 suppresses the numerical oscillation using the BE method twice; and stage 4 is resynchronization using the linear interpolation algorithm, so the computation point is synchronized back to the original time step. At the same time, if a switching action is detected in stage 2, it will return to stage 1 to perform interpolation using the linear interpolation algorithm.

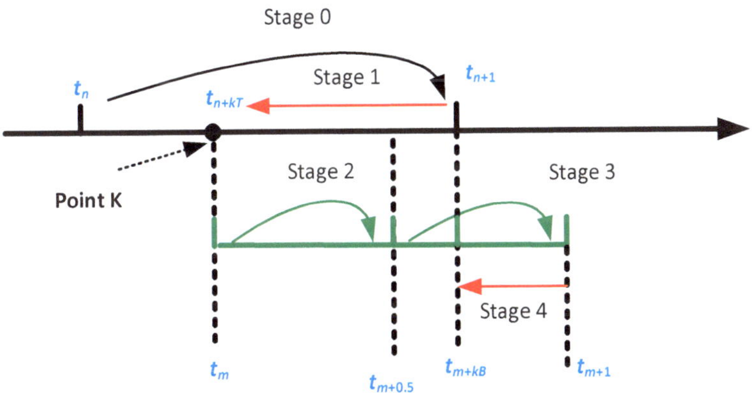

Figure 1. CDA computing framework.

Resynchronization is needed, as all controls and system models need to recognize the effective time-step shift in their local integration scheme. Therefore, two types of interpolation were used, i.e., the interpolation after the TR and the interpolation after the BE method. The former interpolation process was used to obtain the exact time of the switching action in stage 1, and the latter synchronization process was used for the resynchronization computation in stage 4.

2.2. Linear Interpolation after TR Method

EMT-type tools simulate the transient phenomenon of dynamic elements, which can be written in standard differential equation form as,

$$\frac{dy}{dt} = f(t, y) \tag{1}$$

Assuming that interpolation is performed at point K and $t_{n+k_T} = t_n + k_T h$, then, using the TR and linear interpolation algorithms, we have the following formulas:

$$\begin{cases} y_{n+1} = y_n + \frac{h}{2} f(t_n, y_n) + \frac{h}{2} f(t_{n+1}, y_{n+1}) \\ y_{n+k_T} = (1 - k_T) y_n + k_T y_{n+1} \end{cases}, \tag{2}$$

Assuming that y_{n+k_T} is the term to be solved, y_{n+1} is an intermediate term, and the step size is changed to a new one, which is $\widetilde{h} = k_T h$, then the above formulas can be rewritten as the following RK formulas:

$$\begin{cases} \mathrm{F}_1 = f(t_n, y_n) \\ \mathrm{F}_2 = f\left(t_n + \frac{1}{k_T} \widetilde{h}, y_n + \frac{\widetilde{h}}{2k_T} \mathrm{F}_1 + \frac{\widetilde{h}}{2k_T} \mathrm{F}_2\right) \\ y_{n+k_T} = y_n + \frac{\widetilde{h}}{2} \mathrm{F}_1 + \frac{\widetilde{h}}{2} \mathrm{F}_2 \end{cases}. \tag{3}$$

For simplicity, the s-stage RK case is often described using the following Butcher tableau:

$$\begin{array}{c|ccc} c_1 & a_{11} & \cdots & a_{1s} \\ \vdots & \vdots & \ddots & \vdots \\ c_s & a_{s1} & \cdots & a_{ss} \\ \hline & b_1 & \cdots & b_s \end{array} = \begin{array}{c|c} \mathbf{c} & \mathbf{A} \\ \hline & \mathbf{b}^{\mathbf{T}} \end{array}. \tag{4}$$

The Butcher tableau is a useful tool for the Runge–Kutta method, and the accuracy of the algorithm can be calculated according to rooted trees as Formulas (5) to (6), where e is an s-dimensional column vector whose elements are all 1 and C is a diagonal matrix composed of c elements.

$$\text{First-order accuracy}: \ \mathbf{b}^T \cdot \mathbf{e} = 1, \tag{5}$$

$$\text{Second-order accuracy}: \ \mathbf{b}^T \cdot \mathbf{C} \cdot \mathbf{e} = \frac{1}{2}. \tag{6}$$

Formula (2) features the Butcher tableau (7), and it is shown that the Butcher tableau (7) satisfies the condition of (5). Only when $k_T = 1$ is the condition of (6) satisfied. Thus, when $k_T \neq 1$, the formula is a first-order accuracy algorithm, and the use of first-order linear interpolation will reduce the computational accuracy.

$$\begin{array}{c|cc} 0 & 0 & \\ \frac{1}{k_T} & \frac{1}{2k_T} & \frac{1}{2k_T} \\ \hline & \frac{1}{2} & \frac{1}{2} \end{array}. \tag{7}$$

2.3. Linear Interpolation after BE Method

To distinguish the symbols in Section 2.2, m and k_B are used instead, and $m = n + k$, $k_B = 1 - k_T$. As illustrated in Figure 1, the linear interpolation in stage 4 is often performed after calculating the BE method twice, so there are three values that can be used, y_m, $y_{m+0.5}$, and y_{m+1}, while the first point $y_n y_m$ is a discontinuous point; thereby, we obtain the following formulas:

$$\begin{cases} y_{m+0.5} = y_m + \frac{h}{2} f(t_{m+0.5}, y_{m+0.5}) \\ y_{m+1} = y_{m+0.5} + \frac{h}{2} f(t_{m+1}, y_{m+1}) \\ y_{m+k_B} = 2(1 - k_B) y_{m+0.5} + (2k_B - 1) y_{m+1} \end{cases}. \tag{8}$$

Assuming that y_{m+k_B} is the term to be solved, $y_{m+0.5}$ and y_{m+1} are intermediate values, and the step size is changed to a new one, which is $\tilde{h} = k_B h$, then Formula (8) can be rewritten as the RK Formula (9) with the Butcher tableau (10).

$$\begin{cases} \mathrm{F}_1 = f\left(t_m + \frac{1}{2k_B}\tilde{h}, y_m + \frac{\tilde{h}}{2k_B}\mathrm{F}_1\right) \\ \mathrm{F}_2 = f\left(t_m + \frac{1}{k_B}\tilde{h}, y_m + \frac{\tilde{h}}{2k_B}\mathrm{F}_1 + \frac{\tilde{h}}{2k_B}\mathrm{F}_2\right) \\ y_{m+k_B} = y_m + \frac{\tilde{h}}{2k_B}\mathrm{F}_1 + \frac{(2k_B - 1)\tilde{h}}{2k_B}\mathrm{F}_2 \end{cases}, \tag{9}$$

$$\begin{array}{c|cc} \frac{1}{2k_B} & \frac{1}{2k_B} & \\ \frac{1}{k_B} & \frac{1}{2k_B} & \frac{1}{2k_B} \\ \hline & \frac{1}{2k_B} & 1 - \frac{1}{2k_B} \end{array}. \tag{10}$$

Formula (10) satisfies the condition of (5). When $k_B = 1 \pm 1/\sqrt{2}$, the condition of (6) is satisfied. Thus, when $k_B \neq 1 \pm 1/\sqrt{2}$, the formula is a first-order accuracy algorithm.

2.4. Error Analysis of CDA Scheme

The BE method is a first-order accuracy algorithm with a local error of $O(h^2)$. Considering error accumulation, the global error is $O(h)$. The TR and BE methods have different orders of accuracy, resulting in significant differences in their errors. When ignoring the secondary error, under the BE method, a three-phase inductor is equivalent to a parallel three-phase resistor $R_{error} = 2L/h$, and a three-phase capacitor under the BE method is equivalent to a series three-phase resistor $R_{error} = h/2C$. Detailed error calculations are provided in Appendix A.

Therefore, when switching between the TR method and BE method, the network also undergoes the operation of connecting or disconnecting resistors. In large-scale power grid calculations, this error propagates throughout the entire grid. In other words, although the fault is local, the algorithm switch causes the fault dynamics of connecting or disconnecting resistors to appear across the entire simulated power grid.

In summary, the linear interpolation algorithm can only maintain the original computational accuracy; the traditional CDA method can maintain the computational accuracy at the second order only with a fixed step using the first-order interpolation algorithm, which renders the interpolation meaningless. In the case of a variable step size, the CDA method will cause the computation accuracy to be reduced to the first order, resulting in large errors.

3. 3S-DIRK-Based Integral Interpolation Scheme

3.1. 3S-DIRK

Based on the characteristics of the Butcher tableau, the algorithms can be categorized into various forms, e.g., FIRK, ERK, DIRK, SDIRK, EDIRK, and FDIRK [6]. To comply with the current form of the EMT-type program, the following characteristics need to be met:

1. Diagonally implicit RK formula (DIRK): This enables a separate computation for each stage, without the need to generate a large matrix, which has the advantage of being a multistep method.
2. Singly diagonally implicit Runge–Kutta formula (SDIRK): The diagonal elements of the algorithm meeting this characteristic are identical, i.e., all the elements of a_{ii} are identical. When solving this special case of the DIRK algorithm, the Jacobi matrix is guaranteed to remain unchanged at each computational stage, which reduces the computational load of matrix decomposition.
3. First-same-as-last formula (FSAL): In this formula, $a_{si} = b_i (i = 1, \cdots, s)$ and $c_s = 1$, which means that the first stage of a step is the same as the last stage from the end of the previous step. This condition ensures that when solving rigid problems, order reduction will not occur in the algorithm [34].

In order to maintain the structure of the EMT-type program, the above three features must meet. Additionally, the three-stage method discussed below is referred to as 3S-DIRK.

3.2. TR-3S-DIRK in Interpolation Process

TR-3S-DIRK is a variable step-size interpolation integration method based on the TR. Assuming that the interpolation step is $\widetilde{h} = k_T h$ and the new step is $\widetilde{h}$, the new DIRK method with second-order accuracy is derived according to Formulas (4) and (5), and its Butcher tableau is as follows:

$$\begin{array}{c|ccc} 0 & 0 & & \\ \frac{1}{k_T} & \frac{1}{2k_T} & \frac{1}{2k_T} & \\ 1 & \frac{3k_T-1-k_T{}^2}{2k_T} & \frac{k_T-1}{2} & \frac{1}{2k_T} \\ \hline & \frac{3k_T-1-k_T{}^2}{2k_T} & \frac{k_T-1}{2} & \frac{1}{2k_T} \end{array} \tag{11}$$

The above algorithm is a special case of the 3S-DIRK algorithm. The elements in the first row of the Butcher tableau are all 0, which has no computation in the first stage and only requires integration on the basis of the TR. This method is referred to as the TR-3S-DIRK.

The stages of TR-3S-DIRK are as follows:

$$\text{Stage 0: TR}\quad y_{n+1} = y_n + \frac{h}{2}f(t_n, y_n) + \frac{h}{2}f(t_{n+1}, y_{n+1}), \tag{12}$$

$$\text{Stage 1: Interpolation}\quad \begin{aligned} y_{n+k_T} &= y_n + \frac{3k_T - 1 - {k_T}^2}{2}hf(t_n, y_n) \\ &+ \frac{k_T - 1}{2}k_T hf(t_{n+1}, y_{n+1}) + \frac{h}{2}f(t_{n+k}, y_{n+k_T}) \end{aligned}. \tag{13}$$

3.3. BE2-3S-DIRK in Synchronous Process

After events or operations occur, the algorithm switches to the BE method, and the interpolation is BE-based synchronous interpolation. The BE method is itself a low-order algorithm, and the linear interpolation further increases the error. Therefore, it is necessary to improve the computational accuracy using an integration method. Based on two-step BE, similar DIRK methods with second-order accuracy can be derived, and their Butcher tableau is as follows:

$$\begin{array}{c|ccc} \frac{1}{2k_B} & \frac{1}{2k_B} & & \\ \frac{1}{k_B} & \frac{1}{2k_B} & \frac{1}{2k_B} & \\ 1 & \frac{3k_B-1-k_B^2}{k_B} & \frac{1-4k_B+2k_B^2}{2k_B} & \frac{1}{2k_B} \\ \hline & \frac{3k_B-1-k_B^2}{k_B} & \frac{1-4k_B+2k_B^2}{2k_B} & \frac{1}{2k_B} \end{array}. \tag{14}$$

The interpolation algorithm (13) has three advantages:

1. The variable step size can be used for interpolation synchronization;
2. The computational accuracy is improved to the second order, while the first-order error brought by the BE algorithm is corrected;
3. It satisfies L-stability, which can eliminate the numerical oscillation in the computation.

This algorithm requires an integral operation on the basis of using the BE method twice. This method is referred to as BE2-3S-DIRK.

The stages of BE2-3S-DIRK are as follows:

$$\text{Stage 2: BE}\quad y_{m+0.5} = y_m + \frac{h}{2}f(t_{m+0.5}, y_{m+0.5}), \tag{15}$$

$$\text{Stage 3: BE}\quad y_{m+1} = y_{m+0.5} + \frac{h}{2}f(t_{m+1}, y_{m+1}), \tag{16}$$

$$\text{Stage 4: Interpolation}\quad \begin{aligned} y_{m+k_B} &= y_m + (3k_B - 1 - {k_B}^2)hf(t_{m+0.5}, y_{m+0.5}) \\ &+ \frac{1 - 4k_B + 2{k_B}^2}{2}hf(t_{m+1}, y_{m+1}) + \frac{h}{2}f(t_{m+k_B}, y_{m+k_B}) \end{aligned}. \tag{17}$$

4. Application of 3S-DIRK in EMT Program

4.1. Application of 3S-DIRK in DAEs

The proposed algorithm is mainly used for interpolation, and the main algorithm in EMT simulation is the TR algorithm (stage 0). When interpolation is required after a switching operation is detected, a new computational logic is adopted, as shown in Figure 2, with the following stages:

Stage 0: Integrate the TR algorithm, with a normal step, while detecting whether any event occurs; if an event occurs, calculate the coefficient for the interpolation (k) and proceed to stage 1; otherwise, repeat stage 0 until the event occurs or the simulation is finished.

Stage 1: Integrate to pint K using the TR-3S-DIRK algorithm; set the new step coefficient to k_T, as in Formula (12); form a current vector; and solve the network's voltage.

Stage 2: Integrate using the BE algorithm with a 1/2 step.

Stage 3: Integrate using the BE algorithm with a 1/2 step.

Stage 4: Integrate using the BE2-3S-DIRK algorithm and set the new step size coefficient to k_B, $k_T + k_B = 1$, as in Formula (16). If no events occur during the process, the steps are illustrated as in Figure 2a. If any events occur during stage 2–4 at the point K′, change the step size and then go back to Stage 2 as in Figure 2b.

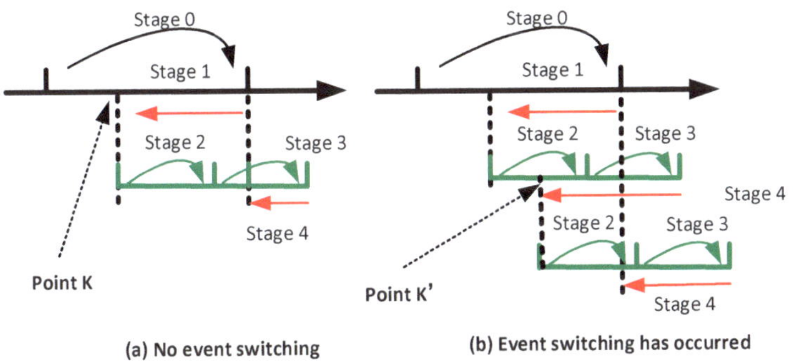

Figure 2. Computational framework of the 3S-DIRK algorithm.

This method can ensure that the value at each step point has second-order accuracy, and the algorithm in the switching process is L-stable and capable of suppressing numerical oscillation. Specifically, the BE2-3S-DIRK method is L-stable and can eliminate the first-order error caused by BE.

4.2. Linear Inductors

The electrical components in the EMT-type program are treated as a parallel circuit of an admittance (G_{equ}) and a current source (I_{hist}), as shown in Figure 3, and the component is converted to the form as:

$$i = G_{equ}u + I_{hist}. \tag{18}$$

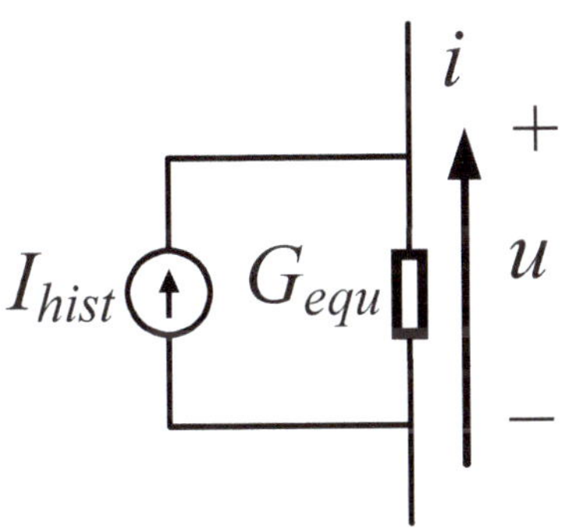

Figure 3. Equivalent circuit of electrical component.

The dynamic equation of the inductor is $L\frac{d}{dt}i = u$. The equivalent admittances of the TR and BE methods are both $G_{equ} = \frac{h}{2L}$.

Assuming that the interpolation coefficients of TR-3S-DIRK and BE2-2S-DIRK are k_T and k_B, respectively, then the formulas of the proposed 3S-DIRK are shown as follows. The derivation processes are explained in Appendix B:

$$\text{TR-3s-FESDIRK}: \quad I_{hist} = i_n + (3k_T - 1 - k_T^2)G_{equ}u_n + k_T(k_T - 1)G_{equ}u_{n+1}, \tag{19}$$

$$\text{BE2-3s-FSDIRK}: \quad I_{hist} = i_m + 2(3k_B - 1 - k_B{}^2)G_{equ}u_{m+0.5} + (1 - 4k_B + 2k_B{}^2)G_{equ}u_{m+1}. \tag{20}$$

4.3. Linear Capacitors

The dynamic equation of the capacitor is $C\frac{d}{dt}u = i$, and the equivalent admittance is $G_{equ} = \frac{2C}{h}$ in both the TR and BE methods.

Similarly, the formulas are derived:

$$\text{TR-3s-FESDIRK}: \quad I_{hist} = G_{equ}u_n + (3k_T - 1 - k_T^2)i_n + k_T(k_T - 1)i_{n+1}, \tag{21}$$

$$\text{BE2-3s-FSDIRK}: \quad I_{hist} = G_{equ}u_m + 2(3k_B - 1 - k_B{}^2)i_{m+0.5} + (1 - 4k_B + 2k_B{}^2)i_{m+1}. \tag{22}$$

In summary, the proposed variable step-size integration algorithm for interpolation can ensure that the equivalent admittances of the elements (G_{equ}) are consistent, and, furthermore, the matrix of the entire system does not need to be modified.

5. Case Study

Two case studies—the Series R-L Circuit and the AC/DC Hybrid Power Systems—are presented, demonstrating that the proposed algorithm retains L-stability, is capable of eliminating numerical oscillations, and ensures the maintenance of second-order computational accuracy.

5.1. Case A: Series R-L Circuit

Voltage chatter occurs when a disturbance is applied at a node to which only the inductor is connected, as shown in Figure 4. In this circuit, the voltage source is an ideal source with zero resistance, a magnitude of 132.79 kV, a frequency of 60 Hz, a ramp-up time of 0.05 s, and an initial phase of 0 degrees. The open resistance and closed resistance are 1×10^{12} ohm and 0.0005 ohm, respectively. The inductance of the inductor is 0.1 H.

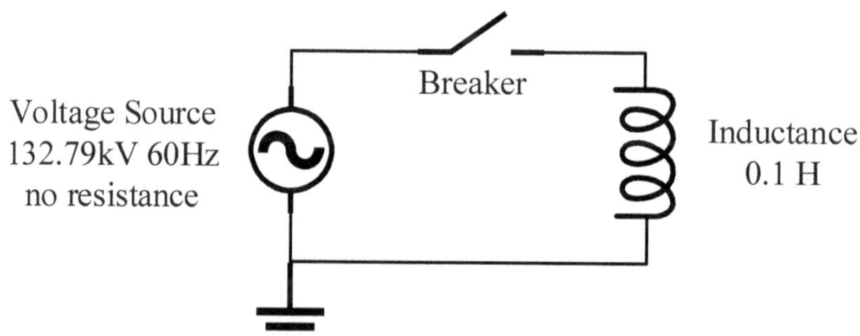

Figure 4. Diagram of series R-L circuit.

Meanwhile, there are two types of breakers: a thyristor and an insulated-gate bipolar transistor (IGBT). The former is a half-controlled type in which the break waits for the current to be reduced to 0 after receiving the off signal, while the latter is the full-controlled type in which the circuit is turned off immediately when the off signal is received. Therefore, the two cases will be discussed separately. In case A1, the switch does not turn off immediately but waits for the current to drop to 0; in case A2, the switch is forced to turn off, i.e., the turn-off action is implemented immediately when the off signal is received.

5.1.1. Case A1: Breaker of Half-Controlled Type

Assuming that the switch is not forced to turn off, the off signal arrives after 0.1 s, but the current at that time is not 0. When the current reaches 0, the time is 0.104167 s, at which point the switch is turned off, so interpolation is required.

As shown in Figure 5, the conventional TR algorithm will cause numerical oscillation, which can only be eliminated through an algorithm with L-stability.

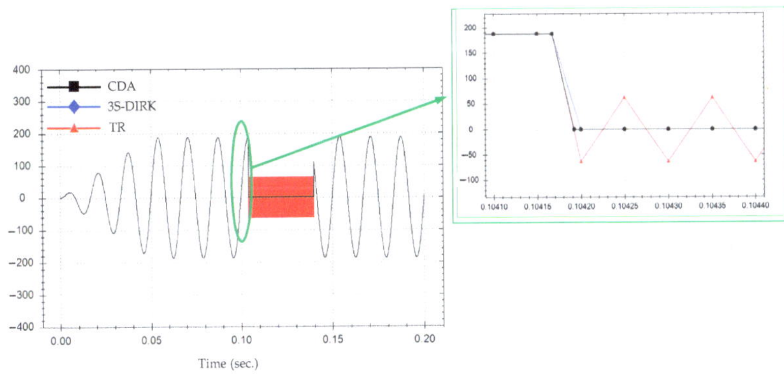

Figure 5. Simulated curve of the voltage across the inductor with half-controlled type breaker.

5.1.2. Case A2: Breaker of Full-Controlled Type

In this case, the switch is forced to turn off when the off signal arrives, at which point the current is zero, and the numerical oscillation is greater.

As shown in Figure 6, both the CDA algorithm and the 3S-DIRK algorithm can eliminate numerical oscillations.

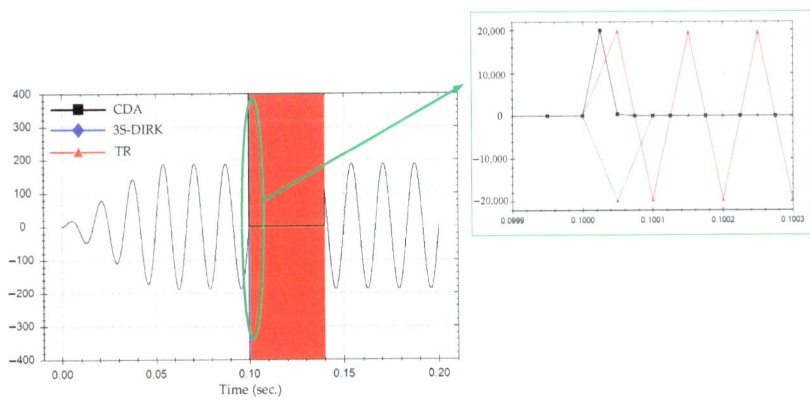

Figure 6. Simulated curve of the voltage across the inductor with full-controlled type breaker.

5.2. Case B: AC/DC Hybrid Power Systems

The CDA method will switch algorithms after events occur and the computation accuracy will be degraded to first order, leading to large errors, especially in the computations for power electronics. The proposed interpolation method in this study uses variable-step integration to maintain the calculation accuracy in the second order, and the errors can be controlled within a small range.

To verify the validity of the proposed 3S-DIRK method, the IEEE9 circuit and the line-commutated, converter-based, high-voltage DC (LCC-HVDC) circuit are combined, that is, both circuits are solved simultaneously, but there is no electrical connection between them. The IEEE 9-bus system serves as a standard test case, comprising three generators, three loads, nine buses, and nine transmission lines, which makes it highly suitable for verifying power system analysis and control algorithms. The LCC-HVDC circuit represents a typical configuration in high-voltage direct current transmission systems, consisting of converters, filters, reactors, DC transmission lines, and control and protection systems [35].

The LCC system starts up at 2 s and withdraws at 4 s, and in this time, the CDA method is put into effect. Meanwhile, the potential error sources are as follows: (1) the BE algorithm, (2) the interpolation algorithm, (3) the number of steps in the BE algorithm, and (4) the step size.

Therefore, it is necessary to set up four cases to separately discuss the impact of these factors:

Case B1: CDA without the interpolation method is performed to show the errors caused by the BE algorithm alone;

Case B2: The standard CDA scheme is employed to show the errors caused by linear interpolation and BE methods;

Case B3: To enlarge the errors caused by the BE algorithm, BE is used five times after the switching action;

Case B4: As discussed in Section 2.4, the errors are affected by the time step size, so the step size is changed to 100 μs from 50 μs.

5.2.1. Case B1: Impact of the BE Algorithm

The computation is implemented without interpolation; the BE algorithm is used twice after the switching action; and the step size is set to 50 μs.

Figure 7 shows that while switching the algorithm, the CDA algorithm causes errors in Bus1-BusA's active power between 2 s and 4 s. Although there are no accidents, a large number of fluctuations occur using CDA method. Meanwhile, during the operation of the LCC, the computational error in the IEEE 9 system is very small using the proposed integral interpolation method.

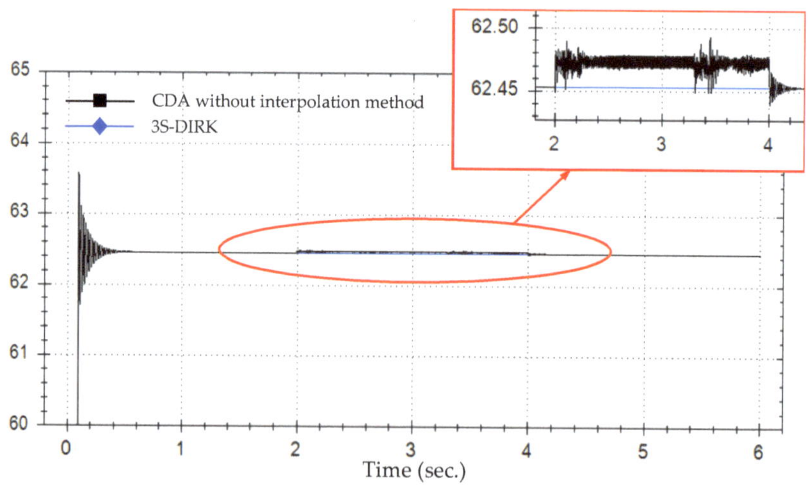

Figure 7. Simulated curve of the active power (MW) of the Bus1-BusA line in case B1.

5.2.2. Case B2: Impact of Interpolation

In this case, interpolation is performed on the basis of the above-described Case B1 example, in which interpolation is performed, and then the BE algorithm is used twice after the switching action, where the step size is set to 50 μs.

Figure 8 shows that in the interpolation process, the CDA error increases. However, the proposed algorithm still performs interpolation after the BE algorithm, but the interpolation coefficient remains at the original step size. The errors caused by the two implementations of the BE algorithm are offset by increasing the computation load, thereby maintaining the computational accuracy in the second order.

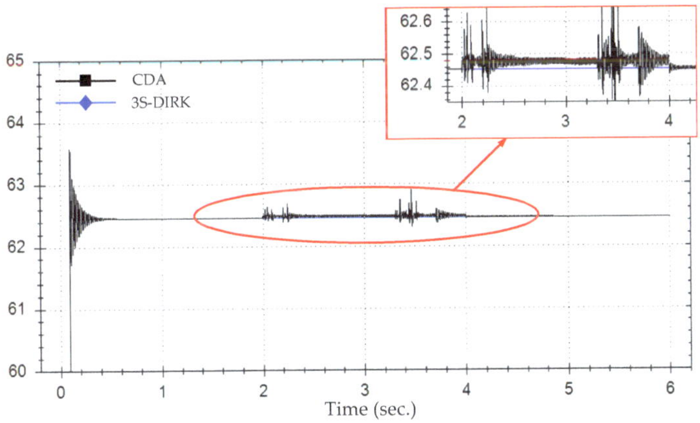

Figure 8. Simulated curve of the active power (MW) of the Bus1-BusA line of Case B2.

5.2.3. Case B3: Impact of the Number of Computations Using the BE Algorithm

Based on the Case B1 example, the number of computations of the BE algorithm is increased to five. Specifically, interpolation is not performed, and then the BE algorithm is used five times after the switching action, in which the step size is set to 50 μs. Because two BE computations are combined into a group in the 3S-DIRK algorithm in this case, three iterations of BE2-3S-DIRK algorithm are performed.

Figure 9 shows that, due to the implementation of multiple rounds of the BE algorithm, the computational error of the CDA is greater. However, the proposed algorithm performs an error correction after every two BE computations, which avoids the increased error derived from the increase in the number of BE computations.

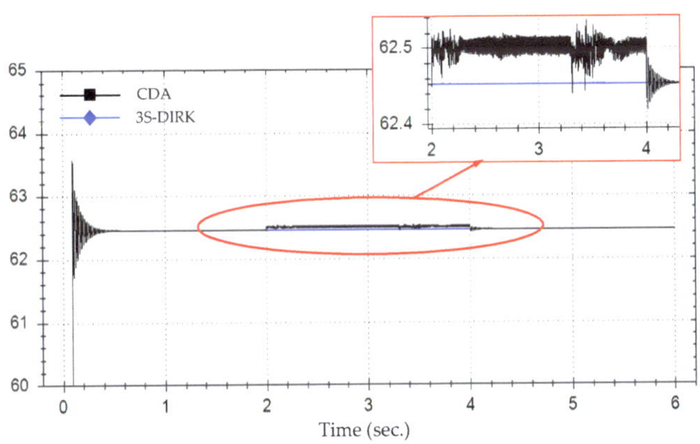

Figure 9. Simulated curve of the active power (MW) of the Bus1-BusA line of Case B3.

5.2.4. Case B4: Impact of the Simulation Step

In this case, the simulation step is changed to 100 μs on the basis of the Case B1 example, in which interpolation is performed; the BE algorithm is then used twice after the switching action, and the step size is set to 50 μs.

Figure 10 shows that the error caused by the BE method is proportional to the step size. However, the proposed algorithm performs well with different step sizes.

To measure the error, the error indicator is defined as the deviation from the stable value within 2 s to 4 s of the active power of the line between Bus 1 and Bus A. The errors are summarized in Table 1.

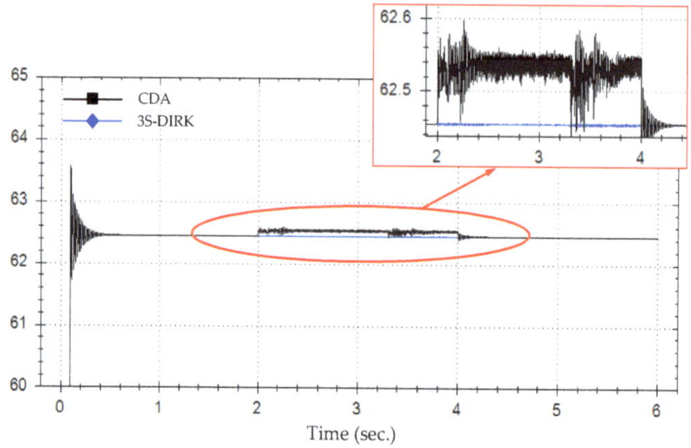

Figure 10. Simulated curve of the active power (MW) of the Bus1-BusA line of Case B4.

Table 1. Error calculation summary table.

Error (MW)	CDA	3S-DIRK
Case B1	0.0404	0.0002
Case B2	0.4454	0.0003
Case B3	0.0831	0.0005
Case B4	0.1454	0.0032

It can be seen that the 3S-DIRK interpolation method proposed in this study can control the errors within a small range, while the conventional CDA method has higher errors, which interferes with the normal computing system due to the power electronics system.

6. Conclusions

In this study, we analyzed the computational accuracy of the CDA method and showed that the interpolation algorithm reduces the computational accuracy to the first order, making it impossible to apply the CDA method to the simulation of large power grids. Using the Butcher tableau, we discussed the accuracy order of the linear interpolation methods and found that linear interpolation algorithms cannot improve the computational accuracy. Therefore, it is impossible to use a linear interpolation algorithm to construct a suitable high-order integration algorithm.

Based on the Runge–Kutta theories, we developed two interpolation algorithms (TR-3S-DIRK and BE2-3S-DIRK) which are used for interpolation after the TR and BE methods. These interpolation algorithms support variable step-size computations while keeping the matrix unchanged and can be used for interpolation and synchronization processes. At the same time, BE2-3S-DIRK can eliminate the first-order error derived from the BE algorithm and increase the computational accuracy to the second order, thereby maintaining the second-order accuracy in the EMT simulations; the BE2-3S-DIRK algorithm is L-stable, which can suppress numerical oscillations.

A new integration scheme was proposed by replacing the linear interpolation of CDA into 3S-DIRK. Then, we derived the formulas for inductors and capacitors and verified that the equivalent admittance of these elements was consistent under different stages, without the need to modify the admittance matrix. We used the Case A examples to verify that the interpolation algorithm retained the L-stability of the original BE algorithm and, thus, is capable of suppressing numerical oscillations. Then, using an example in which the LCC-HVDC system and the IEEE9 system were combined, we verified that the proposed method could maintain second-order computational accuracy throughout the whole computation process.

Moreover, in this study, we provide an algorithm development idea based on the RK architecture, which can be used to further develop new high-order algorithms.

Author Contributions: Conceptualization, X.Y.; methodology, K.S. and K.C.; software, K.C.; validation, H.C.; writing—original draft, K.S. and F.T.; writing—review and editing, X.Y. All authors have read and agreed to the published version of the manuscript.

Funding: This research was financially supported by the "Guangzhou Ultra-Large City New Type Power System Demonstration Zone Special Planning" project, with the number 0301006900220001.

Data Availability Statement: The original contributions presented in the study are included in the article, further inquiries can be directed to the corresponding author.

Conflicts of Interest: Authors Kaiyuan Sun, Kun Chen and Haifeng Cen, ware employed by the company Guangdong Power Grid Co., Ltd. The remaining authors declare that the research was conducted in the absence of any commercial or financial relationships that could be construed as a potential conflict of interest.

Nomenclature

Abbreviations	
EMT	Electromagnetic transient
RK	Runge–Kutta
DIRK	Diagonally implicit RK formula
SDIRK	Singly diagonally implicit Runge–Kutta formula
FSAL	First-same-as-last formula
3S-DIRK	Three-stage diagonally implicit Runge–Kutta
2S-DIRK	Two-stage diagonally implicit Runge–Kutta
TR	Trapezoidal rule
BE	Backward Euler
TR-3S-DIRK	Three-stage diagonally implicit Runge–Kutta methods combined with trapezoidal rule
BE2-3S-DIRK	Three-stage diagonally implicit Runge–Kutta methods combined with backward Euler twice
DAEs	Differential–algebraic equations
ODEs	Ordinary differential equations
TR-BDF2	Trapezoidal method with the second-order backward difference formula
LCC-HVDC	Line-commutated converter-based high-voltage direct current
IGBT	Insulated-gate bipolar transistor
CDA	Critical damping adjustment
Variables	
$f(t,y)$	Function of the differential equation, representing the derivative of y with respect to t
t_n, t_{n+1}	Time at the n-th and n+1-th time step as shown in Figure 1
y_n, y_{n+1}	Value at the n-th and n+1-th time step, respectively
k_T, k_B	Interpolation coefficients in Stage 1 and Stage 4 as in Figure 1
h	Step size
$\widetilde{h}$	New step size
G_{equ}	Equivalent admittance in equivalent circuit
I_{hist}	History current in equivalent circuit
u	Instantaneous voltage
i	Instantaneous current
L	Inductance of inductor
C	Capacitance of capacitor
θ_v	Voltage phase angle
θ_i	Current phase angle
R	Resistance
ω	Angular frequency
$\mathbf{c}$, $\mathbf{b}$	Vector of s size as in Equation (4)
$\mathbf{A}$, $\mathbf{C}$	Matrix of size s x s as in Equation (4)

Appendix A. Error Calculations of BE Method and Interpolation Algorithm

As discussed in Sections 2.2 and 2.3, the BE method and interpolation algorithm is a first-order accuracy algorithm. The TR and BE methods have different orders of accuracy, leading to significant differences in their errors. To illustrate the error in the BE method and linear interpolation, the formulas for single-phase and three-phase inductors are derived and used to compare the numerical error differences resulting from the different orders of accuracy of the two methods. For the sake of convenience, we assume that the interpolation coefficient k equals 0.5.

Appendix A.1. Error in Single-Phase Inductor

We assume that the voltage across the inductor after stabilization is $u(t) = \sqrt{2}u\cdot \sin(\omega t + \theta_v)$ and the current is $i(t) = \sqrt{2}i\cdot \sin(\omega t + \theta_i)$. An error analysis comparison for different algorithms is shown in Table A1.

Table A1. Error analysis for single-phase inductor.

	TR	BE	Interpolation (k = 0.5)
Formula	$i(t) = i(t-h) + \frac{h}{2L}[u(t) + u(t-h)]$	$i(t) = i(t-h) + \frac{h}{L}u(t)$	$i(t) = i(t-h) + \frac{h}{2L}[u(t+h)]$
Magnitude	$u = \frac{2L}{h}i\cdot \tan(\frac{\omega h}{2})$	$u = \frac{2L}{h}i\cdot \sin(\frac{\omega h}{2})$	$u = \frac{2L}{h}i\cdot \sin(\frac{\omega h}{2}) / \cos(\omega h)$
Angle	$\theta_v = \theta_i + \frac{\pi}{2}$	$\theta_v = \theta_i + \frac{\pi}{2} - \frac{\omega h}{2}$	$\theta_v = \theta_i + \frac{\pi}{2} - \frac{\omega h}{2}$

It can be seen that the BE and interpolation methods introduce a first-order deviation in the angle of the inductor element. In terms of magnitude, both algorithms exhibit second-order errors.

Appendix A.2. Error in Three-Phase Inductor

Assuming that, after stabilization, the voltage and current across the inductor are three-phase balanced and the active and reactive power are balanced, then the voltage and current across the three-phase inductor can be expressed as follows:

$$U(t) = \sqrt{\frac{2}{3}}u\cdot \begin{bmatrix} \sin(\omega t + \theta_v) \\ \sin(\omega t + \theta_v - \frac{2}{3}\cdot\pi) \\ \sin(\omega t + \theta_v - \frac{2}{3}\cdot\pi) \end{bmatrix} \tag{A1}$$

$$I(t) = \sqrt{\frac{2}{3}}i\cdot \begin{bmatrix} \sin(\omega t + \theta_i) \\ \sin(\omega t + \theta_i - \frac{2}{3}\cdot\pi) \\ \sin(\omega t + \theta_i - \frac{2}{3}\cdot\pi) \end{bmatrix} \tag{A2}$$

As shown in Table A2, when the secondary error is ignored, the BE method models the three-phase inductor as being equivalent to a parallel three-phase resistor $R_{error} = 2L/h$.

Table A2. Error analysis for three-phase inductor.

	TR	BE	Interpolation (k = 0.5)
Formula	$P = P_0\cos(\omega h) - Q_0\sin(\omega h) + \frac{h}{2L}u^2[1+\cos(\omega h)]$ $Q = Q_0\cos(\omega h) + P_0\sin(\omega h) + \frac{h}{2L}u^2\sin(\omega h)$	$P = P_0\cos(\omega h) - Q_0\sin(\omega h) + \frac{h}{L}u^2$ $Q = Q_0\cos(\omega h) + P_0\sin(\omega h)$	$P = P_0\cos(\omega h) - Q_0\sin(\omega h) + \frac{h}{L}u^2\cos(\omega h)$ $Q = Q_0\cos(\omega h) + P_0\sin(\omega h)$
Active power	$P = 0$	$P = \frac{h}{2L}u^2$	$P = \frac{h}{2L}u^2\cos(\omega h)$
Reactive power	$Q = \frac{h}{2L}u^2\frac{1+\cos(\omega h)}{\sin(\omega h)}$	$Q = \frac{h}{2L}u^2\frac{1+\cos(\omega h)}{\sin(\omega h)}$	$Q = \frac{\Delta t}{2L}u^2\frac{1+\cos(\omega h)}{\sin(\omega h)}\cos(\omega h)$

Appendix A.3. Error in Three-Phase Capacitor

Using the above conditions for derivation, the error for the three-phase capacitor is shown in Table A3.

Table A3. Error analysis of the BE method for three-phase capacitor.

	TR	BE	Interpolation (k = 0.5)
Formula	$P = P_0\cos(\omega h) + Q_0\sin(\omega h) + \frac{h}{2C}i^2[1+\cos(\omega h)]$ $P = P_0\cos(\omega h) + Q_0\sin(\omega h) + \frac{h}{2C}i^2[1+\cos(\omega h)]$	$P = P_0\cos(\omega h) + Q_0\sin(\omega h) + \frac{h}{C}i^2$ $Q = Q_0\cos(\omega h) - P_0\sin(\omega h)$	$P = P_0\cos(\omega h) + Q_0\sin(\omega h) + \frac{h}{C}i^2\cos(\omega h)$ $Q = Q_0\cos(\omega h) - P_0\sin(\omega h)$
Active power	$P = 0$	$P = \frac{h}{2C}i^2$	$P = \frac{h}{2C}i^2\cos(\omega h)$
Reactive power	$Q = -\frac{h}{2C}i^2\frac{1+\cos(\omega h)}{\sin(\omega h)}$	$Q = -\frac{h}{2C}i^2\frac{1+\cos(\omega h)}{\sin(\omega h)}$	$Q = -\frac{h}{2C}i^2\frac{1+\cos(\omega h)}{\sin(\omega h)}\cos(\omega h)$

It can be seen that the BE method for a three-phase capacitor is equivalent to a series connection with a three-phase resistor $R_{error} = h/2C$.

In summary, when switching between the TR and BE methods, the network also undergoes the operation of connecting or disconnecting resistors. In large-scale power grid calculations, this error propagates throughout the entire grid. In other words, although the fault is local, the algorithm switch causes the fault dynamics of connecting or disconnecting resistors to appear across the entire simulated power grid.

Appendix B. Derivation Process of 3S-DIRK of Linear Inductor and Capacitor

Appendix B.1. Linear Inductor

The dynamic equation of the inductor is $L\frac{d}{dt}i = u$. It can be written in standard form as:

$$\begin{cases} \frac{dy}{dt} = f(t,y) \\ y = i \\ f(t,y) = \frac{1}{L}u \end{cases}, \tag{A3}$$

Using Equation (12),

$$i_{n+k_T} = i_n + \frac{3k_T - 1 - k_T^2}{2}h\frac{1}{L}u_n + \frac{k_T - 1}{2}k_T h \cdot \frac{1}{L}u_{n+1} + \frac{h}{2}\frac{1}{L}u_{n+k_T}, \tag{A4}$$

Thus, we obtain the TR-3S-DIRK formulas as a form of Equation (18):

$$\begin{array}{l} i_{n+k_T} = G_{equ}u_{n+k_T} + I_{hist} \\ \begin{cases} G_{equ} = \frac{h}{2L} \\ I_{hist} = i_n + (3k_T - 1 - k_T^2)G_{equ}u_n + k_T(k_T - 1)G_{equ}u_{n+1} \end{cases} \end{array}. \tag{A5}$$

Using Equation (17),

$$i_{m+k_B} = i_m + (3k_B - 1 - k_B{}^2)h\frac{1}{L}u_{m+0.5} + \frac{1 - 4k_B + 2k_B{}^2}{2}h\frac{1}{L}u_{m+1} + \frac{h}{2}\frac{1}{L}u_{m+k_B}, \tag{A6}$$

Thus, we obtain the BE2-3S-DIRK formulas:

$$\begin{array}{l} i_{m+k_B} = G_{equ}u_{m+k_B} + I_{hist} \\ \begin{cases} G_{equ} = \frac{h}{2L} \\ I_{hist} = i_m + 2(3k_B - 1 - k_B{}^2)G_{equ}u_{m+0.5} + (1 - 4k_B + 2k_B{}^2)G_{equ}u_{m+1} \end{cases} \end{array}. \tag{A7}$$

Appendix B.2. Linear Capacitors

The dynamic equation of the capacitor is $C\frac{d}{dt}u = i$. It can be written in standard form as:

$$\begin{cases} \frac{dy}{dt} = f(t,y) \\ y = u \\ f(t,y) = \frac{1}{C}i \end{cases}, \quad \text{(A8)}$$

Using Equation (12),

$$u_{n+k_T} = u_n + \frac{3k_T - 1 - k_T^2}{2} h \frac{1}{C} i_n + \frac{k_T - 1}{2} k_T h \cdot \frac{1}{C} i_{n+1} + \frac{h}{2}\frac{1}{C} i_{n+k_T}, \quad \text{(A9)}$$

Thus, we obtain the TR-3S-DIRK formulas:

$$\begin{aligned} & i_{n+k_T} = G_{equ} u_{n+k_T} + I_{hist} \\ & \begin{cases} G_{equ} = \frac{2C}{h} \\ I_{hist} = G_{equ} u_n + (3k_T - 1 - k_T^2) i_n + k_T(k_T - 1) i_{n+1} \end{cases} \end{aligned} \quad \text{(A10)}$$

Using Equation (17),

$$u_{m+k_B} = u_m + (3k_B - 1 - k_B{}^2) h \frac{1}{C} i_{m+0.5} + \frac{1 - 4k_B + 2k_B{}^2}{2} h \frac{1}{C} i_{m+1} + \frac{h}{2}\frac{1}{C} i_{m+k_B}, \quad \text{(A11)}$$

Thus, we obtain the BE2-3S-DIRK formulas:

$$\begin{aligned} & i_{m+k_B} = G_{equ} u_{m+k_B} + I_{hist} \\ & \begin{cases} G_{equ} = \frac{2C}{h} \\ I_{hist} = G_{equ} u_m + 2(3k_B - 1 - k_B{}^2) i_{m+0.5} + (1 - 4k_B + 2k_B{}^2) i_{m+1} \end{cases} \end{aligned} \quad \text{(A12)}$$

References

1. Mahseredjian, J.; Dinavahi, V.; Martinez, J.A. Simulation tools for electromagnetic transients in power systems: Overview and challenges. *IEEE Trans. Power Del.* **2009**, *24*, 1657–1669. [CrossRef]
2. Ming, Y.; Yongming, Z.; Ziqian, Z. Review of electromagnetic transient simulation algorithms for power system. *Elect. Meas. Instrum.* **2022**, *59*, 10–19.
3. Tanaka, Y.; Baba, Y. Study of a numerical integration method using the compact scheme for electromagnetic transient simulations. *Electr. Power Syst. Res.* **2023**, *223*, 109666. [CrossRef]
4. Subedi, S.; Rauniyar, M.; Ishaq, S.; Hansen, T.M.; Tonkoski, R.; Shirazi, M.; Wies, R.; Cicilio, P. Review of methods to accelerate electromagnetic transient simulation of power systems. *IEEE Access* **2021**, *9*, 89714–89731. [CrossRef]
5. Gear, W. Simultaneous numerical solution of differential-algebraic equations. *IEEE Trans. Circuit Theory* **1971**, *18*, 89–95. [CrossRef]
6. Butcher, J.C. Stability of implicity runge-kutta methods. In *Numerical Methods for Ordinary Differential Equations*, 2nd ed.; John Wiley & Sons Ltd.: New York, NY, USA, 2008; pp. 230–258.
7. Tant, J.; Driesen, J. On the numerical accuracy of electromagnetic transient simulation with power electronics. *IEEE Trans. Power Deliv.* **2018**, *33*, 2492–2501. [CrossRef]
8. Dommel, H.W. Digital computer solution of electromagnetic transients in single- and multiphase networks. *IEEE Trans. Power Appar. Syst.* **1969**, *88*, 734–741. [CrossRef]
9. Pordanjani, S.R.; Mahseredjian, J.; Naïdjate, M.; Bracikowski, N.; Fratila, M.; Rezaei-Zare, A. Electromagnetic modeling of inductors in EMT-type software by three circuit-based methods. *Electr. Power Syst. Res.* **2022**, *211*, 108304. [CrossRef]
10. Wen, H.C.; Ruehli, A.; Brennan, P. The modified nodal approach to network analysis. *IEEE Trans. Circuits Syst.* **1975**, *22*, 504–509.
11. Ametani, A. Electromagnetic transients program: History and future. *IEEJ Trans. Electr. Electron. Eng.* **2021**, *16*, 1150–1158. [CrossRef]
12. Marti, J.R.; Lin, J. Suppression of numerical oscillations in the EMTP. *IEEE Power Eng. Rev.* **1989**, *9*, 71–72. [CrossRef]
13. Alvarado, F.L.; Lasseter, R.H.; Sanchez, J.J. Testing of trapezoidal integration with damping for the solution of power transient problems. *IEEE Trans. Power Appar. Syst.* **1983**, *102*, 3783–3790. [CrossRef]
14. Noda, T.; Takenaka, K.; Inoue, T. Numerical integration by the 2-stage diagonally implicit Runge-Kutta method for electromagnetic transient simulations. *IEEE Trans. Power Deliv.* **2009**, *24*, 390–399. [CrossRef]
15. Noda, T.; Kikuma, T.; Yonezawa, R. Supplementary techniques for 2S-DIRK-based EMT simulations. *Electr. Power Syst. Res.* **2014**, *115*, 87–93. [CrossRef]
16. Olaniyan, A.S.; Bakre, O.F.; Akanbi, M.A. A 2-stage implicit Runge-Kutta method based on heronian mean for solving ordinary differential equations. *Pure Appl. Math. J.* **2020**, *9*, 84–90. [CrossRef]

17. Watson, N.R.; Irwin, G.D. Comparison of root-matching techniques for electromagnetic transient simulation. *IEEE Trans. Power Deliv.* **2000**, *15*, 629–634. [CrossRef]
18. Ye, J.; Xie, J.; Wang, Y.; Zhang, L.; Li, B.; Lv, J.; Li, K.; Jin, S.; Xu, Y.; Ma, W. Electromagnetic transient simulation algorithm for nonlinear elements based on rosenbrock numerical integration method. *IEEE Access* **2024**, *12*, 66981–66992. [CrossRef]
19. Cibik, A.; Eroglu, F.G.; Kaya, S. Analysis of second order time filtered backward Euler method for MHD equations. *J. Sci. Comput.* **2020**, *82*, 38. [CrossRef]
20. Lin, J.; Marti, J.R. Implementation of the CDA procedure in the EMTP. *IEEE Trans. Power Systems* **1990**, *5*, 394–402.
21. Kui, W.; Ying, S. Modeling and simulation of nonlinear component algorithm. *Power Syst. Clean Energy* **2010**, *26*, 34–37.
22. Zou, M.; Mahseredjian, J.; Delourme, B.; Joos, G. On interpolation and reinitialization in the simulation of transients in power electronic systems. In Proceedings of the 14th Power Systems Computation Conference (PSCC 2002), Sevilla, Spain, 24–28 June 2002.
23. Kuffel, P.; Kent, K.; Irwin, G. The implementation and effectiveness of linear interpolation within digital simulation. *Int. J. Electr. Power Energy Syst.* **1997**, *19*, 221–227. [CrossRef]
24. Strunz, K.; Linares, L.; Marti, J.R.; Huet, O.; Lombard, X. Efficient and accurate representation of asynchronous network structure changing phenomena in digital real time simulators. *IEEE Trans. Power Syst.* **2000**, *15*, 586–592. [CrossRef]
25. Strunz, K. Flexible numerical integration for efficient representation of switching in real time electromagnetic transients simulation. *IEEE Trans. Power Deliv.* **2004**, *19*, 1276–1283. [CrossRef]
26. Nzale, W.; Mahseredjian, J.; Fu, X.; Kocar, I.; Dufour, C. Improving Numerical Accuracy in Time-Domain Simulation for Power Electronics Circuits. *IEEE Open Access J. Power Energy* **2021**, *8*, 157–165. [CrossRef]
27. Li, P.; Meng, Z.; Fu, X.; Yu, H.; Wang, C. Interpolation for power electronic circuit simulation revisited with matrix exponential and dense outputs. *Electr. Power Syst. Res.* **2020**, *189*, 106714. [CrossRef]
28. Wang, C.; Fu, X.; Li, P.; Wu, J.; Wang, L. Multiscale Simulation of Power System Transients Based on the Matrix Exponential Function. *IEEE Trans. Power Syst.* **2017**, *32*, 1913–1926. [CrossRef]
29. Kennedy, C.A.; Carpenter, M.H. *Diagonally Implicit Runge-Kutta Methods for Ordinary Differential Equations. A Review*; National Aeronautics and Space Administration: Washington, DC, USA, 2016.
30. Kennedy, C.A.; Carpenter, M.H. Diagonally implicit Runge–Kutta methods for stiff ODEs. *Appl. Numer. Math.* **2019**, *146*, 221–244. [CrossRef]
31. da Silva, C.C.; Lessig, C. Variational symplectic diagonally implicit Runge-Kutta methods for isospectral systems. *BIT Numer. Math.* **2022**, *62*, 1823–1840. [CrossRef]
32. Kamiński, M.; Corigliano, A. Numerical solution of the Duffing equation with random coefficients. *Meccanica* **2015**, *50*, 1841–1853. [CrossRef]
33. Butcher, J.C. Coefficients for the study of Runge-Kutta integration processes. *J. Aust. Math. Soc.* **1963**, *3*, 185–201. [CrossRef]
34. Prothero, A. Robinson, Stability and accuracy of one-step methods for solving stiff systems of ordinary differential equations. *Math. Comp.* **1974**, *28*, 145–162. [CrossRef]
35. Oni, O.E.; Davidson, I.E.; Mbangula, K.N. A review of LCC-HVDC and VSC-HVDC technologies and applications. In Proceedings of the 2016 IEEE 16th International Conference on Environment and Electrical Engineering (EEEIC), Florence, Italy, 7–10 June 2016; IEEE: New York City, NY, USA, 2016; pp. 1–7.

applied sciences

MDPI

Article

Study of the Impact of Acquisition Parameters on Fault Feature Identification Based on Magnetotelluric Modeling

Hui Zhang [1,*] and Fajian Nie [2]

1 College of Geosciences and Engineering, North China University of Water Resources and Electric Power, Zhengzhou 450046, China

2 School of Petroleum Engineering, Yangtze University, Wuhan 430110, China; niefajian@yangtzeu.edu.cn

* Correspondence: zhanghui@ncwu.edu.cn

Abstract: The electromagnetic method is widely used in mineral exploration, geothermal resource exploration, and deep earth structure detection. To improve the resolution and positioning accuracy of magnetotelluric surveys for targets beneath cover, it is essential to utilize forward modeling. This approach allows for a better understanding of the capabilities and limitations of MT in resolving features of different scales. In this paper, we employ forward modeling using the finite element method to simulate a series of continuous fault zones ranging from 100 m to 400 m in width, with varying lengths and dips. These fault zones represent conductive fluid pathways that could be associated with different scales and depths. The forward modeling provides the necessary data and method testing to assess the effectiveness of MT surveys in identifying and resolving such features. Our findings demonstrate that a station spacing of 400 m is optimal for resolving fault zones of various widths. For narrower faults (approximately 100 m), extending the survey line to 12 km or more significantly improves the deep structural inversion of the bounding faults, while an 8 km survey line suffices for wider faults (400 m). However, the vertical extent of these features is less well constrained, with deeper faults appearing broader and inversion depths being notably shallower than actual depths. These results highlight the need for careful interpretation of inversion anomalies, especially when supplementary data from other geophysical methods are lacking. Misinterpretation can lead to inaccurate resource assessments and exploration potential.

Keywords: forward modeling; magnetotelluric; finite element method; NLCG inversion; geothermal exploration

1. Introduction

The magnetotelluric (MT) method is an important geophysical technique that utilizes natural alternating electromagnetic fields to investigate the electrical structure of subsurface media. In recent years, the effectiveness of the MT method has been well-demonstrated in resource exploration and geological disaster detection, especially in mineral imaging and geothermal exploration [1–6]. With the growing complexity of study areas and the diversification of application environments, the MT method faces increasingly complex challenges, necessitating improved data acquisition and more appropriate inversion methods to achieve high-quality images [7–11]. For instance, in geothermal resource exploration, particularly in northern China [12,13], the distribution of thermal storage is mainly controlled by regional faults. Fault zones are crucial for the transportation and distribution of hot water, as they fracture and saturate the deep Ordovician greywacke, creating channels for the upward conduction of Earth's deep thermal energy. Therefore, resolving fault zones is the most critical aspect of thermal exploration in this region. To identify faults beneath a cover, broadband MT surveys and long-period MT surveys can be conducted using high-resolution data to enhance the resolution of deep narrow anomalies [14–16]. However, MT inversion is inherently multi-resolution, and the inversion model often differs significantly from actual geological structures due to the limited resolution of the MT method and the influence of complex subsurface structures [17,18]. Additionally, sparse low-frequency data points are insufficient to resolve deep fault structures [19,20]. Moreover, different survey designs lead to varying interpretation results, affecting anomaly interpretation and geological assessment. To better understand the relationship between inverted anomalies and tectonics, and to improve the resolution and accuracy of target localization beneath cover, it is essential to design an optimal survey system and assess how line length, station spacing, observation bands, and model grid resolution influence the imaging of target tectonics.

This target-oriented survey design can be achieved through forward modeling, allowing for the pre-determination of the subsurface structure and the evaluation of its resolution in MT inversion. Numerical simulation methods for MT forward modeling mainly include the Finite Element Method (FEM) [21,22], the Spectral Method [23], and the Finite Difference Method (FDM) [20,24]. The Finite Element Method is mostly used due to its high computational accuracy, adaptability to complex structures, and unconditional stability. Previous studies have shown that forward modeling plays a crucial role in optimizing survey design and improving the accuracy of data interpretation [25–29]. Dmitriev, V.I. (2010) [30] considered the influence of the coastal effect on bottom magnetotelluric soundings in the shelf zone, using mathematical modeling of the magnetotelluric field into a two-dimensional medium to choose the most effective methods for calculating the fields near the coastal line. Based on magnetotelluric modeling, Zhang, M.L. (2023) [31] applied the compressive sensing theory to MT acquisition and developed an efficient acquisition approach that uses an optimized station layout with sparse irregular patterns. This new acquisition approach can lead to significant savings in both the acquisition cost and operational time while maximizing the information obtained from a given number of stations. Batista, J.D. (2019) [32] applied genetic algorithms and a local methodology integrating the Gauss-Newton and Conjugate Gradient (GNCG) techniques to test one-dimensional inverse modeling of synthetic magnetotelluric data, leading to the development of a hybrid genetic algorithm for MT inversion of one- and two-dimensional synthetic data. In this study, we constructed a geological model consisting of a set of fault zones with varying widths and dips, representing different scales of conductive fluid pathways associated with mineral or geothermal systems. We designed various survey systems for the geological model using two different model grid schemes. For each scenario, we performed finite element forward modeling to generate synthetic data and conducted MT inversion studies to assess the effectiveness of various survey parameters and model mesh designs in interpreting different fault zones in MT detection. Unlike previous studies that focused on uniform or idealized fault models, this work addresses the challenges posed by heterogeneous geological settings, demonstrating that survey length and data acquisition strategies must be tailored to the specific lithological properties of the region. This approach offers a new perspective on optimizing MT surveys for complex subsurface conditions, which is crucial for improving the precision of fault identification in diverse geological environments.

2. MT Forward Modeling and Parameter Setting

2.1. Fault Model Setup

The two-dimensional geological model consists of three faults, labeled F1, F2, and F3, arranged from west to east (Figure 1). These faults originate from the basement and extend to the shallow layers, forming three distinct low-resistance anomaly zones. F1 has an almost vertical dip, while F2 and F3 dip at angles ranging from 65° to 75°, with all three fault zones having the same width. The geological model was digitized onto a grid with two resolutions: 100 m (Scenario 1) and 200 m (Scenario 2), to analyze the inversion and resolution of fault structures in MT at different resolutions. In Scenario 1, the faults are 100 to 200 m wide, while in Scenario 2, they range from 200 to 400 m wide, with resistivity set to 10 ohm-m in both scenarios. The depth and resistivity of the background layers are consistent across both scenarios, as is the contact ratio between the faults and the background layer. The resistivity of the background layer ranges from 100 to 5000 ohm-m, with the highest resistivity at the basement level.

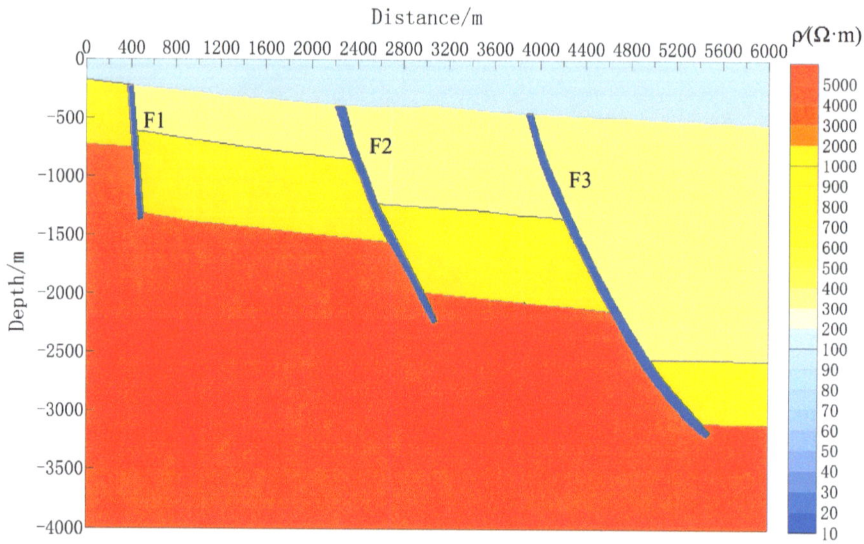

Figure 1. Geological model depicting three faults (F1, F2, and F3) with varying dip angles and resistivity distribution used for forward modeling.

2.2. Finite Element Modeling

The propagation of electromagnetic waves in subsurface media follows Maxwell's equations, which form the basis for deriving magnetotelluric forward modeling equations. In a two-dimensional medium, the electromagnetic field can be divided into two independent wave types: one includes the components E_x, H_y, and H_z, known as transverse electric (TE) polarization mode, while the other includes H_x, E_y, and E_z, referred to as transverse magnetic (TM) polarization mode. When solving the numerical solution of the magnetotelluric field in a two-dimensional medium, Maxwell's equations can be transformed into the following second-order partial differential equations that satisfy E_x and H_x (Equations (1) and (2)) [17,33,34]:

$$\frac{\partial}{\partial y}\left(\frac{1}{i\omega\mu}\frac{\partial E_x}{\partial y}\right)+\left(\frac{1}{i\omega\mu}\frac{\partial E_x}{\partial z}\right)+(\sigma-i\omega\varepsilon)E_x=0 \tag{1}$$

$$\frac{\partial}{\partial y}\left(\frac{1}{\sigma-i\omega\varepsilon}\frac{\partial H_x}{\partial y}\right)+\left(\frac{1}{\sigma-i\omega\varepsilon}\frac{\partial H_x}{\partial z}\right)+i\omega\mu H_x=0 \tag{2}$$

Here, E_x represents the horizontal component of the electric field in the x-direction, H_x is the horizontal component of the magnetic field in the x-direction, σ is the medium's electrical conductivity, ε is the dielectric constant, μ is the magnetic permeability, and $\omega = 2\pi f$ is the angular frequency.

Appropriate boundary conditions must be applied to solve Equations (1) and (2). For example, in TM mode, the top boundary AB is typically set at the surface, where the field value $u = H_x$ is set to 1 unit:

$$u|_{AB}=1 \tag{3}$$

Below the bottom boundary CD, homogeneous rock is assumed at a sufficient depth, where the anomalous field generated by local inhomogeneous rock is negligible on CD. This assumption holds when the lower boundary is deep enough and the medium below is homogeneous, allowing the influence of local inhomogeneities to be ignored. At this depth, the propagation equation for electromagnetic waves below the lower boundary is:

$$u|_{CD} = u_0 e^{-ky} \tag{4}$$

Here, u_0 is a constant, $k = \sqrt{-i\omega\mu\sigma}$ is the propagation coefficient, and σ is the rock's electrical conductivity. Taking the derivative of Equation (4), we get $\frac{\partial u}{\partial z} = -ku$. At the lower boundary $\frac{\partial}{\partial z} = -\frac{\partial}{\partial n}$, the boundary condition at CD is given as:

$$\frac{\partial u}{\partial n} + ku = 0 \tag{5}$$

When the left and right boundaries are sufficiently distant from the inhomogeneity, the anomalous field becomes zero, leading to symmetry in the electromagnetic field:

$$\frac{\partial u}{\partial n} = 0 \tag{6}$$

Based on the above analysis, solving the magnetotelluric field reduces to solving the system of equations formed by partial differential Equation (1) to Equation (2), and the boundary conditions Equations (3)–(6).

$$\begin{array}{ll} \nabla \cdot (\tau \nabla u) + \lambda u = 0 & \in \Omega \\ u = 1 & \in \mathrm{AB} \\ \frac{\partial u}{\partial n} = 0 & \in \mathrm{AD, BC} \\ \frac{\partial u}{\partial n} + ku = 0 & \in \mathrm{CD} \end{array} \tag{7}$$

For the TE polarization mode: $u = E_x$, $\tau = i\omega\mu$, and $\lambda = \frac{1}{\sigma - i\omega\varepsilon}$

For the TM polarization mode: $u = H_x$, $\tau = \frac{1}{\sigma - i\omega\varepsilon}$, and $\lambda = i\omega\mu$.

The finite element method (FEM) is employed to numerically solve this system of equations. To calculate the numerical solution of equation system (7) using the finite element method, we consider its equivalent variational problem:

$$\begin{array}{l} F(u) = \int \left[\frac{1}{2}\tau(\nabla u)^2 - \frac{1}{2}u^2\right] d\Omega \\ u|_{AB} = 1 \\ \delta F(u) = 0 \end{array} \tag{8}$$

The Finite Element Method (FEM) involves dividing the model region Ω into a grid of rectangular elements, with bilinear interpolation applied to approximate the field variables using shape functions. This interpolation method strikes a good balance between computational efficiency and accuracy, which is crucial for capturing the smoothness and continuity of the electromagnetic fields. The principles and algorithms of FEM are extensively covered in the literature [17,33,34], and will not be elaborated here. The system of equations is then assembled, incorporating boundary conditions, and solved for field values u at each node. After obtaining these values, the vertical partial derivative $\frac{\partial u}{\partial z}$ is computed, which in the TM mode corresponds to $\frac{\partial H_x}{\partial z}$. This result is used to calculate the apparent resistivity and impedance phase:

$$\begin{array}{l} Z_{yx} = -\frac{1}{\sigma}\frac{\partial H_x}{\partial z} / H_x \\ \rho_{yx} = \frac{1}{\omega\mu}|Z_{yx}|^2 \\ \phi_{yx} = \arctan\frac{\mathrm{Im}[Z_{yx}]}{\mathrm{Re}[Z_{yx}]} \end{array} \tag{9}$$

Here, Z_{yx} represents the impedance, ρ_{yx} represents the apparent resistivity, and ϕ_{yx} represents the impedance phase in the TM polarization mode. In the case of TM polarization, $\frac{\partial u}{\partial z}$ corresponds to $\frac{\partial E_x}{\partial z}$. Substituting this into Equation (10) yields the apparent resistivity and impedance phase.

$$\begin{aligned} Z_{xy} &= E_x / \left(\frac{1}{i\omega\mu} \frac{\partial E_x}{\partial z} \right) \\ \rho_{xy} &= \frac{1}{\omega\mu} \left| Z_{xy} \right|^2 \\ \phi_{xy} &= \arctan \frac{\mathrm{Im}\left[Z_{xy} \right]}{\mathrm{Re}\left[Z_{xy} \right]} \end{aligned} \tag{10}$$

Here, Z_{xy} represents the impedance, ρ_{xy} represents the apparent resistivity, and ϕ_{xy} represents the impedance phase in the TE polarization mode.

2.3. Forward Modeling Parameters Setting

To assess the ability of various survey parameters to resolve fault zones at different resolutions in MT detection, we devised the following surveys for two model grid scenarios:

(1) The survey length is 14 km, with a frequency band ranging from 0.001 to 1000 Hz, and station spacings of 200 m, 400 m, 600 m, and 800 m, respectively.

(2) The station spacing is 400 m, with a frequency band ranging from 0.001 to 1000 Hz, and survey lengths of 8 km, 10 km, 12 km, and 14 km, respectively.

(3) With a station spacing of 400 m and a survey length of 14 km, the frequency band is divided into two ranges: 0.001 to 1000 Hz for long-period measurements with broadband data, and 0.01 to 100 Hz for relatively narrow-band measurements.

For all survey designs in the two scenarios, forward modeling and inversion parameters remained consistent, except for those comparing survey systems (refer to Table 1), to eliminate discrepancies in accuracy caused by inconsistent parameter settings. In Scenario 1, the horizontal mesh is set to 100 m, while in Scenario 2, it is set to 200 m. The vertical mesh starts at 100 m at the surface and increases logarithmically to a depth of 15 km. The air layer extends to 10,000 m, and a minimum of 100 iterations were performed.

Table 1. Mesh parameters for forward modeling.

Station Spacing (m)	Survey Length (km)	Frequency Band (Hz)	No of Air Layers	No of Vertical Cell	Horizontal Cell Size (Scenario 1) (m)	Horizontal Cell Size (Scenario 2) (m)
200	14	0.001–1000	8	25	100	200
400	14	0.001–1000	8	25	100	200
600	14	0.001–1000	8	25	100	200
800	14	0.001–1000	8	25	100	200
400	12	0.001–1000	8	25	100	200
400	10	0.001–1000	8	25	100	200
400	8	0.001–1000	8	25	100	200
400	14	0.01–100	8	25	100	200

3. MT Synthetic Data Inversion

TE and TM polarized inversions were conducted using the nonlinear conjugate gradient (NLCG) inversion algorithm [35–37] for each scenario and survey design. The NLCG inversion method is widely utilized in MT inversion and offers high resolution and fast computational speed.

The objective function of the inversion calculation is defined as follows:

$$\Psi(m) = (d - F(m))^T V^{-1} (d - F(m)) + \lambda m^T L^T L m \tag{11}$$

where Ψ represents objective function; m represents the model vector; d represents the data vector; F represents forward modeling function; λ represents regularization factor; V represents the covariance matrix with respect to the error vector $e = d - F(m)$; and L represents the two-dimensional differentiation matrix.

The inversion iteration process is:

$$m_0 = given, \tag{12}$$

$$\Psi(m_l + \alpha_l p_l) = \min_{\alpha} \Psi(m_l + \alpha p_l), \tag{13}$$

$$m_{l+1} = m_l + \alpha_l p_l \quad l = 0,1,2\ldots, \tag{14}$$

where α_l represents the search step; p_l represents the search direction in the model space, generated iteratively by the expression:

$$p_0 = -C_0 g_0, \tag{15}$$

$$p_l = -C_l g_l + \beta_l p_{l-1} \quad l = 1,2\ldots, \tag{16}$$

$$\beta_l = \frac{{g_l}^T C_l (g_l - g_{l-1})}{{g_{l-1}}^T C_{l-1} g_{l-1}}, \tag{17}$$

where C_l represents the prior information, $C_l = \boldsymbol{I}$ when no prior information exists.

The number of inversion stations is equal to the number of forward modeling stations. However, all inversions were conducted using a coarser horizontal model mesh compared to that employed for the forward modeling. The model covariance was set to 0.1 for all models, with an RMS misfit of 5%. The results indicate that all inversions achieved an RMS misfit ranging from 0.82 to 1.78 (Table 2). The misfit generally improves as the site spacing and model cell size decrease.

Table 2. Mesh parameters used for inversion of synthetic data.

Station Spacing (m)	Survey Length (km)	Frequency Band (Hz)	No. of Air Layers	No. of Vertical Cell	RMS Misfit (Scenario 1)	RMS Misfit (Scenario 2)
200	14	0.001–1000	8	25	0.82	1.51
400	14	0.001–1000	8	25	1.27	1.38
600	14	0.001–1000	8	25	1.42	1.50
800	14	0.001–1000	8	25	1.64	1.59
400	12	0.001–1000	8	25	1.29	1.40
400	10	0.001–1000	8	25	1.26	1.44
400	8	0.001–1000	8	25	1.26	1.49
400	14	0.01–100	8	25	1.41	1.78

4. Discussion

4.1. Effect of Station Spacing on Fault Imaging

The magnetotelluric inversion results (Figure 2a–d) highlight the effect of station spacing on detecting low-resistance fault zones. The station spacings tested range from 200 m to 800 m, and the results consistently show that even at depths of up to 3 km, the linear structure of the faults can be effectively captured. These findings demonstrate the robustness of the employed inversion method across various station spacings, particularly in delineating fault zones spanning shallow to deep subsurface layers.

However, it is evident that smaller station spacings (200 m and 400 m) provide superior accuracy in defining fault geometries and structures compared to larger spacings (600 m and 800 m). In Scenario 1, fault imaging is more precise when the station spacing is reduced to 200 m or 400 m. In these cases, the spatial locations of the three faults, especially F1 and F3, closely align with their actual positions in the geological model, and their depths are well-matched in the inversion results.

In contrast, when the station spacing increases to 600 m or 800 m, the inversion results begin to diverge from the actual fault positions. Notably, the F1 fault exhibits significant positional deviations, suggesting that larger station spacings reduce the reliability of imaging complex fault geometries. This trend is also apparent in Scenario 2, where the 800 m station spacing causes the F3 fault to show a broad influence area with unclear boundaries, indicating a loss of resolution.

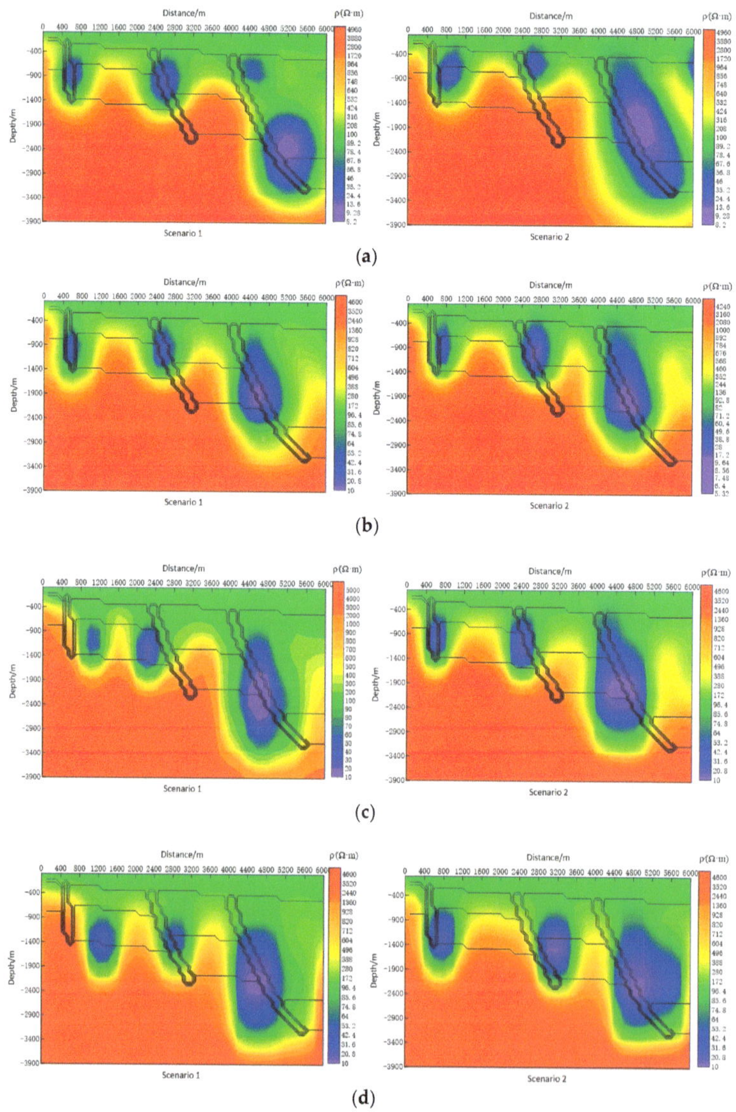

Figure 2. Inversion resistivity profile of different station spacings. (**a**) Station spacing is 200 m. (**b**) Station spacing is 400 m. (**c**) Station spacing is 600 m. (**d**) Station spacing is 800 m.

Additionally, the F2 fault proves challenging to image across all station spacings and scenarios. Regardless of the acquisition parameters, the inversion consistently fails to accurately resolve the depth and position of the F2 fault, suggesting that this fault may either require a denser dataset or be subject to limitations inherent in the inversion algorithm or the characteristics of the fault zone itself.

4.2. *Effect of Survey Length on Fault Imaging*

The inversions (Figure 3a–d) in both scenarios demonstrate that increasing the survey line length enhances the accuracy of fault imaging, particularly for faults F1 and F3. The geological context plays a significant role in these observations, as resistivity contrasts and fault geometries influence the magnetotelluric responses.

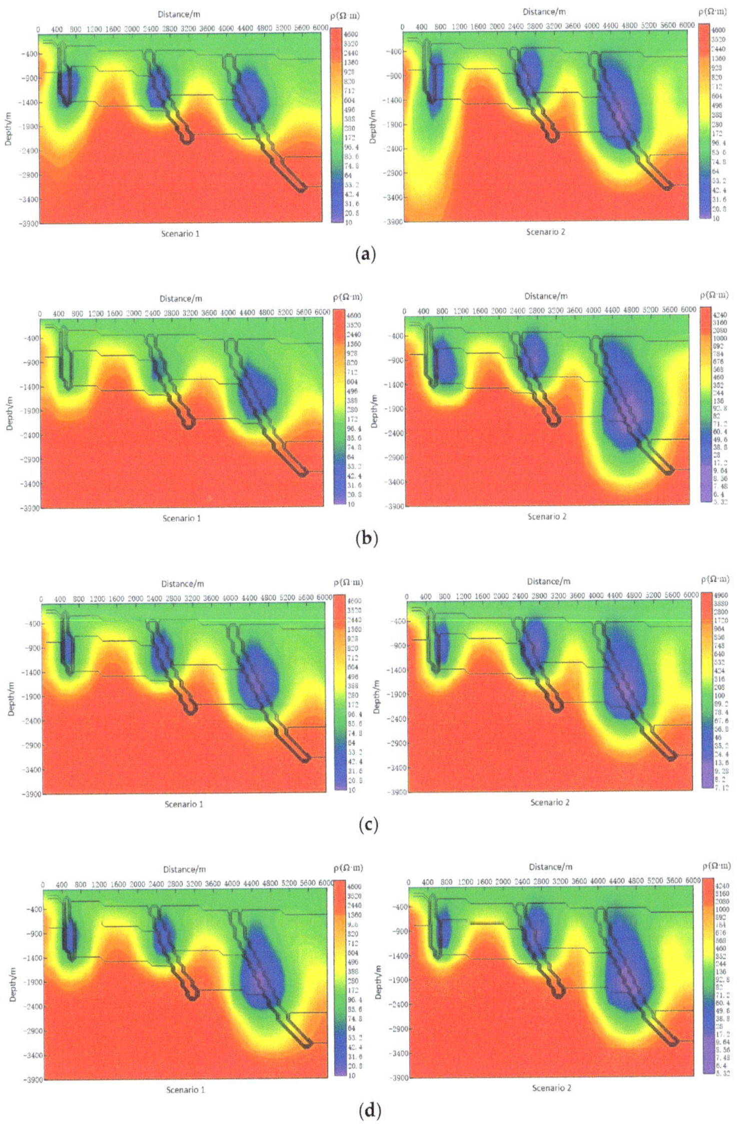

Figure 3. Inversion resistivity profile of different survey length. (**a**) Survey length is 8 km. (**b**) Survey length is 10 km. (**c**) Survey length is 12 km. (**d**) Survey length is 14 km.

In Scenario 2, which features wider faults, the imaging accuracy of F1 improves considerably when the survey line length is extended to 12 km. At this length, the low-resistivity anomalies associated with the F1 fault closely match their true lateral positions in the geological model, and the fault's localization accuracy surpasses that of Scenario 1. This improvement could be attributed to the wider fault, which allows for better signal penetration and easier detection, even in regions with varying geological conditions.

Interestingly, in Scenario 2, the inversion accuracy of the F3 fault is already high at a survey length of 8 km, with no significant improvements when the length is further increased. This suggests that for wider faults with strong resistivity contrasts, a shorter survey line may be sufficient for accurate imaging. The geological implication here is that fault width and the associated lithological properties can reduce the need for extensive survey lengths in certain cases.

In both scenarios, the imaging accuracy of the F2 fault remains consistently poor, regardless of the survey line length. The inversion depth for F2 is always shallower than the actual depth, resulting in smaller-scale low-resistivity anomalies. This persistent issue may be due to geological factors such as lower resistivity contrast or gradational boundaries between the fault and surrounding rock. These characteristics can weaken the electromagnetic response, making it difficult to accurately image the fault using standard MT inversion techniques. This suggests that additional geological information or advanced inversion methods may be required to resolve such features.

4.3. *Effect of Observation Frequency Bands on Fault Imaging*

The inversion results (Figure 4a,b) indicate that, in Scenario 1, the fault zone location and resolution from regular inversion bands are inferior to those from broadband survey data, mainly due to the absence of low-frequency data. The inversion depth of the F3 fault is limited to 2 km, failing to fully capture the depth of the fault. This limitation is not observed in any other survey designs discussed in this paper. Additionally, the image area of the F2 and F3 faults appears significantly broader than their actual dimensions, suggesting an overestimation of fault zone width. The fault depth resolution is notably better in Scenario 2 compared to Scenario 1, particularly for the deeper imaging of the F3 fault, which closely matches the broadband survey data. However, the width of the deeper imaging of the F1 and F3 faults is significantly broader than in the geological model, indicating poor resolution of narrow structures. Among all the survey designs, the F2 fault exhibits the poorest imaging quality, with inversion depths significantly diverging from those in the geological model.

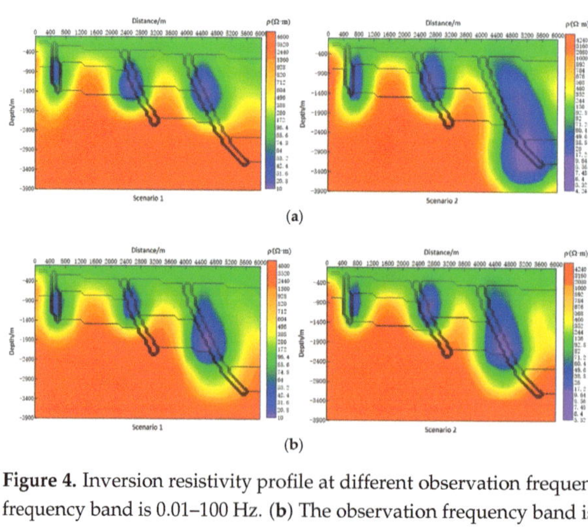

(a)

(b)

Figure 4. Inversion resistivity profile at different observation frequency bands. (**a**) The observation frequency band is 0.01–100 Hz. (**b**) The observation frequency band is 0.001–1000 Hz.

5. Conclusions

A geological model containing various fault zones was constructed, and synthetic data was used for forward modeling and inversion to replicate different survey designs. By analyzing how different survey parameters and model grids interpret fault zones in MT detection, we reached the following conclusions:

(1) Forward modeling results show that a station spacing of 400 m effectively resolves fault zones and transverse conductive fault zones ranging from 100 m to 400 m in width. However, reducing station spacing offers no significant advantage in accurately determining their extent and morphology. In contrast, larger station spacing impacts fault boundary localization and increases deviations from the actual geological model. Specifically, in Scenario 2, when station spacing exceeds 400 m, the imaging of both the F1 and F3 faults deteriorates, with lateral errors increasing from 0.1 km to as much as 1.2 km. These findings emphasize the need for tighter station spacing, particularly in complex geological regions, to achieve accurate fault localization.

(2) In this study's geological model, using a 400 m station spacing, a survey length of 12 km or more yields optimal imaging accuracy for fault boundaries (F1 and F3), especially for narrower faults (Scenario 1), enhancing localization accuracy and inversion depth. For wider faults (Scenario 2), increasing the survey length does not significantly improve the clarity of the deeper faults. In both scenarios, the F1 fault is best imaged with a survey length of 12 km, where the lateral error in fault positioning decreases from 1.0 km to less than 0.3 km. For the F3 fault, a 12 km survey reduces its positional error from 1.5 km to 0.2 km. In contrast, the F2 fault consistently shows a shallower inversion depth, with reduced low-resistivity anomalies and an average depth discrepancy of 1.0 km across all survey lengths.

(3) With a 400 m station spacing, broadband measurements do not significantly improve the clarity of shallow fault structures compared to regular measurements. This limitation may result from insufficient resolution of narrow fault structures in broadband data at a 400 m station spacing. Therefore, acquiring lower frequency and broader bandwidth data through long-period measurements is essential to improve the resolution of deep fault structures.

(4) The inversion results in this study show a shallower inversion depth compared to the true depth, resulting in poor resolution of the faults' vertical extent. Additionally, the inversion results suggest that due to the limitations of MT inversion, resistivity models derived from MT data should be interpreted with caution, especially without support from other geophysical methods, as this may affect resource estimation. For instance, the F2 fault zone appears the smallest in the inversion results, with the resistivity model showing the poorest conductivity, contradicting its true characteristics but not diminishing its exploration potential. The forward modeling in this study has been optimized for a two-dimensional observation system, with simplifications made regarding subsurface structures and conductive materials. However, real-world conditions are likely more complex, requiring consideration of factors such as fault connectivity, variations in fault zones across planes, and the types of conductors involved.

Author Contributions: Conceptualization, H.Z.; methodology, H.Z.; validation, H.Z. and F.N.; formal analysis, F.N.; data curation, F.N.; writing—original draft preparation, H.Z.; writing—review and editing, H.Z. All authors have read and agreed to the published version of the manuscript.

Funding: This research is funded by High-Level Talent Startup Project of North China University of Water Resources and Electric Power.

Data Availability Statement: The experimental data used to support the findings of this study are included in the manuscript.

Conflicts of Interest: The authors declare no conflicts of interest.

References

1. McLeod, J.; Ferguson, I.; Craven, J.; Roberts, B.; Giroux, B. Pre-injection magnetotelluric surveys at the Aquistore CO_2 sequestration site, Estevan, Saskatchewan, Canada. *Int. J. Greenh. Gas Control* **2018**, *74*, 99–118. [CrossRef]
2. Fan, Y.; Chen, X.B.; Tang, J.; Cui, T.F.; Sun, X.Y.; Wang, P.J.; Liu, Z.Y. Three-dimensional modeling of magnetotelluric data from the Hefei-Suqian segment of the Tanlu Fault Zone, Eastern China. *Chin. J. Geophys.* **2022**, *65*, 1336–1353. (In Chinese)
3. Heinson, G.; Didana, Y.; Soeffky, P.; Thiel, S.; Wise, T. The crustal geophysical signature of a world-class magmatic mineral system. *Sci. Rep.* **2018**, *8*, 10608. [CrossRef] [PubMed]
4. Jiang, W.P.; Duan, J.M.; Michael, D.; Andrew, C.; Anthony, S.; Brodie, R.C.; Goodwin, J. Application of Multi-Scale Magnetotelluric Data to Mineral Exploration: An Example from the East Tennant Region, Northern Australia. *Geophys. J. Int.* **2021**, *229*, 1628–1645. [CrossRef] [PubMed]
5. Niu, P.; Han, J.T.; Zeng, Z.F.; Hou, H.S.; Liu, L.J.; Ma, G.Q.; Guan, Y.W. Deep controlling factors of the geothermal field in the northern Songliao basin derived from magnetotelluric survey. *Chin. J. Geophys.* **2021**, *64*, 4060–4074. (In Chinese)
6. Zhang, J.F.; Sun, N.Q.; Liu, Z.L.; Qi, Z.P. Electromagnetic methods in the detection of water hazards in coal mines: A review. *Coal Geol. Explor.* **2023**, *51*, 301–316.
7. Nikiforov, V.M.; Varentsov, I.M.; Shkabarnya, G.N.; Kaplun, V.B.; Zhukovin, A.Y.; Cuong, D.H. Long-period ($10 < T < 20{,}000$ s) Magnetotelluric Studies in Northern Vietnam. *Russ. Geol. Geophys.* **2020**, *61*, 1036–1055.
8. He, J.S. Wide field electromagnetic sounding methods. *J. Cent. South Univ. (Sci. Technol.)* **2010**, *41*, 1065–1072.
9. Cai, J.T.; Chen, X.B.; Fang, L.H.; Qi, H.; Wang, P.J.; Zhang, Y.Y.; Liu, Z.Y.; Xie, X.B.; Zhou, L. Study on the seismotectonic environment in Bengbu city using the dense magnetotelluric array. *Chin. J. Geophys.* **2023**, *66*, 638–658. (In Chinese)
10. Zhang, J.H.; Zhao, G.Z.; Xiao, Q.B.; Xia, Q.B.; Tang, J. Analysis of electric structure of the central Tan-Lu fault zone (Yi-Shu fault zone, 36°N) and seismogenic condition. *Chin. J. Geophys.* **2010**, *53*, 605–611. (In Chinese)
11. Han, J.T.; Kang, J.Q.; Liu, C.; Liu, W.Y.; Zhang, Y.C.; Wang, T.Q.; Guo, Z.Y.; Yuan, T.M.; Liu, L.J. Characteristics of the asthenosphere structure beneath the eastern segment of the Central Asia orogenic belt inferred from a long-period magnetotelluric survey. *Chin. J. Geophys.* **2019**, *62*, 1148–1158. (In Chinese)
12. Qiao, Y.; Zhang, H. Methodology and Application of Deep Geothermal Sounding in Low-Resistance Cover Areas. *Appl. Geophys.* 2023, *Early Access*.
13. Zhang, H.; Sui, S.Q.; Qian, L.R. The application of multiple non-seismic methods to geothermal exploration in Qihe Shandong Province. *Geophys. Geochem. Explor.* **2020**, *44*, 727–733.
14. Patro, P.K. Magnetotelluric Studies for Hydrocarbon and Geothermal Resources: Examples from the Asian Region. *Surv. Geophys.* **2017**, *38*, 1005–1041. [CrossRef]
15. Geiermann, J.; Schill, E. 2-D Magnetotellurics at the geothermal site at Soultz-sous-Forets: Resistivity distribution to about 3000 m depth. *Comptes Rendus Geosci.* **2010**, *342*, 587–599. [CrossRef]
16. Pospeeva, E.V.; Potapov, V.V. Deep Structure of the Junction Zones of the Chuya Tectonic Depression and Its Mountainous Frame in Gorny Altai according to Magnetotelluric Studies. *Russ. Geol. Geophys.* **2021**, *62*, 474–485. [CrossRef]
17. Avdeev, D.B. Three-dimensional electromagnetic modeling and inversion from theory to application. *Surv. Geophys.* **2005**, *26*, 767–799. [CrossRef]
18. Zhou, R.F.; Wand, X.B.; Qin, C.; Xu, Y.C.; Zhang, J.T.; Wang, R. Comprehensive utilization of NLCG and OCCAM in two-dimensional magnetotelluric inversion. *Prog. Geophys.* **2016**, *31*, 2306–2312.
19. Chen, M.S. Analysis of some problems on the detecting depth of frequency electromagnetic sounding. *Coal Geol. Explor.* **2012**, *40*, 67–70.
20. Zhang, H.; Nie, F. Magnetotelluric Forward Modeling Using a Non-Uniform Grid Finite Difference Method. *Mathematics* **2024**, *12*, 2984. [CrossRef]
21. Gallardo, G.E.U.; Ruiz, A.D. High order edge-based elements for 3D magnetotelluric modeling with unstructured meshes. *Comput. Geosci.* **2022**, *158*, 104971. [CrossRef]
22. Klaus, S. Electromagnetic Modeling Using Adaptive Grids—Error Estimation and Geometry Representation. *Surv. Geophys.* **2023**, *45*, 227–314.
23. Tong, X.Z.; Sun, Y.; Zhang, B.Y. An efficient spectral element method for two-dimensional magnetotelluric modeling. *Front. Earth Sci.* **2023**, *11*, 1183150. [CrossRef]
24. Jahandari, H.; Bihlo, A. Forward modelling of geophysical electromagnetic data on unstructured grids using an adaptive mimetic finite-difference method. *Comput. Geosci.* **2021**, *25*, 1083–1104. [CrossRef]
25. Erdogan, E.; Candansayar, M.E. The conductivity structure of the Gediz Graben geothermal area extracted from 2D and 3D magnetotelluric inversion: Synthetic and field data applications. *Geothermics* **2017**, *65*, 170–179. [CrossRef]
26. Zainalpour, A.; Kamali, G.; Moradzadeh, A. Optimal selection of regularization parameter in magnetotelluric data inversion. *Acta Geod. Geophys.* **2022**, *57*, 245–263. [CrossRef]
27. Wang, P.J.; Chen, X.B.; Zhang, Y.Y. Synthesizing magnetotelluric time series based on forward modeling. *Front. Earth Sci.* **2023**, *11*, 1086749. [CrossRef]
28. Tomassi, A.; Trippetta, F.; de Franco, R.; Ruggieri, R. How petrophysical properties influence the seismic signature of carbonate fault damage zone: Insights from forward-seismic modelling. *J. Struct. Geol.* **2023**, *167*, 104802. [CrossRef]

29. Kirkby, A.; Doublier, M.P. Synthetic magnetotelluric modelling of a regional fault network—Implications for survey design and interpretation. *Explor. Geophys.* **2022**, *55*, 617–628. [CrossRef]
30. Dmitriev, V.I.; Mershchikova, N.A. Mathematical Modeling of the Influence of the Coastal Effect on Marine Magnetotelluric Soundings. *Izv.-Phys. Solid Earth* **2010**, *46*, 717–721. [CrossRef]
31. Zhang, M.L.; Li, Y.G. Efficient magnetotelluric data acquisition using irregular stations and compressive sensing reconstruction. *J. Appl. Geophys.* **2023**, *217*, 105169. [CrossRef]
32. Batista, J.D.; Sampaio, E.E.S. Magnetotelluric inversion of one- and two-dimensional synthetic data based on hybrid genetic algorithms. *Acta Geophys.* **2019**, *67*, 1365–1377. [CrossRef]
33. Arun, S.; Rahul, D.; Pravin, K.G.; Israil, M. A MATLAB based 3D modeling and inversion code for MT data. *Comput. Geosci.* **2017**, *104*, 1–11.
34. Mackie, R.L.; Smith, J.T.; Madden, T.R. Three-Dimensional electromagnetic modeling using finite-difference equations-the magnetotelluric example. *Radio Sci.* **1994**, *29*, 923–935. [CrossRef]
35. Rodi, W.; Mackie, R.L. Nonlinear conjugate gradients algorithm for 2D magnetotelluric inversion. *Geophysics* **2001**, *66*, 174–187. [CrossRef]
36. Kelbert, A.; Egbert, G.D.; Schultz, A. Non-linear conjugate gradient inversion for global em induction: Resolution studies. *Geophys. J. Int.* **2008**, *173*, 365–381. [CrossRef]
37. Miensopust, M.P.; Queralt, P.; Jones, A.G. Magnetotelluric 3-D inversion- a review of two successful workshops on forward and inversion code testing and comparison. *Geophys. J. Int.* **2013**, *193*, 1216–1238. [CrossRef]

energies

Article

Aerodynamic Performance and Numerical Validation Study of a Scaled-Down and Full-Scale Wind Turbine Models

Zahid Mehmood *, Zhenyu Wang *, Xin Zhang and Guiying Shen

College of Civil Engineering and Architecture, Zhejiang University, Hangzhou 310058, China; xin.zhang1259@zju.edu.cn (X.Z.); 12312052@zju.edu.cn (G.S.)
* Correspondence: zahidmehmood@zju.edu.cn (Z.M.); wzyu@zju.edu.cn (Z.W.)

Abstract: Understanding the aerodynamic performance of scaled-down models is vital for providing crucial insights into wind energy optimization. In this study, the aerodynamic performance of a scaled-down model (12%) was investigated. This validates the findings of the unsteady aerodynamic experiment (UAE) test sequence H. UAE tests provide information on the configuration and conditions of wind tunnel testing to measure the pressure coefficient distribution on the blade surface and the aerodynamic performance of the wind turbine. The computational simulations used shear stress transport and kinetic energy (SST K-Omega) and transitional shear stress transport (SST) turbulence models, with wind speeds ranging from 5 m/s to 25 m/s for the National Renewable Energy Laboratory (NREL) Phase VI and 4 m/s to 14 m/s for the 12% scaled-down model. The aerodynamic performance of both cases was assessed at representative wind speeds of 7 m/s for low, 10 m/s for medium, and 20 m/s for high flow speeds for NREL Phase VI and 7 m/s for low, 9 m/s medium, and 12 m/s for the scaled-down model. The results of the SST K-Omega and transitional SST models were aligned with experimental test measurement data at low wind speeds. However, the SST K-Omega torque values exhibited a slight deviation. The transitional SST and SST K-Omega models yielded aerodynamic properties that were comparable to those of the 12% scaled-down model. The torque values obtained from the simulation for the full-scale NREL Phase VI and the scaled-down model were 1686.5 Nm and 0.8349 Nm, respectively. Both turbulence models reliably predicted torque and pressure coefficient values that were consistent with the experimental data, considering specific flow regimes. The pressure coefficient was maximum at the leading edge of the wind turbine blade on the windward side and minimum on the leeward side. For the 12% scaled-down model, the flow simulation results bordering the low-pressure region of the blade varied slightly.

Keywords: wind turbine structural aerodynamics; CFD verification; scaled-down modeling; parametric study; turbulence modeling; wind tunnel testing; pressure variations

1. Introduction

Wind power generation has emerged as a primary, sustainable, and renewable power source and is gaining significant importance and economic viability in response to climate change and other power sources [1]. The design and optimization of wind turbine rotors remain challenging and significant owing to the complex wind flow environment [2,3]. The blades and hubs in wind turbines capture the force of the wind, and the captured wind force rotates the wind turbine rotor to convert the wind energy to electrical power. The design and performance of wind turbines rely heavily on the aerodynamic characteristics of the blades. Various research methodologies such as computational fluid dynamics (CFD), wind tunnel testing, the blade momentum method (BMM), and the vortex wake method have been employed in the field of aerodynamics [4]. Both eddy simulation (LES) and time-averaged Navier–Stokes equation (RANS) computational fluid dynamics (CFD) techniques have proven effective for studying wind turbine performance and analyzing the complex flows generated [5]. Recently, a high-fidelity optimization technique was also introduced to leverage the benefit of machine learning to reduce noise and improve aerodynamic efficiency. Geng et al. [6] proposed an advanced framework based on transfer learning and deep reinforcement learning to optimize the design. These methodologies helped to effectively optimize wind turbines to capture the maximum available energy.

In recent years, experimental research on the unsteady aerodynamic testing of the NREL wind turbine at NASA has been conducted at the Ames Wind Tunnel (80 × 120 feet) facility at the NREL, located in the United States of America. Later, researchers obtained simulation results using structural properties and geometries similar to those of the reference wind turbine provided by the NREL. The aerodynamic performance of NREL Phase VI was evaluated numerically using different models, such as the Navier–Stokes equations, wake-free, and blade element theory, under stall conditions for different yaw angle values. Their findings did not show a trend consistent with the experimental data, and a significant difference was observed between the predicted and actual observations [7–9]. Apart from the traditional CFD methods, to simulate complicated fluid dynamics, the lattice Boltzmann method (LBM) has gained popularity, especially in the case of wind turbine rotors. Lyu et al. [10] used advanced near-wall refinement to obtain better fluid flow predictions using a hybrid method in the LBM. This method has advanced the sliding mesh approach for LBM. Lyu et al. [11] applied the sliding method in the LBM as an alternative to traditional CFD approaches for wind turbine rotor simulation to capture complex flow dynamics.

Regardless of these differences, the above-mentioned numerical method provides a cost-effective and quick solution to differential equations by simulating fluid behaviors and wind turbine aerodynamic modeling. Several studies have used computational fluid dynamics (CFD) to investigate the aerodynamic characteristics of wind turbines. Menter et al. [12] presented two equation-based eddy viscosity turbulence models by incorporating the K-Omega and K-ε models near the inner and outer regions of boundary-layer walls. The transport of turbulent shear stress was incorporated into the formulation of turbulent viscosity to enhance the flow separation onset predictions for flows on smooth surfaces with adverse pressure gradients. The SST K-Omega model has been applied to various complex flow fields. It provides more accurate and reliable results than other RANS turbulence models [13]. To evaluate the aerodynamic forces and structural performance of the NREL Phase VI wind turbine, Lee et al. [14] used fluid–structure interaction (FSI). The results prove that this method is highly effective for wind turbine aerodynamic studies. Song and Perot [15] also adopted computational fluid dynamics (CFD) simulations to study surface pressure, torque, and fluid flow behaviors using an NREL Phase VI wind turbine. Wang et al. [16] employed a decomposed method to explore the impact of boundary layer suction (BLS) on the aerodynamic characteristic performance using a horizontal-axis wind turbine. The research findings indicate that the BLS controlled the stability of the flow and increased the pressure differences on the sides of the surface. Zhong and Li [4] investigated the aerodynamic performance and flow field using the biplane geometry of a wind turbine blade. They applied the Spalart–Allmaras model based on a single equation. The results revealed that the power coefficient increased by 23.7% compared to that of the baseline blade. Dias and Camacho [17] utilized a shear stress transport K-Omega simulation model for aerodynamic studies. Their findings inferred that the shear stress transport K-Omega model outcome lies within the standard deviation of the experimental observations for wind velocities ranging from 5 to 15 m/s. The design and optimization of aeroacoustics are also very important for wind turbine rotors. Lee et al. [18] utilized the SST K-Omega turbulence model to investigate the aerodynamics of two horizontal-axis wind turbine blades. The results obtained from the experiments and simulations were consistent at various wind speeds. Sedighi et al. [19] used the shear stress transport K-Omega turbulence model to investigate the aerodynamic behavior of a horizontal-axis wind turbine with a dimpled blade surface. Their findings suggest that the torque and power generation values can be effectively increased by appropriately designing a dimpled blade surface. The effect of a whale hump on the aerodynamic efficiency of an NREL Phase VI wind turbine was successfully investigated by Ke et al. [20]. To study the aerodynamic efficiency, the SST K-Omega turbulence model was applied by varying the tubercle position, wavelength, and amplitude. Their study revealed that the wind turbine performance was enhanced by increasing the lift over the low-pressure side of the blade. Menegozzo et al. [21] evaluated the scenarios of extreme loading on six NREL Phase VI wind turbines by applying the shear stress transport K-Omega simulation model. Based on the IEC-61400-2 standard, extreme operating gust (EOG) was considered at the inlet boundary conditions [22]. Many factors, including the surrounding topography, climate, and turbine surroundings, influence the flow characteristics of wind turbines. Numerical simulations can be used to examine the impact of terrain on wind turbine performance. Previous studies have investigated the effects of various slopes, from flat to very complex terrains, on wind turbine performance using the RANS and LES techniques. Similarly, to obtain further information about the flow conditions surrounding onshore wind turbines, other researchers [23–26] used field measurements to study the effects of several complicated terrain types, including wind profiles in the atmospheric boundary layer (ABL) and turbine wake recovery in hilly and forested areas and wakes, on the performance of multiple wind turbines. A recent experimental study provided a valuable understanding of wind turbine wake flow behavior. A turbulent flow decay region in the wake of two-bladed wind turbines was experimentally investigated for future turbulent modeling [27]. In their studies, the simulation and experimental/field results exhibited consistent trends in the stress, moment, and axial force at the root of the blade. None of the aforementioned turbulence models effectively capture the transition phase from laminar to turbulent flow.

With the evolution of computational technology, alternate models have been devised for wind turbine simulations to encapsulate the flow shift from laminar to turbulent phases. Menter et al. [28] developed a new correlation-based transition model by introducing intermittency (γ) and the momentum-thickness Reynolds number. Two further transport equations for the transition Reynolds number and intermittency (γ) were coupled with the shear stress transport K-Omega model for the transitional shear stress transport four-equation-based model [29–31]. This model has been validated for various engineering applications and can forecast the inherent bypass transition and separation induced in wall-enclosed flows [28,32,33]. In line with other models for engineering applications, this model can estimate fully turbulent flows under free shear stress. Turbulent limiters are introduced in stagnation regions to avoid the overprediction of turbulent kinetic energy [28,34].

Shaik and Subramanian [35] employed the shear stress transport K-Omega model with the gamma–theta transition theory approach coupled with URANS equations. The results obtained from the URANS simulation efficiently captured the aerodynamic characteristics. Lanzafame et al. [29] created a three-dimensional computational fluid dynamics (CFD) approach to estimate the aerodynamic operation of wind turbines. The results of the transitional SST and SST K-Omega turbulence models were optimized using a high-quality mesh. Amiri et al. [36] used the Reynolds stress turbulence (RST) method to predict the structural fatigue life of a large wind turbine. Moshfeghi et al. [37] presented the effect of near-wall mesh spacing on the effective capture of the aerodynamic performance of the NREL Phase VI reference turbine model by applying the gamma–theta model. Based on previous studies, the numerical simulation results exhibit strong agreement for predominantly attached flows characterized by low wind speeds. However, they deviated from the measurements at higher wind speeds over the blades under stall conditions. Other past studies [8,28,29,32,35,37,38] have demonstrated that the transition model provides reliable predictions for boundary layers during turbulent flow evolution from laminar flow. The transition model is the most suitable for studying wind turbine aerodynamic behavior under stall and separated flow conditions.

To date, researchers have focused on full-scale NREL Phase VI wind turbines for fully developed turbulence flow models. The numerical study of the aerodynamic behavior of a full-scale wind turbine and a scaled-down wind turbine still needs to be focused on considering laminar and turbulent flow separation boundary layers. Similarly, the accuracy of the shear stress transport K-Omega and transitional SST models should be used to investigate the geometry of the scaled-down wind turbine model based on the airfoils.

This study investigated two different geometrical models, namely, the NREL Phase VI reference wind turbine and 12% scaled-down models for wind tunnel testing. This study primarily focused on validating computational fluid dynamics turbulence models for a reference wind turbine model. The second objective was to investigate the aerodynamic performance of the scaled-down geometry for future parametric studies of aerodynamic characteristics, which will optimize future wind tunnel testing on scaled-down wind turbine models. The SST K-Omega (two equations) and transitional SST (four equations) turbulence models were employed for the CFD simulations, and the results were analyzed to determine the sensitivity of the model. First, the findings were cross-matched with the unsteady aerodynamic experiment (UAE) test sequence H measurements for the full-scale NREL Phase VI conducted at the NREL, as well as UAE horizontal-axis turbine measurements taken at NASA Ames wind tunnel, an 80-feet by 120-feet facility. Second, the simulation results of the 12% scaled-down wind turbine model were validated using low-speed wind tunnel testing data collected at the Korea Aerospace Research Institute (KARI, 80 in × 120 in). Finally, the aerodynamic performance of the 12% scaled-down model and a comparison of the turbulence models are discussed. The workstation, equipped with 512 GB RAM and a 64 × 2 GHz CPU, was used to perform all the simulations in this study.

The main contributions of this study are as follows:

This is the first study to validate the aerodynamic performance efficiency of a 12% scaled-down model. This will provide valuable insight into the scaled-down model at a reduced computational cost for future wind tunnel optimization. Second, a comparative analysis of both turbulence models (SST K-Omega and transitional SST) on both geometrical models offers a sensitivity analysis of the flow separation between the laminar and turbulent phases. This signifies the progression emphasizing the scaled-down technique compared with previous studies that focused only on the full scale. Finally, a comprehensive pressure coefficient distribution analysis on the blade surface provides new insights for studying the flow separation phenomena in a scaled-down model, which has been left unexplored in past studies. This validation study will facilitate future wind tunnel testing to study the aerodynamics and wake of wind turbines, which is impossible on a full scale, particularly under complex wind conditions.

The remainder of this paper is organized as follows: Section 2 discusses the computational models, meshing techniques, governing equations, computational parameters, and boundary conditions used in this study. Section 3 presents and discusses our results. Finally, Section 4 summarizes the findings and concludes the study.

2. Methodology

2.1. Computational Models

This study employed the NREL Phase VI horizontal-axis turbine geometry for numerical modeling. The reference geometry was stall-regulated, and the rotor was 10 m in diameter. An S809 airfoil was used for blade spans ranging in chord length from 0.25 m to 1 m. The transitional component joined the cylindrical portion of the blade with an airfoil. The details of the reference turbine geometry were presented by Hand et al. [8]. Only half of the turbine rotor was selected for the simulations to avoid an extensively heavy computational domain. The domains used in the simulations comprised two components: the inner rotating domain housing the turbine rotor and the outer stationary domain. Cylindrical domains were created owing to the rhythmic and periodic nature of the wind turbine rotor. The domain consisted of 10 times the diameter of the upstream side, 25 times the diameter of the downstream side, and a radius of 10 times the diameter of the turbine rotor. Similarly, the exact domain ratio was maintained for a 12% scaled-down turbine model, as shown in Figure 1. The upstream, downstream, and radius of the outer domain were selected as 101, 252, and 101 m for the full-scale model and 13, 31, and 13 m for the scaled-down model, respectively.

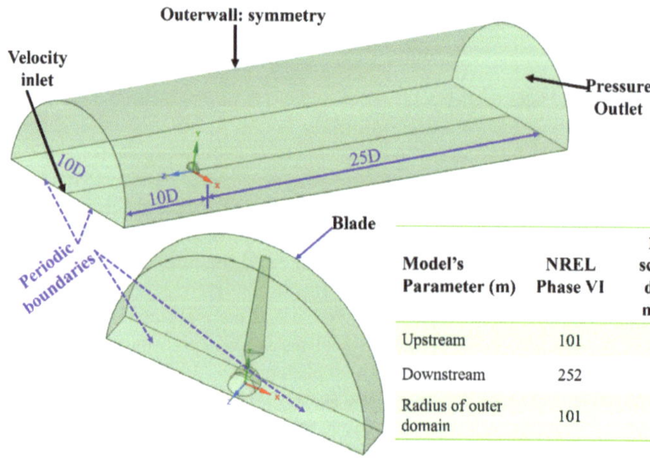

Model's Parameter (m)	NREL Phase VI	12% scaled-down model
Upstream	101	13
Downstream	252	31
Radius of outer domain	101	13

Figure 1. Details of the domain used for the CFD simulation.

2.2. Meshing for CFD Simulation

This study explores the aerodynamic performance of wind turbine models; therefore, a refined mesh is required on blade surfaces. Specifically, a mesh found on both the leading and trailing edges was carefully considered owing to more curvatures in the geometry. A well-crafted unstructured mesh was generated using ICEM computational fluid dynamics based on local and global size parameters. To obtain $Y^+ < 1$, an inflation layer containing 45 prisms with a growth rate of 1.25 and a prism ratio of 1.5 is redistributed at an initial height of 0.000015 mm for the geometry of the reference turbine blade. Similarly, 51 prism layers were generated at an initial height of 0.0000015 mm to analyze the flow dynamics near the blade surface for a scaled-down model. Many attempts have been made to obtain a Y^+ value of less than one based on the initial heights. The Y^+ value is determined as follows:

$$Y^+ = \frac{y_p u_t}{v} \tag{1}$$

where y_p is the distance of the first element from the blade surface, u_t is the friction velocity, and v is the kinematic viscosity [29].

2.3. Governing Equations and Theories of Turbulence

The calculations for the computational simulations were based on RANS and the continuity equation for compressible fluid flow. Shear stress transport (SST) and transitional shear stress transport turbulence models were applied to deal with the problem and compare the results. The following equations for mass and conservation were used [20,39]:

$$\frac{\partial \rho}{\partial t} + \frac{\partial}{\partial x_i}(\rho u_i) = 0 \tag{2}$$

$$\frac{\partial(\rho u_i)}{\partial t} + \frac{\partial(\rho u_i u_j)}{\partial x_i} = -\frac{\partial p}{\partial x_i}\frac{\partial}{\partial x_i}\left[\mu\left(\frac{\partial u_i}{\partial x_j} + \frac{\partial u_j}{\partial x_i} - \frac{2}{3}\delta_{ij}\frac{\partial u_i}{\partial x_i}\right)\right] + \frac{\partial}{\partial x_j}\left(-\rho\overline{u_i{}' u_j{}'}\right) + \rho f_i + F_i \tag{3}$$

where f_i denotes the body force; F_i is a source term; μ is the fluid dynamic viscosity; and the Kronecker delta function is symbolized as δ_{ij}, and it takes a value equal to one when i and j are equal and zero otherwise. The other term $\rho\overline{u_i{}' u_j{}'}$ represents the Reynolds stress tensor, where $u_i{}^{/}$ and $u_j{}^{/}$ represent fluctuating velocities. By applying the Boussinesq theory, the Reynolds stress can be formulated as follows:

$$\rho\overline{u_i{}^{/}u_j{}^{/}} = \mu_t\left(\frac{\partial u_i}{\partial x_j} + \frac{\partial u_j}{\partial x_i} - \frac{2}{3}\delta_{ij}\frac{\partial u_k}{\partial x_k}\right) - \frac{2}{3}\rho k\delta_{ij} \tag{4}$$

Here, k is the turbulent kinetic energy, and the eddy viscosity is expressed as μ_t. The Reynolds stress is correlated to the turbulence, and the term $\rho\overline{u_i{}^{/}u_j{}^{/}}$ is determined using the turbulence model. The transition thickness and intermittency variables used in the equation are as follows [31,38]:

$$\frac{\partial(\rho\gamma)}{\partial t} + \frac{\partial(\rho U_j\gamma)}{\partial x_j} = P_{\gamma1} - E_{\gamma1} + P_{\gamma2} - E_{\gamma2} + \frac{\partial}{\partial x_j}\left[\left(\mu + \frac{\mu_i}{\sigma x_j}\right)\frac{\partial}{\partial x_j}\right] \tag{5}$$

$$\frac{\partial\left(\rho\widetilde{Re\theta t}\right)}{\partial t} + \frac{\partial\left(\rho U_j\widetilde{Re\theta t}\right)}{\partial x_i} = P_{\theta t} + \frac{\partial}{\partial x_i}\left[\sigma_{\theta t}\left(\mu + \frac{\mu_t}{\sigma_\gamma}\right)\frac{\partial\widetilde{Re\theta t}}{\partial x_i}\right] \tag{6}$$

where the time-averaged velocity is represented by U_j, μ is the dynamic viscosity, *air* density is denoted by ρ, and the transition sources are denoted by P_γ and E_γ. The momentum thickness Reynolds number corresponds to $\widetilde{Re\theta t}$, which is used to determine the transition point. γ is the intermittency variable in the equation used to locally initiate the transition.

2.4. Computational Parameters and Boundary Conditions

ANSYS Fluent 2022 R1 was used to perform the 3D unsteady state numerical simulations. The RANS equations were solved using the SST K-Omega and transitional SST models. The SIMPLE convergence algorithm facilitates the coupling of pressure and velocity. A second-order discretization scheme was applied to the convection and viscous terms in the governing equations. For three-dimensional unsteady numerical simulations, convergence was achieved with no appreciable change in the residuals while augmenting the iteration counts. Achieving residual normalization stability required a value of at least 1×10^{-7} for x, y, and z momentum; 1×10^{-6} for dissipation of turbulence and kinetic energy; and 1×10^{-5} for continuity. The velocity inlet was defined as an inlet for the simulation domain, and the outlet was defined as a pressure outlet, no-slip rotating condition for the blade surface, symmetry for the far-field, and 180-degree periodic boundary condition for periodicity, as shown in Figure 1.

NREL Phase VI operates as a stall-regulated turbine with a radius of 5.029 m, as measured at the NASA Ames wind tunnel. For the aerodynamic performance study, various data were gathered during the experimentation of the NREL Phase VI reference turbine [8]. This study adopted the reference turbine test sequence H measurements [7]. Wind velocities in the range of 5–25 m/s were used for the numerical simulations. In this range, the flow behaves differently: at low speed, the flow is fully attached; at medium speed, the flow remains attached to the inner part of the blade, and at high velocities, the flow is completely separated from the wind turbine blade's surface.

2.5. Mesh Sensitivity Analysis

Computational mesh sensitivity and refinement analyses were performed based on three grid types: coarse, medium, and fine. A representative wind flow of 10 m/s with SST K-Omega and a transitional SST model based on the same scheme and topology was used. The total number of element reference model geometries was 19.3, 20.9, and 21.6 million for the three types of grids, respectively. Similarly, the total number of elements for the scaled-down model (12%) was 36.5, 38.6, and 42.5 million, respectively. The results of mesh size and grid sensitivity analyses are presented in Table 1 and shown in Figure 2.

Table 1. Grid sensitivity analysis.

Model Name	Mesh	Mesh Size (Million)	Torque (Nm)	R_C
NREL Phase VI	m_3: Coarse	19.3	1607.68	0.035
	m_2: Medium	20.9	1686.50	
	m_1: Fine	21.6	1689.32	
12% Scaled Down	m_3: Coarse	36.5	0.710664	0.030
	m_2: Medium	38.6	0.834985	
	m_1: Fine	42.5	0.849573	

Figure 2. Meshing for simulations.

The CFD solution must lie within the asymptotic convergence margin to evaluate the uncertainty. The mesh convergence ratio was defined to determine the asymptotic limit of convergence [40].

$$R_c = \frac{k_{12}}{k_{23}} \tag{7}$$

Here, $k_{12} = m_1 - m_2$ and $k_{23} = m_2 - m_3$.

The torque values at the three grids were analyzed, and an $R_c < 1$. Similarly, the percentage variance between the fine and medium-mesh torque values was less than 2%. Therefore, a medium-mesh model was used to perform the simulations in this study.

3. Results and Discussion

In this section, we analyze and validate the torque, thrust force, and pressure coefficients for NREL Phase VI and the 12% scaled-down model. The simulation results were evaluated using experimental data for validation.

3.1. Torque and Thrust Force

The results obtained from the numerical simulations were cross-verified with the experimental measurements of sequence H obtained from a wind tunnel test performed at the NASA Amed wind tunnel for NREL Phase VI. The experimental values are represented by the actual values with bars of uncertainty [40,41]. Similarly, to verify the CFD models for a 12% scaled-down geometry, the torque values were compared with the available experimental data obtained at the Korea Aerospace Research Institute (KARI). The details of the operational parameters were consistent with the experimental work conducted by Cho et al. [41] to validate the numerical simulations. The torque can be defined as follows:

$$T = \frac{1}{2}\rho A_s U_\infty R C_m \tag{8}$$

where ρ is the density (1.225 kg/m^3), A_s is the rotor blade swept area, $U\infty$ is the inlet velocity, R is the blade radius, and C_m is the moment coefficient estimated using CFD code.

As depicted in Figure 3a,b, the simulated findings obtained from the transitional shear stress transport and SST K-Omega turbulence models were validated with experimental data, and there was no considerable difference between wind velocities of 5 m/s and 10 m/s. At 10 m/s, both models overpredicted the torque values compared with the experimental results. The turbulence models provided an exceptional analogy with experimental data in the attached flow region but overestimated at the peak and underestimated the stall condition [42]. At 10 m/s, the maximum torque was observed at lower wind speeds before stalling at higher wind speeds. Above 10 m/s, the transitional SST model yielded more accurate results under stall conditions, up to a wind speed of 25 m/s. However, the simulated results of the transitional SST model fell below the experimental values but remained within the standard deviation measures. For wind speeds between 10 and 16 m/s, the SST K-Omega simulation results were in good agreement with the experimental values. However, after a wind flow velocity of 16 m/s, the full separation of the flow was evident, complex flow separation in the full stall region was not entirely captured in SST K-Omega, and poor performance was observed. The shear stress transport K-Omega model shows inadequacy at higher wind velocities for predicting the separation regions in uncertain flow separation. In general, the shear stress transport K-Omega model cannot correctly estimate the separation point [37]. Both the transitional SST and SST K-Omega models provided satisfactory predictions of the thrust force reasonably accurately based on the experimental data of the reference wind turbine (NREL Phase VI).

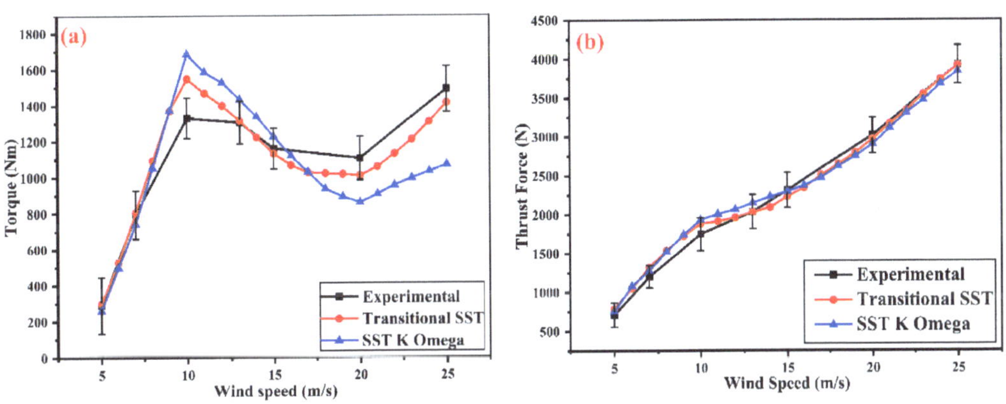

Figure 3. Comparison of torque (**a**) and thrust force (**b**) using CFD and experimental data for NREL Phase VI.

Similarly, the results obtained from the numerical simulation for torque and thrust force for the 12% scaled-down model turbine were validated with experimentally measured values for wind speed of 4 ms–14 m/s. As illustrated in Figure 4a, at wind flow velocities between 4 and 9 m/s, the simulated results of both the transitional SST and SST K-Omega models were in close agreement with the experimental values, even reaching a flow velocity of 10 m/s for the transitional SST model. At wind flow velocities of 10 and 11 m/s, the shear stress transport K-Omega turbulence model overpredicted the torque value. Above 11 m/s, the experimental and simulation results showed a difference of 2–7%. The transitional SST and SST K-Omega turbulence models showed reliable conformity with the experimental trend. As illustrated in Figure 4b, the thrust force calculated from the transitional SST and SST K-Omega turbulence models showed a similar trend, with an increase in flow velocities ranging from 4 m/s to 14 m/s.

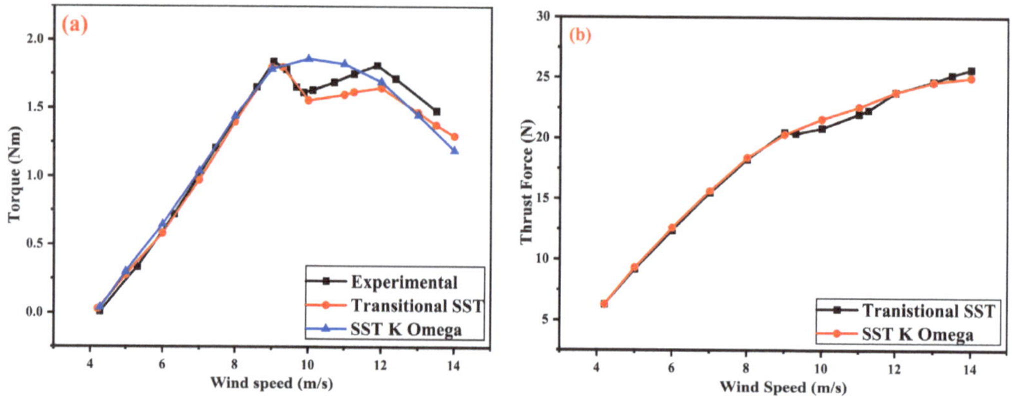

Figure 4. Comparison of torque (**a**) and thrust force (**b**) using CFD and experimental data for the scaled-down model based on NREL Phase VI.

3.2. *Pressure Coefficients*

3.2.1. NREL Phase VI Validation

The validation of the simulation results in conjunction with the NREL Phase VI reference wind turbine was obtained from the experimental measurements of the test sequence at wind flow velocities of 7, 10, and 20 m/s, as illustrated in Figure 5, Figure 6, and Figure 7, respectively. Radial sections were considered for the NREL Phase VI reference turbine at blade spans of 30, 47, 63, 80, and 95%. The coefficient of pressure at any section of the turbine blade is defined as follows [20]:

$$Cp = \frac{P - P_\infty}{\frac{1}{2}\rho\left[U_\infty{}^2 + (r\omega)^2\right]} \tag{9}$$

where $U\infty$ is the free stream velocity (m/s), r is the distance of the blade section from the root in meters, air density is represented as ρ in kilograms/cubic meters, ω is the angular velocity of the rotor in radians per second, the local pressure at any section is represented as P in pascal, and atmospheric pressure is expressed as $P\infty$.

Figure 5 illustrates a comparative analysis of the pressure coefficient of the CFD simulation findings using the transitional SST and SST K-Omega turbulence models with the experimental data at a wind flow velocity of 7 m/s. The comparison showed a consistent trend in the CFD findings with the experimental data values in all sections. A marginal difference was observed between the CFD simulations and experimental results on the suction side at 63%, 80%, and 95% along the blade span, as shown in Figure 5c–e, respectively. The CFD simulation slightly overestimated the values of the blade suction side near the leading edge.

Figure 6 shows a comparison of the pressure coefficient of the CFD simulation results using the transitional SST and SST K-Omega turbulence models with the experimental results at a wind flow velocity of 10 m/s. When the flow velocity was 10 m/s, the results of the CFD simulation were consistent with the experimental measurements, with a slight difference between the simulated results of the transient shear stress transport model and the SST K-Omega model at 30% (Figure 6a) and 47% (Figure 6b) of the blade span, but they were within the standard deviation of the experimental data. At 80% (Figure 6d) of the blade section, there was a slight overestimation of the pressure in the CFD simulation along the suction side. At a blade span of 47%, the CFD results were underestimated when moving toward the trailing edge compared to the experimental data.

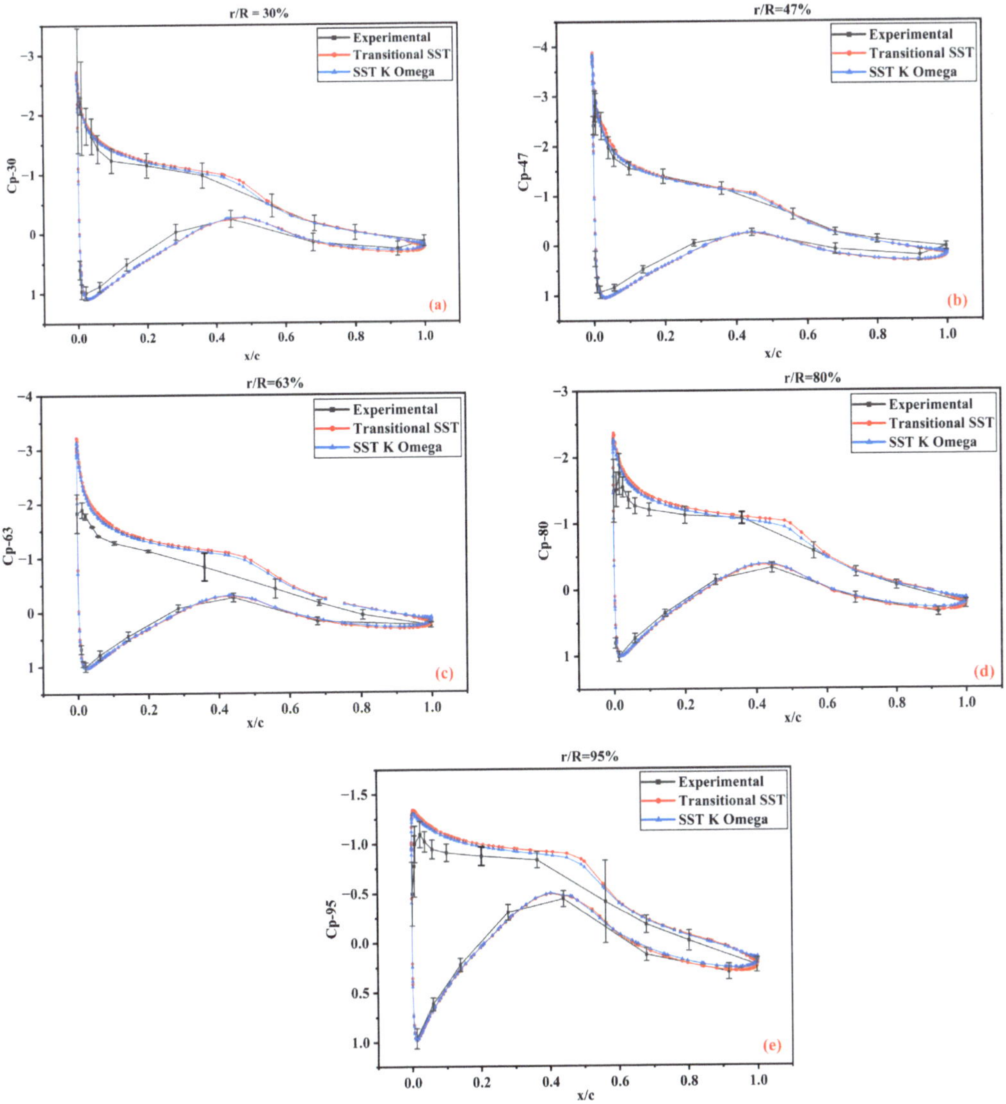

Figure 5. Comparison of the coefficient of pressure of NREL Phase VI obtained from CFD simulations and experimental data at a free-stream velocity of 7 m/s and at different blade sections (**a**–**e**).

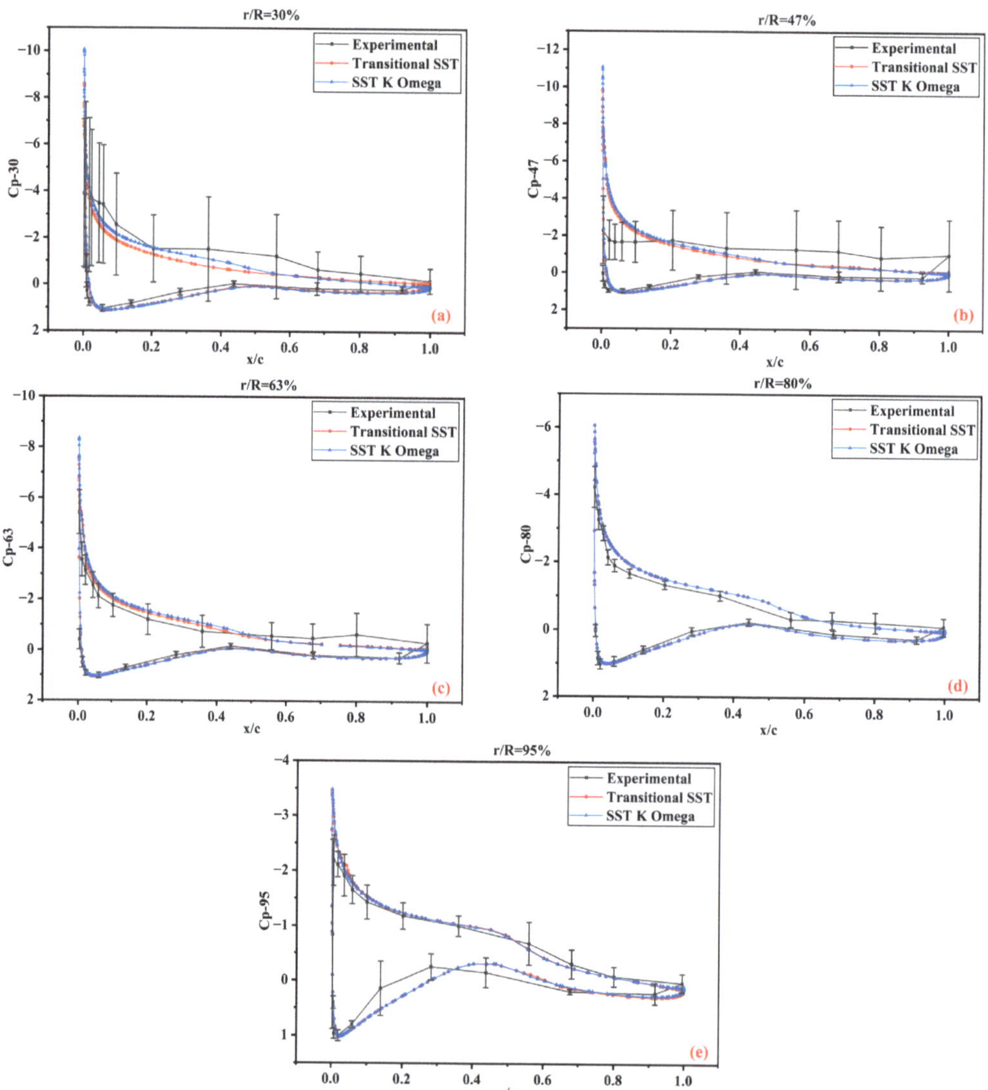

Figure 6. Comparison of the coefficient of pressure of NREL Phase VI obtained from CFD simulations and experimental data at a free-stream velocity of 10 m/s and at different blade sections (**a**–**e**).

Figure 7 shows a comparison of the pressure coefficient of the CFD simulation results derived using the transitional SST and SST K-Omega turbulence models with the experimental results obtained at a wind speed of 20 m/s. When the wind speed was 20 m/s, the simulation results strongly agreed with the experimental results at the front side of the blade while increasing the pressure from extending to the trailing edge. Specifically, on the suction side, there was a distinction between the results obtained from the simulations and experiments. At 30% of the blade span (Figure 7a), the turbulence models underestimated the pressure coefficient values by up to 80% of the chord. At 47% (Figure 7b) of the blade span, the turbulence models underestimated the pressure coefficient value while moving toward the trailing edge. Similarly, different patterns were observed for the blade spans of

63% (Figure 7c), 80% (Figure 7d), and 95% (Figure 7e). However, all the simulation results were within the standard deviation range of the experimental data over the blade span.

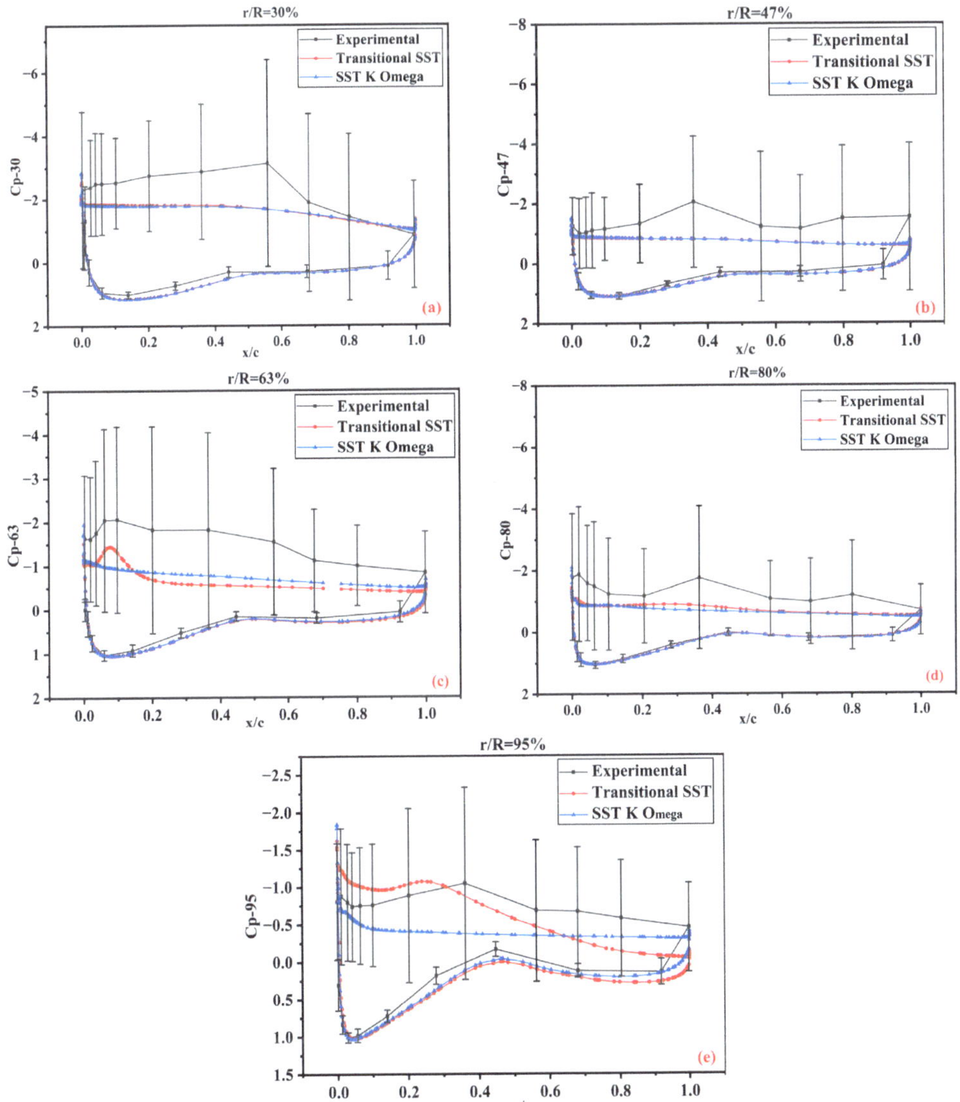

Figure 7. Comparison of the coefficient of pressure of NREL Phase VI obtained from CFD simulations and experimental data at a free-stream velocity of 20 m/s and at different blade sections (**a**–**e**).

Overall, both the transitional SST and SST K-Omega models provided good validation of the front surface of the blade using experimental data. Similarly, both turbulence models provided satisfactory results on the suction side while extending from the leading edge, except for some differences at a high inlet speed of 20 m/s. The trend behavior of the pressure variation at higher wind flow velocities, particularly at the suction side, is more prominent in the transitional SST turbulence model for the detection and localization of turbulent flow separation, as found at 63% and 95% spans of the blade.

3.2.2. The 12% Scaled-Down Model

To the best of our knowledge, the aerodynamic performance analysis of the 12% scaled-down model is yet to be included in future research. The experimental data validated the torque values predicted by the transitional SST and SST K-Omega models. After validating the torque values with the reference experimental data, the coefficients of pressure at various spans of the model wind turbine were determined. The transitional SST and SST K-Omega models were applied to capture the flow variations that reflect the approximation of a full-scale turbine model based on aerodynamic scaling. Matching the tip speed ratio and Mach number in the scaled-down model ensures the replication of the wind and rotational speeds of the real turbine rotor. This section presents a detailed analysis of the pressure distribution accompanying the pressure coefficients of the 12% scaled-down model at wind flow velocities of 7, 9, and 12 m/s. Radial sections were considered for a 12% scaled-down wind turbine model with blade spans of 30, 47, 63, 80, and 95%. Figure 8, Figure 10, and Figure 12 show the results for the pressure coefficients. Figure 9, Figure 11, and Figure 13 show the variation in the pressure contours obtained from the transitional shear stress transport and shear stress transport K-Omega turbulence models, respectively.

At Wind Speed = 7 m/s

Figure 8 shows a comparison of the pressure coefficients predicted by the turbulence models for the selected wind velocity of 7 m/s. The simulation results at a wind speed of 7 m/s showed no marked deviation from the transitional shear stress transport model and the SST K-Omega model at the blade cross-section. Approximately 50% to 85% of the chord length and up to 80% (Figure 8d) of the blade span, there is a deviation in the results of the transitional shear stress transport model from the SST K-Omega model while moving toward the trailing edge, specifically at the suction side. At 95% (Figure 8e) of the blade span, the pressure coefficient values predicted by SST K-Omega were higher than those predicted by the transitional SST model in the suction region extending from the leading edge. After the midpoint of the chord on the front side of the blade, the transitional SST model yielded slightly higher values. The transitional SST model predicted the maximum pressure value on the blade span, except at a radial distance of 95% (Figure 8e). The pressure value was −968.2 pascal for the shear stress transport K-Omega model and −954.9 pascal for the transitional SST model located on the blade's suction side. The pressure contours obtained from both turbulence models are shown in Figure 9.

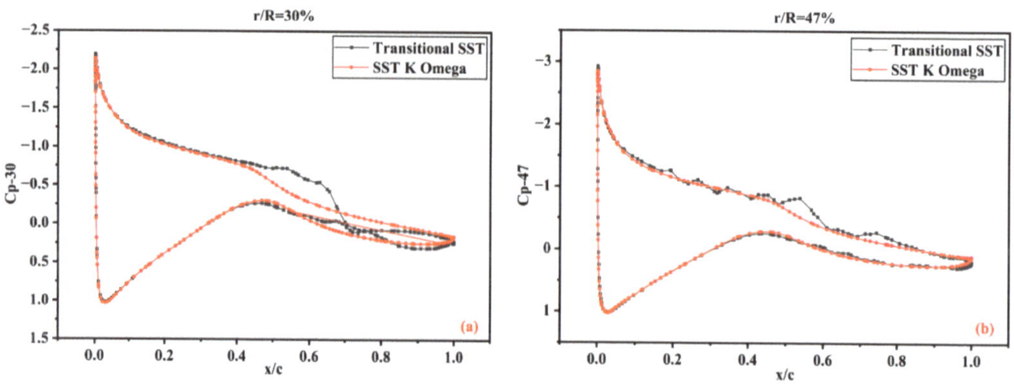

Figure 8. *Cont.*

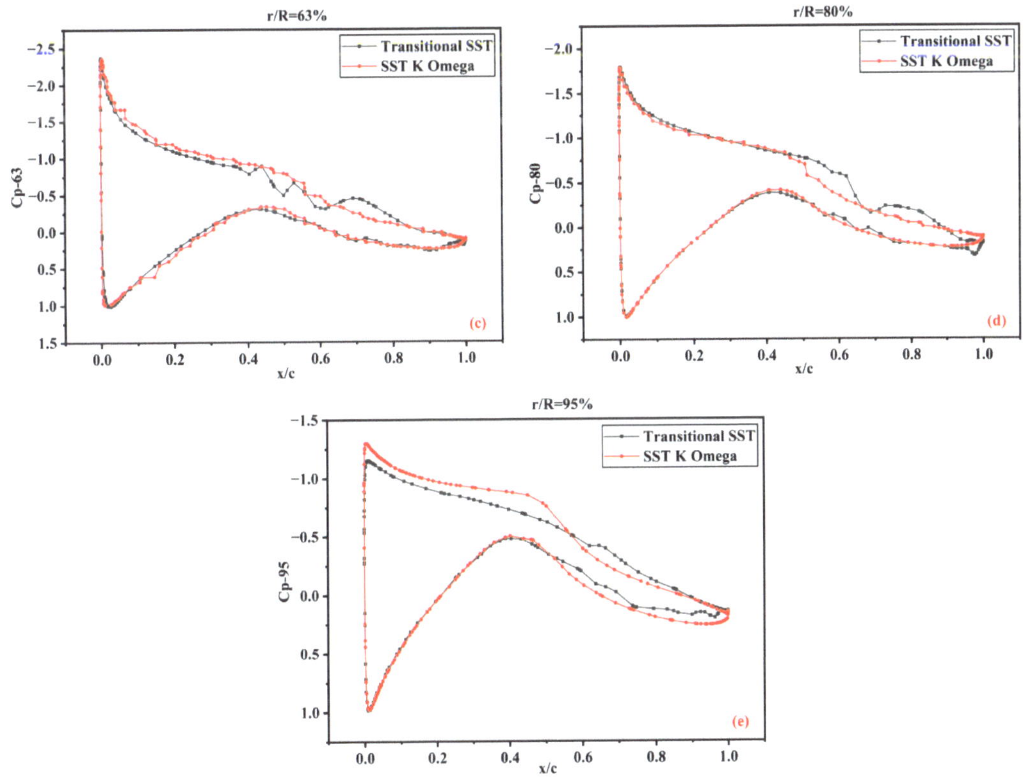

Figure 8. Comparison of pressure coefficients obtained from transitional SST and SST K-Omega turbulence models at a wind velocity of 7 m/s and at different blade sections (**a–e**).

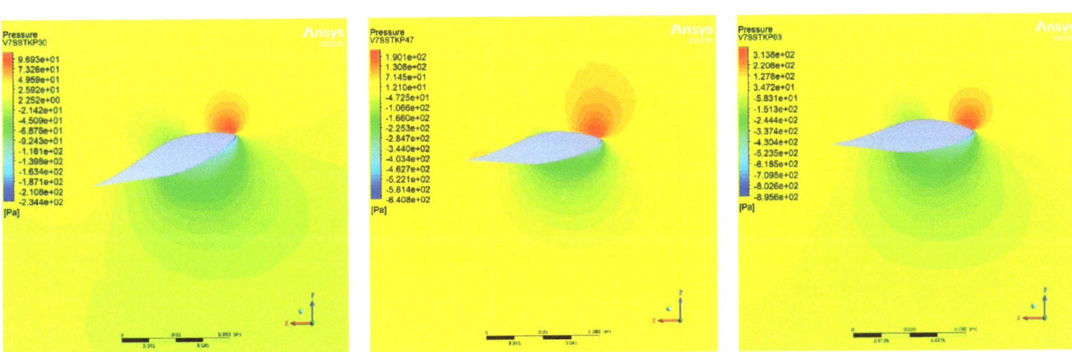

Figure 9. *Cont.*

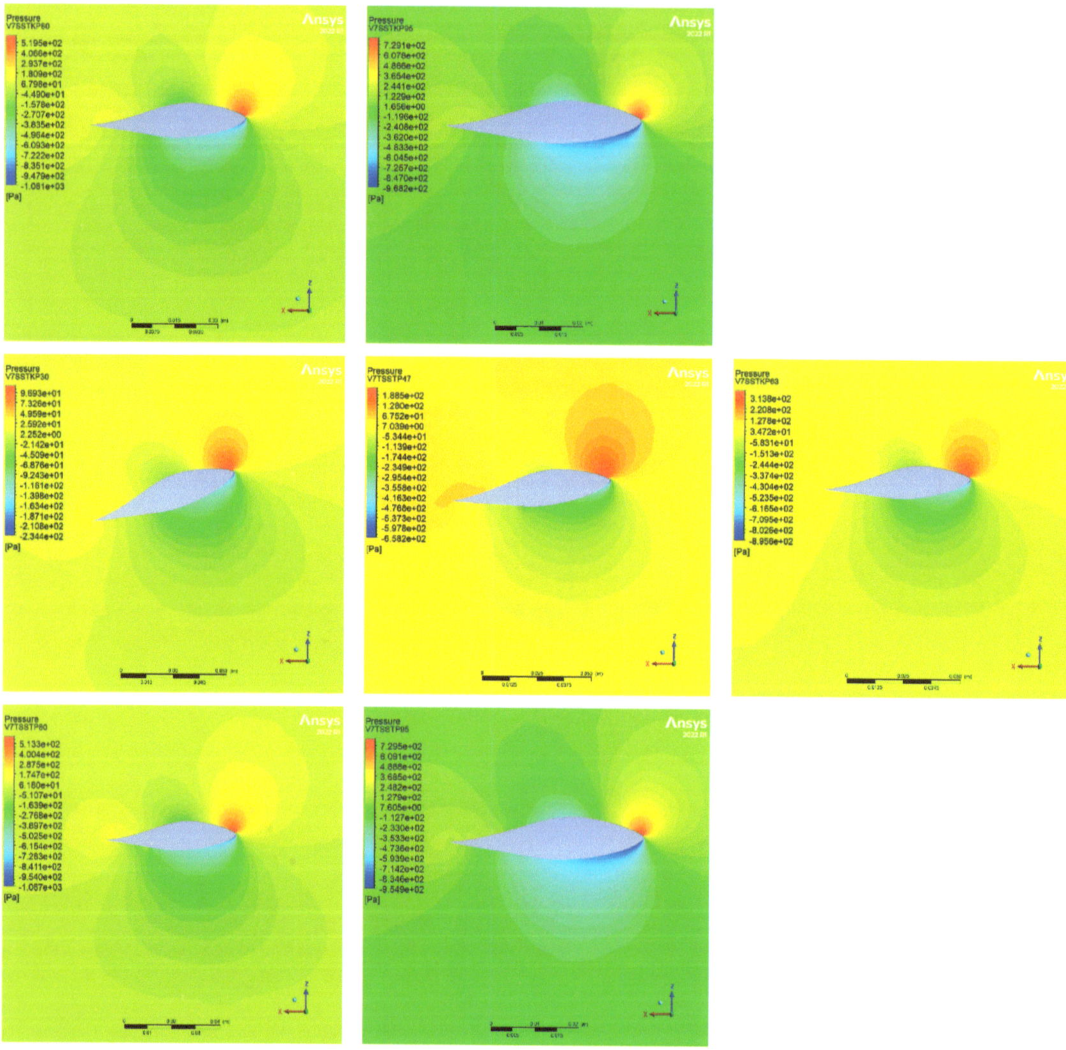

Figure 9. Pressure variation contours at different radial locations of the blade by SST K-Omega (graph with title SSTKP) and transitional SST (graph with title TSSTP) models at a wind velocity of =7 m/s.

At Wind Speed = 9 m/s

Figure 10 shows a comparison of the pressure coefficients predicted by both turbulence models at a selected wind flow velocity of 9 m/s. The numerical simulation results of the transitional SST model and SST K-Omega exhibit similar trends along the blade span. At a radial distance of 63% (Figure 10c), the transitional SST model captured the flow pattern better than SST K-Omega on the suction side of the blade covering from the leading edge to the trailing edge. Similarly, at a blade position of r/R = 95% (Figure 10e), local pressure was predicted well in the vicinity of the trailing edge in the transitional SST model in the suction region of the blade. The contours of the pressure distribution at 9 m/s of wind speed at different points along the blade are presented in Figure 11. The values of pressure obtained from the transitional SST model were higher at all span locations except at r/R = 47% (Figure 10b) radial distance. The maximum pressure value predicted by SST K-Omega was greater on both sides of the blade at a radial distance of 47%. (Figure 10b).

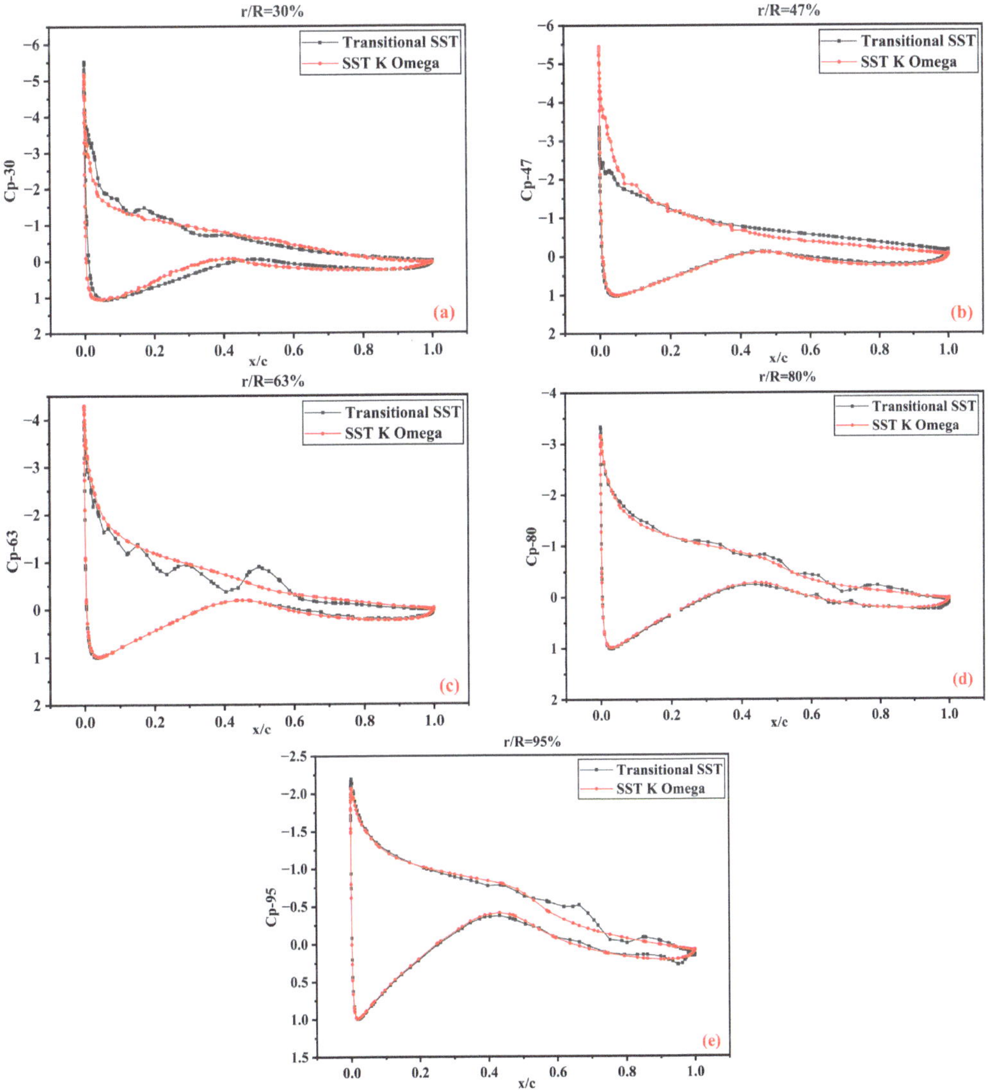

Figure 10. Comparison of pressure coefficients obtained from transitional SST and SST K-Omega turbulence models at a wind speed of 9 m/s and at different blade sections (**a**–**e**).

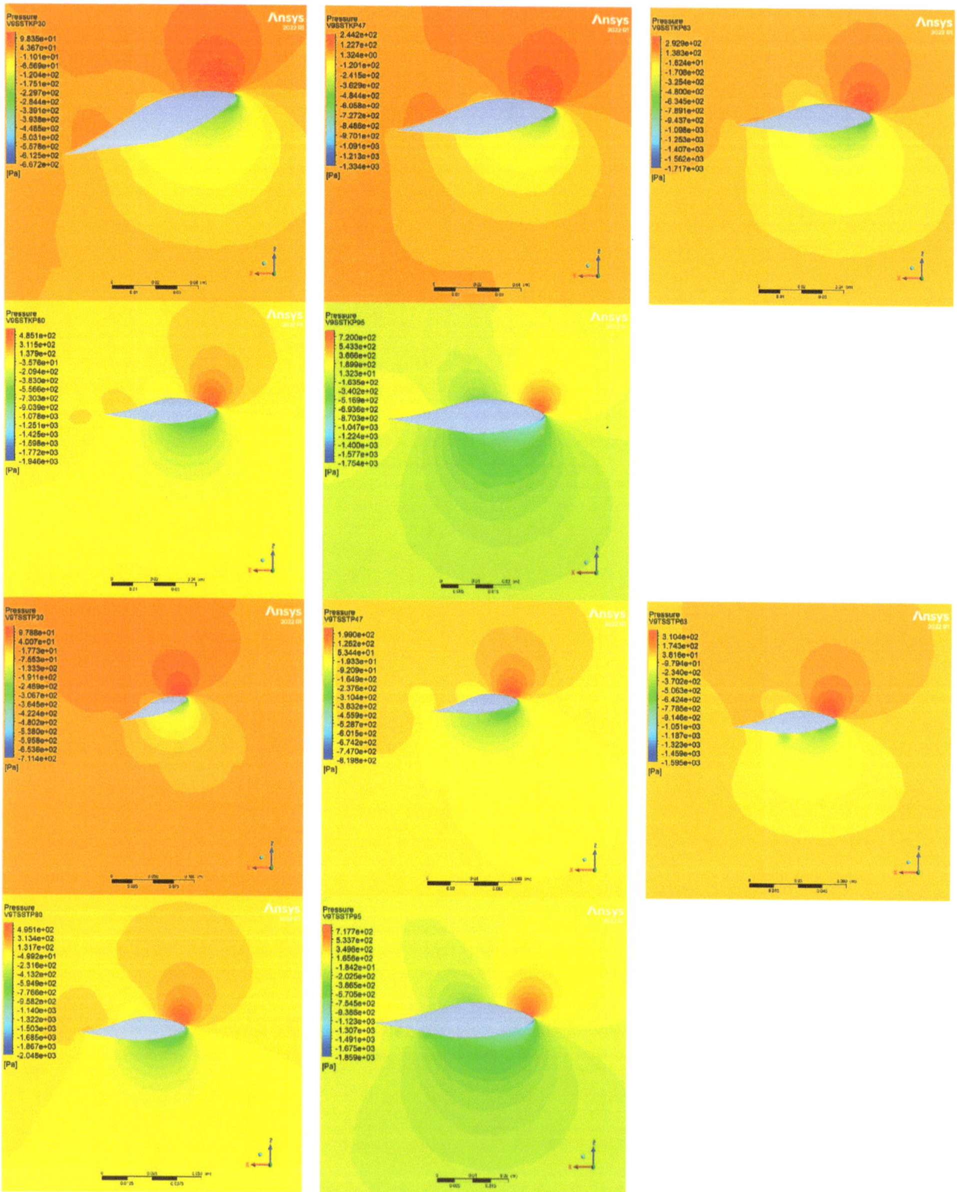

Figure 11. Pressure variation contours at different radial locations of the blade using SST K-Omega (graph with title SSTKP) and transitional SST (graph with title TSSTP) models at a wind speed of 9 m/s.

At Wind Speed = 12 m/s

The highest wind speed selected for the 12% scaled-down model was 12 m/s for the pressure analysis. Figure 12 shows a comparison of the pressure coefficients predicted by both turbulence models at a specified wind speed of 12 m/s. Figure 13 shows the contours of the pressure distribution on the front and suction sides of the blade for the SST K-Omega and transitional SST turbulence models, respectively. When the wind speed reached 12 m/s, the predictions of the pressure coefficient by both turbulence models were equal, except for some deviations along the suction region of the blade given by the transitional SST turbulence model. At r/R = 30%, the transitional SST model captured the flow separation more precisely than the SST K-Omega model. Similarly, at r/R = 80% (Figure 12c) and 95% (Figure 12e) of the blade span, the transitional SST model more precisely estimated the flow phenomena along the suction side of the blade. The SST K-Omega model provided the maximum pressure value in the suction regions at 95% of the blade radial span. The maximum pressure predicted by the transitional SST model was 715.2 pascal at r/R = 95% (Figure 12e). A comparison of the simulation results confirmed that the suction pressure increased, covering the distance from the bottom of the blade to its tip. The maximum suction was observed at r/R = 80% (Figure 12c) at wind flow speeds of 7 and 9 m/s. At high speeds, the maximum suction pressure was observed at r/R = 95% (Figure 12e), adjacent to the trailing edge of the blade. Similarly, the highest pressure and negative suction values were observed near the leading edge. The main concentration of suction pressure was adjacent to the leading edge, and the flow separated while moving toward the trailing edge of the blade. The pressure contours confirmed the reattachment close to the end of the blade tip. The maximum pressure observed was 927 pascals on the front side of the blade near the leading edge at a representative speed of 12 m/s.

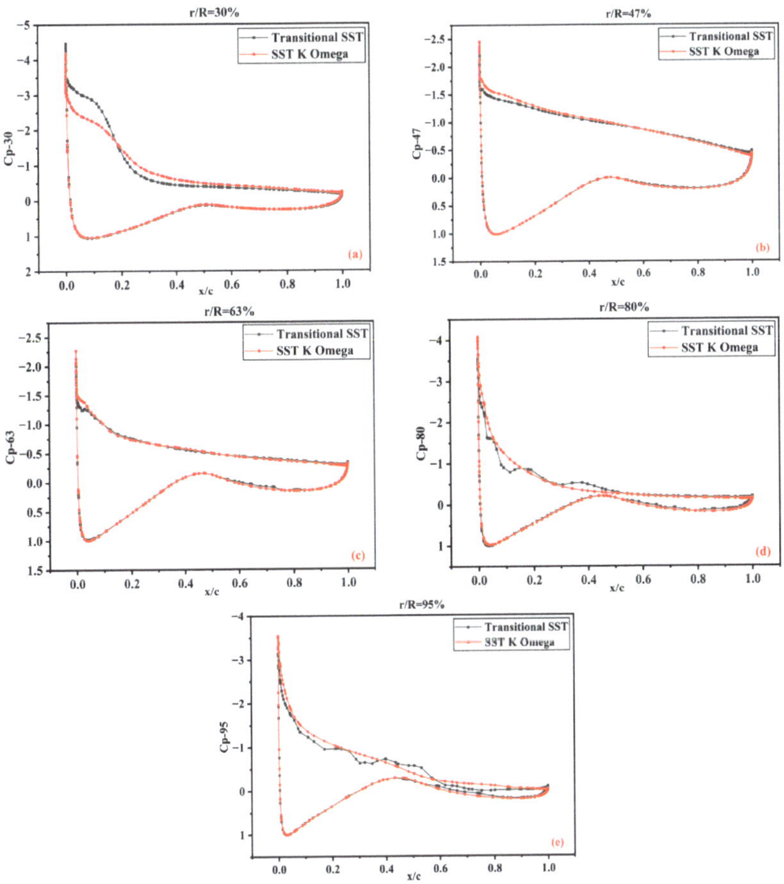

Figure 12. Comparison of pressure coefficients obtained from SST K-Omega and transitional SST models at a wind speed of 12 m/s and at different blade sections (**a**–**e**).

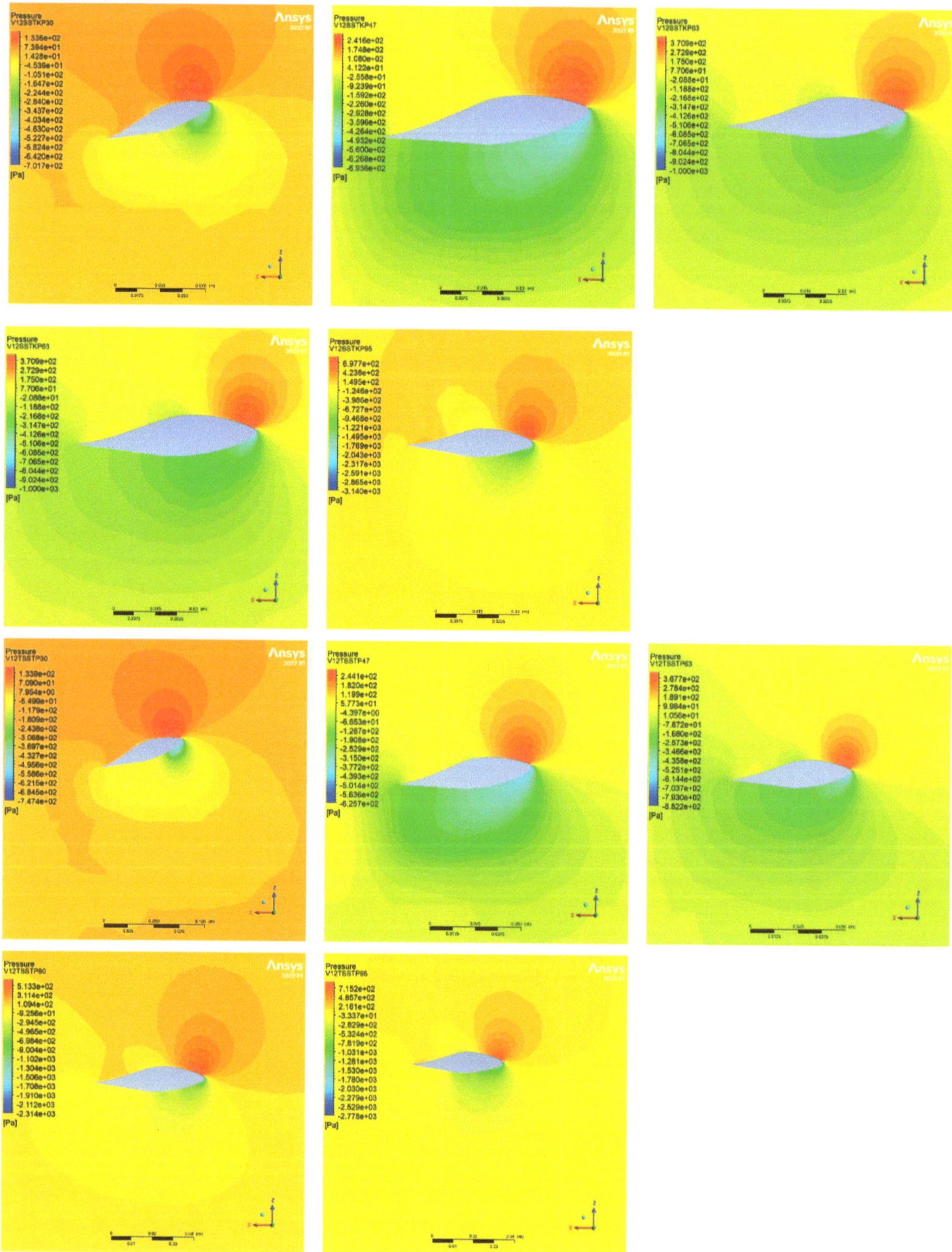

Figure 13. Pressure variation contours at different radial locations of the blade using SST K-Omega (graph with title SSTKP) and transitional SST (graph with title TSSTP) models at a wind speed of 12 m/s.

4. Conclusions

This study validated the characteristics and aerodynamic performance of full-scale NREL Phase VI and 12% scaled-down models using the transitional SST and SST K-Omega turbulence models. The wind speeds varied from 5 to 25 m/s and 4 to 14 m/s, respectively. The representative wind speeds selected for this study were 7, 10, and 20 m/s for the full-scale model and 7, 9, and 12 m/s for the scaled-down model. Based on the results of the numerical simulations, the following conclusions were drawn:

1. For wind flow velocities in the range of 5–10 m/s, the numerical simulation results of the transitional SST and SST K-Omega models matched well with the experimental data of the torque and thrust force. At a wind velocity of 10 m/s, both turbulence models overestimated the torque and thrust force values. Likewise, the torque values estimated by both turbulence models for a 12% scaled-down model up to a wind speed of 9 m/s displayed excellent trends with the experimental results. At a wind speed of 9 m/s, flow separation from the blade surface was observed, and the numerical solutions of the transitional SST and SST K-Omega deviated from the experimental data. Overall, the transitional SST model matched reasonably well with the experimental trend and slightly underestimated the torque values after a wind speed of 10 m/s, with a difference of less than 6%.
2. The pressure coefficients predicted by the transitional SST and SST K-Omega models show satisfactory agreement with the experimental data, except for some deviations at high speeds, but lie within the standard deviation of the experimental data.
3. The numerical results for the pressure coefficient curves obtained from the turbulence models were similar. The variation primarily occurred on the suction side, extending from the base to the blade tip. In the boundary layer region, the transitional SST model captured the flow phenomena better than the shear stress transport K-Omega model in the transition from the laminar to turbulent layers. Likewise, advancing from the leading edge to the trailing edge of the blade, the transitional SST model demonstrated an improved predicted flow separation. The simulation results of this study were limited to the specific flow regimes considered in the analysis.

5. Future Work

In the future, using the scaled-down model, the dynamic interaction of wind turbine blades, nacelles, and towers will be investigated to study their impact on performance and structural stability. Advanced measurement techniques and instrumentation will be developed to study fine-scale flow features and load dynamics.

Author Contributions: Conceptualization, Z.M. and Z.W.; methodology, Z.M.; software, Z.M.; validation, Z.M., Z.W. and X.Z.; formal analysis, Z.M. and X.Z.; investigation, Z.M. and G.S.; resources, Z.W.; data curation, Z.W.; writing—original draft preparation, Z.M.; writing—review and editing, Z.M. and G.S.; visualization, Z.M. and G.S.; supervision, Z.W.; project administration, Z.W. and X.Z.; funding acquisition, Z.W. All authors have read and agreed to the published version of the manuscript.

Funding: This study was financially supported by the National Natural Science Foundation of China (Grant Nos. 51779224 and 51579221) and the Basic Public Welfare Research Program of Zhejiang Province (Grant No. LHZ19E090002).

Data Availability Statement: The original contributions presented in the study are included in the article, further inquiries can be directed to the corresponding author.

Acknowledgments: The authors thank the National Renewable Energy Laboratory (NREL), United States Department of Energy, for providing experimental data to validate and verify the numerical results.

Conflicts of Interest: The authors declare no conflicts of interest.

References

1. Panwar, N.L.; Kaushik, S.C.; Kothari, S. Role of renewable energy sources in environmental protection: A review. *Renew. Sustain. Energy Rev.* **2011**, *15*, 1513–1524. [CrossRef]
2. Chehouri, A.; Younes, R.; Ilinca, A.; Perron, J. Review of performance optimization techniques applied to wind turbines. *Appl. Energy* **2015**, *142*, 361–388. [CrossRef]
3. Veers, P.; Bottasso, C.L.; Manuel, L.; Naughton, J.; Pao, L.; Paquette, J.; Robertson, A.; Robinson, M.; Ananthan, S.; Barlas, T.; et al Grand challenges in the design, manufacture, and operation of future wind turbine systems. *Wind Energy Sci.* **2023**, *8*, 1071–1131 [CrossRef]
4. Zhong, J.; Li, J. Aerodynamic performance prediction of NREL phase VI blade adopting biplane airfoil. *Energy* **2020**, *206*, 118182. [CrossRef]
5. Ye, M.; Chen, H.-C.; Koop, A. Verification and validation of CFD simulations of the NTNU BT1 wind turbine. *J. Wind. Eng. Ind. Aerodyn.* **2023**, *234*, 105336. [CrossRef]
6. Geng, X.; Liu, P.; Hu, T.; Qu, Q.; Dai, J.; Lyu, C.; Ge, Y.; Akkermans, R.A. Multi-fidelity optimization of a quiet propeller based on deep deterministic policy gradient and transfer learning. *Aerosp. Sci. Technol.* **2023**, *137*, 108288. [CrossRef]
7. Schreck, S.J. *IEA Wind Annex XX: HAWT Aerodynamics and Models from Wind Tunnel Measurements*; Final Report; National Renewable Energy Laboratory (NREL): Golden, CO, USA, 2008.
8. Hand, M.M.; Simms, D.A.; Fingersh, L.J.; Jager, D.W.; Cotrell, J.R.; Schreck, S.; Larwood, S.M. *Unsteady Aerodynamics Experiment Phase VI: Wind Tunnel Test Configurations and Available Data Campaigns*; National Renewable Energy Laboratory (NREL): Golden, CO, USA, 2001.
9. Giguère, P.; Selig, M.S. *Design of a Tapered and Twisted Blade for the NREL Combined Experiment Rotor*; National Renewable Energy Laboratory (NREL): Golden, CO, USA, 1999.
10. Lyu, C.; Liu, P.; Hu, T.; Geng, X.; Qu, Q.; Sun, T.; Akkermans, R.A.D. Hybrid method for wall local refinement in lattice Boltzmann method simulation. *Phys. Fluids* **2023**, *35*, 017103. [CrossRef]
11. Lyu, C.; Liu, P.; Hu, T.; Geng, X.; Sun, T.; Akkermans, R.A. A sliding mesh approach to the Lattice Boltzmann Method based on non-equilibrium extrapolation and its application in rotor flow simulation. *Aerosp. Sci. Technol.* **2022**, *128*, 107755. [CrossRef]
12. Menter, F.R. Two-equation eddy-viscosity turbulence models for engineering applications. *AIAA J.* **1994**, *32*, 1598–1605. [CrossRef]
13. Wilcox, D.C.; Suga, K. *Turbulence Modelling for CFD*; DCW Industries Inc.: La Cañada, CA, USA, 1993; p. 460.
14. Lee, K.; Huque, Z.; Kommalapati, R.; Han, S.-E. Fluid-structure interaction analysis of NREL phase VI wind turbine: Aerodynamic force evaluation and structural analysis using FSI analysis. *Renew. Energy* **2017**, *113*, 512–531. [CrossRef]
15. Song, Y.; Perot, J.B. CFD Simulation of the NREL Phase VI Rotor. *Wind. Eng.* **2015**, *39*, 299–309. [CrossRef]
16. Wang, P.; Liu, Q.; Li, C.; Miao, W.; Yue, M.; Xu, Z. Investigation of the aerodynamic characteristics of horizontal axis wind turbine using an active flow control method via boundary layer suction. *Renew. Energy* **2022**, *198*, 1032–1048. [CrossRef]
17. Dias, M.M.G.; Ramirez Camacho, R.G. Optimization of NREL phase VI wind turbine by introducing blade sweep, using CFD integrated with genetic algorithms. *J. Braz. Soc. Mech. Sci. Eng.* **2022**, *44*, 52. [CrossRef]
18. Lee, M.-H.; Shiah, Y.C.; Bai, C.-J. Experiments and numerical simulations of the rotor-blade performance for a small-scale horizontal axis wind turbine. *J. Wind. Eng. Ind. Aerodyn.* **2016**, *149*, 17–29. [CrossRef]
19. Sedighi, H.; Akbarzadeh, P.; Salavatipour, A. Aerodynamic performance enhancement of horizontal axis wind turbines by dimples on blades: Numerical investigation. *Energy* **2020**, *195*, 117056. [CrossRef]
20. Ke, W.; Hashem, I.; Zhang, W.; Zhu, B. Influence of leading-edge tubercles on the aerodynamic performance of a horizontal-axis wind turbine: A numerical study. *Energy* **2022**, *239*, 122186. [CrossRef]
21. Menegozzo, L.; Monte, A.D.; Benini, E.; Benato, A. Small wind turbines: A numerical study for aerodynamic performance assessment under gust conditions. *Renew. Energy* **2018**, *121*, 123–132. [CrossRef]
22. International Electrotechnical Commission. *Wind Turbines–Part 2: Design Requirements for Small Wind Turbines*; Technical Report 61400-2; International Electrotechnical Commission: Geneva, Switzerland, 2013.
23. Radünz, W.C.; Sakagami, Y.; Haas, R.; Petry, A.P.; Passos, J.C.; Miqueletti, M.; Dias, E. Influence of atmospheric stability on wind farm performance in complex terrain. *Appl. Energy* **2021**, *282*, 116149. [CrossRef]
24. Seim, F.; Gravdahl, A.R.; Adaramola, M.S. Validation of kinematic wind turbine wake models in complex terrain using actual windfarm production data. *Energy* **2017**, *123*, 742–753. [CrossRef]
25. Subramanian, B.; Chokani, N.; Abhari, R. Aerodynamics of wind turbine wakes in flat and complex terrains. *Renew. Energy* **2016**, *85*, 454–463. [CrossRef]
26. Zhao, F.; Gao, Y.; Wang, T.; Yuan, J.; Gao, X. Experimental Study on Wake Evolution of a 1.5 MW Wind Turbine in a Complex Terrain Wind Farm Based on LiDAR Measurements. *Sustainability* **2020**, *12*, 2467. [CrossRef]
27. Liu, Z.; Liu, P.; Guo, H.; Hu, T. Experimental investigations of turbulent decaying behaviors in the core-flow region of a propeller wake. *Proc. Inst. Mech. Eng. Part G J. Aerosp. Eng.* **2020**, *234*, 319–329. [CrossRef]
28. Menter, F.R.; Langtry, R.B.; Likki, S.R.; Suzen, Y.B.; Huang, P.G.; Völker, S. A Correlation-Based Transition Model Using Local Variables—Part I: Model Formulation. *J. Turbomach.* **2004**, *128*, 413–422. [CrossRef]
29. Lanzafame, R.; Mauro, S.; Messina, M. Wind turbine CFD modeling using a correlation-based transitional model. *Renew. Energy* **2013**, *52*, 31–39. [CrossRef]

30. Sihag, S.; Kumar, M.; Tiwari, A.K. CFD Validation and Aerodynamic Behaviour of NREL Phase VI Wind Turbine. *IOP Conf. Ser. Mater. Sci. Eng.* **2022**, *1248*, 012063. [CrossRef]
31. Ji, B.; Zhong, K.; Xiong, Q.; Qiu, P.; Zhang, X.; Wang, L. CFD simulations of aerodynamic characteristics for the three-blade NREL Phase VI wind turbine model. *Energy* **2022**, *249*, 123670. [CrossRef]
32. Langtry, R.; Gola, J.; Menter, F. Predicting 2D Airfoil and 3D Wind Turbine Rotor Performance using a Transition Model for General CFD Codes. In Proceedings of the 44th AIAA Aerospace Sciences Meeting and Exhibit, Reno, NV, USA, 9–12 January 2006.
33. Blocken, B.; van Druenen, T.; Toparlar, Y.; Andrianne, T. Aerodynamic analysis of different cyclist hill descent positions. *J. Wind. Eng. Ind. Aerodyn.* **2018**, *181*, 27–45. [CrossRef]
34. Kato, M. The Modeling of Turbulent Flow Around Stationary and Vibrating Square Cylinders. In Proceedings of the 9th International Symposium on Turbulent Shear Flows, Kyoto, Japan, 16–18 August 1993.
35. Shaik, M.; Subramanian, B. Computational investigation of NREL Phase-VI rotor: Validation of test sequence-S measurements. *Wind Eng.* **2023**, *47*, 973–994. [CrossRef]
36. Amiri, M.M.; Shadman, M.; Estefen, S.F. URANS simulations of a horizontal axis wind turbine under stall condition using Reynolds stress turbulence models. *Energy* **2020**, *213*, 118766. [CrossRef]
37. Moshfeghi, M.; Song, Y.J.; Xie, Y.H. Effects of near-wall grid spacing on SST-K-ω model using NREL Phase VI horizontal axis wind turbine. *J. Wind. Eng. Ind. Aerodyn.* **2012**, *107–108*, 94–105. [CrossRef]
38. Younoussi, S.; Ettaouil, A. Numerical Study of a Small Horizontal-Axis Wind Turbine Aerodynamics Operating at Low Wind Speed. *Fluids* **2023**, *8*, 192. [CrossRef]
39. Pope, S.B. *Turbulent Flows*; Cambridge University Press: Cambridge, UK, 2000.
40. Stern, F.; Wilson, R.; Shao, J. Quantitative V&V of CFD simulations and certification of CFD codes. *Int. J. Numer. Methods Fluids* **2006**, *50*, 1335–1355.
41. Cho, T.; Kim, C.; Lee, D. Acoustic measurement for 12% scaled model of NREL Phase VI wind turbine by using beamforming. *Curr. Appl. Phys.* **2010**, *10* (Suppl. S2), S320–S325. [CrossRef]
42. Hartwanger, D.; Horvat, A. 3D Modelling of a Wind Turbine Using CFD. In Proceedings of the NAFEMS UK Conference "Engineering Simulation: Effective Use and Best Practice", Cheltenham, UK, 10–11 June 2008.

energies

MDPI

Article

Simulating Water Invasion Dynamics in Fractured Gas Reservoirs

Yueyang Li [1,2], Enli Zhang [1], Ping Yue [2,*], Han Zhao [1], Zhiwei Xie [2,*] and Wei Liu [1]

[1] Research Institute of Exploration and Development, Southwest Oilfield Company, PetroChina, Chengdu 610500, China
[2] State Key Laboratory of Reservoir Geology and Development, Southwest Petroleum University, Chengdu 610500, China
* Correspondence: 201331010025@swpu.edu.cn (P.Y.); 201911000121@stu.swpu.edu.cn (Z.X.); Tel.: +86-130-8804-0285 (P.Y.); +86-177-1104-9070 (Z.X.)

Abstract: The Longwangmiao Formation gas reservoir in the Moxi block of the Sichuan Basin is a complex carbonate reservoir characterized by a low porosity and permeability, strong heterogeneity, developed natural fractures, and active water bodies. The existence of natural fractures allows water bodies to easily channel along these fractures, resulting in a more complicated mechanism and dynamic law of gas-well water production, which seriously impacts reservoir development. Therefore, a core-based simulation experiment was designed for oil–water two-phase flow. Three main factors influencing the water production of the gas reservoir, namely fracture permeability, fracture penetration, and water volume multiple, were analyzed using the orthogonal test method. The experimental results showed that the influences of the experimental parameters on the recovery factor and average water production can be ranked as water volume multiple > fracture penetration > fracture permeability, with the influence of the water volume multiple being slightly greater than that of the other two parameters. It provides a certain theoretical basis for water control of the gas reservoir.

Keywords: marine carbonate reservoir; fractured gas reservoir; water invasion law; orthogonal experiment

1. Introduction

Pore structures in carbonate reservoirs consist of three basic types: fracture vugs, fractures, and pores [1]. Fractures play a significant role in reservoirs. On the one hand, the existence of fractures improves the seepage conditions; on the other hand, edge-bottom water channeling along fractures can cause premature flooding in production wells, leading to adverse effects on gas recovery rate [2–6]. Fracturing or drilling processes can also affect the water output of gas well [7,8]. The Longwangmiao Formation gas reservoir in the Moxi block of the Sichuan Basin is characterized by developed faults and fractures, with the reservoir type being primarily the fracture type. There exists well-developed bottom water in the gas field, with active water body energy in local areas. Some wells of the gas reservoir have encountered water production issues in the initial stage of operation, which have severely restricted efficient development of the reservoir. Therefore, it is necessary to research the water invasion law for the fractured gas reservoir.

Currently, there is relatively little research on the water invasion mechanism in naturally fractured gas reservoirs, and most studies are conducted through numerical simulation [9–15]. In physical experimental studies, glass models or cast thin sections are utilized to carry out microscopic water flooding experiments to understand the micro-seepage mechanism [16–21], while the studies on macroscopic water invasion laws and development effects are mainly conducted through long-core physical experiments [22–24]. Shen Weijun [22] studied water invasion mechanisms under the conditions of various fracture widths using a whole-fracture core experiment, but did not consider the different degrees of fracture penetration. He Jialin [23] took into account fracture penetration, but more cores were required for his experiment. However, the gas well of the target reservoir was approximately 4000 m deep, resulting in high costs for core sampling and difficulties in obtaining complete carbonate long cores. Given this situation, a rational experimental method was designed. Under the core-limited conditions, it is possible to carry out an efficient and reliable experiment with a small number of core samples by optimizing the experimental sequence and combining the core cutting and fracturing, along with the orthogonal test method. This allows for analyses and a sensitivity study on fracture permeability, fracture penetration, and water volume multiple.

2. Experimental Objective

A simulation experiment of water invasion was conducted using carbonate cores with fractures that were obtained by artificially fracturing full-diameter reservoir core samples to simulate the depletion of fractured carbonate gas reservoirs with edge-bottom water. It aimed to evaluate the impacts of fracture permeability, fracture penetration, and water body volume on the water invasion law of gas wells, thereby providing a theoretical basis for water control.

3. Experimental Setup and Procedure

3.1. Experimental Setup

In accordance with the experimental requirements, the experimental setup for the study on water invasion mechanism was designed as shown in Figure 1. This setup consisted of a confining pressure pump, core holder, pressure gauge, flowmeter, back pressure controller, and so on. The core holder is a critical component in a core displacement device, which was positioned vertically to simulate a gas reservoir. The intermediate container was connected to the bottom of core holder and filled with formation water to simulate bottom water. The back pressure controller and the gas flowmeter were used for production regulation.

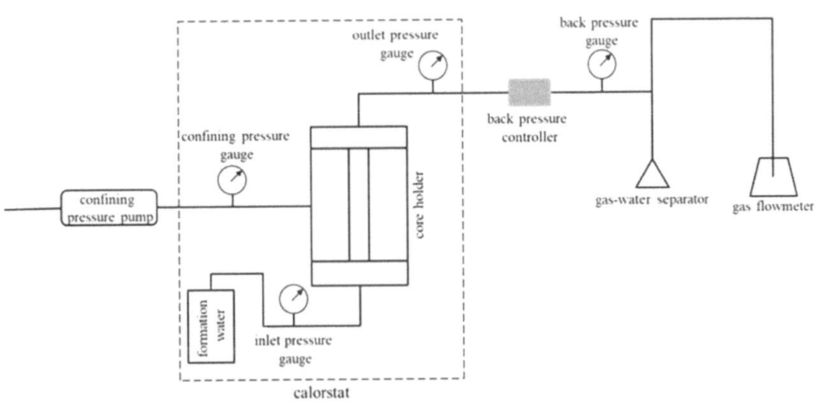

Figure 1. Experimental setup for water invasion mechanism.

3.2. Experiment Conditions

3.2.1. Experimental Temperature and Pressure

The experimental temperature was set at 142 °C according to the reservoir temperature of the Longwangmiao Formation gas reservoir.

The experimental internal pressure and confining pressure were 75 MPa and 85 MPa, respectively, according to the formation pressure of the Longwangmiao Formation gas reservoir.

3.2.2. Experimental Fluids

Natural gas and formulated formation water were chosen as the experimental fluids. The natural gas was sampled on site (Table 1), while the formation water for experimental use was prepared in the laboratory based on an analysis report of the field water samples (Table 2).

Table 1. Natural gas composition.

Component	Molar Percentage (%)	Component	Molar Percentage (%)
O_2 + Ar	0.0000	iC_4H_{10}	0.1670
He	0.0092	nC_4H_{10}	0.2111
H_2	0.0985	iC_5H_{12}	0.0920
N_2	0.5265	nC_5H_{12}	0.0776
CO_2	4.1874	C_6H_{14}	0.1313
H_2S	0.0218	C_7H_{16}	0.1115
CH_4	89.6571	C_8H_{18}	0.0653
C_2H_6	3.7082	C_9H_{20}	0.0279
C_3H_8	0.9044	$C_{10}H_{22}$	0.0032
relative density = 0.6399		compression factor = 0.9970	
hydrogen sulfide = 313 g/m^3		carbon dioxide = 82,271 mg/m^3	
molecular weight = 18.58 g/mol			

Table 2. Formation water salinity.

Density	pH Value	HCO_3^- (mg/L)	Cl^- (mg/L)	SO_4^{2-} (mg/L)	Ca^{2+} (mg/L)
1.055	6.1	433.1	44,792.1	751.5	5010
Mg^{2+} (mg/L)	**K^+/Na^+(mg/L)**	**Total Anion (mg/L)**	**Total Cation (mg/L)**	**Surin Classification**	**Total Mineralization (mg/L)**
608	22,854.4	45,976.7	28,662.3	chloride calcium type	74,639.0

3.2.3. Experimental Cores

Carbonate cores that met the requirements were selected from full-diameter reservoir core samples. Artificial fractures were created to obtain the experimental fractured carbonate cores. In this study, cylindrical carbonate cores with a length of 18.23 cm and diameter of 9.92 cm were utilized, which had a porosity of 4.2% and permeability of 0.97 mD under an overburden pressure of 75 MPa. The artificial fractures were created by splitting the cores in the middle and filling them with quartz.

3.3. *Experimental Scheme Design*

A sensitivity analysis was conducted in this experiment using three parameters, fracture permeability, fracture penetration, and water volume multiple, to determine their effects on water invasion in gas wells. The upper and lower limits of the experimental parameters were set according to the actual situation of the gas reservoir. An orthogonal design method was adopted for the experimental schemes to carry out the corresponding physical simulation tests. The schemes are detailed in Table 3, where the fracture permeability refers to the permeability under an overburden pressure of 75 MPa.

Table 3. Design of orthogonal experimental schemes.

Scheme No.	Fracture Permeability (mD)	Fracture Penetration (%)	Water Volume Multiple
1	50	100	5
2	50	75	10
3	50	50	30

Table 3. *Cont.*

Scheme No.	Fracture Permeability (mD)	Fracture Penetration (%)	Water Volume Multiple
4	100	75	30
5	100	50	5
6	100	100	10
7	200	50	10
8	200	100	30
9	200	75	5

3.4. Experimental Procedure

In line with the purpose of the physical simulation experiment for the depletion of a bottom-water carbonate gas reservoir and the schematic diagram of the experimental setup, the specific experimental procedure was as follows:

(1) Create artificial fractures in a core sample according to the fracture permeability and penetration in one experimental scheme.
(2) Install the artificially fractured core into the core holder and apply a confining pressure up to 85 MPa.
(3) Saturate the core model with gas, ensuring that the gas saturation is complete once the reservoir pressure is balanced at 75 MPa.
(4) Fill the high-pressure resistance intermediate container with the simulated formation water and pressurize it to 75 MPa.
(5) Connect the water in the intermediate container with the gas-saturated core model.
(6) Utilize the constant temperature chamber to heat both the core holder and the intermediate container and maintain it at a constant temperature of 142 °C.
(7) Gradually reduce the internal pressure at intervals of 2 Mpa; record the average gas and water production rates, as well as the pressure at water breakthrough for each 2 MPa pressure drop. Terminate the experiment when the internal pressure reaches 5 MPa.

The core samples were replaced with samples with different fracture parameters and water volume multiples in the intermediate container were used; the above steps were repeated until all the experimental schemes were executed.

4. Experiment Result Analysis

The simulation experiment to study the effect of fracture permeability, fracture penetration, and water volume multiple on the water production mechanism and water breakthrough performance was carried out using the orthogonal test method. Five indicators were taken into account in the simulation, including the recovery factor, average water–gas ratio, pressure at water breakthrough, average gas production, and average water production. The results are shown in Table 4.

Table 4. Experimental results of the dynamic simulation of water invasion mechanism.

Scheme No.	Parameter Value			Experimental Results				
	Fracture Permeability (mD)	Fracture Penetration (%)	Water Volume Multiple	Recovery Factor f	Average Water–Gas Ratio (mL/mL)	Pressure at Water Breakthrough (MPa)	Average Gas Production (mL)	Average Water Production (mL)
1	50	100	5	0.69	0.0035	40	790	2.8
2	50	75	10	0.72	0.0040	42	750	3
3	50	50	20	0.73	0.0054	46	700	3.8
4	100	75	20	0.73	0.0049	50	740	3.6
5	100	50	5	0.77	0.0032	40	750	2.4

Table 4. *Cont.*

Scheme No.	Parameter Value				Experimental Results			
	Fracture Permeability (mD)	Fracture Penetration (%)	Water Volume Multiple	Recovery Factor *f*	Average Water–Gas Ratio (mL/mL)	Pressure at Water Breakthrough (MPa)	Average Gas Production (mL)	Average Water Production (mL)
6	100	100	10	0.75	0.0045	49	840	3.8
7	200	50	10	0.72	0.0029	48	980	2.8
8	200	100	20	0.64	0.0049	52	970	4.8
9	200	75	5	0.66	0.0036	46	950	3.4

4.1. Analysis of Water Production Types

A double-logarithmic coordinate was adopted for displaying the simulation results of the above experimental schemes, with the water–gas ratio on the vertical axis and the recovery factor on the horizontal axis, to obtain the diagnostic curves of the water production patterns (see Figure 2). The upward segment after water breakthrough can intuitively reflect the water invasion law. The slope of the straight line of the upward segment, denoted by α, was used to characterize the rate at which the water–gas ratio increased with the recovery factor. Referring to the water production classification standard of Cheng Youyou [14], the water production was divided into three patterns based on the α value: Pattern 1 ($1 \leq \alpha < 2$), Pattern 2 ($2 \leq \alpha < 3$), and Pattern 3 ($\alpha > 3$).

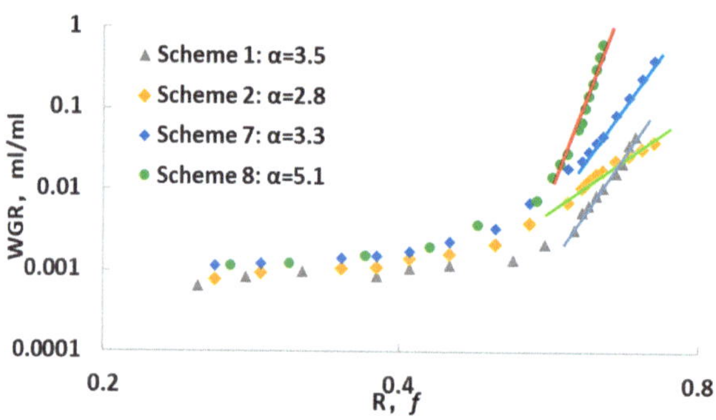

Figure 2. Diagnostic curves of water production patterns.

When $1 \leq \alpha < 2$, it indicates a water invasion performance in a fracture-free reservoir, representing the normal dynamic characteristics of water cresting and water coning.

When $2 \leq \alpha < 3$, it indicates the development of semi-penetrating or low-angle fractures that do not directly connect with the wellbore or bottom water, indicating a fractured carbonate reservoir.

When $\alpha > 3$, it indicates the existence of fully penetrating large fractures, large vugs, or penetrative fracture networks; bottom water directly connects with the production well through fractures and vugs, leading to rapid water invasion and water channeling during development [14].

By analyzing the diagnostic curves of the water production pattern of each designed experimental scheme, the following conclusions were drawn.

Due to the presence of fractures in all the schemes, the slopes were relatively steep, indicating that the water production patterns basically belong to Pattern 2 and Pattern 3.

The experimental results showed that the greater the fracture penetration, the fracture permeability, and the water volume multiple, the steeper the slopes in the water invasion diagnostic curves and the lower the recovery factor. This indicates that the greater the fracture penetration, the fracture permeability, and the water volume multiple, the more severe the water invasion and water channeling.

The comparison of Scheme 1 with Scheme 2 showed that Scheme 1 had a larger water volume multiple (10 times for Scheme 1 and 5 times for Scheme 2), a lower penetration rate (75% for Scheme 1 and 100% for Scheme 2), and a higher α value (3.5 for Scheme 1 and 2.8 for Scheme 2), indicating that the water volume had a greater impact on water invasion than the fracture penetration.

4.2. Analysis of Effects of Experimental Parameters on Water Invasion Indicators

The key experimental parameters included fracture permeability, fracture penetration, and water volume multiple. The water invasion indicators mainly included the recovery factor, average water–gas ratio, pressure at water breakthrough, average gas production, and average water production. The influence of each experimental parameter on the water invasion indicators was evaluated by conducting a range analysis on the results of the orthogonal test. According to the principle of the orthogonal test, the larger the range value, the greater the influence of the parameters. A summary of the results is shown in Table 5.

Table 5. Summary of the evaluation results for the dynamic simulation experiment of water invasion mechanism.

Evaluation Indicator	Parameter	Fracture Permeability K, mD		Fracture Penetration G, %		Water Volume Multiple M	
		Value	Result	Value	Result	Value	Result
Recovery Factor R, *f*	Mean value 1	50	0.713	50	0.740	5	0.707
	Mean value 2	75	0.735	63	0.721	7.5	0.725
	Mean value 3	100	0.750	75	0.703	10	0.730
	Mean value 4	150	0.721	87	0.697	15	0.720
	Mean value 5	200	0.673	100	0.693	20	0.700
	Range		0.077		0.047		0.030
Average Water–Gas Ratio WGR, mL/mL	Mean value 1	50	0.0043	50	0.0038	5	0.0034
	Mean value 2	100	0.0042	75	0.0041	10	0.0038
	Mean value 3	200	0.0038	100	0.0043	20	0.0051
	Range		0.0005		0.0005		0.0017
Pressure at Water Breakthrough P, MPa	Mean value 1	50	42.7	50	44.7	5	42.0
	Mean value 2	100	46.3	75	46.0	10	46.3
	Mean value 3	200	48.7	100	47.0	20	49.3
	Range		6.0		2.3		7.3
Average Gas Production Q_g, mL	Mean value 1	50	747	50	810	5	830
	Mean value 2	75	760	63	811	7.5	847
	Mean value 3	100	777	75	813	10	857
	Mean value 4	150	860	87	835	15	837
	Mean value 5	200	967	100	867	20	803
	Range		220		57		54
Average Water Production Q_w, mL	Mean value 1	50	3.20	50	3.00	5	2.87
	Mean value 2	100	3.27	75	3.33	10	3.20
	Mean value 3	200	3.67	100	3.80	20	4.07
	Range		0.47		0.80		1.20

4.2.1. Recovery Factor

Based on the range value, the three experimental parameters were ranked in order of their influence on the recovery factor: water volume multiple > fracture penetration > fracture permeability; however, the strengths of the influence were not very different. When impacted by the fractured water breakthrough, the recovery factor was relatively high with a fracture permeability of 100 mD, fracture penetration of 50%, and moderate water volume multiple of 10 times. A fracture permeability that is too low limits the gas flow, and both a great fracture penetration and a large water volume multiple are not conducive to improving the recovery rate (Figure 3).

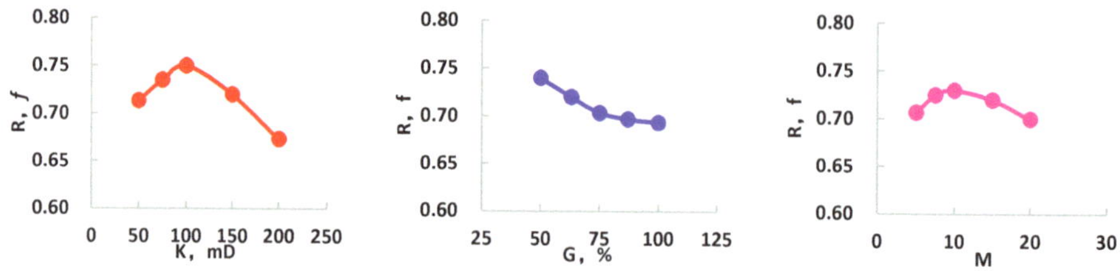

(a) Fracture permeability (b) Fracture penetration (c) Water volume multiple

Figure 3. Effects of experimental parameters on recovery factor (R).

4.2.2. Average Water–Gas Ratio

Based on the range value, the three parameters were ranked in order of their influence on the average water–gas ratio: fracture penetration > water volume multiple > fracture permeability. Within the selected range of mean values, the water volume multiple and the fracture permeability were comparable in terms of influence on the average water–gas ratio, whereas the fracture penetration had a significantly greater influence than the other parameters. Under the impact of fractured water breakthrough, the greater the fracture penetration and the water volume multiple, the higher the average water–gas ratio (Figure 4).

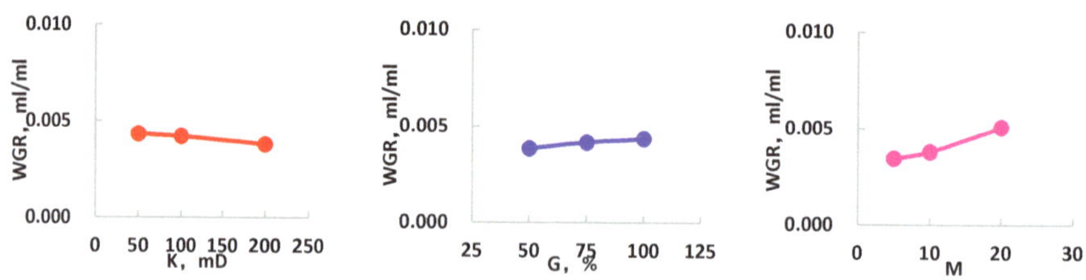

(a) Fracture permeability (b) Fracture penetration (c) Water volume multiple

Figure 4. Effects of experimental parameters on average water–gas ratio (WGR).

4.2.3. Pressure at Water Breakthrough

Based on the range value, the three parameters were ranked in order of their influence on the pressure at water breakthrough: fracture permeability > water volume multiple > fracture penetration. The pressure at water breakthrough corresponds to the timing of water breakthrough. Within the selected range of mean values, the influence of fracture permeability on the pressure was much greater than that of the other two parameters. Under the impact of fractured water breakthrough, the greater the permeability, the fracture penetration, and the water volume multiple, the higher the pressure point at water breakthrough and the earlier the water breakthrough time (Figure 5).

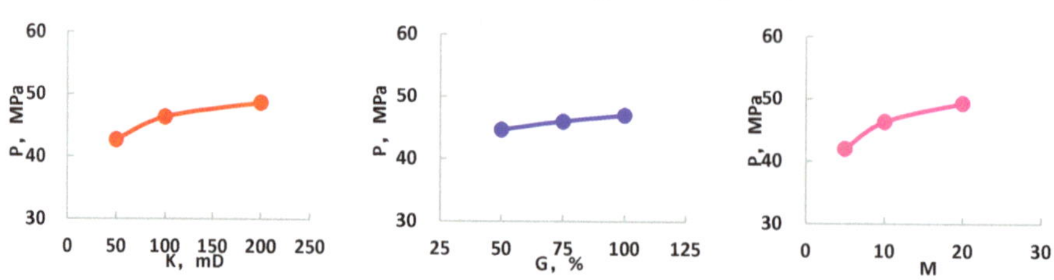

(a) Fracture permeability (b) Fracture penetration (c) Water volume multiple

Figure 5. Effects of experimental parameters on pressure at water breakthrough.

4.2.4. Average Gas Production

Based on the range value, the three parameters were ranked in order of their influence on the average gas production: fracture permeability > fracture penetration > water volume multiple. The fracture permeability had the greatest influence. Driven by the fractured bottom water, the average gas recovery rate in the initial stage before water breakthrough was positively correlated with the fracture permeability. The greater the fracture penetration, the larger the average gas production; the average gas recovery rate was relatively high when the water volume multiple was moderate, but its impact was limited (Figure 6).

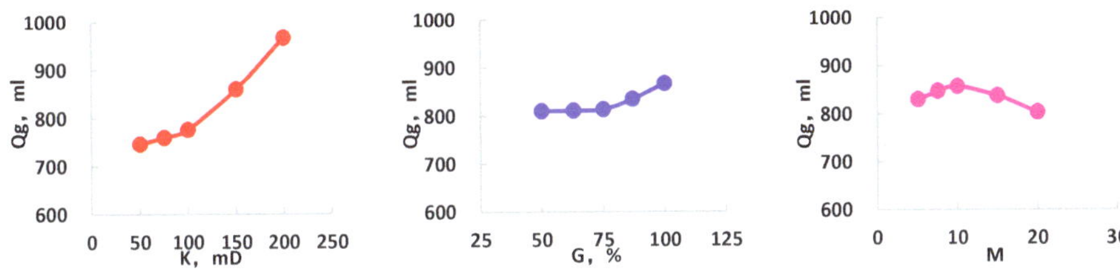

(**a**) Fracture permeability (**b**) Fracture penetration (**c**) Water volume multiple

Figure 6. Effects of experimental parameters on average gas production (Q_g).

4.2.5. Average Water Production

Based on the range value, the three parameters were ranked in order of their influence on the average water production: water volume multiple > fracture penetration > fracture permeability. The water volume multiple had the greatest influence. Driven by the fractured bottom water, the average water production rate after water breakthrough exhibited a positive correlation with the fracture permeability, fracture penetration, and water volume multiple. The higher the parameter values, the greater the average water production (Figure 7).

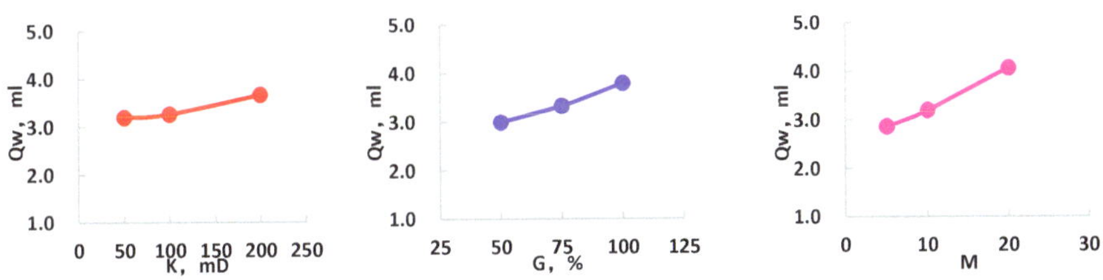

(**a**) Fracture permeability (**b**) Fracture penetration (**c**) Water volume multiple

Figure 7. Effects of experimental parameters on average water production (Q_w).

5. Conclusions

(1) By optimizing the experimental sequence and utilizing the orthogonal test method, it is possible to carry out more effective and comprehensive experimental research with a smaller number of core samples.

(2) The core size was sufficient to obtain the diagnostic curve pattern of water production. The greater the fracture permeability, the fracture penetration, and the water volume multiple, the larger the slope of diagnostic curves and the more severe the water invasion and water channeling.

(3) The experimental results showed that the effects of the experimental parameters on the recovery factor and average water production can be ranked as water volume multiple > fracture penetration > fracture permeability, with the influence of the water volume multiple being slightly greater than that of the other two parameters. For the average water–gas ratio, the fracture penetration exhibited a stronger sensitivity and

more significant influence; in the case of the water breakthrough time and average gas production, the fracture permeability had a stronger sensitivity and greater influence.

(4) For fracture-pore-type gas reservoirs, if there are through-fractures between the gas well and the edge and bottom water during the development process, the edge and bottom water will rapidly advance into the wellbore along the fractures. The advancing speed of the fracture water invasion front is closely related to the reservoir permeability and edge and bottom water. There was a positive correlation between the size of the water body, that is, the higher the reservoir permeability and the larger the water body, the faster the front advance speed, which will lead to a significant decrease in gas well production and final recovery volume.

Author Contributions: Conceptualization, methodology, validation, formal analysis, project administration, and funding acquisition, P.Y. and Y.L.; software, data curation, visualization, and writing—original draft preparation, Z.X.; investigation, H.Z.; resources, W.L.; writing—review and editing, E.Z.; supervision, P.Y. All authors have read and agreed to the published version of the manuscript.

Funding: This project was supported by the Key Program of the National Natural Science Foundation of China (No. U24B200683) and the Scientific Research Foundation of the Education Department of Sichuan Province, China (No. 2024NSFSC2012).

Data Availability Statement: The original contributions presented in this study are included in the article. Further inquiries can be directed to the corresponding author.

Conflicts of Interest: Authors Yueyang Li, Enli Zhang, Han Zhao, Wei Liu were employed by Research Institute of Exploration and Development, Southwest Oilfield Company, PetroChina. The remaining authors declare that the research was conducted in the absence of any commercial or financial relationships that could be construed as a potential conflict of interest.

References

1. Zhang, B.; Zheng, R.; Liu, H.; Lei, W.; Chen, R. Characteristics of Carbonate Reservoir in Callovian-Oxfordian of Samandepe Gasfield, Turkmenistan. *Acta. Geol. Sin.* **2010**, *84*, 117–126.
2. Cheng, Y.; Mu, L.; Zhu, E.; Zhang, P.; Guo, C.; Leng, Y.; Wei, Z.; Chen, Y.; Xing, Y.; Cheng, M.; et al. Water producing mechanisms of carbonate reservoirs gas wells: A case study of the Right Bank Field of Amu Darya, Turkmenistan. *Pet. Explor. Dev.* **2017**, *44*, 89–96. [CrossRef]
3. Song, H.; Cheng, H.; Yuan, F.; Cheng, L.; Yue, P. Prediction and Application of Drilling-Induced Fracture Occurrences under Different Stress Regimes. *Processes* **2024**, *12*, 1874. [CrossRef]
4. Sun, Z. Production characteristics and the optimization of development schemes of fractured gas reservoir with edge or bottom water. *Pet. Explor. Dev.* **2002**, *29*, 69–71. [CrossRef]
5. Hu, Y.; Li, X.; Wan, Y.; Jiao, C.; Xu, X.; Guo, C.; Jing, W. The experimental study of water invasion mechanism in fracture and the influence on the development of gas reservoir. *Nat. Gas. Geo.* **2016**, *5*, 910–917. [CrossRef]
6. Tang, J.R. *Gas Reservoir Engineering Technology*; Petroleum Industry Press: Beijing, China, 2011.
7. Qiang, L.; Li, Q.; Wang, F.; Wu, J.; Wang, Y. The Carrying Behavior of Water-Based Fracturing Fluid in Shale Reservoir Fractures and Molecular Dynamics of Sand-Carrying Mechanism. *Processes* **2024**, *12*, 2051. [CrossRef]
8. Li, Q.; Li, Q.; Han, Y. A Numerical Investigation on Kick Control with the Displacement Kill Method during a Well Test in a Deep-Water Gas Reservoir: A Case Study. *Processes* **2024**, *12*, 2090. [CrossRef]
9. Zhang, L.; He, W. Single well water invasion model for fractured bottom water gas reservoirs. *Nat. Gas. Ind.* **1994**, *14*, 48–50.
10. Tao, L.; Yong, F. Numerical simulation study on water coning of large-fractured gas well with finite volume method. *J. Southwest Pet. Univ. Sci. Technol. Ed.* **2001**, *23*, 19–22. [CrossRef]
11. Xu, X.; Wan, Y.; Chen, Y.; Hu, Y.; Mei, Q.; Jiao, C.-Y. Physical simulation of water invasion and water control for the fractured water-bearing gas reservoirs. *Nat. Gas. Geosic.* **2019**, *30*, 1508–1518. [CrossRef]
12. Hu, Y.; Le, P.; Guo, C.; Chen, P.; Xiao, Y.; Qu, S.; Wang, X. Experimental Study on Water Invasion in Full-Diameter Cores from Fractured Carbonate Reservoirs. *Xinjiang Pet Geo.* **2023**, *4*, 479–484.
13. Zhang, X.; Zhang, L.; Li, Y.; Zhou, Q. A new method for water invasion performance prediction in the early stage of fractured and water-carrying gas reservoir development. *J. Southwest Pet. Univ. Sci. Technol. Ed.* **2007**, *29*, 82–85. [CrossRef]
14. Peng, X.; Du, Z. Percolation model and numeral simulation of bottom water gas reservoirs with big fractures. *Nat. Gas. Ind.* **2004**, *24*, 116–119. [CrossRef]
15. Wang, X.; Tian, J.; Zhu, G.; Liang, B. A 3D numerical simulation of gas-water biphase for fractured gas reservoirs with low permeability. *China Offshore Oil Gas* **2010**, *22*, 168–171. [CrossRef]

16. Persoff, P.; Pruess, K. Two-phase flow visualization and relative permeability measurement in natural rough-walled rock fractures. *Water Resour.* **1995**, *31*, 1175–1186. [CrossRef]
17. Zhou, K.; Li, N.; Zhang, Q.; Tang, X. Experimental research on gas-water two phase flow and confined gas for mation mechanism. *Nat. Gas. Ind.* **2002**, *22*, 122–125.
18. Fan, H.; Zhong, B.; Li, X.; Liu, Y.; Yang, H.; Feng, X.; Zhang, Y. Studies on water invasion mechanism of fractured-watered gas reservoir. *Nat. Gas. Geo.* **2012**, *23*, 1179–1184.
19. Hu, Y.; Li, X.; Wan, Y.; Lu, J.; Zhu, H.; Zhang, Y.; Zhu, Q.; Yang, M.; Niu, L. Physical simulation on gas percolation in tight sandstone. *Pet. Explor. Dev.* **2013**, *40*, 580–584. [CrossRef]
20. Fang, F.; Li, X.; Gao, S.; Xue, H.; Zhu, X.; Liu, H.; An, W.; Li, C. Visual simulation experimental study on water invasion rulesof gas reservoir with edge and bottom water. *Nat. Gas. Geo.* **2016**, *27*, 2246–2252. [CrossRef]
21. Wang, L.; Yang, S.; Peng, X.; Deng, H.; Li, L.; Meng, Z.; Kun, Q.; Wang, Q. Visual experiments on the occurrence characteristics of multi-type reservoir water in fracture-cavity carbonate gas reservoir. *Acta. Pet. Sin.* **2018**, *39*, 686–696. [CrossRef]
22. Shen, W.; Li, X.; Liu, X.; Lu, J.; Jiao, C. Physical simulation of water influx mechanism in fractured gas reservoirs. *J. Cent. South Univ. Sci. Technol.* **2014**, *45*, 3283–3287.
23. He, G. Experimental study on EOR of gas reservoir with different penetrating level. *Pet. Ind. Tech.* **2016**, *23*, 79.
24. Li, C. Recovering technology of fractured and watered gas reservoirs. *Nat. Gas. Ind.* **2003**, *23*, 123–126. [CrossRef]

applied sciences

MDPI

Article

Flow-Induced Strength Analysis of Large Francis Turbine Under Extended Load Range

Xingping Liu [1], Xingxing Huang [2,3], Weijiang Chen [1] and Zhengwei Wang [2,*]

[1] Power Station Management Bureau, Y.R. Wanjiazhai Water Multi-Purpose Dam Project Co., Ltd., Linyi 036412, China; liuxingping502854@foxmail.com (X.L.); chenweijiang0303@foxmail.com (W.C.)

[2] Department of Energy and Power Engineering, Tsinghua University, Beijing 100084, China; xingxing.huang@gmx.ch

[3] Future Energy Research Institute, S.C.I. Energy (Swiss), 8706 Zurich, Switzerland

* Correspondence: wzw@mail.tsinghua.edu.cn

Abstract: To meet the load requirements of the power grid, the hydroelectric power plants need to extend the operational load range of the turbine units, which are often operated under off-design operating conditions. This new challenge significantly changes the flow characteristics of the hydro turbine units. Strong vibrations and high stresses caused by pressure pulsations at various loads directly lead to severe damage to the runner blades, threatening the safe operation of the hydropower unit. In this study, the detailed flow dynamics analysis under three loading conditions of a large-scale Francis turbine, i.e., 33.3%, 66.6%, and 100% of the Francis turbine's rated power, is investigated with computational fluid dynamics (CFD) calculations. The pressure files at different operating conditions are adopted to carry out the corresponding flow-induced strength analysis of the Francis runner prototype. The pressure distributions and flow velocity distributions at these three typical operating conditions are studied, and the maximum stress of the runner gradually increases with the power output of the turbine, but it is only around one-third of the yield stress of the runner material. It reveals that the runner is safe to operate in the extended operation range from a 33.3% to 100% of the rated power load. The analysis approach in this work can be applied to other hydraulic machinery including Francis turbines, pumps and pump–turbines.

Keywords: Francis turbine; extended load range; multiple operating conditions; flow dynamics; flow-induced strength; CFD and FEM

1. Introduction

The global energy crisis, characterized by rising energy demand, resource depletion, and environmental concerns, necessitates a transition towards more sustainable energy sources. Hydropower, as one of the most important and technically mature renewable energy sources that can be developed on a large scale, plays a pivotal role in this transformation and increases sustainable energy generation [1–3]. Francis turbines can operate efficiently in a wide range of current and head conditions and are widely used in hydropower power plants as one of the most important baseload generating units for power systems.

Francis turbine units remain the mainstay of hydropower generation worldwide due to their ability to adapt to a wide range of heads and their high efficiency. Their adaptability and reliability have contributed significantly to the global push towards renewable energy. Photovoltaic and wind turbines as the mainstay of new energy have the disadvantage of intermittency and volatility, with power generation output fluctuating according to

https://doi.org/10.3390/app15052422

changes in weather and time. The instability of photovoltaic and wind power requires large hydropower units as a regulating power source to balance the grid and ensure the stability and reliability of the power supply. In order to cope with load fluctuations after a huge amount of photovoltaic and wind power is connected to the grid, hydropower units, especially large Francis turbines, are required to operate in the load range to balance the grid and ensure the stability and reliability of the power supply.

Changes in electricity consumption lead to continuous fluctuations in the load demand on the grid for hydroelectric power plants throughout the day. The hydropower units in the power plants must operate frequently under a broad range of loads under off-design operating conditions in order to meet the power grid's requirements for absorbing new energy sources, such as photovoltaic and wind power. This has resulted in severe issues with the turbine runners and has drastically altered the operating characteristics of the units [4,5]. Francis turbine runners are susceptible to unstable flow when they operate under off-design conditions for an extended period. This can result in intense vibrations in the runner blades and a reduction in power generation efficiency [6,7]. Francis turbines can experience flow-induced cracks and damage in severe instances [8–10]. Figure 1 shows damage to the runner blade of a large prototype Francis turbine.

Figure 1. Flow-induced damage of the runner blade of a large Francis turbine [11].

Scholars worldwide have conducted a large number of in-depth studies on the unsteady flow characteristics of Francis turbines when they are operated outside of their rated operating conditions. The flow distribution in the Francis runner flow channel under partial loads is irregular, causing cavitation on the runner blades . Cavitation erosion decreases turbine performance, increases hydraulic losses, and causes material loss on the runner blades [12,13]. The runner blades of Francis turbines may develop pits and cracks as a result of prolonged cavitation erosion [14,15]. The time- and frequency-domain properties in the vaneless region of Francis units under various load conditions were monitored and simulated [16].

At part-load conditions, the Francis turbine unit can experience draft tube pressure pulsations, leading to severe hydraulic vibrations in the Francis turbine unit and threatening its stability [17,18]. Experimental and numerical studies of swirling flow and flow instability in Francis turbines under different loads have been carried out to capture the transient rotating flow characteristics in the draft tube [19]. Through the Francis turbine model test, it is possible to directly observe the complex flow characteristics of the draft tube vortex ropes under different operating conditions [20].

The physical mechanisms of unsteady flow and flow-induced structural vibration and stress in Francis turbine units under different operating conditions can be further elucidated by performing field measurements of pressure pulsation and stress on a prototype Francis runner [19,21,22]. To gain a deeper understanding of the prototype unit's vibration mechanism underwater, the vibration measurements of a simplified hydraulic turbine model have been carried out using novel distributed fiber optic sensing technology to monitor vibrations and stresses [23]. The underwater structural vibrations of Francis turbine runners were measured and simulated to evaluate the model behavior of the runners considering the added-mass effects of the surrounding water [24]. Studies involving field measurements and numerical simulations have shown that unsteady flow-induced stresses in Francis runners are closely related to pressure pulsation levels, rotational speed, and the geometric configuration of the turbine [25].

According to the results of a study into the runner blade breakdown of a 200 MW Francis turbine, cavitation damage from operations at low part loads combined with severe high-cycle fatigue caused the blade to crack [26]. Another investigation discovered that cracks on the runner blades of a prototype Francis turbine were produced by operating load conditions deviating from the rated condition [10]. Some Francis units may also be damaged as a result of excessive dynamic stresses within the runner and flow-induced structural resonance under off-design conditions [8]. According to some experts, the Francis runner's blades cracked because of the uneven weld surface, which was exacerbated by the weld filler material's hydrogen embrittlement [9]. The development of cracks in Francis turbine runners is a complicated issue that is impacted by several variables. One of the most frequent conclusions is that turbulence, particularly in off-design operating conditions, is a major factor in causing these cracks, even though the precise causes may differ.

Many researchers have studied the flow dynamics and structural vibration of Francis turbine runners by computational fluid dynamics (CFD), finite element method (FEM) and field measurements, respectively. However, the research for the comprehensive analysis of large prototype Francis turbines with full three-dimensional fluid–structure interaction (FSI) methods is still relatively limited [27]. In order to capture the flow-induced dynamic behavior more accurately and to avoid serious flow-induced damage in Francis turbines, the flow-induced strength analysis of a 183 MW prototype Francis turbine is carried out in this study following the workflow shown in Figure 2.

First, a geometrical model of the Francis turbine unit's fluid and structural domains is created. Second, three operating loads are chosen to perform the CFD analysis of the Francis turbine unit's entire flow passage in order to thoroughly examine the pressure distribution and the flow distribution characteristics: 33.3%, 66.6%, and 100% of the rated power. Ultimately, the finite element model of the Francis runner is mapped with the pressure data calculated from the CFD analysis, and an in-depth investigation of the runner's stress distribution is conducted.

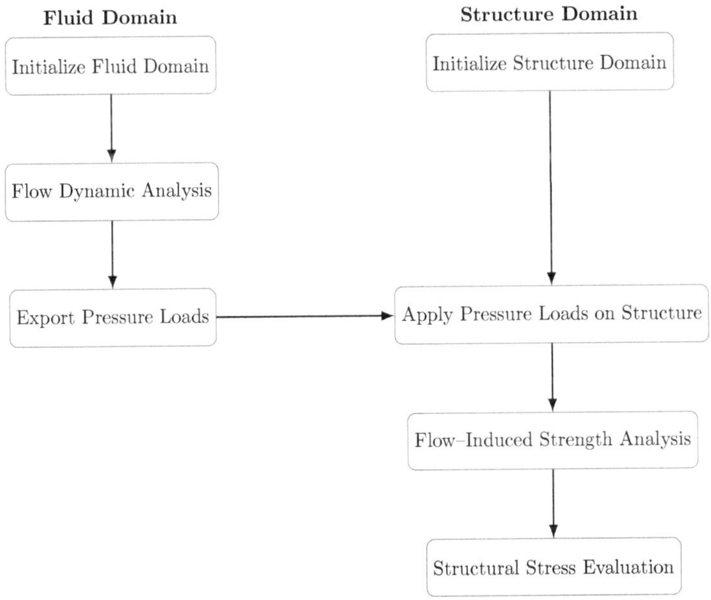

Figure 2. The workflow of flow-induced strength analysis of the Francis turbine in this study.

2. Methodology and Mathematical Physics Equations

2.1. Fluid Dynamics Governing Equations

The governing equations for turbulent fluid flow in the hydraulic Francis turbine units are based on the fundamental laws of mass, momentum, and energy conservation. Since the water in hydraulic turbines is thought to be incompressible and at a constant temperature, the Reynolds-averaged Navier–Stokes (RANS) equations can be used to explain the fundamental laws of mass and momentum conservation:

$$\frac{\partial \overline{u_i}}{\partial t} + \frac{\partial(\overline{u_i u_j})}{\partial x_j} = -\frac{1}{\rho}\frac{\partial \overline{p}}{\partial x_i} + \frac{\partial}{\partial x_j}[\nu(\frac{\partial \overline{u_i}}{\partial x_j} + \frac{\partial \overline{u_j}}{\partial x_i})] - \frac{\partial \overline{u_i' u_j'}}{\partial x_i} + \overline{f_i} \tag{1}$$

$$\frac{\partial \overline{u_i}}{\partial x_i} = 0 \tag{2}$$

where i and j are indicators which fall in $1,2,3$, x_i and x_j are the Cartesian coordinate directions, $\overline{u}_i$ and $\overline{u}_j$ are the time-averaged terms of the fluid velocity, and $\overline{u}_i'$ and $\overline{u}_j'$ are time-fluctuating terms of the fluid velocity. Time is represented by t, fluid density by ρ, time-averaged pressure by $\overline{p}$, kinematic viscosity by ν, Reynolds stress tensor by $\overline{u_i'u_j'}$, and time-averaged external forces by f_i.

The RANS equations introduce an additional unknown quantity, the Reynolds stress tensor $\overline{u_i'u_j'}$, which accounts for the effects of turbulent fluctuations on the mean flow field [28]. To close the RANS equations, various turbulence models are introduced, which parameterize the Reynolds stress tensor in terms of the mean flow properties and turbulence quantities. The k-ε model is based on transport equations for the turbulent kinetic energy k and its dissipation rate ε, and assumes an isotropic eddy viscosity to relate Reynolds stresses to mean flow gradients [29]. Utilizing transport equations for k and the turbulent kinetic energy dissipation rate ω, the k-ω model improves predictions in near-wall regions and for flows with adverse pressure gradients [30]:

$$k = \frac{1}{2}\overline{u'_i u'_j} \tag{3}$$

$$\varepsilon = \nu\overline{(\frac{\partial u'_i}{\partial x_k}\frac{\partial u'_i}{\partial x_k})} \tag{4}$$

$$\omega = \frac{\varepsilon}{k} \tag{5}$$

These turbulent models utilize the Boussinesq hypothesis, introducing the turbulent viscosity ν_T to express the Reynolds stress tensor as

$$\overline{u'_i u'_j} = -\nu_T(\frac{\partial \overline{u_i}}{\partial x_j} + \frac{\partial \overline{u_j}}{\partial x_i}) + \frac{2}{3}k\delta_{ij} \tag{6}$$

The Shear Stress Transport (SST) model is a hybrid approach that combines the advantages of the k-ω model in near-wall regions with the k-ε model in the free-flow regions, providing enhanced accuracy in separating flows [31]. The SST model is widely used for the turbulent flow simulations for hydraulic turbines and therefore also adopted in this investigation:

$$\frac{\partial \omega}{\partial t} + \frac{\partial}{\partial x_j}(\overline{u}_j\omega) = \frac{\partial}{\partial x_j}\left[(\nu + \frac{\nu_T}{\sigma_{\omega_2}})\frac{\partial \omega}{x_j}\right] + 2(1 - F_1)\frac{1}{\sigma_{\omega_2}\omega}\frac{\partial k}{\partial x_j}\frac{\partial \omega}{\partial x_j} - \beta_2\omega^2 + \alpha\frac{\omega}{k}P_k \tag{7}$$

$$\frac{\partial k}{\partial t} + \frac{\partial(\overline{u}_j k)}{\partial x_j} = \frac{\partial}{\partial x_j}\left[(\nu + \frac{\nu_T}{\sigma_{k2}})\frac{\partial k}{\partial x_j}\right] - \beta' k\omega + P_k \tag{8}$$

The mixing function F_1 in Equation (7) is expressed as

$$F_1 = \tanh(\arg_1^4) \tag{9}$$

$$\arg_1 = \min\left[\max\left(\frac{\sqrt{k}}{\beta'\omega y}, \frac{500\nu}{y^2\omega}\right), \frac{4k}{\sigma_{\omega_2}CD_{k\omega}y^2}\right] \tag{10}$$

$$CD_{k\omega} = \max\left(2\rho\frac{1}{\sigma_{\omega_2}\omega}\frac{\partial k}{\partial x_i}\frac{\partial \omega}{\partial x_i}, 10^{-10}\right) \tag{11}$$

where y is the distance to the nearest wall.

The turbulent kinetic energy generation term P_k in Equation (8) can be written with the strain rate S_{ij} as

$$P_k = 2\nu_T S_{ij}\frac{\partial \overline{u}_i}{\partial x_j} \tag{12}$$

where S_{ij} is the strain rate.

$$S_{ij} = \frac{1}{2}(\frac{\partial \overline{u_i}}{\partial x_j} + \frac{\partial \overline{u_j}}{\partial x_i}) \tag{13}$$

The turbulent viscosity ν_T used in the above equations is defined as

$$\nu_T = \frac{\alpha_1 k}{\max(\alpha_1\omega, SF_2)} \tag{14}$$

$$F_2 = \tanh(\arg_2^2) \tag{15}$$

$$\arg_2 = \max\left(\frac{2\sqrt{k}}{\beta' \omega y}, \frac{500\nu}{y^2 \omega}\right) \tag{16}$$

The other parameters are $\alpha_1 = 0.31$, $\alpha_2 = 0.44$, $\beta' = 0.09$, $\beta_2 = 0.0828$, $\sigma_{k_2} = 1$, and $\sigma_{\omega_2} = 1/0.856$.

2.2. Fluid-Induced Structural Dynamics Governing Equations

With the calculated results by the fluid dynamics analysis, the fluid-induced force $\left\{F^f\right\}$ acting on the structure surfaces S of the pumped-storage unit can be obtained by

$$\left\{F^f\right\} = \iint_S p\, dS \tag{17}$$

Taking the flow-induced force $\left\{F^f\right\}$ applied on the pumped-storage unit into account, the fluid-induced structural dynamics equation can be expressed as

$$[M^s]\{a\} + [D^s]\{v\} + [S^s]\{d\} = \{F^s\} + \left\{F^f\right\} \tag{18}$$

where $[M^s]$, $[D^s]$, and $[S^s]$ denote the mass, damping, and stiffness matrices of the structures for the pumped-storage unit, respectively. $\{a\}$, $\{v\}$, and $\{d\}$ represent the acceleration, velocity, and displacement vectors of the unit structures, respectively. $\{F^s\}$ is the external force acting on the unit structures but excludes the flow-induced force $\left\{F^f\right\}$.

The structural displacement $\{d\}$ of the unit structures can be obtained by solving Equation (18), and the flow-induced stress $\{\sigma\}$ of the structures for the pumped-storage unit can be obtained by the following equation:

$$\{\sigma\} = [E^s]\left[B^d\right]\{d\} \tag{19}$$

where $[E^s]$ and $[B^d]$ are the elasticity matrix and the strain–displacement matrix of the structures for the pumped-storage unit.

With the maximum, middle, and minimum principal stresses (σ_1, σ_2 and σ_3), the equivalent stress $\sigma_{eq.}$ can be calculated by Equation (20) to investigate the flow-induced stress behavior of the structures for the pumped-storage unit, including the head-cover bolts and bottom-ring bolts:

$$\sigma_{eq.} = \sqrt{\frac{(\sigma_1 - \sigma_2)^2 + (\sigma_2 - \sigma_3)^2 + (\sigma_3 - \sigma_1)^2}{2}} \tag{20}$$

3. Specifications of the Francis Turbine Unit and Investigated Operating Conditions

The hydropower plant of this study has six large-scale Francis turbine generation units, each rated at 183 MW. The turbine units' rated head and maximum head are 68 m and 80 m. The rated discharge and the rotating speed of the turbine are 294 m^3s^{-1} and 100 rpm, respectively. The Francis turbine unit has 20 stay vanes and 20 guide vanes, and the turbine runner has 13 blades. The rotating direction of the turbine runner is clockwise in the top view. The parameter specifications of the Francis turbine unit are listed in Table 1.

The three-dimensional geometrical model of the water passage of the Francis turbine prototype is shown in Figure 3. The inlet and outlet of the turbine flow passages are also illustrated in the figure. The spiral case is the Francis turbine unit's flow intake structure, and its spiral design helps to maintain an even distribution of flow velocity and pressure, thereby minimizing energy loss as the water enters the stay vane flow passage. Stay vanes within the stay ring, positioned radially between the spiral case and the guide vanes, provide structural support to the spiral case and other structures of the turbine unit. The guide vanes can be adjusted in their opening angle to regulate water flow and finely control the power output on the turbine. The Francis runner is the core component of the turbine. When the water flows through the Francis runner, its potential and kinetic energy are converted into the mechanical energy of the rotor system. The draft tube, located at the outlet of the runner, discharges water from the runner to the downstream and converts the remaining kinetic energy of water exiting the runner into pressure energy. This increases the overall efficiency of the turbine by reducing energy losses. The water in the crown chamber and band chamber between the runner and the surrounding stationary structures balances the pressure and thrust of the runner. Labyrinth seals in the crown and belt chambers help to reduce water leakage of the runner passage.

Table 1. The specifications of the Francis turbine unit.

Specification	Value	Unit
Rated power	183	MW
Rated head	68	m
Maximum head	80	m
Rated discharge	294	m^3s^{-1}
Rotating speed	100	rpm
Stay vane number	20	-
Guide vane number	20	-
Runner blade number	13	-

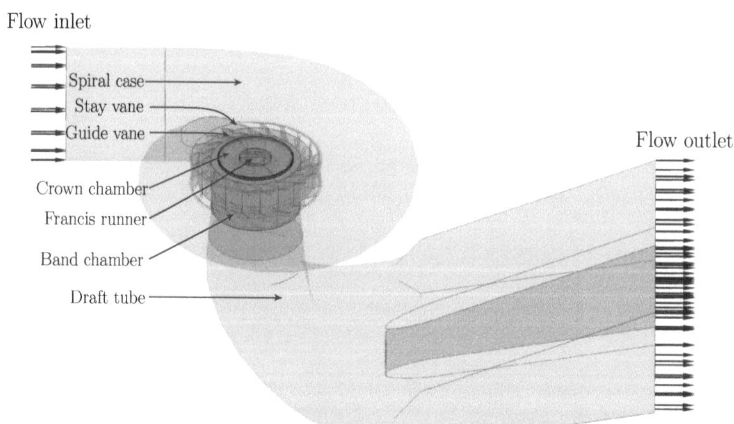

Figure 3. The geometrical model of the water passage of the Francis runner unit.

The power plant operation statistics show that the Francis turbine units often operate at 74 m, close to the maximum head. To meet the requirements of the grid load regulation, the turbine units of the power plant have to operate under off-design operating conditions for a long period and even extend their operation to the low load zone. In this study, three operating conditions of 33.3%, 66.6%, and 100% of the rated power are adopted for the CFD simulation and subsequent flow-induced strength analysis. The operating parameters of the selected operating conditions are analyzed based on site measurements (Table 2).

To perform the CFD simulation of the turbine unit, the fluid domain computational model of the Francis turbine is generated with high-quality tetrahedral and hexahedral meshes. The meshes near the structure wall surfaces of key flow domains such as runner blades and guide vanes are refined. There are five mesh layers near the wall surfaces, with the first mesh height of 0.1 mm and an expansion ratio of 1.2. Three groups of meshes are generated for the fluid domain of the Francis turbine unit, with 7.3 million, 14.2 million, and 25.8 million elements, respectively, and the mesh sensitivity assessment is performed to ensure the reliability and convergence of the calculation results. The deviation of the power output of the Francis turbine at 100% of load calculated by the first mesh group from the target 183 *MW* is around 10%, which cannot be used. The power output of the second and third mesh groups is close to the measured power onsite. However, the simulation time with the third mesh group is much longer than that with the second mesh group. According to the computing resources and requirements, the second mesh group with 14.2 million elements is selected to ensure a reasonable balance between calculation accuracy and efficiency (Table 3).

Table 2. Operating conditions of the Francis turbine unit.

Operating Condition (OP)	Guide Vane Opening (%)	Flow Discharge (%)	Power Output (%)
OP1	46	42	33.3
OP2	67	68	66.6
OP3	100	100	100

The meshes of different fluid domains including the spiral case, stay vanes, guide vanes, Francis runner, crow and band chambers, and draft tube of the turbine unit, and the entire fluid domain of the turbine unit are presented in Figures 4 and 5.

Table 3. Meshes of different fluid domains of the Francis turbine unit.

Fluid Domain	Element Amount ($\times 10^6$)	Element Type
Spiral case and stay vane	10.0	Hexahedral and Tetrahedral
Guide vane	1.60	Hexahedral
Crown chamber and band chamber	0.35	Hexahedral
Francis runner	1.62	Hexahedral and Tetrahedral
Draft tube	0.63	Hexahedral and Tetrahedral
Total	14.2	Hexahedral and Tetrahedral

The flow domains of the spiral case, stay vane, guide vane, draft tube, and crown and band chambers are set up in the stationary frame, and the runner flow domain is located in the rotating frame with the rated rotating speed of 100 rpm, which allows to simulate the centrifugal and Coriolis forces and other rotational effects on the fluid flow. Interface surfaces are specified between the runner fluid domain and adjacent fluid domains to transfer data between the rotating frame and the stationary frame.

The spiral case inlet and the draft tub outlet are defined with the field-measured flow discharge and static pressure values from site campaigns in the power plant. All surfaces of the fluid domain connected to the turbine unit structures are designated as no-slip walls. Automatic wall treatment is selected for the wall surfaces. The total physical time in the simulation is set to 3.6 s, corresponding to 10 rotation revolutions for the Francis runner, and the residual threshold of the CFD simulations is specified as 10^{-6}. It can ensure the convergence of the CFD simulation and avoid unnecessary calculation overhead caused by an excessively long calculation time. Finally, 720 steps per revolution is set up to fully capture the complex turbulent flow in the turbine flow domains.

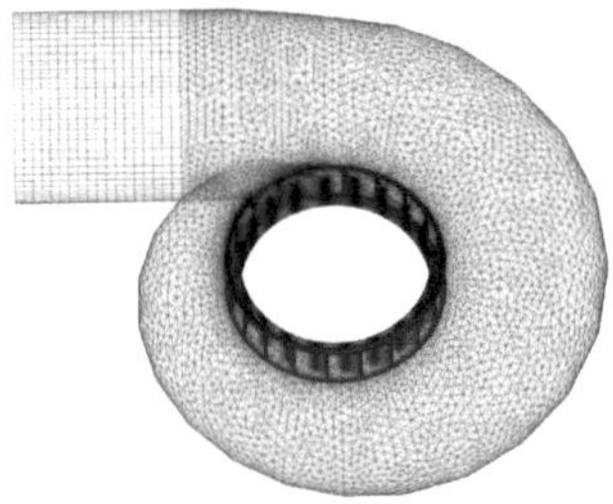

(**a**) Flow domains of the spiral case and stay vanes

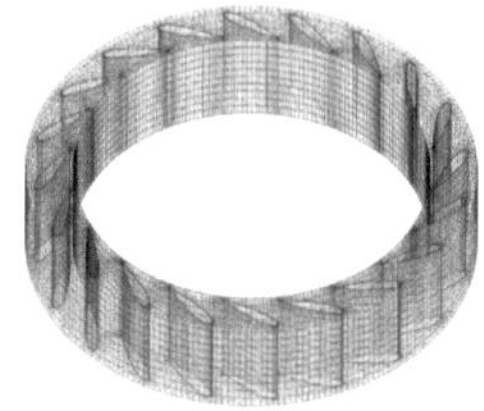

(**b**) Flow domain of the guide vanes

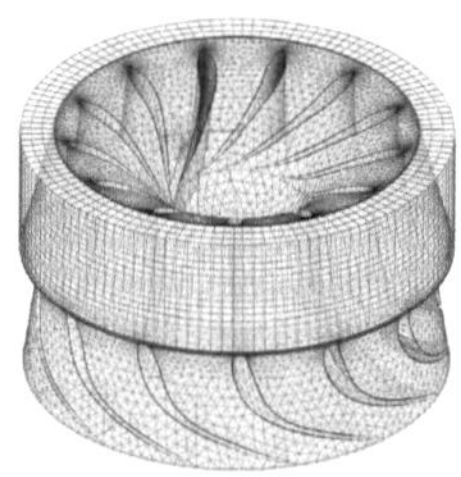

(**c**) Flow domain of the Francis runner

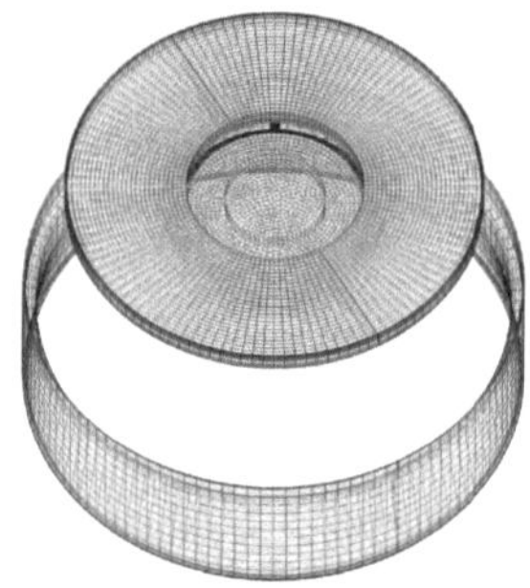

(**d**) Flow domain of the crow and band chambers

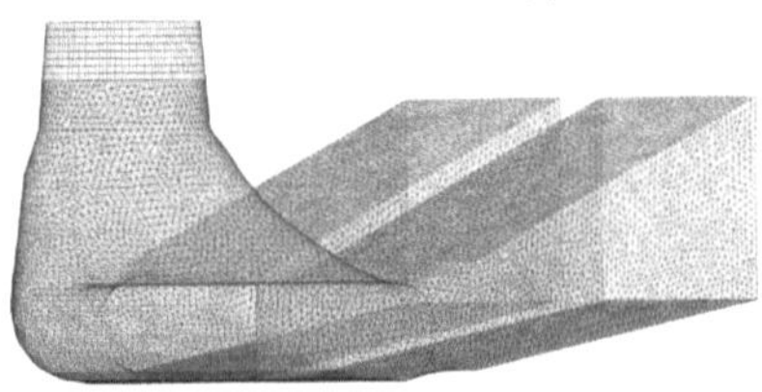

(**e**) Flow domain of the draft tube

Figure 4. Meshes of different fluid domains of the Francis turbine unit.

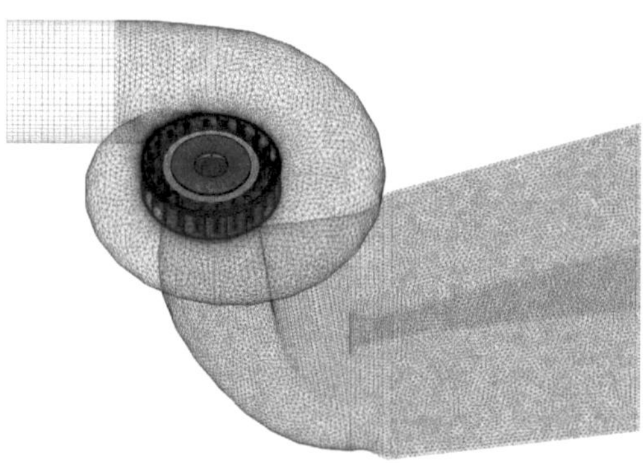

Figure 5. The fluid domain mesh of the Francis turbine.

4. Results and Discussions of Flow-Induced Strength Analysis

4.1. Flow Pressure Distributions and Velocity Distributions of the Francis Turbine Unit

Under three different operating loads of 33.3%, 66.6%, and 100% of the rated power, the simulation results of the power output of the Francis turbine approach the measured ones, which verifies the reliability of the CFD calculation results. In order to better compare the calculation results of the Francis turbine under different operating conditions, the pressure and velocity of the turbine in this study are normalized according to the flowing formulae:

$$p^* = \frac{p}{\rho_f g H} \tag{21}$$

$$\mathbf{u}^* = \frac{\mathbf{u}}{u_{max.}} \tag{22}$$

where g is the acceleration of gravity, H is the operating head of the Francis turbine unit, and $u_{max.}$ is the maximum velocity value of the three load cases.

The normalized pressure distribution in the flow domain of the Francis turbine unit at three different operating conditions, 33.3%, 66.6%, and 100% of the rated power, are demonstrated in Figures 6, 7 and 8, respectively.

The results reveal that at 33.3%, 66.6%, and 100% loads of the rated power, the pressure of the Francis turbine flow domains decreases from the spiral case to the draft tube in general. The pressure in the spiral case is the highest in the flow domain, where the hydraulic energy is dominated by pressure energy. In the runner flow domain, the pressure drops significantly across the runner due to the pressure energy being transferred from the fluid to the rotating blades as kinetic energy. After the runner runner outlet, the fluid flows into the draft tube, which acts as a diffuser to recover part of the kinetic energy into pressure energy. The pressure in the draft tube is the lowest in the entire flow passages of the Francis turbine unit.

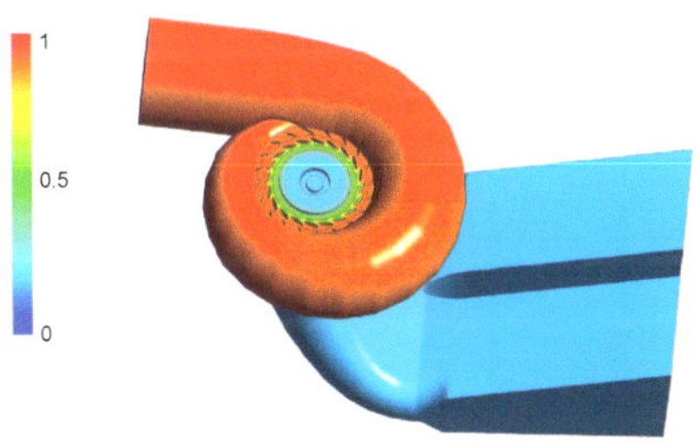

Figure 6. Normalized pressure distribution of the Francis turbine unit (OP1, 33.3% rated power load).

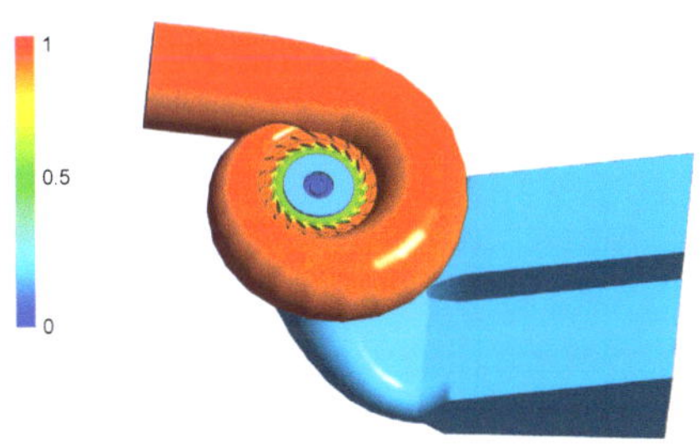

Figure 7. Normalized pressure distribution of the Francis turbine unit (OP2, 66.6% rated power load).

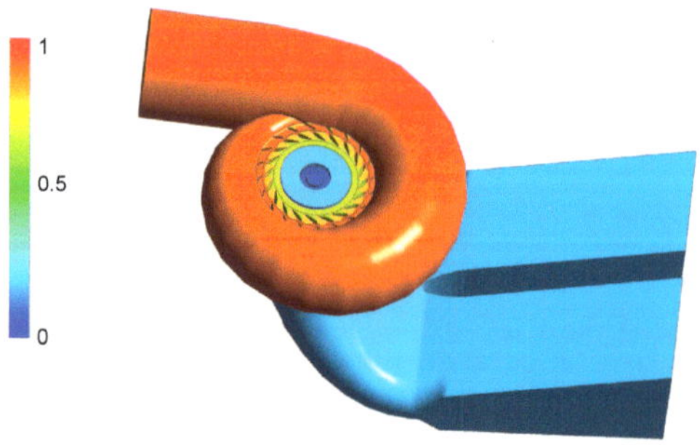

Figure 8. Normalized pressure distribution of the Francis turbine unit (OP3, 100% rated power load).

Under various loads, the flow rate and flow direction of the water entering the runner can be controlled by adjusting the guide vane openings. Since the guide vane opening angles are significantly different at the three loads, the local pressure distributions in the guide vane domains also vary with each other. At a 33.3% rated power load, the guide vane opening is relatively small, and the water flow is ejected to the runner blade fluid channels from the narrow gaps between adjacent guide vanes. There is a significant difference in pressure between the upstream and downstream of the guide vanes. The pressure upstream of the guide vane is higher, and the pressure downstream of the guide vane is lower. With the guide vane opening increases at a 66.6% rated power load, the discharge increases accordingly to generate more power. And the pressure level of the flow between the guide vanes increases. At a 100% rated power load, the guide vanes are fully opened, and the pressure gradually decreases from the spiral case toward the guide vane outlet. The pressure gradient in the guide vane region is less than that at 33.3% and 66.6% loads of rated power. The normalized flow velocity distributions of the Francis turbine flow domain at 33.3%, 66.6%, and 100% loads of rated power are shown in Figures 9–11, respectively.

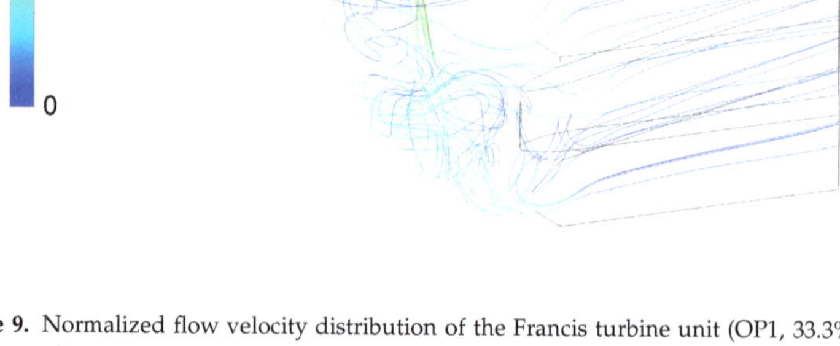

Figure 9. Normalized flow velocity distribution of the Francis turbine unit (OP1, 33.3% load of rated power).

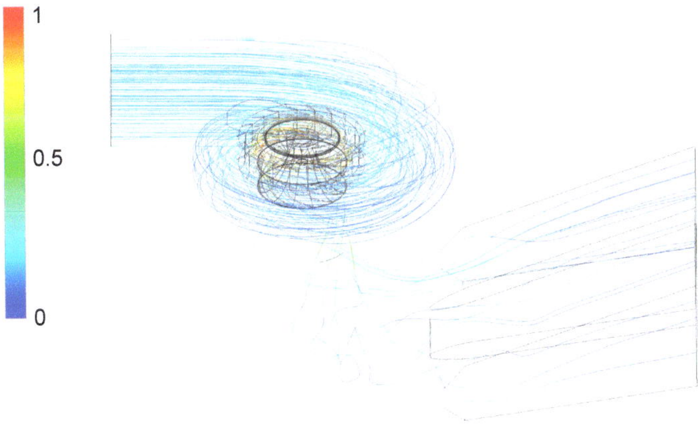

Figure 10. Normalized flow velocity distribution of the Francis turbine unit (OP2, 66.6% load of rated power).

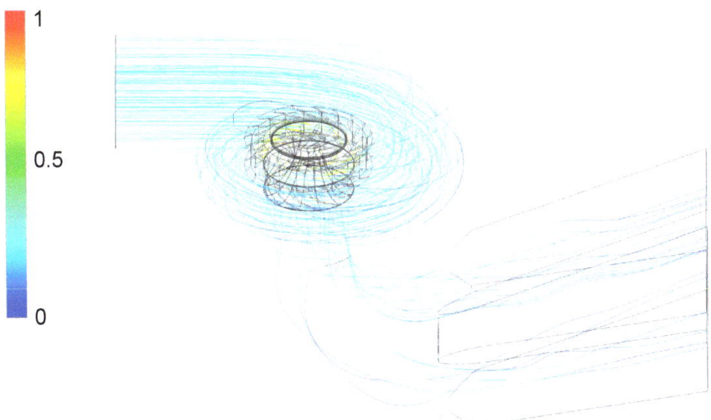

Figure 11. Normalized flow velocity distribution of the Francis turbine unit (OP3, 100% load of rated power).

The flow streamlines of the Francis turbine under different operating conditions from 33.3% to 100% of the rated power load have obvious differences. At the 33.3% rated power load, the flow in the spiral case is relatively stable, but the flow between guide vanes and in the vaneless region is obviously irregular. Due to the small opening of the guide vanes and the narrow water flow passages in between, the fluid velocity increases when passing through the gap between adjacent guide vanes, forming a high shear zone with high speed locally. The fluid streamline distribution in the draft tube is disordered and asymmetric with obvious vortex and backflow. The loss of fluid kinetic energy increases, resulting in a reduction in turbine efficiency. At the 66.6% rated power load, the flow in the spiral case remains stable. With a large guide vane opening, the flow in the guide vane flow domain and the runner flow domain is improved. The streamlines gradually become regular and the flow velocity of the high speed ring in the vaneless region is decreased. In the runner blade flow channel, the streamlines are relatively smooth, and the energy transfer is more efficient. The streamline distribution is more regular than that at the 33.3% rated power load, and the vortex and backflow are weakened but not completely eliminated.

When the power output reaches 100% of the rated power load, the flow streamlines of the turbine flow passage are distributed more regularly, and the energy transfer is the highest. The flow velocity distribution in the spiral case is very uniform and provides stable inflow conditions to the guide vane region. The flow lines are distributed regularly among the guide vanes, and the water gradually accelerates as it passes through the guide vanes, forming a uniform flow field into the runner flow domain. The high-speed ring in the vaneless region disappears. The streamlines flow smoothly along the blade surface inside the Francis runner without obvious separation or turbulence, and the kinetic energy is gradually converted into mechanical energy. The streamlines in the draft tube are of high regularity. The flow velocity decreases evenly, and backflow and vortex phenomena are almost disappeared. At the 100% rated power load, the flow distributions in all flow domains are optimized to the best state, and the turbine efficiency and stability reach the highest level.

4.2. Flow-Induced Strength Analysis of the Francis Runner

In order to evaluate the strength of the Francis runner under three different operating conditions, the flow-induced strength analysis of the runner needs to be performed by combining the above CFD simulations with finite element analysis (FEA). The data of the pressure distributions on the runner at 33.3%, 66.6%, and 100% loads of rated power can be exported from the CFD analysis performed in the previous section and mapped on the finite element of the Francis runner as surface loads.

High-quality tetrahedral elements are used to create the Francis runner's finite element model (Figure 12). The Francis turbine runner blades' geometrical curvature varies significantly, particularly at the fillets, which are likely to become locations of concentrated stress. To accurately record the stress values, the mesh at the runner blades' fillets is improved and refined. The Francis runner is subjected to the boundary conditions, which include the fixed restrictions, rotating speed, and gravity acceleration.

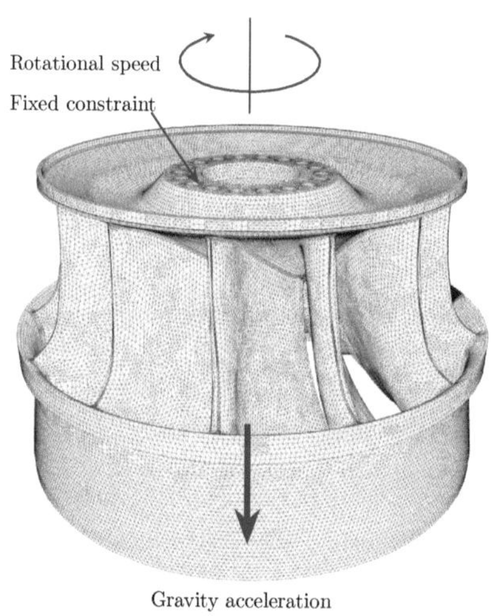

Figure 12. Finite element model of the Francis turbine for the flow-induced stress analysis.

The Francis runner in this study is made of stainless steel, and the material parameters of the runner are summarized in Table 4 and include a density of 7700 kg m^{-3}, a Young's modulus of 2.1 × 10^5 MPa, a Poisson's ratio of 0.3, and a yield stress of 550 MPa. This value of yield stress ensures that the Francis runner can withstand operational loads under normal and transient conditions without permanent deformation.

Table 4. Parameters of the hydro-generator unit.

Parameter	Value	Unit
Density	7700	kg m^{-3}
Young's modulus	2.1 × 10^5	MPa
Poisson's ratio	0.3	-
Yield stress	550	MPa

The pressure files across the entire flow passage of the Francis turbine unit at three different loads can be exported from the CFD analysis discussed in the previous section. This CFD-derived pressure data can then be accurately mapped onto the nodes of the finite element model to facilitate the subsequent flow-induced structural stress analysis (Figure 13).

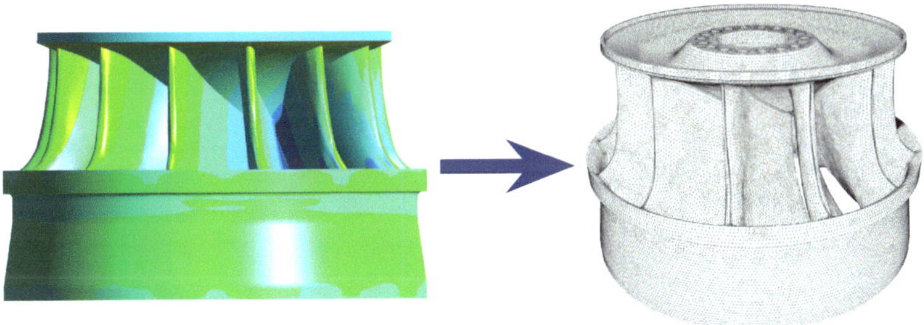

Figure 13. Mapping the CFD-derived pressure profile to the finite element model of the Francis turbine.

As shown in Figure 14, the maximum stresses of the runner at 33.3%, 66.6%, and 100% loads of rated power are located in the fillets between the blade trailing edge and the crown.

When the Francis runner is running stably at a rated speed, its load is determined by the hydraulic torque acting on the runner. As the load of the Francis unit increases, the opening of the unit's guide vanes increases accordingly; therefore, the water flow entering the runner flow passage and the runner hydraulic torque increases. Eventually, the value of the runner hydraulic torque is equal to the value of the counter load torque on the turbine shaft so that a new torque equilibrium is formed in the Francis turbine unit. As shown in Figure 15, the maximum stress of the runner gradually increases with the power from 33.3% to 100% of the rated power load.

As the hydraulic torque on each runner blade increases with the load, the flow-induced stress of the runner blades also increases accordingly. The maximum stress values of the runner at the 33.3% rated power load and the 66.6% rated power load are 0.45 and 0.74 times the stress at the 100% rated power load, which is around 1/3 of the yield stress of runner material. Therefore, the runner is safe during the extended operation range from 33.3% to 100% of the rated power load. As the service life of the Francis turbine increases, the performance of the materials used in the turbine runner may degrade, especially for materials that are susceptible to cavitation erosion or corrosion. To ensure the safe and efficient operation of the units, it is still recommended to strengthen the inspection of the runners to prevent problems before they occur.

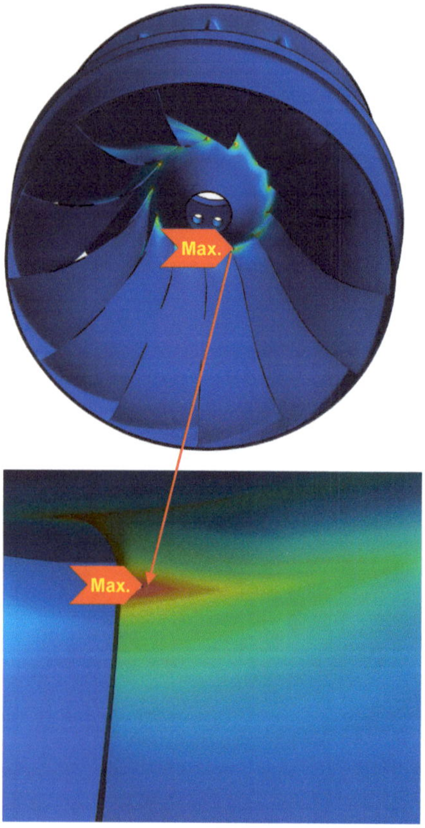

Figure 14. Comparison of the normalized maximum stress of the Francis runner under three different loads.

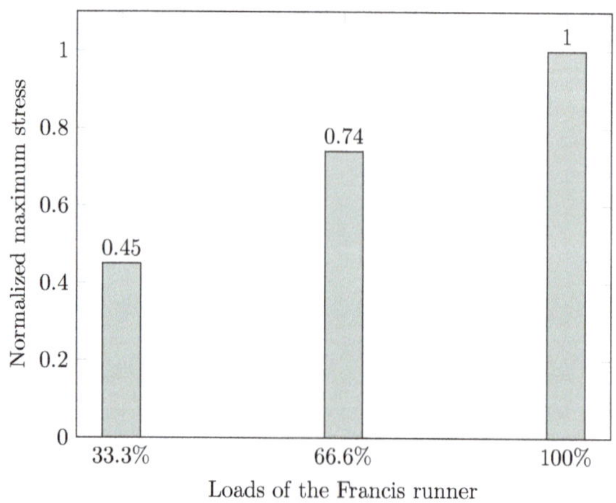

Figure 15. Normalized stress of the Francis runner under three different loads.

The Francis runner's fatigue and cavitation resistance deteriorate with deteriorating material, making the runner blades more vulnerable to stress concentration at off-design conditions. This can hasten damage and impact the turbine's stability and service life. To address these issues, the next plan is to measure the flow characteristics and strength behavior of the Francis turbine under about 60 different operating conditions, ranging from 0 to 100% loads of rated power. Then, with the help of the corresponding flow-induced strength simulations, the flow pattern and flow-induced stress characteristics under these operating conditions will be thoroughly analyzed, and effective vibration monitoring and control measures will be implemented to guarantee the turbine's dependability and efficiency after long operating times.

5. Conclusions

In the study, the geometrical model CFD calculation models of the entire fluid domain including the spiral case, stay vanes, guide vanes, turbine runner, crow and band chambers, and draft tube of the turbine unit were constructed. CFD simulations and the flow-induced strength analysis under three operating conditions in the extended load range from the 33.3% rated power load to the full load have been performed.

The Francis turbine flow domain pressure drops from the spiral case to the draft tube under 33.3%, 66.6%, and 100% loads of rated power, according to the CFD calculation results. Nevertheless, the guide vane region's local pressure distributions change with various loads. The pressure differential between the guide vanes' upstream and downstream is evident for the 33.3% and 66.6% loads of rated power, with the upstream pressure being higher than the downstream pressure. The pressure gradient in the guide vane region is lower at the 100% rated power load when the guide vanes are fully opened than when it is at 33.3% and 66.6% loads of rated power.

The flow velocity distributions of the turbine under three operating conditions vary from one another. The flow in the spiral case is relatively stable for all loads. However, the flows in the guide vane region and the vaneless zone are strongly related to the guide vane openings. The flow forms a high-speed ring in the vaneless region at the 33.3% rated power load due to the small guide vane opening. The fluid velocity in the draft tube at the 33.3% and 66.6% rated power loads is asymmetric and disordered with the vortex and backflow. At the 100% rated power load, the flow of the whole turbine flow domain is more regular and stable. The high-speed ring in the vaneless region and the backflow and vortex in the draft tube disappear.

According to the Francis runner's flow-induced strength analysis, the runner's maximum stress progressively rises as the power increases from 33.3% to 100% of the rated power load. Therefore, the maximum stress on the runner blades increases with the Francis runner's load. Under 100% of the rated power load, the analyzed runner's maximum stress is 2.2 and 1.4 times that under 33.3% and 66.6% of the rated power load, respectively, but it is only about one-third of the runner material's yield stress. This runner is safe to operate throughout the extended range of 33.3% to 100% of the rated power load. Because of the Francis runner's distinctive design, the flow-induced structural stress on the runner blades varies nonlinearly with the runner load.

The study provides an in-depth analysis of flow-induced stresses of a large Francis turbine over an extended load range. By integrating CFD and FEA approaches, this study fills a critical gap in understanding the operating limits of large Francis turbines to ensure their safe operation under varying load conditions. The analysis approach proposed in this study can be applied to other hydraulic machinery including Francis turbines, pumps, and pump–turbines.

Author Contributions: Conceptualization, X.H.; methodology, X.H.; investigation, X.L. and X.H.; validation, W.C.; writing—original draft preparation, X.H. and X.L.; writing—review and editing, X.H. and W.C.; supervision, Z.W. All authors have read and agreed to the published version of the manuscript.

Funding: This research and the APC were funded by the project "Study on Stable Operation of Units in Wanjiazhai Hydropower Station" of Wanjiazhai Water Multi-Purpose Dam Project Co., Ltd. Grant number: WJZ-ZB-2022-027.

Institutional Review Board Statement: Not applicable.

Informed Consent Statement: Not applicable.

Data Availability Statement: The original contributions presented in this study are included in the article. Further inquiries can be directed to the corresponding author.

Conflicts of Interest: The authors Xingping Liu and Weijiang Chen are employed by the company Power Station Management Bureau, Y.R. Wanjiazhai Water Multi-Purpose Dam Project Co., Ltd. The authors declare that this study received funding from Wanjiazhai Water Multi-Purpose Dam Project Co., Ltd. The funder was not involved in the study design, collection, analysis, interpretation of data, the writing of this article or the decision to submit it for publication. The remaining authors declare that the research was conducted in the absence of any commercial or financial relationships that could be construed as potential conflicts of interest.

Abbreviations

The following abbreviations are used in this manuscript:

CFD	Computational fluid dynamics
FE	Finite element
FEA	Finite element analysis
FSI	Fluid–structure interaction
OP	Operating condition
SST	Shear stress transport

References

1. Bayazıt, Y. The effect of hydroelectric power plants on the carbon emission: An example of Gokcekaya dam, Turkey. *Renew. Energy* **2021**, *170*, 181–187. [CrossRef]
2. Paraschiv, L.S.; Paraschiv, S. Contribution of renewable energy (hydro, wind, solar and biomass) to decarbonization and transformation of the electricity generation sector for sustainable development. *Energy Rep.* **2023**, *9*, 535–544. [CrossRef]
3. Rodríguez-Pérez, A.M.; Rodríguez-Gonzalez, C.A.; López, R.; Hernández-Torres, J.A.; Caparrós-Mancera, J.J. Water Microturbines for Sustainable Applications: Optimization Analysis and Experimental Validation. *Water Resour. Manag. Int. J. Publ. Eur. Water Resour. Assoc. (EWRA)* **2024**, *38*, 1011–1025. [CrossRef]
4. Ambec, S.; Crampes, C. Electricity provision with intermittent sources of energy. *Resour. Energy Econ.* **2012**, *34*, 319–336. [CrossRef]
5. Rehman, S.; Al-Hadhrami, L.M.; Alam, M.M. Pumped hydro energy storage system: A technological review. *Renew. Sustain. Energy Rev.* **2015**, *44*, 586–598. [CrossRef]
6. Aakti, B.; Amstutz, O.; Casartelli, E.; Romanelli, G.; Mangani, L. On the performance of a high head Francis turbine at design and off-design conditions. *J. Physics Conf. Ser.* **2015**, *579*, 012010. [CrossRef]
7. Seidel, U.; Mende, C.; Hübner, B.; Weber, W.; Otto, A. Dynamic loads in Francis runners and their impact on fatigue life. In *Proceedings of the IOP Conference Series: Earth and Environmental Science*; IOP Publishing: Bristol, UK, 2014; Volume 22, p. 032054.
8. Flores, M.P.; Urquiza, G.; Rodriguez, J. A. A Fatigue Analysis of a Hydraulic Francis Turbine Runner. *World J. Mech.* **2012**, *2*, 28–34. [CrossRef]
9. Frunzăverde, D.; Muntean, S.; Mărginean, G.; Câmpian, V.; Marşavina, L.; Terzi, R.; Şerban, V. Failure analysis of a Francis turbine runner. *IOP Conf. Ser. Earth Environ. Sci.* **2010**, *12*, 012115. [CrossRef]
10. Adhikari, M.P.; Poudel, L. Fatigue Analysis of Francis Turbine Runner as a Result of Flow-induced Stresses. In *Proceedings of IOE Graduate Conference, 2019-Winter*; Institute of Engineering, Tribhuvan University: Kirtipur, Nepal, 2019; Volume 7, pp. 7–13.

11. Liu, P.; Huang, X.; Yang, T.; Wang, Z. Flow-Induced Fatigue Damage of Large Francis Turbines Under Multiple Operating Loads. *Appl. Sci.* **2024**, *14*, 12003. [CrossRef]
12. Gordon, J.L. Determining turbine runner metal loss caused by cavitation erosion. *Int. Water Power Dam Constr.* **1991**, *43*, 21–24.
13. Escaler, X.; Egusquiza, E.; Farhat, M.; Avellan, F.; Coussirat, M. Detection of cavitation in hydraulic turbines. *Mech. Syst. Signal Process.* **2006**, *20*, 983–1007. [CrossRef]
14. Celebioglu, K.; Altintas, B.; Aradag, S.; Tascioglu, Y. Numerical research of cavitation on Francis turbine runners. *Int. J. Hydrogen Energy* **2017**, *42*, 17771–17781. [CrossRef]
15. Tiwari, G.; Shukla, S.; Prasad, V.; Patel, V.k. Numerical Investigation of Hydrodynamics of the Cavitating Turbulent Flow in a Low Head Prototype Francis Turbine. *AIP Conf. Proc.* **2020**, *2273*, 50006. [CrossRef]
16. Kobro, E. *Measurement of Pressure Pulsations in Francis Turbines*; Norwegian University of Science and Technology: Trondheim, Norway, 2010.
17. Pasche, S.; Gallaire, F.; Avellan, F. Origin of the synchronous pressure fluctuations in the draft tube of Francis turbines operating at part load conditions. *J. Fluids Struct.* **2019**, *86*, 13–33. [CrossRef]
18. Nicolet, C.; Zobeiri, A.; Maruzewski, P.; Avellan, F. Experimental investigations on upper part load vortex rope pressure fluctuations in francis turbine draft tube. *Int. J. Fluid Mach. Syst.* **2011**, *4*, 179–190. [CrossRef]
19. Sotoudeh, N.; Maddahian, R.; Cervantes, M.J. Investigation of Rotating Vortex Rope formation during load variation in a Francis turbine draft tube. *Renew. Energy* **2020**, *151*, 238–254. [CrossRef]
20. Ciocan, G.D.; Iliescu, M.S.; Vu, T.C.; Nennemann, B.; Avellan, F. Experimental study and numerical simulation of the FLINDT draft tube rotating vortex. *J. Fluids Eng.* **2007**, *129*, 146–158. [CrossRef]
21. Huang, X.; Ou, H.; Huang, H.; Wang, Z.; Wang, G. Flow-Induced Stress Analysis of a Large Francis Turbine Under Different Loads in a Wide Operation Range. *Appl. Sci.* **2024**, *14*, 11782. [CrossRef]
22. Cao, D.; Dodge, M.; Zhou, Y.; Liu, J. Stress measurement of runner blade for a Francis prototype turbine. *J. Physics Conf. Ser.* **2024**, *2752*, 012215. [CrossRef]
23. Yao, C.; Huang, X.; Zhu, P.; Brès, C.S.; Thévenaz, L. Preliminary Study of Distributed Fiber Optic Sensing Technologies in Hydraulic Machinery. In *Proceedings of the Light, Energy and the Environment 2018 (E2, FTS, HISE, SOLAR, SSL), Singapore, 5–8 November 2018*; Optica Publishing Group: Washington, DC, USA, 2018; p. EM2A.3. [CrossRef]
24. Lais, S.; Liang, Q.; Henggeler, U.; Weiss, T.; Escaler, X.; Egusquiza, E. Dynamic Analysis of Francis Runners—Experiment and Numerical Simulation. *Int. J. Fluid Mach. Syst.* **2009**, *2*, 303–314. [CrossRef]
25. Zhang, D.; Quan, Q.; Huang, X.; Wang, Z.; Wang, B.; Xiao, Y. Transient Flow-Induced Stress Investigation on a Prototype Reversible Pump–Turbine Runner. *Energies* **2024**, *17*, 3026. [CrossRef]
26. Luna-Ramírez, A.; Campos-Amezcua, A.; Dorantes-Gómez, O.; Mazur-Czerwiec, Z.; Muñoz-Quezada, R. Failure analysis of runner blades in a Francis hydraulic turbine—Case study. *Eng. Fail. Anal.* **2016**, *59*, 314–325. [CrossRef]
27. Huang, X.; Chen, L.; Wang, Z.; Li, H.; Chen, S.; Hu, K.; Li, C.; Qiu, L. Stress Characteristic Analysis of Pump-Turbine Head Cover Bolts during Load Rejection Based on Measurement and Simulation. *Energies* **2022**, *15*, 9496. [CrossRef]
28. Chou, P.Y. On velocity correlations and the solutions of the equations of turbulent fluctuation. *Q. Appl. Math.* **1945**, *3*, 38–54. [CrossRef]
29. Jones, W.P.; Launder, B.E. The prediction of laminarization with a two-equation model of turbulence. *Int. J. Heat Mass Transf.* **1972**, *15*, 301–314. [CrossRef]
30. Menter, F. Zonal Two Equation k-ω Turbulence Models For Aerodynamic Flows. *AIAA Paper* **1993**, *1993*, 2906. [CrossRef]
31. Menter, F.R. Two-equation eddy-viscosity turbulence models for engineering applications. *AIAA J.* **1994**, *32*, 1598–1605. [CrossRef]

MDPI AG
Grosspeteranlage 5
4052 Basel
Switzerland
Tel.: +41 61 683 77 34

MDPI Books Editorial Office
E-mail: books@mdpi.com
www.mdpi.com/books

www.ingramcontent.com/pod-product-compliance
Lightning Source LLC
Chambersburg PA
CBHW041425210926
51850CB00013B/3358